Gächter / Müller
Plastics Additives Handbook

Plastics Additives Handbook

Stabilizers, Processing Aids, Plasticizers, Fillers, Reinforcements, Colorants for Thermoplastics

Edited by
R. Gächter and H. Müller

with contributions by Dr. H. Andreas, Drs. A.L. Berg, K. Berger, Dipl.-Ing. K-D. Böhme, Dr. A.W. Bosshard, Dr. W. Brotz, Dr. W. Damm, Dr. H.W. Finck, Dr. F. Gugumus, Dr. E. Herrmann, Dipl.-Ing. H. Hurnik, Ir. J. Jansen, Dr. H. Jenkner, Dr. R. Leimgruber, Dr. J. Lorenz, Dr. G. Menzel, Dr. H. Müller, Dr. H.P. Schlumpf, Dr. K. Schwarzenbach, Ing. W. Sommer

HANSER PUBLISHERS

Hanser Publishers, Munich Vienna New York

Distributed in the United States of America by
Macmillan Publishing Company, New York
and in Canada by
Collier Macmillan Canada, Inc., Ontario

Distributed in USA by
Scientific and Technical Books
Macmillan Publishing Company
866 Third Avenue, New York, N.Y. 10022

Distributed in Canada by
Collier Macmillan Canada, Inc.
1200 Eglington Ave. E, Suite 200,
Don Mills, Ontario M3C 3N1 Canada

Distributed in all other countries by
Carl Hanser Verlag
Kolbergerstraße 22
D-8000 München 80

CIP-Kurztitelaufnahme der Deutschen Bibliothek

Plastics additives handbook: stabilizers, processing aids, plasticizers, fillers, reinforcements, colorants for thermoplastics / ed. by R. Gächter and H. Müller. With contributions by H. Andreas ... Transl. and rev. ed. – Munich; Vienna; New York: Hanser, 1987.
Dt. Ausg. u. d. T.: Taschenbuch der Kunststoff-Additive
ISBN 3-446-15072-2

NE: Gächter, Reinhard [Hrsg.]; Andreas, Holger [Mitverf.]

ISBN 3-446-15072-2 Carl Hanser Verlag München Wien
ISBN 0-02-947911-8 Macmillan Publishing Company, New York

Library of Congress Catalog Card Number 83-062289

Printed in Germany by Universitätsdruckerei H. Stürtz AG, Würzburg

Preface

The vigor of the plastics industry remains unbroken but there has been a shift in emphasis with regard to technical development. Today front stage is occupied by the development of plastics and plastics systems, taylor-made for specific applications.

Along with copolymers and physical blends of various plastics, additives are enjoying a key position within this new trend in development. They permit the use of plastics in applications where the plastics material as such would have had small chance of success.

Additives are accepted today as full-fledged partners of plastics. In this sense, this Additives Handbook is a logical and long overdue complement to the trade literature. The Additives Handbook is a reference book having as objective to summarize the state of the art achieved in this area. Next to this somewhat static point of view, attempts have been made to present an outlook of the future, whenever relevant technical trends could be discerned at this time. It is hoped in this way to stimulate further, in depth collaboration among plastics producers, plastics processors, and additives manufacturers. This in turn should lead to long term meaningful problem solutions, thereby further strengthening the position of plastics materials in technology and the economy in general.

In view of the diversity of the subject matter, it was deliberately decided not to have one single author treat the vast topic Additives. Instead, a number of authors qualified by their professional activity, are attempting to describe how additives contribute to rendering thermoplastic polymers – the scope of the Handbook is limited to them – capable of fulfilling their manifold purposes.

The individual chapters of the Handbook are structured essentially in an introductory outline of the technical/commercial aspects of the problem, followed by a short synopsis of the corresponding physical and chemical fundamentals. The subsequent main part deals with the various products or product groups and their use in individual plastics. The already mentioned future outlook forms the concluding part.

The penultimate chapter of the book is concerned with toxicological problems faced daily by additives producers and users alike. The increasingly more severe requirements imposed by the authorities on additives for approval of their use in plastics in contact with food, exert profound influence on the future development in this area. A short résumé of this problem complex appeared, therefore, appropriate.

Already two German editions have been published of this Additives Handbook, which has become an indispensable reference to all those who are professionally connected with the complex world of additives for thermoplastic polymers. Readers' responses have clearly shown that it helps fill a real gap in the trade literature.

The present English version complies with numerous wishes from all sectors of the plastics industry. Apparently a broad and comprehensive survey of additives is missing from the extensive English literature on plastics. The English text corresponds to a great extent to the latest German edition, only minor changes and linguistic adaptations have been made by the authors of the various chapters.

In this connection we should like to thank particularly Dr. *Peter Klemchuk* for his valuable contributions in preparing a technically and grammatically correct translation.

Basel, Switzerland, November 1984. R. Gächter
H. Müller

Preface to the Second Edition

The Plastics Additives Handbook has clearly filled an important gap in the plastics technology literature. Only this fact can explain that the First Edition sold out in just two years. In this short time interval there have been relatively few innovations and, therefore, it was not deemed necessary to revise the text. However, the indexes of trade names and suppliers at the end of each chapter were updated.

The Editors hope that this Second Edition will appeal to all plastics technologists who have not, as yet, become acquainted with this Handbook.

Basel, July 1987 The Editors

Contents

4 PVC Stabilizers

Dr. Holger Andreas, Bensheim

5 Plasticizers

Ing. Werner Sommer, Frankfurt/M.

6 Lubricants and Related Auxiliaries for Thermoplastic Materials

Dr. Walter Brotz, Gersthofen

11 Flame Retardants for Thermoplastics
Dr. Herbert Jenkner, Köln

12 Antistatic Agents
Dr. H.W. Finck, Basel

13 Fluorescent Whitening Agents
Kurt Berger, Basel

14 Biostabilizers
Dr. Joachim Lorenz, Bensheim

16 Organic Peroxides as Cross-Linking Agents

Drs. A.L. Berg, Amersfoort

17 Nucleating Agents for Partly Crystalline Polymers

Ir. Joost Jansen, Geleen

18 Aspects of Plastics Additives Relative to Industrial Hygiene and Food Quality

Dr. Roland Leimgruber, Itingen

19 Compound Development and Compounding of Thermoplastics

Dr. Gerhard Menzel, Marl

TECHNICAL SERVICE AND DEVELOPMENT

The Rohm and Haas philosophy of service demands that research, development and technical support be maintained at the same high level as its world-wide manufacturing capabilities. To provide the key link between customers, technical service, sales, production and research, the company has an established product development and technical service laboratory complex in Valbonne, Southern France. Here is located a specially equipped facility designed to provide support and interaction with the PVC processing industry to ensure the continuing development of additives to meet the special needs of the European market. Further development work is conducted at the company's two manufacturing plants for modifiers in Lauterbourg, France and Grangemouth, Scotland.

Rohm and Haas maintains its basic "new product" research programme at its main research laboratories located in the U.S.A. at Spring House and Bristol, Pa. Through a regular interchange of information of both a technical and a market-orientated nature, new product and application development is able to take advantage of input not only from the European and American markets, but from other regions of the world as well.

Innovation and improvements to the extensive Paraloid product range are part of a continuous process, thus wherever PVC is to be processed, Rohm and Haas will be seeking to provide additives and technical support to optimise the effective use of this, the most versatile of thermoplastic materials.

PROCESSING AIDS

Rigid PVC is more difficult to process than other large volume thermoplastic materials such as polyethylene or polystyrene. Its non-homogeneous morphology results in difficult fusion and low melt strength. It also suffers from a narrow processing temperature range between fusion and degradation and shows a tendency to stick to processing equipment.

For rigid PVC formulations these difficulties have been overcome by the development of special processing skills on the part of PVC fabricators, combined with the use of Paraloid acrylic processing aids developed by Rohm and Haas.

PARALOID PROCESSING AIDS improve the processing of PVC through :-

- Faster, more controlled fusion
- Greater homogenization of the melt
- Elimination of plateout
- Increased melt strength and elongation
- Elimination of surface defects

These features result in numerous benefits to the PVC processor, both during manufacturing operations:

- Less sensitivity to fluctuations in processing conditions
- Broader processing temperature range
- More consistent melt characteristics
- Less downtime for cleaning
- Higher yields
- Easier use of lower cost powder blends

. . . . and in the finished product

- Improved surface quality
- Improved sparkle and clarity
- Improved physical properties

Depending on the actual process being used, these benefits are manifested in many different ways:

CALENDERING

- Higher and more uniform output from compounding extruders or intensive mixers
- Lower scrap rates due to less critical processing conditions, and quicker establishment of saleable quality production
- Film and sheet products with improved end-use fabrication properties, including improved thermoformability and higher consumer appeal

EXTRUSION

- Improved control of dimensional tolerances due to more consistent melt flow
- Lower scrap rates due to more rapid establishment of equilibrium conditions and elimination of melt fracture
- Increased output rates, resulting in lower unit costs.

BLOW MOULDING

- Stronger, more consistent components through improved parison strength and better welds
- More uniform weight control
- More sparkling appearance and greater consumer appeal

INJECTION MOULDING

- Lower reject rates due to reduction of splash
- Higher quality parts, due to less obvious, stronger weld lines, and improved appearance
- Improved performance of parts due to lower moulding stress

PROCESSING AIDS

PARALOID K 120 N

Versatile, all purpose processing aid, the original standard of the industry. Usually chosen for medium and high K-value PVC and for use in plasticised PVC formulations.

PARALOID K 125

All purpose, high efficiency processing aid with similar properties to Paraloid K 120 N.

LUBRICATING - PROCESSING AID

PARALOID K 175

This product offers strong external lubricating action supplemented with acrylic processing aid functionality. This unique polymeric lubricant greatly facilitates PVC processing but without the detrimental side effects of conventional lubricants. The outstanding features of Paraloid K 175 are:

- Reduced sticking even at high temperatures
- Improved dynamic heat stability
- No impairment of clarity
- All the benefits of an acrylic processing aid

The key lubricating feature of Paraloid K 175 gives production cost savings through longer production runs between clean-ups of calender rolls or extrusion dies, higher output at higher operating temperatures, and lower scrap rates. Formulating versatility is gained especially in clear applications, resulting in easy processing of calendered film and blow moulded bottles. Paraloid K 175 is also effective in less critical industrial applications: for example it considerably broadens the range of injection speeds and temperatures tolerable in injection moulding; it can improve the binding of high filler levels in low cost formulations for pipe and other products; it eliminates lubricant migration in film and sheet giving improved print and ink acceptance.

MBS IMPACT MODIFIERS

Impact modifiers are included in rigid PVC formulations to achieve toughness level required for many commercial applications, for example:-

BOTTLES:-

FOOD – Mineral water, vegetable oils, etc

NON FOOD – Cosmetics, household products, etc.

FILM AND SHEET:

– Blister packs, furniture films, etc.

INDUSTRIAL – Building Products:

– Pipes, profiles, injection mouldings, etc.

For each application Rohm and Haas offers a complete range of MBS products, both KM and BTA series, with specific properties detailed below.

PARALOID KM SERIES

PARALOID KM 608 A

- Crystal clarity and sparkle
- Freedom from stress-whitening
- Good impact strength
- Excellent solvent resistance

Specially developed for clear film and sheet applications requiring maximum esthetics. Eliminates loss of consumer appeal caused by stress whitening developed through rough handling of packaged goods. Permits cold fabrication of blister packs, overlays, furniture films, etc., without whitening. Imparts unsurpassed solvent resistance to films to be used for tapes and packaging. The standard of the industry for the most demanding sheet and film applications.

PARALOID KM 611

- High impact strength
- Outstanding resistance to stress crazing by vegetable oils and other aggressive liquids
- Good clarity
- Excellent processability

A most versatile general-purpose impact modifier widely used in all major processes and applications; specially suited for calendered films and oil bottles. Unique stress-crazing resistance broadens range of PVC for packaging applications. In extrusion provides an excellent balance of processability and impact strength in the finished products.

PARALOID KM 653

- Outstanding impact efficiency
- Unsurpassed low temperature impact strength
- Low contribution to melt viscosity

Optimized for maximum impact efficiency, especially useful at low temperatures. A valuable additive at low concentrations for conduit, pipe and pipe fittings and other industrial goods. Outstanding toughness/melt viscosity balance for injection moulding compounds.

MBS IMPACT MODIFIERS

PARALOID BTA SERIES

PARALOID BTA III F

- Extremely low contribution to taste and odour
- High resistance to breakage
- Excellent clarity and high surface gloss
- Absence of colour reversal and fluorescence
- Good colour and thermal stability in calcium-zinc and tin stabilised PVC compounds

An impact modifier offering an outstanding balance of properties for PVC bottles, particularly for the packaging of mineral water and other products sensitive to taste and odour contribution. Also suitable for calendered film intended for food packaging applications.

PARALOID BTA III FF1

- Improved transparency
- Higher impact strength efficiency
- Improved thermal stability and better early colour
- Free flowing powder characteristics
- Extremely low contribution to taste and odour
- High resistance to breakage
- Excellent clarity and high surface gloss
- Absence of colour reversal and fluorescence
- Good colour and thermal stability in calcium-zinc and tin stabilised PVC compounds

Developed especially for the production of PVC bottles for mineral water and the packaging of taste-sensitive products. Its inherent free-flowing powder characteristics make it more easily handled in automated processing equipment, or for applications where performance improvements above Paraloid BTA III F are required.

PARALOID BTA III OR

- Very high resistance to stress-cracking in presence of vegetable oils and many other aggressive materials
- Good clarity and surface gloss
- Good colour and thermal stability in calcium-zinc and tin stabilised PVC compounds
- Low contribution to taste and odour

Specially developed impact modifier to minimise stress-cracking in PVC bottles used for packaging of vegetable oils and several other aggressive products.

PARALOID BTA III N2

- Excellent transparency
- High impact strength
- Low degree of odour and taste transfer
- Excellent rheological performance under critical calendering conditions
- Good thermal stability
- Low contribution to crease whitening

An Impact modifier designed for use in calendering and extrusion of PVC film and sheet for all applications particularly food packaging, requiring a high level of transparency, and extruded PVC applications requiring high levels of impact strength efficiency. Also suitable for PVC bottles intended for food and non food packaging applications.

PARALOID BTA III K3

- Complete absence of stress whitening, even after repeated flexing and creasing
- Good impact resistance, particularly to falling-dart impact
- Glass-like transparency combined with high surface gloss and sparkle
- Efficient dispersion and absence of gel formation
- Exceptionally low "plate out" level
- Good thermal stability and absence of colour development
- Good thermoforming properties
- Ease of handling and low level of dustiness

A modifier ideally suited for calendered film applications requiring high levels of gloss and clarity and absence of crease whitening, such as blister packs, adhesive tapes, metallised films, laminates and overlays.

ACRYLIC IMPACT MODIFIERS

Unfortunately, MBS modifiers cannot maintain their toughness and efficiency under prolonged outdoor exposure. To satisfy the demands of the rapidly growing building products market, PVC formulators require impact modifiers which can preserve appearance and toughness through long-term weathering and exposure.

Rohm and Haas, the world leaders in acrylic chemistry, has developed impact modifier technology based on durable acrylic elastomers to meet this need.

PARALOID KM 323B

- High impact strength
- Excellent weatherability
- Easy, non-critical processing
- Little effect on stiffness or other physical properties

Proven performance in outdoor PVC applications requiring long term retention of toughness and appearance such as window-profiles, door and door frames cladding, rainwater goods etc. Allows complete freedom in the formulation of PVC compounds because it is effective with almost all types of PVC resin and any desired toughness level can easily be achieved. Easy processing and effective performance over a wide range of conditions. Strength/stiffness properties of PVC are maintained after modification, as needed for building products.

PARALOID KM 334

- Outstanding impact strength, especially at low temperatures
- Excellent weatherability and impact retention
- Wide processing range
- Little effect on stiffness or other physical properties

A higher efficiency all acrylic impact modifier combining the benefits of Paraloid KM 323B with improved low temperature toughness. Building product applications containing Paraloid KM 334 exhibit excellent retention of impact strength and appearance after prolonged outdoor exposure. Paraloid Acrylic Impact Modifiers have also been found to be effective in improving the properties of thermoplastics other than PVC.

SPECIALITY MODIFIERS

PARALOID KM 318 F

PVC FOAM MODIFIER

- Combination of processibility, lubrication and impact
- Unsurpassed foam stability, cell uniformity
- Low foam density
- Excellent surface quality
- High impact strength and resistance to nail-splitting

Specially developed modifier for PVC foam compounds used in extruding profiles for construction, mouldings, siding, pipe, fittings and panels.

PARALOID KF 710

GLOSS CONTROL AND ANTI-BLOCKING AGENT

- Stable matt finish in PVC and ABS
- Excellent blocking resistance; good slip and resistance to soiling
- High contact clarity
- No effect on impact strength or other performance properties

A fine-grind powder for easy dispersibility in soft PVC compounds or with low shear processes.

FOOD REGULATIONS

The following summarises the conformance of Rohm and Haas Paraloid Processing Aids and Impact Modifiers for PVC with certain important regulations which govern food packaging and food contact applications. Before adopting these products on a commercial scale however, prospective users are advised that it is their responsibility to abide by the regulations concerning the finished product they are manufacturing.

CONFORMANCE OF PARALOID PROCESSING AIDS AND IMPACT MODIFIERS WITH SELECTED FOOD PACKAGING REGULATIONS		COUNTRY AND REGULATION											
		AUSTRIA	BELGIUM - Royal Decree 8/25/1976	CZECHOSLOVAKIA	FRANCE - Booklet 1227	GERMANY - BGA Recommendations	ITALY - Min. Decree 3/21/1973	NETHERLANDS	POLAND	SPAIN - Min. Decree 2886/1976	SWITZERLAND	UNITED KINGDOM	U.S.A. - FDA 178 3790
Processing Aids	- K 120 N	■	■	■	■	■	■	■	■	■	■	■	■
	- K 125	■	■		■	■	■		■	■	■	■	■
Lub./Processing Aid	- K 175	■	■	■	■	■	■	■	■	■	■	■	■
MBS Impact Modifiers	- KM 608A	■	■		■	■	■	■		■	■	■	■
	- KM 611	■	■	■	■	■	■	■	■	■	■	■	■
	- KM 653	■	■		■	■	■		■	■	■	■	■
	- BTA III F	□	■		■	■	■	■		□	■	■	■
	- BTA III FFI	□	■		■	■	■	■		□	■	■	■
	- BTA III OR	□	■		■	■	■	■		□	■	■	■
	- BTA III N2	□	■		■	■	■	■		□	■	■	■
	- BTA III K3		■		■	■	■	■					
Acrylic Impact Modifiers	- KM 323B		■		■		■	■		■	■		
	- KM 334		■		■		■	■					
Speciality Modifiers	- KM 318 F		■		■	■	■			■	■		
	- KF 710		■		■	■	■			■			

□ = Approvals in progress

For the latest information on the conformance of Paraloid Modifiers with selected food packaging regulations please contact your nearest Rohm and Haas Sales Office.

SAFE HANDLING

Safe handling of Paraloid Modifiers requires no special precautions beyond those which constitute good clean housekeeping and sensible principles of safety, health and enviromental control on a commercial scale. The table below summarises good working practice and contains helpful recommendations. Further information on the safe handling of Paraloid Modifiers can be obtained by contacting you nearest Rohm and Haas Sales Office.

SAFE HANDLING PROCEDURE	
Avoid electrostatic discharge	When transferring to or from containers, all equipment should be grounded to avoid electrostatic discharge.
Clean up spillages promptly	These materials are finely divided and (like talcum powder) can be very slippery when spilt on the ground. Always clean up promptly with a damp brush or vacuum cleaner, and use sweepings as landfill or destroy. Regular cleaning of surfaces is recommended to prevent build-up of dust.
Minimise dust hazards	Because of the materials' dusty nature, use exhaust ventilation and wear dust mask as needed when handling in quantity. In the event of breathing excessive dust, or getting it in the eyes, remove into fresh air, rinse eyes and nasal passages with warm water. Use protective clothing to prevent skin irritation. Avoid dust clouds and prevent flammable concentrations. Use dust-tight electrical equipment, including lighting fittings.
Fire-fighting	Can form explosive dust cloud mixtures. Combustible, but products of combustion are no more hazardous than those of wood or paper. Fight fire with water mist, but avoid use of high velocity water streams. Dry chemicals and carbon dioxide also effective. Wear self contained breathing apparatus in accordance with British Standard 4667 Part 2 (1974).

STORAGE

All Paraloid Modifiers should be stored in a cool, dry, well ventilated place, protected from direct exposure to sunlight.

IMPORTANT NOTICE

Our recommendations for the use of the products listed in this brochure are based on tests believed to be reliable. The standard quality of the materials and their adherence to our published specifications, if any, are guaranteed. In the event material is not of specified quality or does not meet specifications, the seller will replace the material as its sole obligation.

Since the use of these products is beyond the control of the manufacturer, no guarantee, expressed or implied, is made as to such use. The manufacturer expressly disclaims responsibility for effects incidental to such use, handling or possession, or the results obtained, whether or not in accordance, or claimed to be in accordance, with the directions.

1 Antioxidants

Dr. Kurt Schwarzenbach, Ciba-Geigy, Basel/Switzerland

1.1 Introduction

Virtually all polymeric materials, both of synthetic or natural origin, undergo oxidation reactions with oxygen. Technically it is important to distinguish whether such oxidation reactions are taking place as purely thermal processes at usually elevated temperatures, or with the assistance of light (mainly ultraviolet). The topic of this chapter is the thermal oxidation of polymers.

Oxidation can manifest itself in every stage of the life cycle of a polymer, i.e. during manufacturing and storage of the material, or during processing and end use.

Plastics materials are very different from each other as far as their inherent resistance to oxidation is concerned. As examples, polystyrene and polymethyl methacrylate are quite stable, also at normal processing temperatures. Highly unsaturated polymers are much more sensitive to oxidation. Differences in stability against oxidation may arise not only from the broad variety of chemical structures of polymers, but, with a given polymer, also from such differences as crystallinity, morphology or nature and amount of catalyst residues.

The typical manifestations of oxidation are summarized by the term aging phenomena. They are very dependent on the polymer, how it is processed and fabricated, and its end use. On the one hand, the polymer appearance may be changed. Above all, discoloration, especially yellowing, is a frequently experienced phenomenon (typical for styrenic polymers, polyamides, and polycarbonates). Also important are changes in surface appearance (hardening, chalking, cracking, loss of gloss) or loss of transparency. On the other hand, and often more important, is the loss of mechanical properties, such as impact strength, flexural strength, tensile strength and elongation since the loss of the physical properties of a polymer impairs the usefulness of articles manufactured therefrom.

Fundamentally, there are various means available to retard thermal oxidation (methods for stabilization):

- Structural modification of the polymer, e.g. by copolymerization with vinyl group containing antioxidants [28],
- reaction of the end groups to a stable derivative; this procedure is common mainly for polyacetals,
- addition of antioxidants.

Addition of antioxidants is the most commonly used method of stabilization. Antioxidants are chemical entities which retard oxidation and thus the aging of the polymer. Usually, they are effective at concentrations of a fraction of one percent. It is desirable to add the antioxidants as early as possible in the life cycle of a polymer, e.g. before the drying stage of the frequently powdery material. The practical usefulness of a number of commercial polymers is possible only through their efficient stabilization.

1.2 Economical situation

The growth in volume and the economic importance of plastics additives have followed directly the growth of the plastics industry. The growth of plastics has been accompanied by higher demands on the service life and quality of plastics goods. Because of this, the last 20 years have seen great efforts in research and development of plastics additives.

The field of antioxidants and its technical and economic importance are summarized in two reviews, published in 1976 and 1979 [1, 2]. They contain all antioxidant patents granted in the United States between 1972 and 1978, covering the main fields of application, such as in plastics, lubricants, elastomers, foodstuffs etc. The list contains roughly 500 patents, 40% of which are held by the 5 most active companies.

Some figures of antioxidant consumption in plastics applications (excluding thermal stabilizers for PVC) are listed in table 1.

Table 1. Antioxidants for thermoplastics, consumption in metric tons.

	Western Europe (t)	USA (t)	Japan (t)	Others*) (t)	Total (t)
1977	15000	14500	5900	8400	43800
1980	18300	17900	8100	11500	55800

*) without Eastern Countries

The figures correspond to an average annual growth rate of 6%. The forecast until 1985 is of the same order of magnitude.

The distribution of the total consumption in Western Europe among the main countries is given in table 2.

Table 2. Share of antioxidant consumption by country in Western Europe.

West Germany	25%	Great Britain	19%
France	17%	Spain	4%
Italy	17%	Others	2%
Benelux	16%		

The 55800 metric tons consumed in 1980 was used for the stabilization of some 25 million metric tons of thermoplastic resins (excluding PVC). Over 80% of the total amount were used in five polymers, as illustrated in table 3.

Table 3. Share of different thermoplastics on antioxidant consumption.

Plastics material	Share (%)
Acrylonitrile-butadiene-styrene copolymers (ABS)	28
Impact polystyrene (IPS)	15
Polypropylene (PP)	27
High density polyethylene (HDPE)	8
Low density polyethylene (LDPE)	8
Others	14

The distribution of the total 1980 consumption among the most important chemical structures is shown in table 4. Figures are given both as totals and individually for the two most important antioxidant consuming polymers. It can be readily concluded from the figures in table 4 that over 50% of the phenols of high molecular weight were used in polypropylene and that about 50% of the consumed BHT was used in ABS.

Table 4. Distribution of antioxidant consumption among the different chemical classes (1980).

Chemical class	Consumption			
	In PP (t)	In ABS (t)	In other plastics (t)	Total (t)
2,6-Ditert.butyl-p-cresol	1900	6700	4800	13400
Phenols of medium and high molecular weight	7500	2900	3600	14000
Phosphites	1500	4700	8300	14500
Thioethers	4250	1350	4900	10500
Amines	–	–	1700	1700
Others	–	–	1700	1700
Total consumption	15150	15650	25000	55800

1.3 Autoxidation and mechanisms of antioxidant action

1.3.1 Non-inhibited autoxidation

The reaction of organic compounds with molecular oxygen is called autoxidation, because such reactions are taking place "automatically" [3], whenever organic materials are exposed to the atmosphere. Frequently, the presence of small amounts of impurities is essential. Autoxidation is characterized by two features: autocatalysis and inhibition by additives. These features are generally typical for radical reactions, which, in turn, are almost invariably chain reactions.

Kinetic investigations of the autoxidation reaction were first performed by *Bolland* and *Gee* [4]. They reacted low molecular weight, saturated hydrocarbons in the liquid phase with oxygen. The results were rationalized by a mechanism, which can be depicted as follows:

INITIATION

$$RH \longrightarrow R^{\cdot} \qquad (1)$$

PROPAGATION

$$R^{\cdot} + O_2 \longrightarrow ROO^{\cdot} \qquad (2)$$

$$ROO^{\cdot} + RH \longrightarrow ROOH + R^{\cdot} \qquad (3)$$

CHAIN BRANCHING

$$ROOH \longrightarrow RO^{\cdot} + {}^{\cdot}OH \qquad (4)$$

$$2\,ROOH \longrightarrow RO^{\cdot} + ROO^{\cdot} + H_2O \qquad (5)$$

TERMINATION

$$2\,ROO^{\cdot} \longrightarrow \text{inactive products} + O_2 \qquad (6)$$

$$R^{\cdot} + ROO^{\cdot} \longrightarrow ROOR \qquad (7)$$

$$2\,R^{\cdot} \longrightarrow R{-}R \qquad (8)$$

According to present knowledge, the same sequence of reactions occurs also in autoxidations of polymeric substances. This means that they consist of a starting reaction, generating free radicals ($R^{\cdot}$), propagating reactions, yielding reaction products (ROOH) and in a series of termination reactions leading to radical destruction.

The details of the mechanisms of the initiating reaction (1) are not yet fully understood. Usually, it is assumed that the oxidation initiating polymer radical $R^{\cdot}$ is formed by heat, light or by mechanical stress or by reactions with radicals originating from foreign sources.

Considering the two basic reactions of propagation, it is obvious that the two radicals $R^{\cdot}$ and $ROO^{\cdot}$ (alkyl radicals and peroxy radicals) are the high energy key intermediates. In the kinetic treatment of autoxidation it is assumed that a stationary state is reached in which the concentrations of $R^{\cdot}$ and $ROO^{\cdot}$ are constant. (2) is a very fast reaction. At reasonably high oxygen pressure the concentration of the peroxy radicals is far greater than the concentration of the alkyl radicals. Consequently, the main termination reaction under such conditions is (6).

The key reactions of chain branching, (4) and (5), are both decay reactions of hydroperoxides. Such reactions are characterized by high activation energies. Consequently, they are becoming significant only at elevated temperatures, i.e. over 120 °C. The newly formed radicals $RO^{\cdot}$ and ${}^{\cdot}OH$ are capable to react with RH just as peroxy radicals and thus can initiate new reaction chains:

$$RO^{\cdot} + RH \longrightarrow ROH + R^{\cdot} \qquad (9)$$

$${}^{\cdot}OH + RH \longrightarrow H_2O + R^{\cdot} \qquad (10)$$

The autocatalytic character of the autoxidation is implied by such kinetic chain branching reactions.

The decay of hydroperoxides is strongly catalyzed by metal ions. The most active catalysts are such metals which form two ions of similar stability and whose oxidation numbers differ by one (e.g. Fe^{+2}/Fe^{+3}; Co^{+2}/Co^{+3}; Mn^{+2}/Mn^{+3}; Cu^{+1}/Cu^{+2}).

$$ROOH + Me^{+} \longrightarrow RO^{\cdot} + Me^{+2} + OH^{-} \qquad (11)$$

$$ROOH + Me^{+2} \longrightarrow ROO^{\cdot} + Me^{+} + H^{+} \qquad (12)$$

$$2\,ROOH \longrightarrow ROO^{\cdot} + RO^{\cdot} + H_2O \qquad (5)$$

Thus, the overall effect of metal ions consists of a catalysis of the formally bimolecular process (5), which produces radicals able to trigger new radical chains.

The effect of autoxidation on a polymer may be cross-linking or chain splitting. Cross-linking is easy to explain by the termination reactions (7) and (8). Cleavage of the polymer chain may occur according to one of the following reactions:

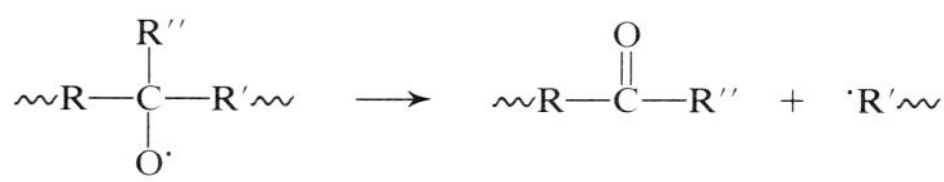

(13)

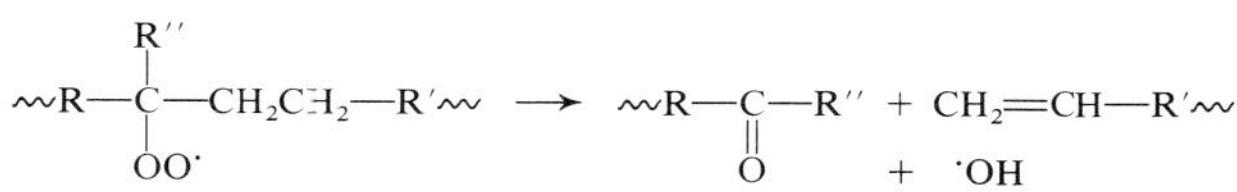

(14)

1.3.2 Inhibited autoxidation

The effectiveness of antioxidants is based on the fact that they are able to influence the outlined processes in different ways, dependent on their chemical structure. Normally, two large, basic groups of antioxidants are distinguished: First, chain terminating or primary antioxidants and second, hydroperoxide decomposers or secondary antioxidants, frequently called synergists. The majority of primary antioxidants are sterically hindered phenols or secondary aromatic amines. They are capable of undergoing fast reactions with peroxy radicals, therefore, they are frequently called radical scavengers.

An example of such a reaction sequence is given by the following scheme:

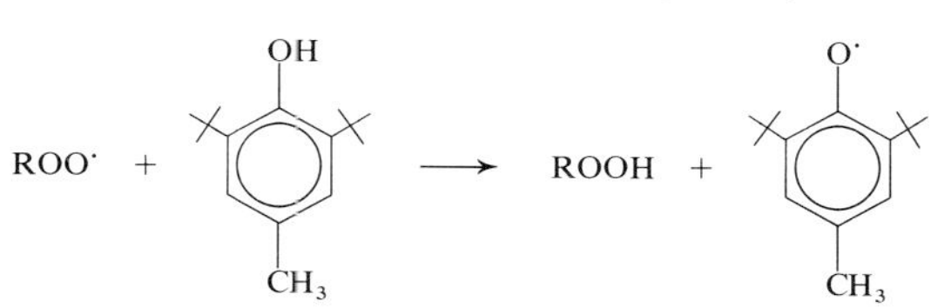

(15)

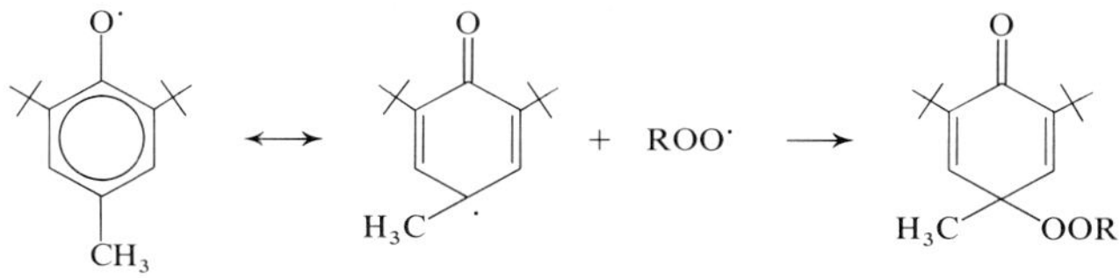

(16)

According to the above scheme, stabilization is achieved by the fact that reaction (15) competes with reaction (3), transforming the reactive peroxy radical into the much less reactive phenoxy radical, which, in turn, is capable to react with a second peroxy radical according to (16). There is abundant literature where the isolation of the generated peroxycyclohexadienones (16) is described, together with numerous other oxidation products from various antioxidants [5 to 9].

The thermal stability of the peroxycyclohexadienones (16) is limited, their decomposition leads to new reaction chains also at temperatures below 150 °C. The phenoxy radicals (15) do not initiate new radical chains at the normal temperatures of use and testing, however, this propagation reaction becomes possible at high temperatures. This means that the effectiveness of sterically hindered phenols is decreasing with increasing temperature.

The secondary antioxidants are essentially sulfur compounds (mostly thioethers and esters of thiodipropionic acid) or triesters of phosphorous acid (phosphites). Both have the ability to react with hydroperoxides to yield non-radical products following heterolytic mechanisms. The reaction of phosphites to phosphates is an example [10]:

$$P(OR)_3 + R'OOH \longrightarrow OP(OR)_3 + R'OH \tag{17}$$

With phosphites though, there is evidence for accompanying homolytic reactions with peroxy radicals [11, 12].

Thioethers react with hydroperoxides in a first stage to yield sulfoxides and alcohols:

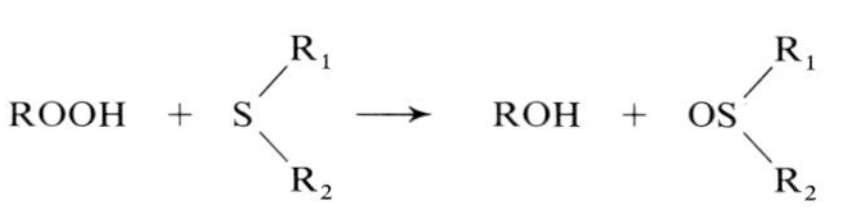

(18)

Sulfoxides themselves yield on further oxidation even more powerful hydroperoxide decomposers than the original sulfides, in that they are able to destroy several equivalents of hydroperoxides. This catalytic effect is explained by the intermediate occurrence of sulfenic acids and sulfur dioxide [13 to 15]. The reaction with hydroperoxides is a good explanation for the fact that the effectiveness of secondary antioxidants is particularly evident whenever the substrate is already preoxidized to an appreciable extent; i.e. when it contains hydroperoxides.

Frequently, synergistic effects are observed when primary and secondary antioxidants are combined, this means that the effectiveness of the combination exceeds the effectiveness of the individual components. Such synergistic effects are very often explained with the concept of the simultaneous occurrence of the radical reactions and the non-radical hydroperoxide decomposition.

Detailed reviews on the present knowledge of autoxidation and inhibition of oxidation reactions have been published repeatedly [16 to 19].

1.4 Chemical structures

An extremely valuable survey of additives, patents and the chemical structures proposed therein is found in the book by *J. Voigt* [20]. A recently published extensive catalogue of stabilizers for polyolefins lists approximately 500 compounds and the 3000 corresponding trade names, half of the compounds being antioxidants [21]. The tabulated compounds are not all commercially available.

Table 5. Important commercial antioxidants for thermoplastics.

Antioxidant group	Chemical class	Number of representatives
Primary antioxidants	Sterically hindered phenols	34
	Secondary aromatic amines	3
Secondary antioxidants	Thioethers	4
	Phosphites and phosphonites	7
	Zinc dibutyldithiocarbamate	1
Synergistic mixtures of primary and secondary antioxidants		5
Total		54

Table 5 depicts the distribution of the most important compounds among their chemical structures.

The greatest diversity is found in the class of the sterically hindered phenols. Most of them are commercially available as relatively pure chemicals. Although aromatic amines are often more powerful antioxidants than phenols, their application is limited to vulcanized elastomers. Their staining properties prohibit their use in most thermoplastics.

Phenols may be further classified according to their particular structure, or according to their molecular weight. An important goal of antioxidant research of the past twenty years was to provide polyphenols of high molecular weight and low volatility (table 6).

Table 6. Important sterically hindered phenols, classification according to chemistry and molecular weight.

Chemical class	Number of products in indicated range of molecular weight		
	<300	300 to 600	>600
Thiobisphenols	–	2	–
Alkylidene-bisphenols	–	9	3
Alkylphenols	2	3	–
Hydroxybenzyl compounds	–	1	5
Acylaminophenols	–	3	–
Hydroxyphenylpropionates	–	1	5

Phosphites and phosphonites are mostly aromatic in nature, e.g. substituted triphenylphosphites. The reason for the use of aromatic compounds is their more favorable hydrolytic stability, compared to aliphatic esters of phosphorous acid. Some representatives are solid at room temperature. This facilitates their manufacture, purification and handling.

The majority of the thioethers are esters of thiodipropionic acid with fatty alcohols.

The compounds of the tables 5 and 6 are represented by the following chemical structures:

a) Thiobisphenols, e.g.:

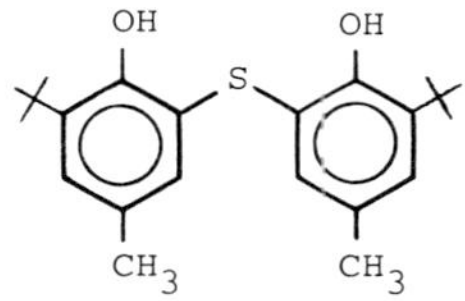

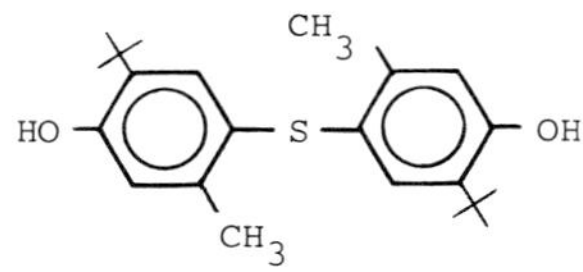

b) Alkylidene-bisphenols, e.g.:

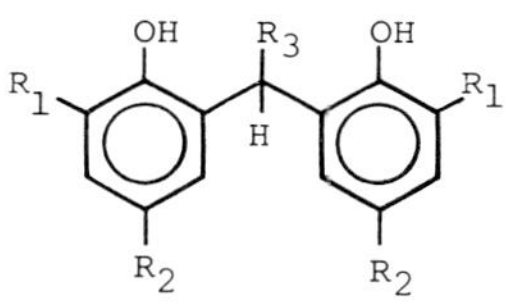

R_1	R_2	R_3
tert. butyl	tert. butyl	—H
tert. butyl	tert. butyl	—CH_3
tert. butyl	sec. butyl	—CH_3
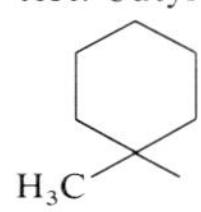	—CH_3	—H
tert. butyl	—C_2H_5	—H
tert. butyl	—CH_3	—H
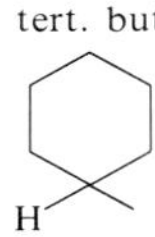	—CH_3	—H
—C_9H_{19}	—CH_3	—H

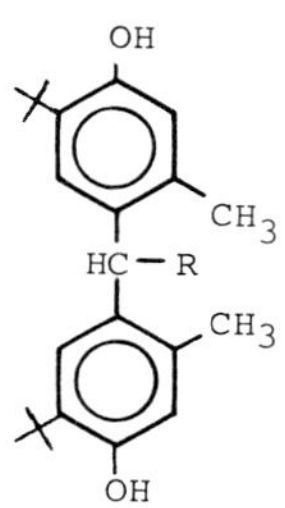

R
—C_3H_7
—CH_2—CH—CH_3 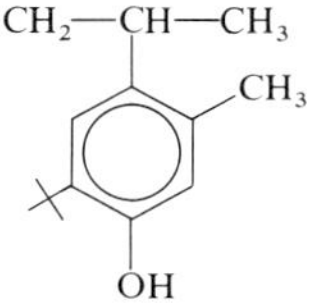

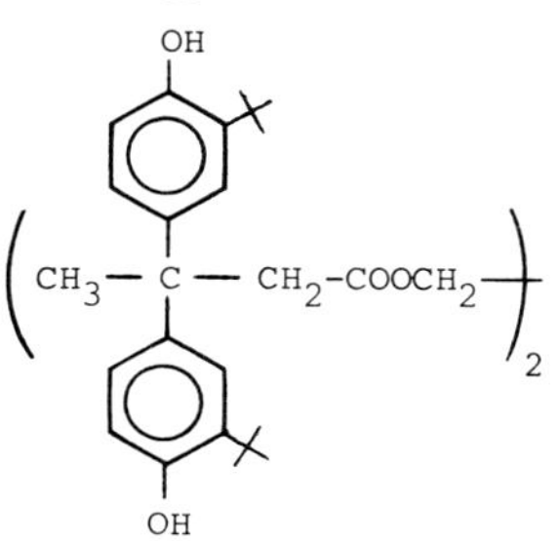

c) Alkylphenols, e. g.:

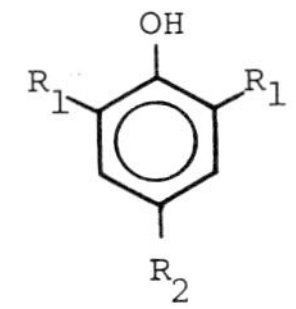

R_1	R_2
tert. butyl	$-CH_3$
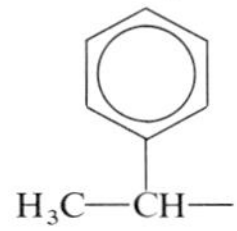	$-C_9H_{19}$
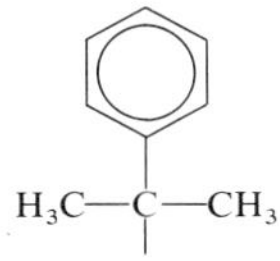	$-C_9H_{19}$

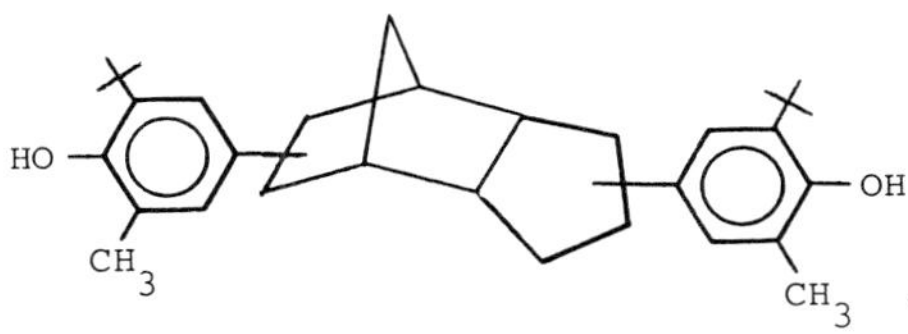

d) Hydroxybenzyl compounds, e. g.:

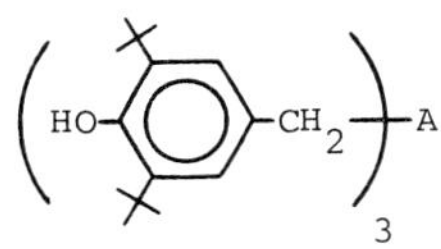

A
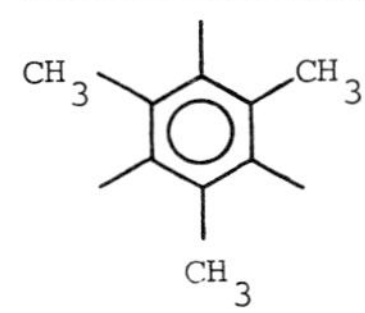
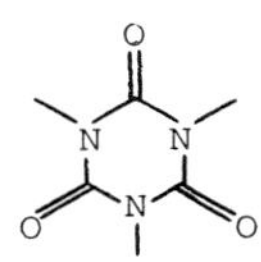

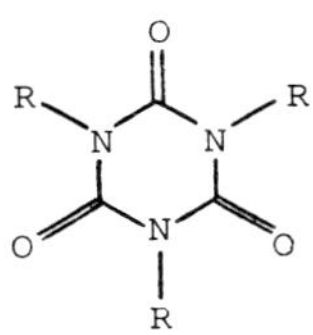

R =

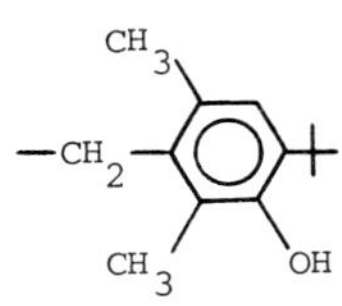

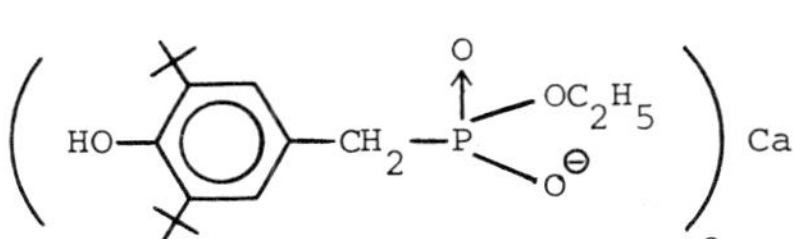

e) Aminophenols, e.g.:

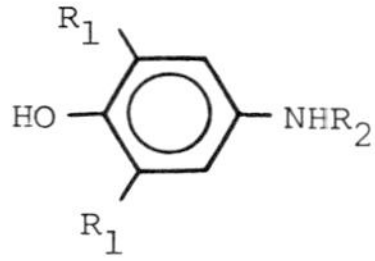

R_1	R_2
—H	$—COC_{11}H_{23}$
—H	$—COC_{17}H_{35}$
tert. butyl	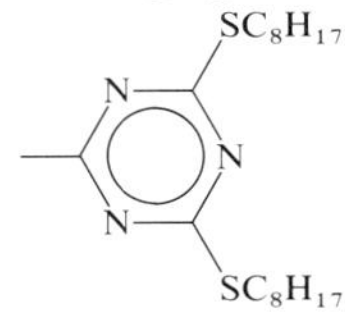

f) Hydroxyphenylpropionates, e.g.:

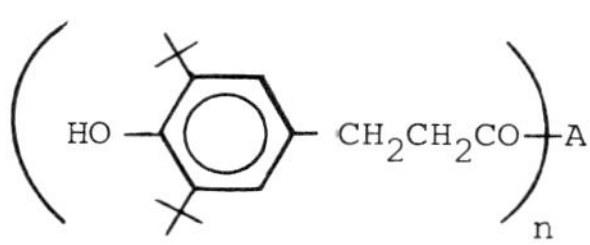

n	A
1	$—OC_{18}H_{37}$
2	$—O(CH_2)_6O—$
2	$—NH—(CH_2)_6—NH—$
2	$—O(CH_2)_2—S—(CH_2)_2O—$
3	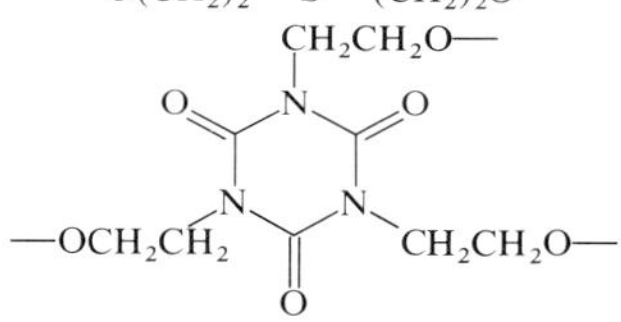
4	$C—(CH_2O)_4$

g) Secondary aromatic amines, e.g.:

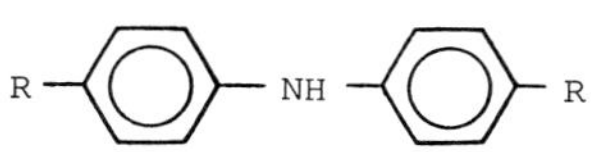

R
tert. octyl
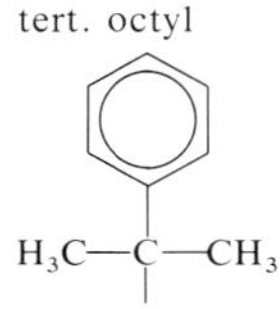

h) Thioethers, e.g.:

$CH_2CH_2—COOR$
|
S
|
$CH_2CH_2—COOR$

R
$—C_{12}H_{25}$
$—C_{13}H_{27}$
$—C_{18}H_{37}$

$C_{18}H_{37}—SS—C_{18}H_{37}$

i) Phosphites and phosphonites, e.g.:

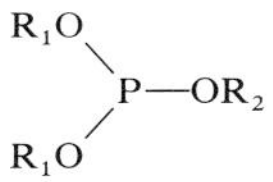

R_1	R_2
—phenyl	—n-decyl
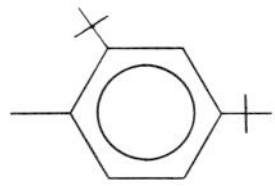	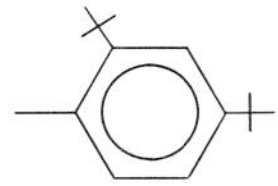
—nonylphenyl	—nonylphenyl

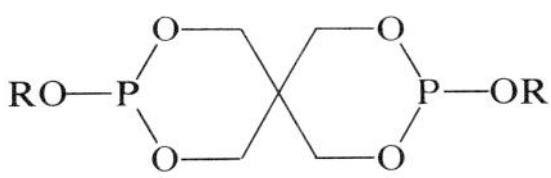

R
$—C_{18}H_{37}$
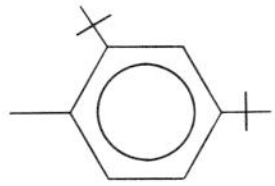

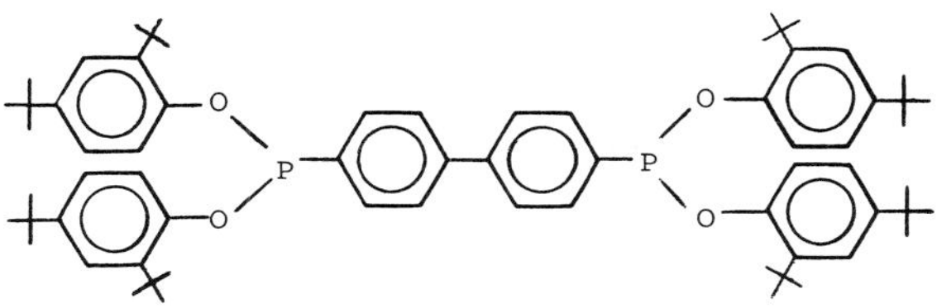

It is an obvious approach to combine the effectiveness of primary and secondary antioxidants by combining the two structural elements in one and the same molecule [22]. It is noteworthy that there are very few such compounds among the most important commercial products. Of the 34 phenols and polyphenols in table 6 only five contain a sulfur atom.

1.5 Physical and chemical requirements

Besides its main function of inhibiting the oxidation of polymeric materials, an antioxidant must fulfill a whole series of important ancillary properties which are usually determining factors in assessing the practicability of a compound. Depending on the polymer to be stabilized and its end use, one or the other ancillary property will be particularly important. Apart from cost considerations, ancillary properties may be divided into three groups: chemical, physical and toxicological. A special chapter deals with the latter group (cf. chapter 18).

Toxicological properties may be of crucial importance in the choice of an antioxidant because for many applications plastics and additives need specific toxicological approvals. Food packaging may serve as an example. The conditions for toxicological approval are different from country to country. Information on the extent to which approval is granted in different plastics is provided by the additives manufacturers.

1.5.1 Chemical ancillary properties

Discoloration

An antioxidant has to be colorless and should contribute as little as possible to discoloration of the substrate during long term use. The discoloration caused by aromatic amines is the main reason for the fact that antioxidants of this type are rarely used in thermoplastics (see table 4). Some discoloration (usually yellowing) may be observed also with sterically hindered phenols. Chemically, this is explained with the formation of oxidation products of the antioxidants [5 to 9]. In substrates, which tend less to discoloration on aging, such as polyolefins and polyacetals, discoloration is most readily seen. In plastics, such as polyurethanes, polycarbonates and styrenics, the discoloration of the substrate itself is superimposed by possible discoloration caused by any additive. Depending on the conditions of aging, a yellowing effect may be caused mainly by the substrate or by the additive. Such environmental conditions may be exposure to heat, irradiation, alkali or industrial waste gases, leading to what is commonly known as "gasfading".

Antioxidant-caused discoloration during heat treatment (processing; oven aging) is generally not very important, in discoloring substrates it is superimposed by the inherent discoloration of the substrate.

The effects observed are somewhat more complicated under conditions of artificial or outdoor weathering. The reaction products of substituted phenols with peroxy radicals show rather low light fastness. Consequently, no discoloration is usually observed under intense irradiation (direct sunlight, Xenotest). The situation is quite different under conditions of indirect daylight, usually behind glass (ambient daylight). During some weeks or months some yellowing may build up which disappears on short and intense irradiation, yet it is still cause of inconvenience.

A remedy for many types of discoloration is very often the use of a phosphite or a thioether.

Thermal stability

An antioxidant has to survive the conditions of manufacturing and processing. The commercially available materials all fulfill this requirement of short term heat stability at temperatures of up to 300 to 320 °C to a satisfactory degree.

Hydrolytic stability

Some commercial antioxidants are esters of carboxylic acids. Ester groups normally do not hydrolyze under conditions encountered by a polymer. Phosphites, on the other hand, are much more sensitive towards hydrolysis. This point has to be taken into account in storing and using phosphites. To facilitate the industrial use of phosphites, three measures can be taken:

- Use of aromatic phosphites of high purity, which are inherently more resistant towards hydrolysis than aliphatic phosphites,
- improving storage stability by admixing a small amount of a base,
- blending with a water repellent wax or any suitable hydrophobic compound.

1.5.2 Physical ancillary properties

Handling

Commercial antioxidants are usually free flowing powders with melting points of up to 250 °C or liquids of medium to low viscosity. To ensure storage and transportation stability, melting or softening points should not be below 50 °C. With rising requirements concerning industrial hygiene, dust-free preparations are preferred.

Solubility and emulsifiability

Whenever the antioxidant is added during the manufacturing step of the polymer, it has to be introduced into the process in a suitable physical form. For some processes the antioxidant has to be soluble in the monomer or in the polymerization solvent. If the antioxidant is preferably added as a liquid, coadditives (e.g. a liquid phosphite or thioether) may serve as the solvent. In the ABS emulsion polymerization process, the preferred method is to add the antioxidant as an emulsion. Products with high melting points are less suitable in this instance.

Volatility

In the final stages of polymer manufacture (drying, monomer devolatilization), an added antioxidant should not be lost through volatilization. Many antioxidants, mainly the long term heat stabilizers, are added during processing of the polymer, they should not volatilize or cause unwanted odor. Most commercial antioxidants fulfill this requirement.

Extraction resistance

Resistance to extraction is generally a very important requirement. Typical applications for plastics materials calling for extraction resistance of the additives are outdoor or domestic applications (e.g. washing machine parts). In addition, extraction resistance is one condition for toxicological approval [23, 24].

Recent publications indicate that efforts are being made to further improve the extraction stability of additives [25]:

- Synthesis of oligomeric antioxidants with molecular weights of 3000 and above, see e.g. [26].
- Chemical bonding of the antioxidant to the polymer by a grafting process [27].
- Incorporation of antioxidants into the polymer backbone by copolymerization with a suitable monomer [28].

However, these attempts have not led to widespread practical use so far.

Compatibility and migration

With some polymers it is occasionally observed that additives tend to bloom on the surface of the material. This is especially true for low density polyethylene and polyurethanes. Such processes usually require a few weeks or months. Blooming is observed when the additives form a supersaturated solution in the polymer and have sufficiently high diffusion rates to migrate to the surface. It can be shown, that diffusion coefficients are dependent on the polymer (glass transition temperature), its morphology (crystallinity, orientation), and on the additive itself. A rough relation states that the susceptibility for blooming is about 1:10:100 for PP, HDPE and LDPE. Quantitative measurements of diffusion coefficients of additives in polyolefins and polyurethanes are the subject of numerous publications [29 to 36].

1.6 Testing of antioxidants

The key problem is the choice of appropriate testing methods designed to yield quantitative data, which are relevant to practice and which assess the effectiveness of an antioxidant in a reasonable period of time.

1.6.1 Investigations with model systems for polymers

There are many investigations which have been carried out on model systems and for which practical considerations were of little importance. A frequently employed method is the kinetic study of the oxidation of model compounds and the determination of the effect of added antioxidants. Model compounds are low molecular weight species containing the structural functionality of a specific polymer. For example:

- Ethylene dibenzoate, diethyleneglycol dibenzoate and 2-hydroxyethyl benzoate for linear polyesters [37],
- low molecular weight amides for polyamides [38],
- branched low molecular weight hydrocarbons for polyolefins,
- cumene, tetralin or decalin as compounds in which the oxidation to the corresponding hydroperoxides and their reaction products can easily be followed.

Another method is the study of the reaction of antioxidants with defined hydroperoxides (ROOH) or peroxy radicals ($ROO^{\cdot}$). A frequently employed hydroperoxide is cumene hydroperoxide. Peroxy radicals are generated in situ, they are for instance formed by homolysis of compounds yielding C-radicals in the presence of oxygen [13, 39 to 41].

Oxygen uptake measurement is still an important method for the quantitative assessment of the effectiveness of antioxidants. Both liquid or dissolved model compounds or the polymer itself may be used as the substrate [39].

These chemical methods have provided very valuable information on the mechanism of autoxidation and its inhibition. However, the antioxidant properties essential for practical purposes cannot always be estimated accurately. One main reason presumably is the fact that the differences in viscosity and diffusion between a model and a polymer can be considerable.

1.6.2 Testing methods related to actual practice

The assessment of the effectiveness of an antioxidant by testing methods related to practice consists almost exclusively of performing accelerated aging tests at elevated temperature. The quality of the test specimens is very important and has to be checked carefully. This means specifically:

- The substrate being tested must not be preoxidized and should be prestabilized as little as possible.
- The distribution of the additives must be homogeneous throughout the test specimen.
- Series of test specimens must consist of identical entities (morphology, dimension, thermal history).

Measurement of the change of one or several physical properties yields quantitative data indicating the damage sustained by the polymeric material during aging. Physical properties should be measured in accordance with the appropriate standards.

As mentioned earlier, oxidative deterioration can occur during all stages of the life cycle of a polymer, i.e. during manufacturing, processing and end use. In principle it is advisable to add the antioxidants as early as possible, i.e. whenever possible during polymerization or polycondensation. This ensures uniform distribution of the additive and consequently optimum protection against aging phenomena during recovery, drying and pelletizing.

The effectiveness of antioxidants under *processing conditions* has to be assessed on the polymer melt. The usual procedure consists of multiple extrusions or prolonged residence times in a hot cylinder or in an injection molding machine. By far the most important properties of the polymer which are measured under such conditions are the change of the melt flow index and possibly yellowing. For certain substrates Brabender plastograms give a good indication of the stability of the polymer melt. Short term treatment in a forced draft oven in the neighborhood of the processing temperatures is suitable for styrenic polymers, the most important criterion being yellowing.

The effectiveness of antioxidants under *conditions of actual use* (i.e. as long term heat stabilizers) is estimated almost exclusively by means of oven aging tests at temperatures below the melting point of the polymer. Relatively high temperatures yield rapid results, but these are generally not very significant in actual practice. Optimally, oven aging tests should be carried out as close to use temperature as practical.

Table 7. Testing of thermoplastics and criteria.

Phase	Test	Criteria
Manufacture	Laboratory polymerization or polycondensation	Yellowing, melt flow index, mechanical properties
Processing	Heat treatment of the melt; e.g. multiple extrusions	Yellowing, melt flow index, mechanical properties
End use	Oven aging (if necessary irradiation)	cf. table 8

Table 8. Important test criteria for different thermoplastics on oven aging.

Criteria	Poly-ole-fins	Sty-renics	Poly-amides	Poly-ure-thanes	Poly-car-bon-ates	Poly-ace-tals	Thermo-plastic elastomers	PBTP
Embrittlement	xx		x					
Yellowing	x	xx	x	xx	xx	x	x	
Tensile strength Elongation Tensile stress at yield	x	x	xx	x		x	xx	x
Impact strength	x	xx	x		x	x		x
Flexural strength	x	x						
Solution viscosity			x		x			
Weight loss						xx		

xx Main criterion

The test criterion for a polyolefin, e.g. polypropylene, is the induction time to the beginning of the visually detectable desintegration (brittleness). In addition, the examination of discoloration during oven aging is also significant. Because unstabilized polyolefins (polyacetals are another good example) do not discolor, any discoloration has to be attributed to oxidation products generated by the additives. In contrast, styrenic polymers (as examples) are strongly discoloring under conditions of aging, and the ability of the antioxidant to inhibit such discoloration is one of the most important test criteria. The same statement holds for polyurethanes and polycarbonates.

The changes in mechanical properties as a function of oven aging time are the most generally applied test criteria. Impact strength, elongation and flexural strength are crucial for styrenic polymers, whereas tensile strength, elongation and tensile stress at yield are important for polyolefins, polyamides, polyurethanes and thermoplastic elastomers. Whenever the measurement of solution viscosities is considered to be relevant, its determination is widespread (e.g. for polyamides and polycarbonates). Measuring the weight loss of polyacetals is specific to this substrate [42]. To the list of criteria may be added gloss, hardness and appearance of the surface; electrical, spectroscopic and thermoanalytic properties.

1.7 Chemical analysis of antioxidants

The analytical determination of antioxidants in plastics is an important part of every production control. Suitable methods are provided by the additives suppliers. The methods should be specific, simple and quick to carry out and sensitive down to the ppm-range.

Chromatographic methods are used almost exclusively for antioxidants. The semi-quantitative thin layer chromatography is widely replaced by vapor phase chromatography today. Antioxidants with high molecular weight are estimated using high pressure liquid chromatography [43].

1.8 Stabilization of selected plastics

The scope of this book does not permit extensive treatment of the state of the art of the stabilization of thermoplastic materials; this section is intended merely as a description of some typical examples.

1.8.1 Polypropylene (PP)

Polypropylene is quite thermostable in the absence of oxygen. However, it is characterized by a marked sensitivity towards oxidation. Consequently, polypropylene is stabilized before it reaches the processor. The antioxidants are added at least partially during the manufacturing process and at the latest during pelletizing.

The most important requirements concerning the ancillary properties of an antioxidant for polypropylene are:

- No tendency to discolor the substrate,
- compatibility and resistance against extraction,
- low volatility.

1.8.1.1 Stabilization of polypropylene during processing

Polypropylene is normally processed at temperatures between 220 and 280 °C. It will degrade under these conditions if insufficiently stabilized. This degradation is, among other things, manifested by a rise in melt flow index.

Antioxidant systems in technical use are composed of processing stabilizers, long term heat stabilizers, calcium- or zinc-stearate, and synergists if necessary.

Typical processing stabilizers are:

- 2,6-Ditert.butyl-p-cresol (butylated hydroxy toluene, BHT),
- phosphites and phosphonites, e.g.:
  tetrakis-(2,4-ditert.butylphenyl)-4,4′-biphenylylenediphosphonite,
  distearyl-pentaerythrityl-diphosphite,
  tris-(nonylphenyl)-phosphite,
  tris-(2,4-ditert.butylphenyl)-phosphite,
  bis-(2,4-ditert.butylphenyl)-pentaerythrityl-diphosphite.

These additives are commonly added in concentrations between 0.05 and 0.25%. The effectiveness of different stabilizer systems is best assessed by multiple extrusions. The melt flow index of the extruded material is determined after each pass (figure 1) [44].

It should be mentioned that in commercial polypropylenes, phosphorus compounds are always used together with a sterically hindered phenol [45 to 47].

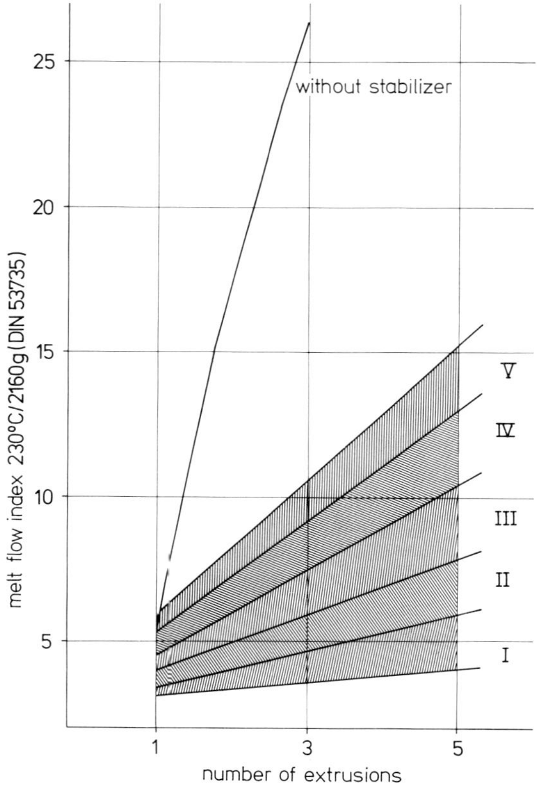

Fig. 1. Change of melt flow index of polypropylene during multiple extrusion (single screw extruder, 20 mm, L/D = 20, 100 rpm, temperature at the die 260 °C).

I: tetrakis-(2,4-ditert.butylphenyl)-4,4′-biphenylylenediphosphonite or bis-(2,4-ditert.butylphenyl)-pentaerythrityl-diphosphite (0.1%), combination phosphite (0.1%) + long term heat stabilizer (0.05%);

II: tetrakis-(2,4-ditert.butylphenyl)-4,4′-biphenylylenediphosphonite or bis-(2,4-ditert.butylphenyl)-pentaerythrityl-diphosphite (0.05%), BHT (0.1%), combination phosphite (0.05%) + long term heat stabilizer (0.05%), combination BHT (0.05%) + long term heat stabilizer (0.05%);

III: like II at half the concentrations;

IV: long term heat stabilizers without additional processing stabilizer (0.1%);

V: long term heat stabilizers at 0.05%.

The values of the change in melt flow index are quite dependent on the particular polymer. However, the ranking in effectiveness of individual formulations will not change significantly from one polymer to the other.

If the requirements for processing stability are not exceedingly high, it is sufficient to use the high molecular weight phenols alone. In contrast to BHT, the extraction stability of phosphorus compounds is excellent, in addition, they suppress occasional slight discoloration.

1.8.1.2 Long term heat stabilizers for polypropylene

At elevated temperatures and with access of air, polypropylene disintegrates to powdery oxidation products. This process is characterized by well reproducible induction periods, commonly called oven life times. End points are easily detected even visually. The disintegration begins mostly at the edges and corners of the test specimen. The oven life times are characteristic for a particular test specimen and are reported usually in days. They are a function of a number of variables. The most important of them are the following:

- Stabilizer/stabilizer system,
- stabilizer concentration,
- specimen thickness,
- test temperature.

The dependence on a number of further experimental parameters was published by *Forsman* [48].

The most important long term heat stabilizers for polypropylene are phenols of medium (300 to 600) and especially high (600 to 1 200) molecular weight, which are frequently used together with thioethers as synergists, e.g. dilauryl thiodipropionate, DLTDP, or distearyl thiodipropionate, DSTDP, or dioctadecyl disulfide:

- Octadecyl 3-(3,5-ditert.butyl-4-hydroxyphenyl)-propionate,
- pentaerythrityl tetrakis-3-(3,5-ditert.butyl-4-hydroxyphenyl)-propionate,
- 1,3,5-tris-(3,5-ditert.butyl-4-hydroxybenzyl)-isocyanurate,
- 1,1,3-tris-(5-tert.butyl-4-hydroxy-2-methylphenyl)-butane,
- 1,3,5-tris-(3,5-ditert.butyl-4-hydroxybenzyl)-mesitylene,
- ethyleneglycol bis-[3,3-bis-(3′-tert.butyl-4′-hydroxyphenyl)-butyrate].

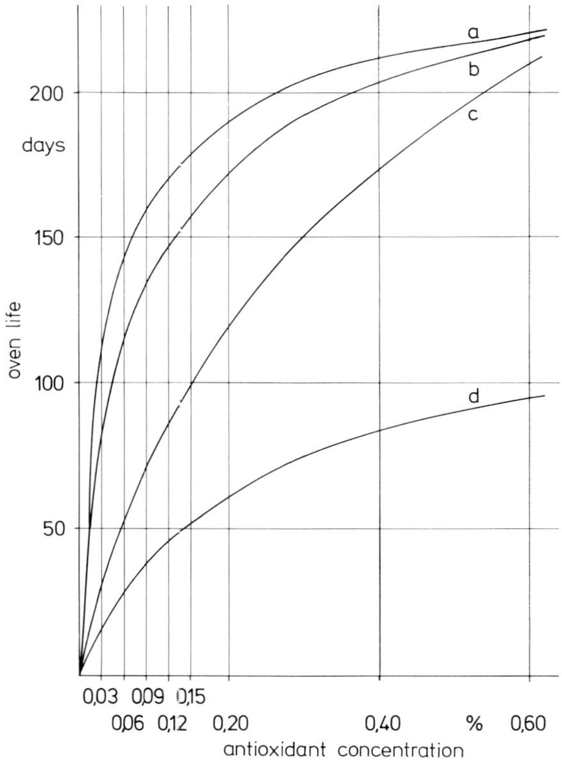

Fig. 2. Oven life of compression molded 1 mm polypropylene plaques at 135 °C. Effectiveness of selected long term heat stabilizers as function of concentration.

a: pentaerythrityl tetrakis-3-(3,5-ditert.butyl-4-hydroxyphenyl)-propionate,

b: 1,3,5-tris-(3,5-ditert.butyl-4-hydroxybenzyl)-mesitylene,

c: 1,3,5-tris-(3,5-ditert.butyl-4-hydroxybenzyl)-isocyanurate,

d: 1,1,3-tris-(5-tert.butyl-4-hydroxy-2-methylphenyl)-butane.

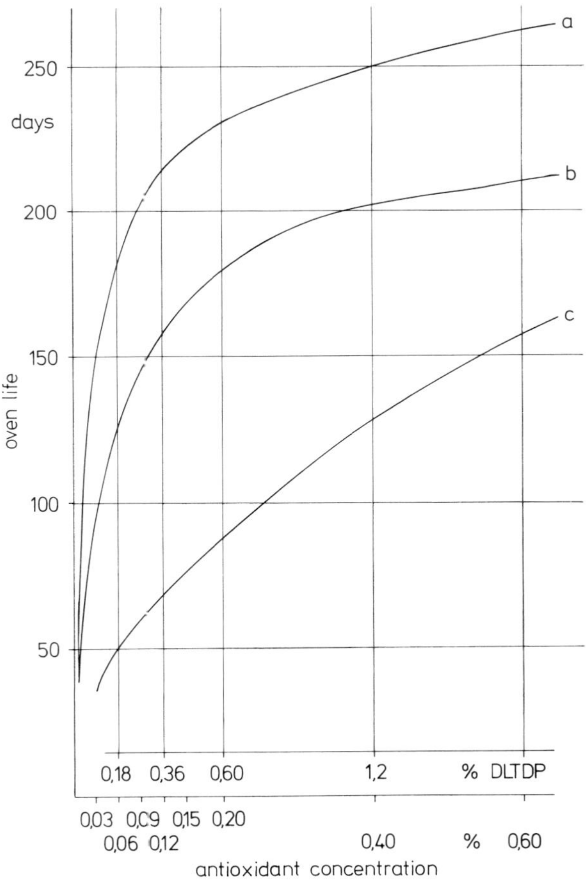

Fig. 3. Oven life of compression molded 1 mm polypropylene plaques at 135 °C. Effectiveness of selected long term heat stabilizers in combination with the triple amount of DLTDP as function of concentration.

a: pentaerythrityl tetrakis-3-(3,5-ditert.butyl-4-hydroxyphenyl)-propionate,

b: 1,1,3-tris-(5-tert.butyl-4-hydroxy-2-methylphenyl)-butane,

c: octadecyl 3-(3,5-ditert.butyl-4-hydroxyphenyl)-propionate.

1.8.1.2.1 Stabilization of polypropylene, thick cross sections

The curves in figures 2 and 3 give an indication of the order of magnitude of oven life times. The curves only hold for a polypropylene of a particular kind and origin. Different polypropylenes can differ distinctly from one another.

Naturally, oven life times are temperature dependent [49, 50]. Within the testing range of 125 to 150 °C, a temperature increase of 10 °C diminishes the times by a factor of roughly 2.5 (for a specimen thickness

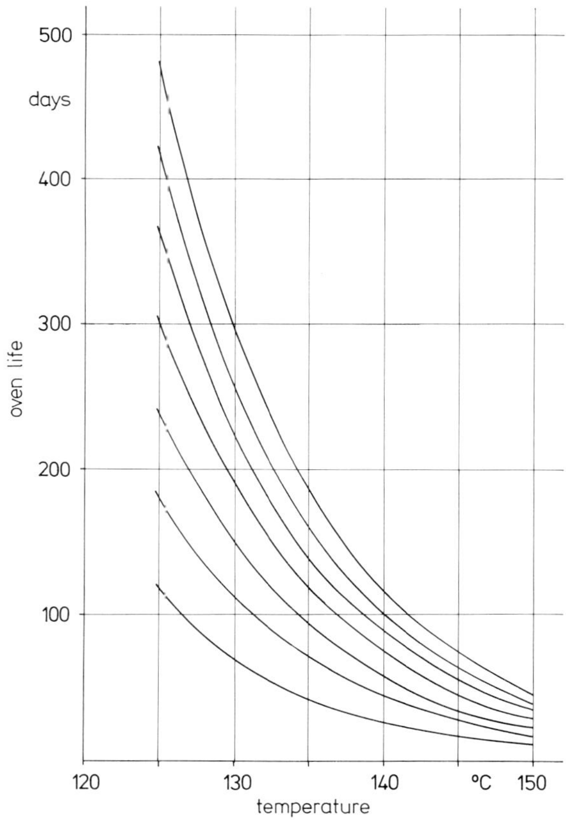

Fig. 4. Temperature dependence of oven life for polypropylene. One distinct curve applies to a test specimen of defined thickness and stabilization.

of 0.5 to 3 mm, figure 4). This has to be regarded as a rough rule of thumb. There are two main reasons, which can cause deviations:

- During rather long oven aging times, antioxidants of low and medium molecular weight may become lost through volatilization.
- *De Jonge* [51] has published experimentally determined oven life times down to temperatures of 80 °C, which show that the temperature dependence (and thus, the apparent activation energy) is decreasing with falling temperature. This has to be expected considering reactions (4) and (5) outlined on p. 5.

The dependence of the oven life times on the thickness of the test specimen [49, 50] is shown in figure 5. The figures may be summarized by two rules of thumb:

- In the thickness range of over 1 mm, oven life times increase by a factor of about 1.3 for every millimeter increase in thickness.
- For thin structures (thicknesses of 60 microns and below, i.e. stretched tapes), oven life times are practically independent of thickness.

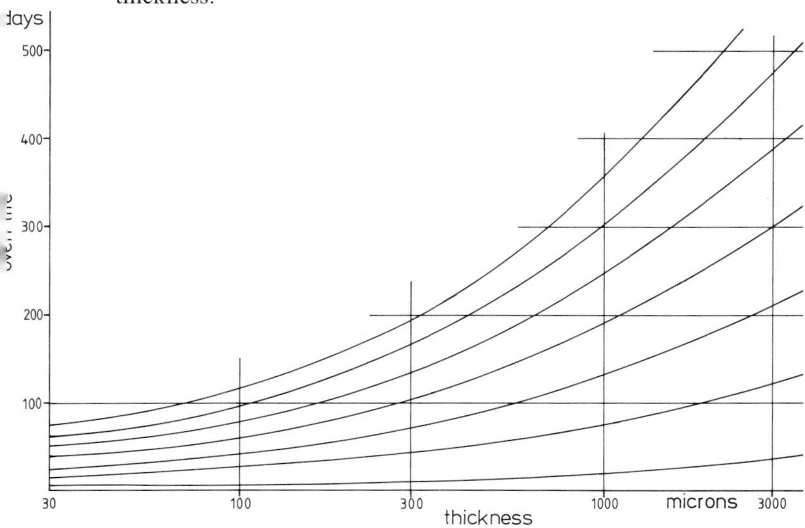

Fig. 5. Dependence of oven life on sample thickness for polypropylene. One distinct curve applies to a test specimen of defined stabilization at constant temperature (logarithmic scale).

The curves in figures 4 and 5 are based on a large number of test results [52]. Differences between individual polymers and stabilizer systems were averaged for a better overview. Figures 4 and 5 make it possible to extend literature data [53 to 57] to a range of temperatures and thicknesses which facilitates the comparison of data of different origin; examples follow in figures 6 and 7. The basis for both is curve *d* in figure 2 [1,1,3-tris-(5-tert.butyl-4-hydroxy-2-methylphenyl)-butane]. Figure 6 is an extension to different temperatures with the aid of figure 4, and figure 7 is an extension to different thicknesses with the aid of figure 5.

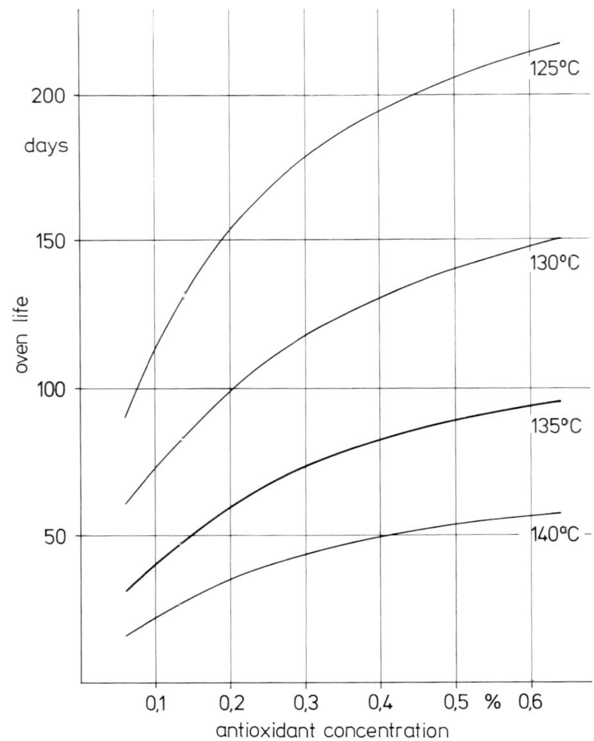

Fig. 6. Oven life of compression molded 1 mm polypropylene plaques. Effectiveness of 1,1,3-tris-(5-tert.butyl-4-hydroxy-2-methylphenyl)-butane as function of concentration and temperature.

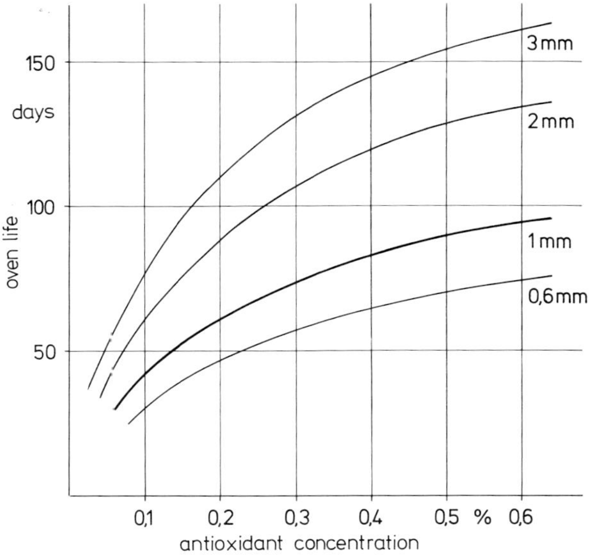

Fig. 7. Oven life of compression molded, 1 mm polypropylene plaques at 135 °C. Effectiveness of 1,1,3-tris-(5-tert.butyl-4-hydroxy-2-methylphenyl)-butane as function of concentration and sample thickness.

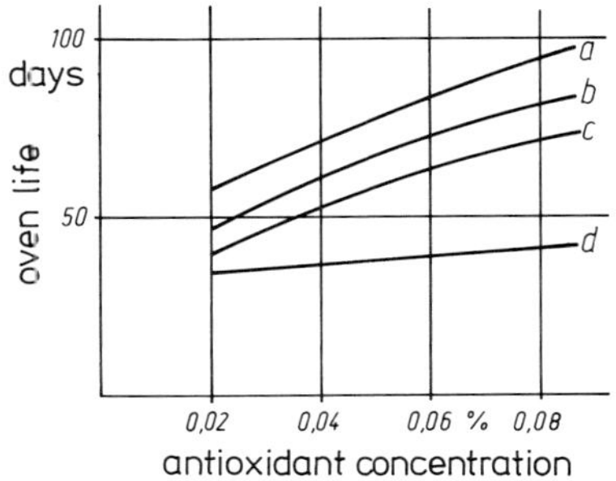

Fig. 8. Oven life of polypropylene stretched tapes (thickness: 30 microns) at 100 °C. Effectiveness of selected long term heat stabilizers as function of concentration.

a: 1,3,5-tris-(3,5-ditert.butyl-4-hydroxybenzyl)-isocyanurate,
b: pentaerythrityl tetrakis-3-(3,5-ditert.butyl-4-hydroxyphenyl)-propionate,
c: 1,3,5-tris-(3,5-ditert.butyl-4-hydroxybenzyl)-mesitylene,
d: octadecyl 3-(3,5-ditert.butyl-4-hydroxyphenyl)-propionate.

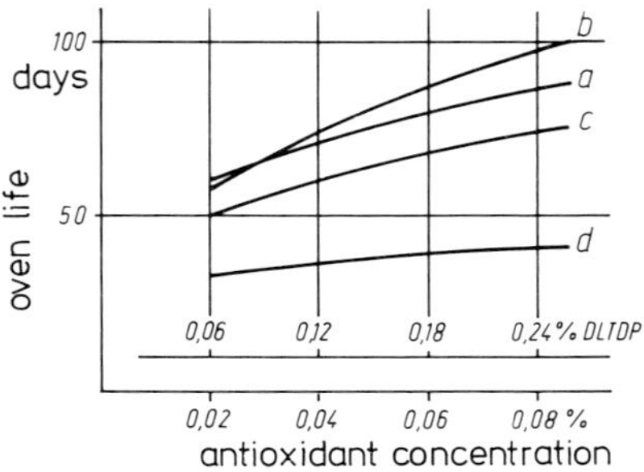
100
days
oven life
50
b
a
c
d
0,06
0,12
0,18
0,24% DLTDP
0,02
0,04
0,06
0,08 %
antioxidant concentration

Fig. 9

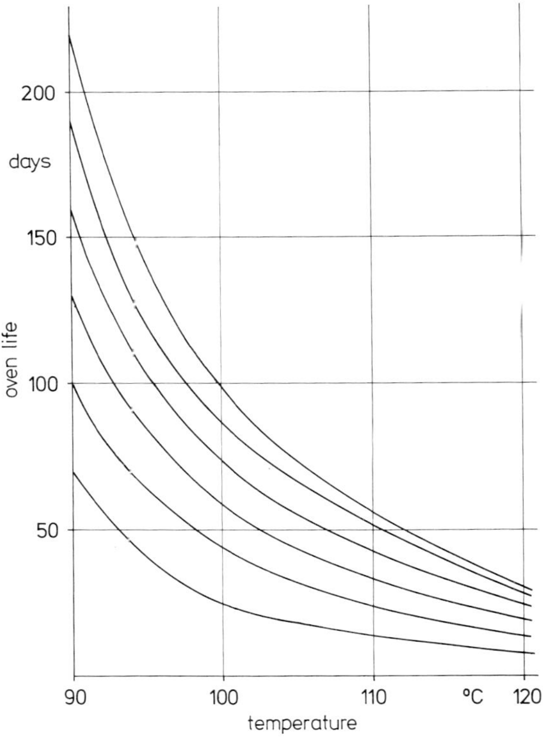
200
days
150
oven life
100
50
90
100
110
°C
120
temperature

Fig. 10

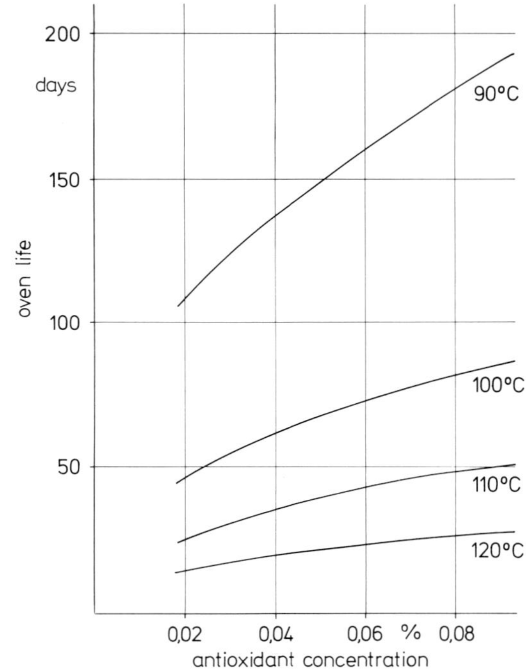

Fig. 11. Oven life of polypropylene stretched tapes. Effectiveness of pentaerythrityl tetrakis-3-(3,5-ditert.butyl-4-hydroxyphenyl)-propionate as function of concentration and temperature.

↖

Fig. 9. Oven life of polypropylene stretched tapes (thickness: 30 microns) at 100 °C. Effectiveness of selected long term heat stabilizers in combination with the triple amount of DLTDP as function of concentration.

a: 1,3,5-tris-(3,5-ditert.butyl-4-hydroxybenzyl)-isocyanurate,

b: pentaerythrityl tetrakis-3-(3,5-ditert.butyl-4-hydroxyphenyl)-propionate,

c: 1,3,5-tris-(3,5-ditert.butyl-4-hydroxybenzyl)-mesitylene,

d: octadecyl 3-(3,5-ditert.butyl-4-hydroxyphenyl)-propionate.

Fig. 10. Temperature dependence of oven life of polypropylene stretched tapes. One distinct curve applies to a test specimen of defined stabilization.

←

1.8.1.2.2 Stabilization of polypropylene, thin cross sections

Figures 2 to 7 are representations of oven life times of specimens with cross sections of half a millimeter and above. These are usually manufactured by injection molding or extrusion. Correspondingly, figures 8 to 11 represent oven life times of stretched tapes with a thickness of 30 microns. The variables are again temperature and antioxidant concentration, visually observable disintegration is the test criterion.

As mentioned earlier, oven life times may vary significantly, depending on the origin of the polypropylene, therefore, the curves shown are indications of the effectiveness of the antioxidants relative to each other.

The dependence of the oven life times of stretched tapes on temperature is similar to test specimens with thick cross sections. A temperature increase of 10 °C results in a life time reduction by a factor of about 2.2 within the temperature range of 90 to 125 °C, see figure 10.

Figure 12 is a representation of data for oven life times of polypropylene fibers; figure 10 may be used for an estimate of temperature dependence.

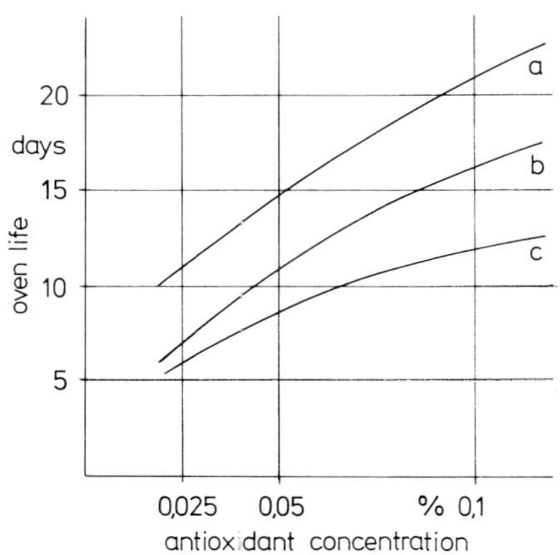

Fig. 12. Oven life of polypropylene fibers at 110 °C. Effectiveness of selected long term heat stabilizers as function of concentration.

a: 3,5-ditert.butyl-4-hydroxybenzyl-phosphonic acid monoethyl ester, calcium salt,

b: 1,3,5-tris-(3,5-ditert.butyl-4-hydroxybenzyl)-isocyanurate,

c: ethyleneglycol bis-[3,3-bis-(3′-tert.butyl-4′-hydroxyphenyl)-butyrate].

1.8.2 High density polyethylene (HDPE)

High density polyethylenes are less sensitive to oxidation, compared to polypropylene, so that generally lower stabilizer concentrations are sufficient. As in polypropylene, antioxidants can be added during a suitable stage of manufacture or during pelletizing.

The requirements concerning color properties, compatibility, volatility and general stability are the same as for polypropylene.

1.8.2.1 Stabilization of HDPE during processing

HDPE is processed at temperatures between 180 and 280 °C. Material manufactured according to Ziegler-Natta processes shows mostly an increase of the melt flow index on processing (which is a result of chain scission, similar to polypropylene), whereas HDPE manufactured according to Phillips processes is predominantly cross-linking, which results in a decrease of the melt flow index. The antioxidants in technical use are the same as for polypropylene. Phenols of medium and high molecular weight are also active as long term heat stabilizers. Concentrations between 0.03 and 0.15% are usual. The ranking in effectiveness of various additives is evident from table 9. Comparison with figure 1 is possible (corresponding data for polypropylene). The increase of the melt flow index of a particular HDPE, polymerized by Ziegler-Natta catalysts, was recorded during multiple extrusions (25 mm one screw extruder,

Table 9. Effectiveness of antioxidants as processing stabilizers in HDPE (according to [58 to 60]).

	I	II	III	IV	V
Ziegler type	Phosphonite 0.1%	Phosphonite 0.05%	BHT 0.1%	Phosphites 0.1%	Long term heat stabilizers (phenols) 0.1%
Phillips type	Phosphonite 0.1%	BHT 0.1%	Long term heat stabilizers (phenols) 0.1%	Phosphonite 0.05%	Phosphites 0.1%

L/D = 20, maximum temperature 240 °C). The test for a Phillips type HDPE consisted of prolonged treatment of the polymer melt in a melt flow index determination unit. The decrease of the melt flow index was recorded as a function of time (230 °C, load 0.325 kp) [58 to 62].

1.8.2.2 Long term heat stabilizers for HDPE

The thermooxidative stability of HDPE is determined basically in much the same manner as of polypropylene. Likewise, the end point is defined as the time until embrittlement occurs (oven life time). It may be determined by a simple manual bending test.

Similar or identical products are used as stabilizers as in polypropylene. The concentrations are distinctly lower and rarely exceed 0.1% (figures 13, 14). The use of costabilizers (thioethers, phosphites) is customary also in HDPE. The curves in the figures apply to the test conditions and materials used [55].

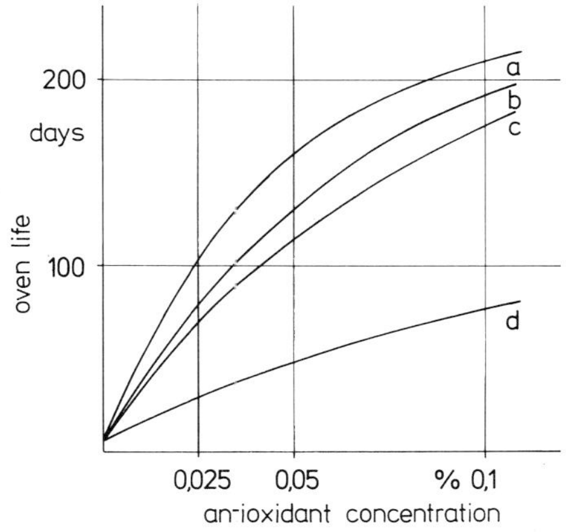

Fig. 13. Oven life of compression molded HDPE (thickness: 0.625 mm) at 125 °C. Effectiveness of selected long term heat stabilizers as function of concentration [55].

a: pentaerythrityl tetrakis-3-(3,5-ditert.butyl-4-hydroxyphenyl)-propionate,

b: 1,3,5-tris-(3,5-ditert.butyl-4-hydroxybenzyl)-isocyanurate,

c: octadecyl 3-(3,5-ditert.butyl-4-hydroxyphenyl)-propionate,

d: 1,1,3-tris-(5-tert.butyl-4-hydroxy-2-methylphenyl)-butane.

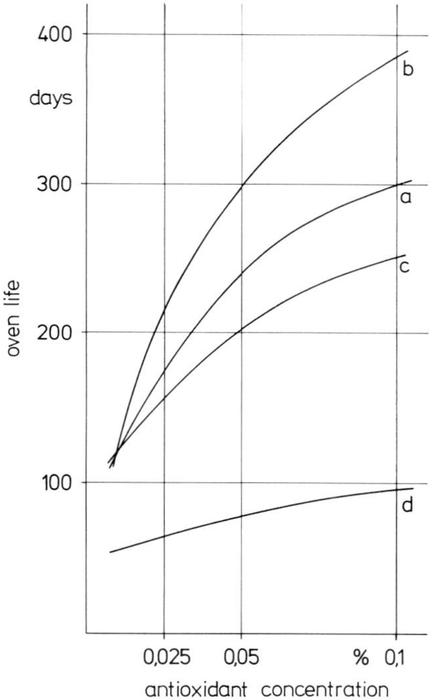

Fig. 14. Oven life of compression molded HDPE (thickness: 0.625 mm) at 125 °C. Effectiveness of selected long term heat stabilizers as function of concentration (base stabilization: 0 25% DLTDP [55]).

a: pentaerythrityl tetrakis-3-(3,5-ditert.butyl-4-hydroxyphenyl)-propionate,

b: 1,3,5-tris-(3,5-ditert.butyl-4-hydroxybenzyl)-isocyanurate,

c: octadecyl 3-(3,5-ditert.butyl-4-hydroxyphenyl)-propionate,

d: 1,1,3-tris-(5-tert.butyl-4-hydroxy-2-methylphenyl)-butane.

1.8.2.3 Processing stabilizers for HDPE of high molecular weight (HMWPE)

High density polyethylenes with molecular weights of up to 1 million and above have gained considerably in importance during the last years.

Because of their interesting mechanical properties such polyethylenes are suitable for the manufacturing of big containers and of paperlike

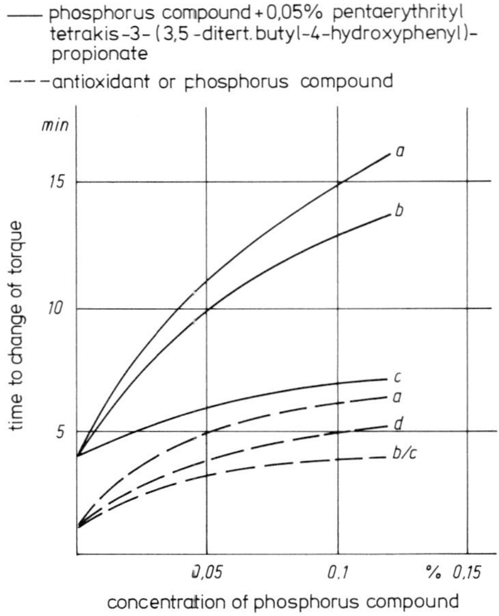

Fig. 15. Effectiveness of phosphorus compounds in high molecular weight HDPE (Phillips type, Brabender plastograph, 220 °C, 50 rpm).

a: tetrakis-(2,4-ditert.butylphenyl)-4,4′-biphenylylenediphosphonite,

b: tris-(2,4-ditert.butylphenyl)-phosphite,

c: distearyl-pentaerythrityl-diphosphite,

d: pentaerythrityl tetrakis-3-(3,5-ditert.butyl-4-hydroxyphenyl)-propionate.

film. Such high molecular weight resins demand processing temperatures between 220 and 260 °C and comparatively long residence times. Thus, good processing stability is particularly important, since only slight cross-linking or polymer chain degradation is sufficient to make it impossible to manufacture perfect articles.

As mentioned earlier, combinations of phenolic antioxidants with phosphites or phosphonites play a vital role in the processing stabilization of polypropylene. To this day such combinations represent practically the only possibility for the processing stabilization of high molecular weight HDPE.

Recording of torque as a function of time, temperature and rotation speed during plasticization in a Brabender plastograph has proved to be a useful test. This kind of testing is of little use for polypropylene and HDPE, but is quite suitable for high molecular weight HDPE (Phillips type). The experimental arrangement allows the determination of induction times until the onset of cross-linking, which is indicated by a sharp rise in torque. This, in turn, is an indication of the effectiveness of added stabilizers [58].

As can be seen from figure 15, remarkable synergistic effects are observed by combining pentaerythrityl tetrakis-3-(3,5-ditert.butyl-4-hydroxyphenyl)-propionate with phosphorus compounds. The effectiveness of pentaerythrityl tetrakis-3-(3,5-ditert.butyl-4-hydroxyphenyl)-propionate alone (curve *d*) is roughly valid also for other hindered phenols. Similarly, pentaerythrityl tetrakis-3-(3,5-ditert.butyl-4-hydroxyphenyl)-propionate may be substituted by another hindered phenol in the combinations (solid curves), and the resulting effects are of the same order of magnitude.

1.8.3 Low density polyethylene (LDPE)

LDPE is a material which is extensively used for the manufacture of films. During processing, which is carried out at temperatures of around 200 °C, cross-linking and thus formation of gel can occur through oxidation. Such gel particles are visible in the film as agglomerates, known to the processor as "fish eyes" or "arrow heads". The compatibility of the additives is an important point. To avoid blooming, the practical concentrations are limited, since the compatibility of any additive in LDPE is considerably lower than in other polyolefins. The processing stabilizers customary for polypropylene are also used in low density polyethylene. Normal concentrations seldom exceed 0.1%.

Quantitative data have been generated by periodic measurement of the melt flow index during prolonged residence times of the polymer melt in the cylinder of the measuring device at 220 °C (figure 16) [58 to 60].

1.8.4 Linear low density polyethylene (LLDPE)

In recent years LLDPE has rapidly gained significance. It is used mainly for the production of films, very frequently in blends with low density polyethylene. The recommended processing stabilizers consist of the systems commonly used for polypropylene, namely, combinations of a phosphite and a long term heat stabilizer in overall concentrations up to 0.1%. The determination of the melt flow index after multiple extrusion is a useful test method.

The effectiveness of long term heat stabilizers is assessed in the usual manner, i.e. by oven aging techniques at temperatures not exceeding 120 °C. The end point may be determined by a simple manual bending test. Typical oven life times at 120 °C amount to 40 to 50 days for 1 mm plaques stabilized with 0.03% pentaerythrityl tetrakis-3-(3,5-ditert.butyl-4-hydroxyphenyl)-propionate or 0.03% octadecyl 3-(3,5-ditert.butyl-4-hydroxyphenyl)-propionate.

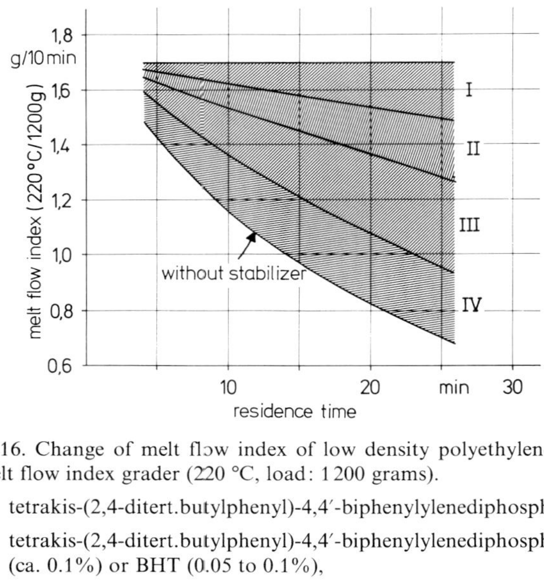

Fig. 16. Change of melt flow index of low density polyethylene during aging in a melt flow index grader (220 °C, load: 1200 grams).

I: tetrakis-(2,4-ditert.butylphenyl)-4,4′-biphenylylenediphosphonite (>0.1%),

II: tetrakis-(2,4-ditert.butylphenyl)-4,4′-biphenylylenediphosphonite (ca. 0.1%) or BHT (0.05 to 0.1%),

III: tetrakis-(2,4-ditert.butylphenyl)-4,4′-biphenylylenediphosphonite (0.05%) or higher molecular weight phenols (0.03 to 0.05%),

IV: distearyl-pentaerythrityl-diphosphite (0.05 to 0.1%).

1.8.5 Cross-linked polyethylene

For primary cable insulation, cable jackets and pipes manufactured from LDPE, medium to high molecular weight grades are used which frequently are cross-linked after processing. There are essentially three methods for cross-linking low density polyethylene:

- Cross-linking with peroxides,
- cross-linking with β- or γ-rays,
- grafting of the polymer with vinyl siloxanes and subsequent hydrolysis to Si—O—Si-units (cf. chapter 16).

For these technical applications, long term heat stabilization is of primary importance, since life times of up to 50 years are required (usually at elevated temperatures with short-time peaks up to 100 °C for cables). The antioxidants have to be extremely compatible and resistant to extraction.

Some typical antioxidants customary for insulation are mentioned below. The simultaneous use of metal deactivators has become increasingly important (cf. chapter 2 on metal deactivators). Color properties are of importance only in that they should not interfere in the color coding of communication wire with pigments.

For communication cable insulation:

- 4,4′-Thiobis-(3-methyl-6-tert.butylphenol),
- 2,2′-thiobis-(4-methyl-6-tert.butylphenol),
- 2,2′-methylenebis-(4-methyl-6-α-methylcyclohexylphenol),
- 1,1,3-tris-(5-tert.butyl-4-hydroxy-2-methylphenyl)-butane,
- 2,2′-thiodiethyl-bis-3-(3,5-ditert.butyl-4-hydroxyphenyl)-propionate,
- pentaerythrityl tetrakis-3-(3,5-ditert.butyl-4-hydroxyphenyl)-propionate,
- dilauryl thiodipropionate as synergist;

for power cable insulation:

- Pentaerythrityltetrakis-3-(3,5-ditert.butyl-4-hydroxyphenyl)-propionate,
- polymeric 2,2,4-trimethyl-1,2-dihydroquinoline,
- 4,4′-thiobis-(3-methyl-6-tert.butylphenol),
- 2,2′-thiodiethyl-bis-3-(3,5-ditert.butyl-4-hydroxyphenyl)-propionate,
- distearyl thiodipropionate as synergist.

Table 10 gives an indication of the effectiveness of a few antioxidant formulations which are customary for power cable insulations.

Table 10. Oven life of peroxide cross-linked LDPE (thickness 1.65 mm) at 150 °C. Time to 50% retained elongation.

Stabilizer		Oven life time (days)
Without stabilizer		1
Polymeric 2,2,4-trimethyl-1,2-dihydroquinoline	0.5%	47
4,4′-Thiobis-(3-methyl-6-tert.butylphenol)	0.3%	50
	0.5%	85
2,2′-Thiobis-(4-methyl-6-tert.butylphenol)	0.3%	47
	0.5%	85
2,2′-Thiobis-(4-methyl-6-tert.butylphenol) + distearyl thiodipropionate	0.3% 0.6%	65

1.8.6 Thermoplastic polyolefin elastomers

Thermoplastic polyolefin elastomers have gained inroads in two main applications:

- Automotive, e.g. bumper covers and fender extensions,
- cable primary insulation and jackets.

Such articles typically have long lives. Their stabilization should be carried out following the recommendations for polypropylene with high stability requirements.

1.8.7 Crystal polystyrene (PS)

Heat treatment of polystyrene in an inert environment results in thermal degradation which manifests itself in a decrease of the solution viscosity and the formation of monomeric styrene and oligomers. Under oxidative conditions unmodified polystyrene belongs to the more stable polymers, so that for many applications the addition of an antioxidant is not required. Nevertheless, repeated processing may lead to oxidative damage of the material, e.g. during the manufacturing of articles which are foamed and subsequently vacuum formed, where considerable amounts of scrap have to be recycled. The oxidative damage leads to an increase of the melt flow index, and to embrittlement of the material. Stabilization is effected by the addition of octadecyl 3-(3,5-ditert.butyl-4-hydroxy-

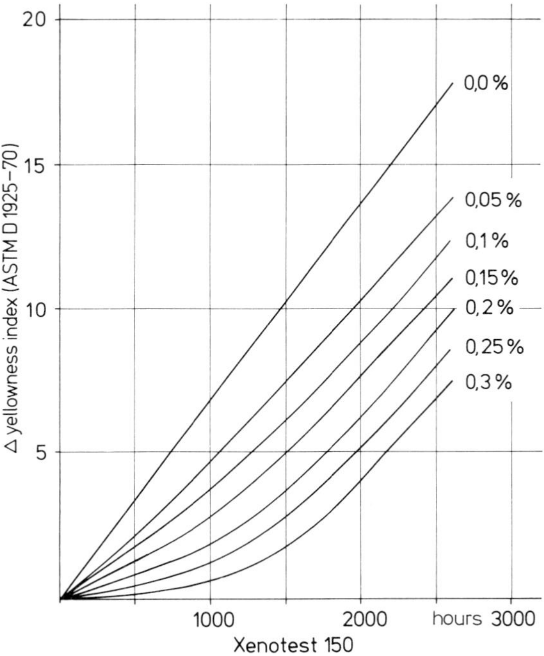

Fig. 17. Yellowing of injection molded polystyrene on u.v. irradiation. Effectiveness of octadecyl 3-(3,5-ditert.butyl-4-hydroxyphenyl)-propionate as function of concentration (Δ yellowness index: difference between irradiated and original samples).

phenyl)-propionate at concentrations of up to 0.15%, if necessary in combination with phosphites or phosphonites to improve color.

However, crystal polystyrene is rather sensitive towards photooxidation, which leads to yellowing and embrittlement. Monomeric styrene definitely promotes the yellowing effect. Octadecyl 3-(3,5-ditert.butyl-4-hydroxyphenyl)-propionate imparts to polystyrene a certain improvement in light stability when used alone and shows synergistic effects with u.v.-absorbing light stabilizers (figures 17 and 18).

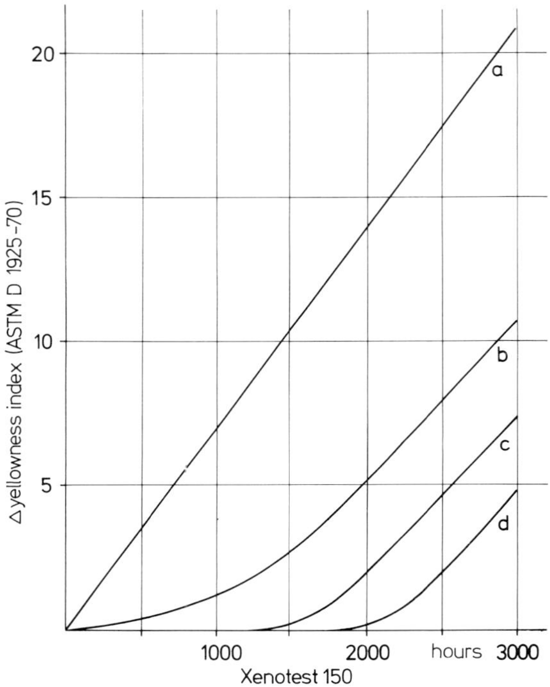

Fig. 18. Yellowing of injection molded polystyrene on u. v. irradiation (△ yellowness index: difference between irradiated and original samples).

a: without stabilizer,

b: octadecyl 3-(3,5-ditert.butyl-4-hydroxyphenyl)-propionate (0.25%),

c: 2-(2′-hydroxy-5′-methylphenyl)-benzotriazole (0.25%),

d: octadecyl 3-(3,5-ditert.butyl-4-hydroxyphenyl)-propionate (0.1%) + 2-(2′-hydroxy-5′-methylphenyl)-benzotriazole (0.15%).

1.8.8 Styrene-acrylonitrile copolymers (SAN)

In contrast to unmodified crystal polystyrene, SAN discolors already during heat treatment under inert conditions. During repeated processing a distinct rise in melt flow index is observed which is an indication of decrease of molecular weight [63]. These phenomena are also a consequence of non-oxidative chemical reactions, thus, the effectiveness of added antioxidants is limited. Yellowing during injection molding (230 to 250 °C) is suppressed to a certain extent by the addition of a phosphite (e.g. tris-[2,4-ditert.butylphenyl]-phosphite), possibly in combination with a phenolic antioxidant (e.g. octadecyl 3-[3,5-ditert.butyl-4-hydroxyphenyl]-propionate). With a total concentration of the antioxidants of 0.1 to 0.2% the yellowness values (yellowness index according to ASTM D 1925–70) of stabilized and unstabilized materials differ by 2 to 5 points after processing.

1.8.9 Impact polystyrene (IPS)

Copolymers of styrene and butadiene are more sensitive to oxidation compared to unmodified polystyrene. This sensitivity is a consequence of the double bonds in the polybutadiene component and manifests itself in yellowing and loss of mechanical properties of the polymer.

In impact polystyrene the following antioxidants or their mixtures are used in total concentrations of 0.1 to 0.25%:

- BHT,
- octadecyl 3-(3,5-ditert.butyl-4-hydroxyphenyl)-propionate,
- 1,1,3-tris-(5-tert.butyl-4-hydroxy-2-methylphenyl)-butane,
- dilauryl thiodipropionate,
- TNPP (only occasionally).

1.8.9.1 Addition of the antioxidants

The former procedure in general use consisted in adding antioxidants to the finished polymer during the pelletizing step, together with lubricants, pigments and possibly other additives.

Recently it has become more and more common to add the antioxidants already at the beginning of the polymerization. This means in practice that the addition is made to the solution of polybutadiene in styrene. With this procedure the following advantages are achieved:

- Accurate metering and homogeneous distribution are facilitated,
- the sensitive rubber phase is protected against oxidation or excessive cross-linking already at an early stage of manufacture.

As a result, polymers are obtained which have improved properties during processing and end use (figure 19). It is a prerequisite for this tech-

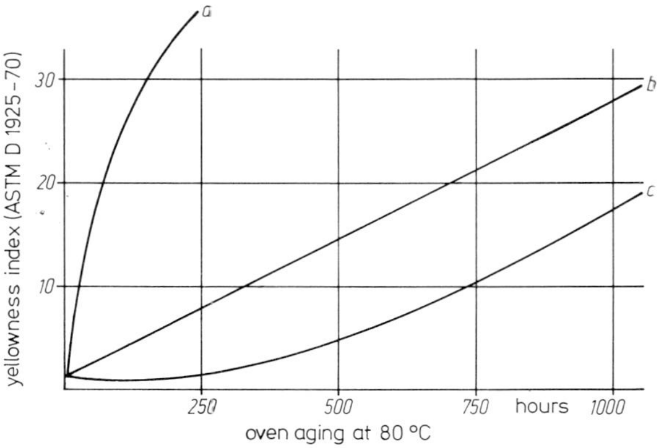

Fig. 19. Yellowing of impact polystyrene (7% polybutadiene) on oven aging at 80 °C. Stabilizer: 0.1% octadecyl 3-(3,5-ditert.butyl-4-hydroxyphenyl)-propionate.

a: without stabilizer,

b: stabilizer added after polymerization,

c: stabilizer added before polymerization.

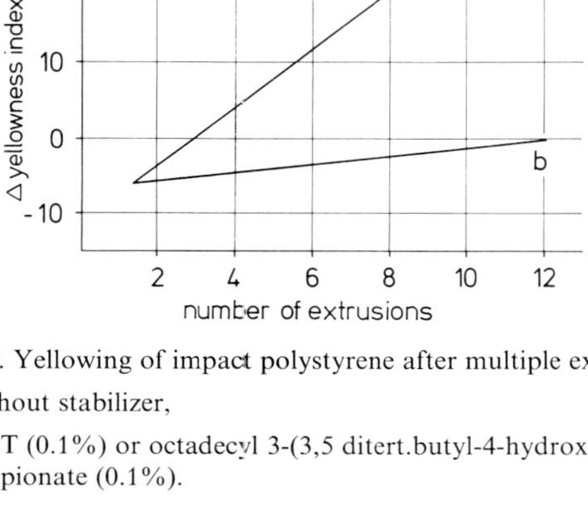

Fig. 20. Yellowing of impact polystyrene after multiple extrusion at 250 °C.

a: without stabilizer,

b: BHT (0.1%) or octadecyl 3-(3,5 ditert.butyl-4-hydroxyphenyl)-propionate (0.1%).

nique that the employed antioxidants have no unfavorable effect on the molecular parameters of the polymer, such as extent of grafting and molecular weight. Furthermore, the antioxidants should have sufficiently high molecular weights in order to minimize losses during the monomer recovery step and avoid contamination of the recycled monomers. For this reason, BHT has lost significance as an antioxidant for IPS due to its considerable volatility and its susceptibility to form colored oxidation products.

1.8.9.2 Thermooxidative properties of impact polystyrene

Impact polystyrene is processed at temperatures between 220 and 260 °C. Oxidative deterioration manifests itself mainly in yellowing as well as in a loss of impact strength and elongation. This is especially true on repeated processing, i.e. when scrap is recycled [64] (figure 20).

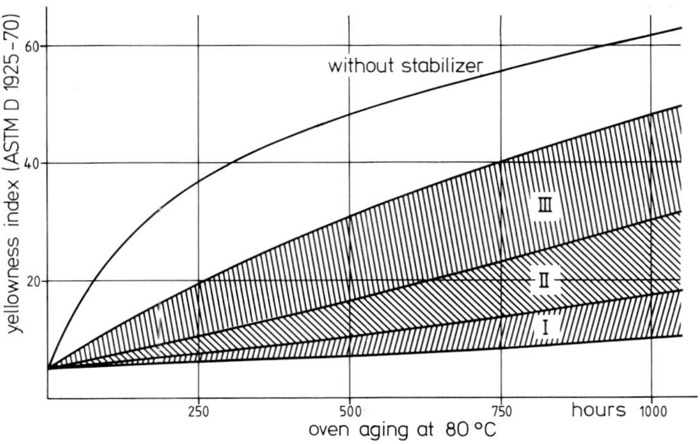

Fig. 21. Yellowing of impact polystyrene (7% polybutadiene) on oven aging at 80 °C (antioxidant added after polymerization).

I: BHT (~0.2%) or octadecyl 3-(3,5-ditert.butyl-4-hydroxyphenyl)-propionate (~0.2%),

II: BHT (~0.1%) or octadecyl 3-(3,5-ditert.butyl-4-hydroxyphenyl)-propionate (~0.1%) or 1,1,3-tris-(5-tert.butyl-4-hydroxy-2-methylphenyl)-butane (~0.2%),

III: BHT (~0.07%) or octadecyl 3-(3,5-ditert.butyl-4-hydroxyphenyl)-propionate (~0.07%) or 1,1,3-tris-(5-tert.butyl-4-hydroxy-2-methylphenyl)-butane (~0.1%).

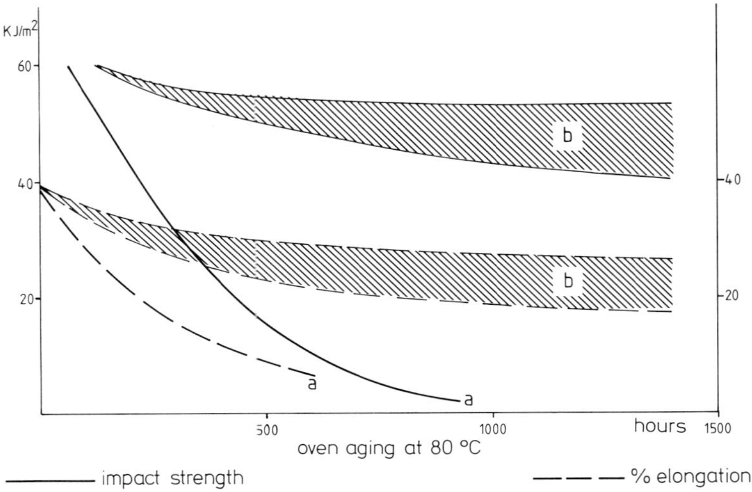

Fig. 22. Change of impact strength and elongation of impact polystyrene as function of aging time at 80 °C (antioxidants added after polymerization).

a: without stabilizer,

b: BHT (0.07 to 0.15%) or octadecyl 3-(3,5-ditert.butyl-4-hydroxyphenyl)-propionate (0.07 to 0.15%) or 1,1,3-tris-(5-tert.butyl-4-hydroxy-2-methyl-phenyl)-butane (0.1 to 0.2%).

The oven aging of impact polystyrene is conducted most commonly either at 80 °C or at 150 to 160 °C. The thermooxidative stability of impact polystyrene with various stabilizers is given in figures 21 and 22. The protection of the polymer against yellowing approximately parallels the preservation of impact strength and elongation. In more theoretical publications the rate of appearance of infrared carbonyl absorption is used as the criterion of oxidative stability and the effectiveness of antioxidants [65].

1.8.10 Acrylonitrile-butadiene-styrene copolymers (ABS)

ABS is comparable to polypropylene in the consumption of antioxidants (cf. table 3). Like impact polystyrene, ABS is sensitive to oxidation caused by the unsaturation of the elastomeric component. However, the antioxidant concentrations normally used in ABS are substantially higher than those used in impact polystyrene. There are two reasons for this:

- The weight fraction of the oxidatively sensitive elastomer component can be as high as 30%, compared to 10% in IPS.
- Articles made of ABS have high value and long life.

1.8.10.1 Stabilization of ABS during manufacturing

The two main processes for the manufacture of ABS are the following:

- Graft polymerization of styrene and acrylonitrile on polybutadiene latex, blending with a separately manufactured SAN latex, the resulting blend is coagulated and dried.
- Separate manufacture, isolation, and drying of the graft polymer and the SAN and blending of the two components at the compounding stage.

Both processes require the drying of powdery polymers which are extremely sensitive to oxidation. Drying is carried out at 100 to 150 °C. If the stabilization is insufficient or if the antioxidants are unevenly distributed, or the protection from oxygen inadequate, deterioration of useful properties results or, in extreme cases, even ignition during drying may occur. Thus, antioxidants have to be added before the coagulation step, normally in emulsified form, sometimes in solution. The primary antioxidants are frequently used together with a synergist.

Most important primary antioxidants:

- BHT,
- 2,2′-methylenebis-(4-ethyl-6-tert.butylphenol),
- 2,2′-methylenebis-(4-methyl-6-tert.butylphenol),
- 2,2′-methylenebis-(4-methyl-6-cyclohexylphenol),
- 2,2′-methylenebis-(4-methyl-6-nonylphenol),
- octadecyl 3-(3,5-ditert.butyl-4-hydroxyphenyl)-propionate,
- 1,1,3-tris-(5-tert.butyl-4-hydroxy-2-methylphenyl)-butane.

Synergists:

- Tris-(nonylphenyl)-phosphite,
- dilauryl thiodipropionate.

These antioxidants are either liquids or show comparatively low melting points. This is an important prerequisite for the formation of stable emulsions, whereby TNPP often is used as a solvent for the primary antioxidants. Furthermore, low volatility and low steam volatility are important advantages.

The thermooxidative stability of stabilized ABS powders, as present during the drying stage, is best assessed using differential thermal analysis

Table 11. Differential thermal analysis of ABS in air; heating rate 2.5 °C/min; sample size 400 mg [67, 68].

1% Stabilizer	Oxidation temperature (°C)	
	ABS: 13% Polybutadiene	ABS: 45% Polybutadiene
Without stabilizer	158	128
BHT	192	177
Octadecyl 3-(3,5-ditert.butyl-4-hydroxyphenyl)-propionate	197	193
2,4-Bis-octylmercapto-6-(3,5-ditert.butyl-4-hydroxyanilino)-s-triazine	200	198
TNPP	202	*)

*) not determined

Table 12. Differential thermal analysis of ABS (30% polybutadiene) in oxygen; 170 °C; sample size 5 mg [52].

Stabilizer		Induction period to maximum of exotherm (min)
Without stabilizer		<1
BHT	0.25%	15
BHT + dilauryl thiodipropionate	0.25% } 0.5%	24
Octadecyl 3-(3,5-ditert.butyl-4-hydroxyphenyl)-propionate	0.25%	32
Octadecyl 3-(3,5-ditert.butyl-4-hydroxyphenyl)-propionate + dilauryl thiodipropionate	0.25% } 0.5%	47

[66 to 68]. The measurements can be carried out isothermally or dynamically. In the first instance induction times are measured until the occurrence of an exothermic reaction. Dynamic measurements consist of raising the temperature of the test samples at a constant rate (2.5 to 8 °C/minute is normal) and determining the temperature at which an exothermic reaction begins. Experiments may be carried out in air or in an atmosphere of pure oxygen. Table 11 shows some oxidation temperatures as observed in dynamic experiments. The exotherm temperatures are essentially dependent on stabilization, sample size, weight fraction of polybutadiene in the ABS used and on the heating rate. Table 12 lists a few induction times for isothermal experiments at 170 °C. The effectiveness of the synergist is pronounced.

1.8.10.2 Stabilization of ABS during processing and end use

The main application of ABS is injection molded articles. The antioxidants introduced during the manufacturing of the material generally impart sufficient thermooxidative stability during processing and end use. However, extrusion grades (for pipes, sheets) normally require more effi-

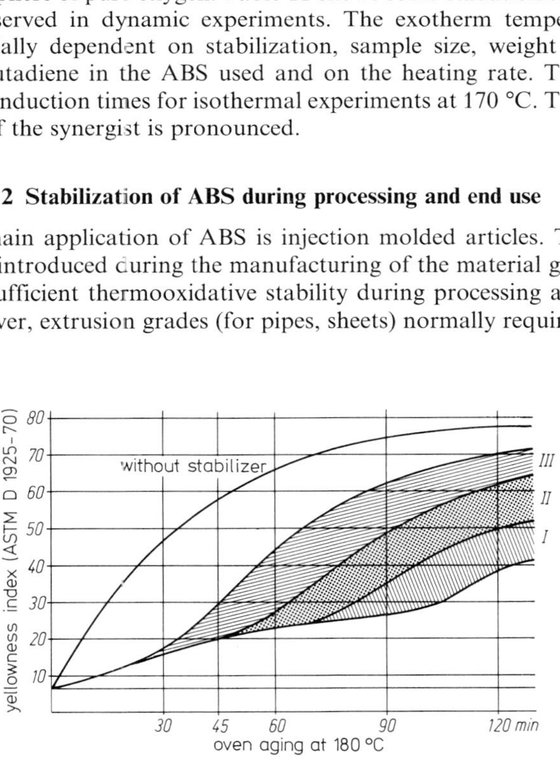

Fig. 23. Yellowing of ABS (30% polybutadiene) on oven aging at 180 °C.

I: methylenebisphenol (~0.6%) or octadecyl 3-(3,5-ditert.butyl-4-hydroxyphenyl)-propionate (~0.8%),

II: methylenebisphenol (~0.45%) or octadecyl 3-(3,5-ditert.butyl-4-hydroxyphenyl)-propionate (~0.6%),

III: methylenebisphenol (~0.3%) or octadecyl 3-(3,5-ditert.butyl-4-hydroxyphenyl)-propionate (~0.4%).

cient stabilization, depending on the process. The antioxidant can be added as an emulsion, during manufacturing to the latex, or later on, during pelletizing. In addition to maintaining color, an important task of the supplementary antioxidant is to retard the loss of impact strength. To this end, a number of antioxidants of medium molecular weight are in use, which belong to the chemical classes of thiobisphenols, methylenebisphenols and hydroxyphenylpropionates. Some of these antioxidants are also used in combination with DLTDP.

The testing of the effectiveness of antioxidants is normally done by oven aging at temperatures of 180 to 200 °C. Figure 23 shows slightly simplified yellowness values, in that the yellowness index before oven aging is 7 throughout. Practically, some differences between antioxidants of about 2 to 5 points after extrusion are usual. The methylenebisphenols contribute to discoloration, whereas octadecyl 3-(3,5-ditert.butyl-4-hydroxyphenyl)-propionate hardly shows this property.

The loss of impact strength of ABS during oven aging at 180 to 200 °C can only be assessed satisfactorily with compression molded test specimens. Injection molded parts show too much shrinking at these temperatures.

Oxidation phenomena on ABS can also be quantified using infrared reflection spectroscopy (ATR; attenuated total reflection). The appearance of the carbonyl absorption (>C=O) and the disappearance of the absorption attributed to the double bond (cis-C=C) are followed simultaneously kinetically [69, 70]. Investigations in the laboratories of CIBA-GEIGY have shown that there is a satisfactory correlation between discoloration and carbonyl buildup.

1.8.11 Styrene based thermoplastic elastomers

Styrene-Butadiene-Styrene (SBS), Styrene-Isoprene-Styrene-Block-Copolymers (SIS)

The thermoplastic elastomers discussed here contain unsaturated soft segments, hence, they are as sensitive to oxidation as graft copolymers of styrene on polybutadiene. Main applications are: shoe soles, ski boots, gaskets. The properties of the material are adjusted to fit a particular application by compounding with polystyrene, naphthenic oils and fillers (calcium carbonate, silica).

1.8.11.1 Stabilization of styrene based thermoplastic elastomers during manufacture

Styrene-based thermoplastic elastomers are manufactured by solution polymerization processes in aliphatic hydrocarbons. Autoxidation during the finishing steps (stripping, drying) manifests itself by a rise in melt flow index and discoloration of the raw polymer. The antioxidant is added to the polymer solution before finishing, hence, it has to be soluble in the polymerization solvent. A number of hindered phenols are used in practice. They are frequently used together with TNPP in a total concentration of around 0.5%.

Primary antioxidants:

- BHT,
- 1,3,5-tris-(3,5-ditert.-butyl-4-hydroxybenzyl)-mesitylene,
- octadecyl 3-(3,5-ditert.butyl-4-hydroxyphenyl)-propionate.

Synergist:

- Tris-(nonylphenyl)-phosphite.

BHT may be lost partially during finishing and drying. The phenols of higher molecular weight are used mainly because of their low volatility. They have the further advantage of protecting the material also during processing and end use. Consequently, test methods have been proposed which include not only the observation of Brabender plastograms at 150 to 200 °C, but also oven aging at 60 to 80 °C.

1.8.11.2 Long term stabilization of styrene based thermoplastic elastomers

Oven aging at 70 to 80 °C has proved to be useful for the estimation of the long term aging properties of styrene-based thermoplastic elastomers and their compounds. The most important test criteria are tensile strength (figure 24, table 13), elongation (figure 25) and discoloration.

The long term heat stabilization of such thermoplastic elastomers and their compounds is achieved mostly with phenolic antioxidants of high molecular weight, possibly in combination with DLTDP in total concentrations up to 0.8%.

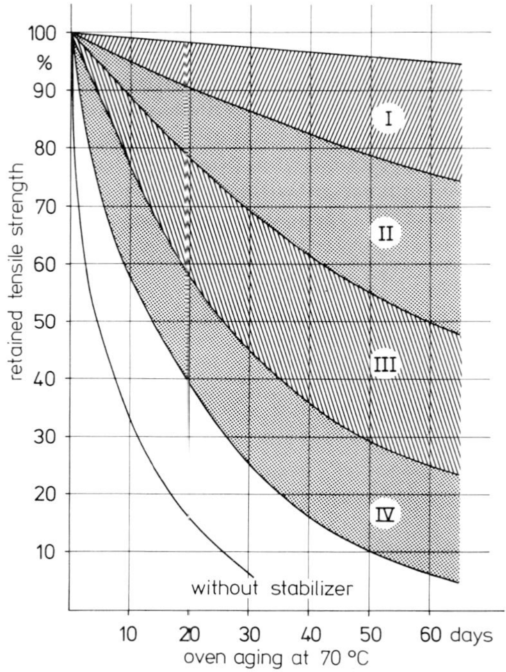

Fig. 24. Tensile strength of linear styrene-butadiene-styrene block copolymer (SBS, thickness: 0.5 mm) on oven aging of tensile bars at 70 °C.

I: 2,4-bis-octylmercapto-6-(3,5-ditert.butyl-4-hydroxyanilino)-s-triazine (0.1%), 2,2′-methylenebis-(4-methyl-6-tert.butylphenol) (0.2 to 0.3%), hydroxyphenylpropionates (up to 1%);

II: 2,2′-methylenebis-(4-methyl-6-tert.butylphenol) (0.1 to 0.2%), di-(3-tert.butyl-4-hydroxy-5-methylphenyl)-dicyclopentadiene (0.3 to 0.5%), pentaerythrityl tetrakis-3-(3,5-ditert.butyl-4-hydroxyphenyl)-propionate (0.5 to 0.8%);

III: di-(3-tert.butyl-4-hydroxy-5-methylphenyl)-dicyclopentadiene (0.2 to 0.3%), pentaerythrityl tetrakis-3-(3,5-ditert.butyl-4-hydroxyphenyl)-propionate (0.3 to 0.5%), pentaerythrityl tetrakis-3-(3,5-ditert.butyl-4-hydroxyphenyl)-propionate + DLTDP 1:1 (0.3 to 0.5%);

IV: pentaerythrityl tetrakis-3-(3,5-ditert.butyl-4-hydroxyphenyl)-propionate (0.1 to 0.3%).

Table 13. Tensile strength of styrene-isoprene-styrene block copolymer (SIS, films; 0.5 mm) after 7 days oven aging at 70 °C. Antioxidant concentration: 1% [71].

Antioxidant	Tensile strength in % of initial
Without antioxidant	0.5
2,4-Bis-octylmercapto-6-(3,5-ditert.butyl-4-hydroxy-anilino)-s-triazine	90
Pentaerythrityl tetrakis-3-(3,5-ditert.butyl-4-hydroxyphenyl)-propionate	85
Butyl zimate	85
2,2′-Methylenebis-(4-methyl-6-tert.butylphenol)	80
Mixture of various alkylated 2,2′-methylenebisphenols	60
Polymeric 2,2,4-trimethyl-1,2-dihydroquinoline	50
1,3,5-Tris-(3,5-ditert.butyl-4-hydroxybenzyl)-mesitylene	30

Primary antioxidants:

- 1,3,5-Tris-(3,5-ditert.butyl-4-hydroxybenzyl)-mesitylene,
- di-(3-tert.butyl-4-hydroxy-5-methylphenyl)-dicyclopentadiene,
- 2,4-bis-octylmercapto-6-(3,5-ditert.butyl-4-hydroxyanilino)-s-triazine,
- pentaerythrityl tetrakis-3-(3,5-ditert.butyl-4-hydroxyphenyl)-propionate,
- octadecyl 3-(3,5-ditert.butyl-4-hydroxyphenyl)-propionate,
- 2,2′-methylenebis-(4-methyl-6-tert.butylphenol),
- 2,2′-methylenebis-(4-methyl-6-nonylphenol),
- polymeric 2,2,4-trimethyl-1,2-dihydroquinoline.

Secondary antioxidants:

- Dilauryl thiodipropionate,
- zinc dibutyldithiocarbamate ("butyl zimate").

Butyl zimate is used primarily in the application of thermoplastic elastomers for adhesives [71]. Methylenebisphenols and aromatic amines are very effective products, however, they have the disadvantage of contributing to discoloration of the substrate. Therefore, their use in light colored articles is questionable.

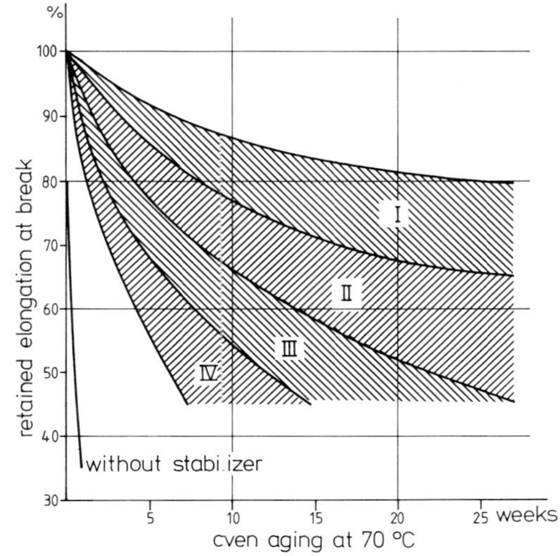

Fig. 25. Retained elongation of a SBS compound*) (thickness: 0.5 mm) on oven aging at 70 °C. The change of tensile strength shows a very similar behavior.

I: 2,4-bis-octylmercapto-6-(3,5-ditert.butyl-4-hydroxyanilino)-s-triazine (0.1%), 2,2′-methylenebis-(4-methyl-6-tert.butylphenol) (0.2 to 0.3%), pentaerythrityl tetrakis-3-(3,5-ditert.butyl-4-hydroxyphenyl)-propionate (0.4 to 0.6%);

II: 2,2′-methylenebis-(4-methyl-6-tert.butylphenol) (0.1 to 0.2%), pentaerythrityl tetrakis-3-(3,5-ditert.butyl-4-hydroxyphenyl)-propionate (0.2 to 0.3%);

III: pentaerythrityl tetrakis-3-(3,5-ditert.butyl-4-hydroxyphenyl)-propionate (0.1 to 0.2%), di-(3-tert.butyl-4-hydroxy-5-methylphenyl)-dicyclopentadiene (0.3 to 0.5%);

IV: octadecyl 3-(3,5-ditert.butyl-4-hydroxyphenyl)-propionate (0.1 to 0.2%), di-(3-tert.butyl-4-hydroxy-5-methylphenyl)-dicyclopentadiene (0.2 to 0.3%).

---

*) Formulation: SBS (linear type) 100 parts, naphthenic oil 70 parts, crystal polystyrene 30 parts.

1.8.12 Polyamide 6 and polyamide 6.6

Discoloration of polyamides during processing can be suppressed to a certain extent by the addition of phosphites, e.g. tris-(2,4-ditert.butyl-phenyl)-phosphite, in concentrations of 0.2 to 0.4%. The decrease of the yellowness index amounts to 5 to 10 points.

The stabilization of polyamides is mainly a matter of long term stabilization. Three main groups of stabilizers have become known:

Copper salts, especially in combination with halogen and/or phosphorus compounds (e.g. copper acetate + potassium iodide/phosphoric acid).

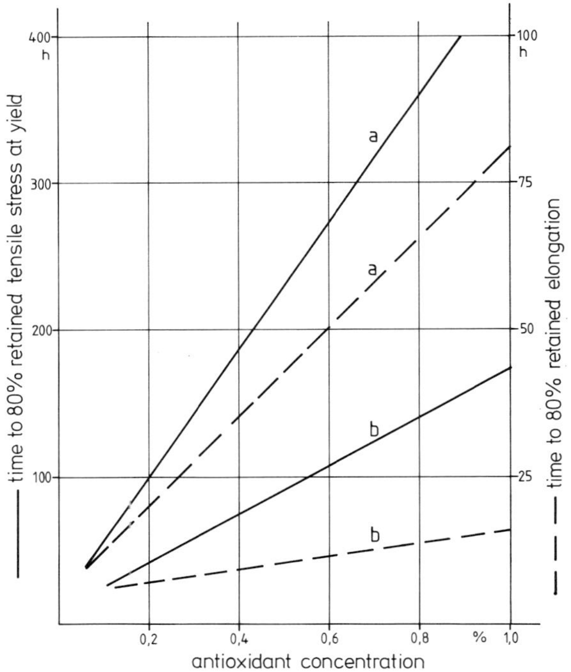

Fig. 26. Oven life of polyamide 6.6 on aging at 150 °C as function of antioxidant concentration (time to 80% retained tensile stress at yield and elongation; sample thickness: 1 mm).

a: N,N′-hexamethylenebis-3-(3,5-ditert.butyl-4-hydroxyphenyl)-propion-amide,

b: 1,1,3-tris-(5-tert.butyl-4-hydroxy-2-methylphenyl)-butane.

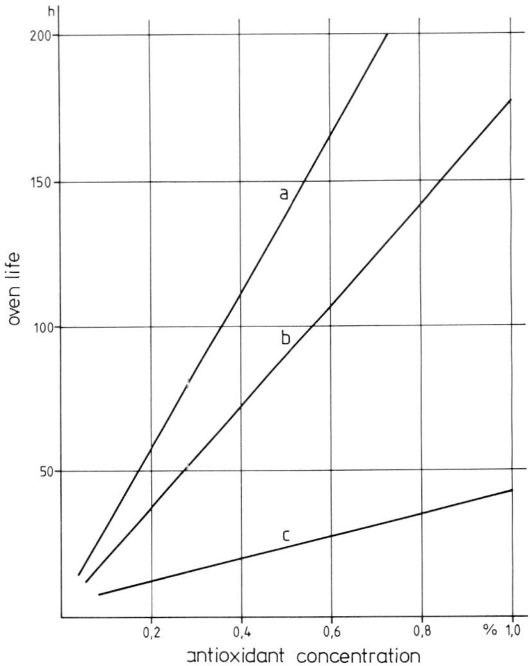

Fig. 27. Oven life of polyamide 6 (thickness: 2 mm) on aging at 160 °C as function of antioxidant concentration.

Antioxidant: N,N′-hexamethylenebis-3-(3,5-ditert.butyl-4-hydroxyphenyl)-propionamide.

a: time to 80% of initial tensile strength,

b: time to 80% of initial tensile stress at yield,

c: time to 50% of initial elongation.

Such systems are effective already in very low concentrations (10 to 50 ppm of copper; approx. 1000 ppm of halogen), although they are extracted quite easily with water and cause discoloration of the substrate.

Aromatic amines are of considerable technical importance for polyamides (e.g. N,N′-dinaphthyl-p-phenylenediamine or N-phenyl-N′-cyclohexyl-p-phenylenediamine). Next to the copper-halogen systems, aromatic amines belong to the most effective stabilizers. They are used in rather high

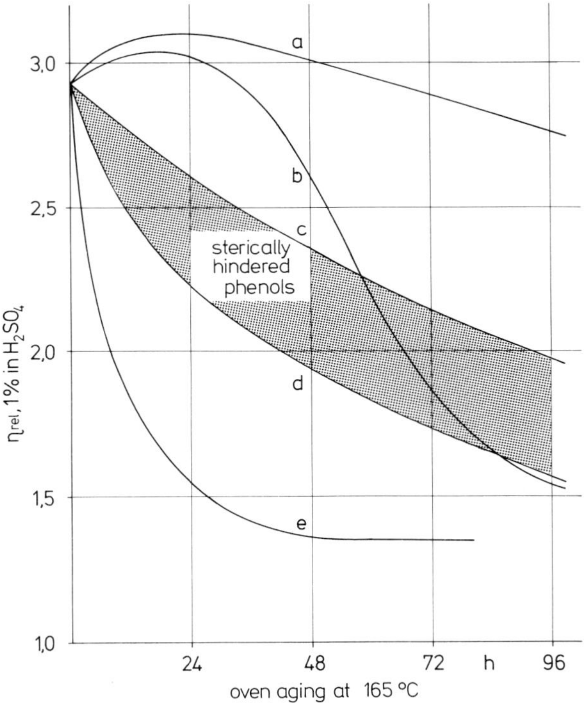

Fig. 28. Decrease of solution viscosity of polyamide 6 on oven aging at 165 °C (thickness: 0.3 mm, antioxidant concentration: 0.5%).

a: 300 ppm Cu as acetate + 1400 ppm jodide as KJ,

b: N,N′-dinaphthyl-p-phenylenediamine,

c: N,N′-hexamethylenebis-3-(3,5-ditert.butyl-4-hydroxyphenyl)-propionamide,

d: 1,3,5-tris-(3,5-ditert.butyl-4-hydroxybenzyl)-mesitylene,

e: without stabilizer.

concentrations: from 0.5 to 2%. They have strongly discoloring properties and are, therefore, used mainly for technical articles which tolerate discoloration.

Hindered phenols do not show the abovementioned disadvantages. They are the material of choice whenever good oxidation stability has to be coupled with good color stability and possibly food approval of the end article. The most important phenolics are the following:

- N,N′-hexamethylenebis-3-(3,5-ditert.butyl-4-hydroxyphenyl)-propionamide,
- 1,1,3-tris-(5-tert.butyl-4-hydroxy-2-methylphenyl)-butane,
- 1,3,5-tris-(3,5-ditert.butyl-4-hydroxybenzyl)-mesitylene,
- BHT.

The antioxidants can be added already during polycondensation, normal concentrations range from 0.3 to 0.7%.

The effectiveness of antioxidants is assessed by oven aging (figures 26 and 27). Criteria are mainly mechanical properties (tensile stress at yield, tensile strength, elongation, impact strength), but also discoloration followed as function of aging time. In such investigations, the water content of the test specimens at the time of the mechanical measurement has to be considered. The values in figures 28 and 29 hold for dry samples, as found directly after oven aging. Water content of a few percent leads to longer oven life times and improved absolute values, especially for

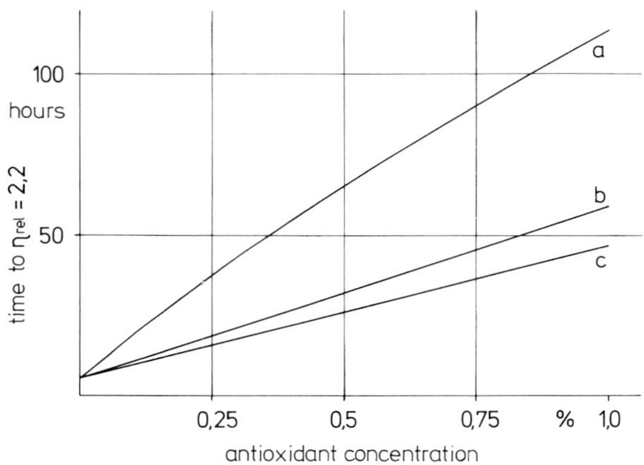

Fig. 29. Oven life of polyamide 6 (thickness: 0.3 mm) at 165 °C. Effectiveness of selected antioxidants as function of concentration. Failure criterion: decrease of solution viscosity η_{rel} from 2.95 to 2.20.

a: N,N′-hexamethylenebis-3-(3,5-ditert.butyl-4-hydroxyphenyl)-propionamide,

b: 1,1,3-tris-(5-tert.butyl-4-hydroxy-2-methylphenyl)-butane,

c: 1,3,5-tris-(3,5-ditert.butyl-4-hydroxybenzyl)-mesitylene.

the elongation. The pick-up of water is brought about by conditioning the test specimens in a defined climate during a few weeks.

Oven life times have also been determined as a function of temperature [72]. For phenolic antioxidants the life times are extended by a factor of 2.1 to 2.3 for every 10 °C reduction in temperature, independent on the test criterion. Thus, a reduction in temperature of 60 °C will result in an extension of life time by a factor of roughly 100. Stabilization with copper is somewhat irregular: the life time extension is less than a factor of 2 here with the consequence that at testing temperatures around 100 to 120 °C copper systems and phenolic antioxidants become comparable in effectiveness.

The determination of solution viscosities as a function of aging time is a frequently used method in addition to the determination of mechanical properties and discoloration (figures 28 and 29).

In polyamides, discoloration does not appear at the same rate as the loss of mechanical properties. Yellowing is observed already after short aging periods, but tensile strength and elongation are hardly affected in the same time period.

1.8.13 Polyamide 11 and polyamide 12

For the stabilization of these polyamides basically the same methods and criteria are employed as for the polyamides 6 and 6.6 (figures 30 to 32). Copper-halogen systems and aromatic amines show the same good effectiveness and also the same disadvantages. The technically employed phenolic antioxidants are used in rather high concentrations of up to about 1%, namely:

- 4-Hydroxy-lauric-anilide,
- 4-hydroxy-stearic-anilide,
- N,N′-hexamethylenebis-3-(3,5-ditert.butyl-4-hydroxyphenyl)-propionamide,
- pentaerythrityl tetrakis-3-(3,5-ditert.butyl-4-hydroxyphenyl)-propionate.

1.8.14 Polymethylene terephthalates (PETP and PBTP)

Polymethylene terephthalates are relatively stable against autoxidation. Occasionally, phenolic antioxidants are used to improve long term stability. The goal is improvement of impact strength and color after thermal treatment. Test temperatures of 150 to 180 °C are usual. An antioxidant of recognized effectiveness is pentaerythrityl tetrakis-3-(3,5-ditert.butyl-4-hydroxyphenyl)-propionate.

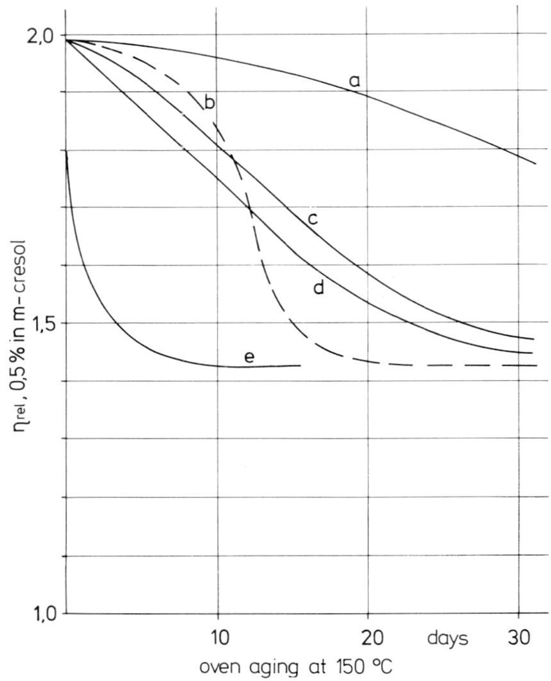

Fig. 30. Decrease of solution viscosity of polyamide 12 on oven aging at 150 °C. (Thickness: 1.0 mm; antioxidant concentration: 0.5%).

a: 300 ppm Cu as acetate + 1400 ppm jodide as KJ,

b: 4,4′-ditert.octyl-diphenylamine,

c: pentaerythrityl tetrakis-3-(3,5-ditert.butyl-4-hydroxyphenyl)-propionate or
N,N′-hexamethylenebis-3-(3,5-ditert.butyl-4-hydroxyphenyl)-propionamide,

d: lauric acid 4-hydroxy-anilide or stearic acid 4-hydroxy-anilide,

e: without stabilizer.

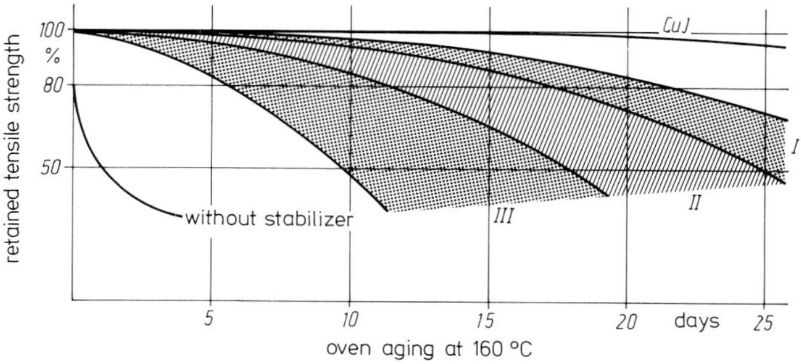

Fig. 31. Change of tensile strength of polyamide 11 (thickness: 1 mm) on oven aging at 160 °C (list of stabilizers cf. figure 32).

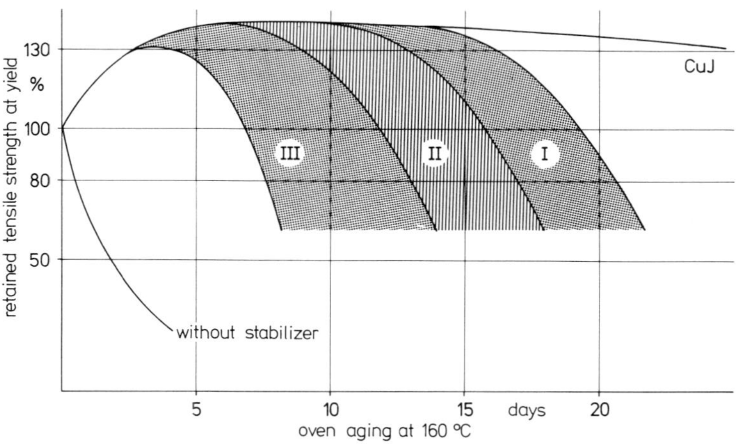

Fig. 32. Change of tensile stress at yield of polyamide 12 (thickness: 1 mm) on oven aging at 160 °C.

CuJ: 300 ppm Cu as acetate + 1400 ppm jodide as KJ;

I: pentaerythrityl tetrakis-3-(3,5-ditert.butyl-4-hydroxyphenyl)-propionate (0.5 to 1.0%), N,N′-hexamethylenebis-3-(3,5-ditert.butyl-4-hydroxyphenyl)-propionamide (0.5 to 1.0%);

II: pentaerythrityl tetrakis-3-(3,5-ditert.butyl-4-hydroxyphenyl)-propionate or N,N′-hexamethylenebis-3-(3,5-ditert.butyl-4-hydroxyphenyl)-propionamide (0.2 to 0.5%), lauric acid 4-hydroxy-anilide or stearic acid 4-hydroxy-anilide (0.5 to 1.0%);

III: lauric acid 4-hydroxy-anilide or stearic acid 4-hydroxy-anilide (0.2 to 0.5%).

1.8.15 Thermoplastic polyester elastomers

Thermoplastic polyester elastomers contain readily oxidizable polyether soft segments. Stabilization usually is necessary, since the material is used for the manufacturing of long-life articles [73]. Essentially two antioxidants are in use in concentrations of up to 1% [74]:

- 4,4′-Di-(α,α-dimethylbenzyl)-diphenylamine,
- N,N′-hexamethylenebis-3-(3,5-ditert.butyl-4-hydroxyphenyl)-propionamide

They can be added during pelletizing, or even better, already during polycondensation. When added during polycondensation, N,N′-hexamethylenebis-3-(3,5-ditert.butyl-4-hydroxyphenyl)-propionamide is partly chemically bound to the polymer because of its amide structure. The antioxidant thus becomes highly stable to extraction. Furthermore, a material with improved color is obtained [75].

The effectiveness of the antioxidants during oven aging at 140 to 180 °C is assessed either by determination of the viscosity number (in m-cresol) as a function of time or by a manual bending test. Breakage of the specimen on bending by 180 °C defines an end point which is sufficiently reproducible (table 14).

Table 14. Oven life of thermoplastic polyester elastomer at 149 °C. Effectiveness of selected antioxidants as function of concentration. Failure criterion: manual bending.

Antioxidant	Oven life time at indicated antioxidant concentration			
	0% (days)	0.2% (days)	0.6% (days)	1.0% (days)
N,N′-hexamethylenebis-3-(3,5-ditert.butyl-4-hydroxy-phenyl)-propionamide	1	8	16	21
4,4′-Di-(α,α-dimethylbenzyl)-diphenylamine	1	11	36	80

1.8.16 Thermoplastic polyurethane elastomers

The properties and the oxidative stability of this class of polymers is determined by the length and the structure of the linear polyether or polyester soft segments and also by the isocyanate. The stability of polyether urethanes against autoxidation is distinctly lower compared to that of polyester urethanes, but the latter are less stable to hydrolysis.

Antioxidants may be used in polyurethanes for rather different purposes:

- Injection molding grades may be stabilized against loss of mechanical properties and against discoloration [76].
- In polyurethane elastomeric fibers, antioxidants are occasionally used as gasfading inhibitors.
- Frequently antioxidants are used in combination with light stabilizers to improve light stability. Such combinations normally show synergistic effects (cf. chapter 3 on light stabilizers).

Another application of antioxidants in polyurethanes is their addition to polyols, above all polyether polyols, which are used in the manufacture of polyurethane foams. Actually, they have two functions: first, they stabilize the polyols against autoxidation during storage, second, they prevent occasionally occurring discoloration in the center of the material during foaming ("scorching" as the result of the strongly exothermic reaction and the low heat conductivity of the foams). The following materials are in use:

Hindered phenols, e.g.

- BHT,
- octadecyl 3-(3,5-ditert.butyl-4-hydroxyphenyl)-propionate,
- pentaerythrityl tetrakis-3-(3,5-ditert.butyl-4-hydroxyphenyl)-propionate;

aromatic amines, e.g.

- 4,4′-ditert.octyl-diphenylamine;

as well as their combinations.

Aromatic amines can be employed only in very limited concentrations (250 to at most 500 ppm) because of their discoloring properties.

The effectiveness of antioxidants in polyols can be assessed using dynamic differential thermal analysis (DSC, DTA). The temperature of the exothermic reaction is determined at a constant heating rate of the test

Table 15. Oxidation temperatures of a polyether polyol; differential scanning calorimetry in oxygen; heating rate: 8 °C/min, additive concentration: 0.05%.

Antioxidant	Oxidation temperature (°C)
Without antioxidant	130
BHT	143
Octadecyl 3-(3,5-ditert.butyl-4-hydroxyphenyl)-propionate	144
4,4′-Ditert.octyl-diphenylamine	165
4,4′-Ditert.octyl-diphenylamine + BHT (1:1)	161
4,4′-Ditert.octyl-diphenylamine + Octadecyl 3-(3,5-ditert.butyl-4-hydroxyphenyl)-propionate (1:1)	165

sample (table 15). To estimate the tendency of a given polyurethane foam formulation to scorching, the use of micro wave ovens was recently recommended as a means of thermal treatment of the foamed test specimen.

1.8.17 Polyacetals (POM)

Polyoxymethylenes show a marked tendency to thermal depolymerization with loss of formaldehyde. To prevent thermal depolymerization, polyoxymethylenes are structurally modified; the two possibilities are acetylation to block the reactivity of the end groups or copolymerization with cyclic ethers, e.g. ethylene oxide.

Furthermore, polyacetals are also sensitive towards autoxidation, which invariably leads to depolymerization as a result of chain scission. The released formaldehyde is very likely to be oxidized to formic acid which, in turn, catalyzes depolymerization. Such damage manifests itself essentially in weight loss, which is typical for polyacetals.

The stabilizer systems are invariably composed of a hindered phenol with a costabilizer. The following are in use:

Hindered phenols:

- 2,2′-Methylenebis-(4-methyl-6-tert.butylphenol),
- 1,6-hexamethylenebis-3-(3,5-ditert.butyl-4-hydroxyphenyl)-propionate,

- pentaerythrityl tetrakis-3-(3,5-ditert.butyl-4-hydroxyphenyl)-propionate;

Costabilizers.

A great number of nitrogen-containing organic compounds are described in the literature as costabilizers for polyacetals, e.g. dicyandiamide, melamine, terpolyamides, urea and hydrazine derivatives. The effectiveness of such compounds is based on their ability to react with formaldehyde and to neutralize acids formed by oxidation, especially formic acid. In addition to nitrogen compounds, salts of long chain fatty acids are also used as acid acceptors, examples are calcium stearate, calcium ricinoleate or calcium citrate.

The practical concentrations are between 0.1 and 0.5% for the phenolic antioxidant and between 0.1 and 1.0% for the costabilizer.

The test for stabilizers used in polyacetals consists in thermal treatment of the material at 120 to 150 °C and also at temperatures of 220 to 230 °C. The lower temperatures give an indication of the effectiveness of antioxidant systems as long term heat stabilizers. The testing is carried out on injection molded or compression molded specimens. Test criteria are mainly color and weight loss, occasionally also tensile strength.

Testing at high temperatures can be carried out either by direct heat exposure of the polymer in granular form in open vessels or by thermogravimetric analysis. Such testing above the melting point of the polymer represent the methods assessing processing stability (tables 16 and 17).

Table 16. Thermogravimetric analysis of polyacetal in air at 220 °C. Sample size: 10 mg; failure criterion: time to 10% weight loss.

Antioxidant 0.2%	Time to 10% weight loss	
	Homopolymer (costabilizer: 0.5% polyamide) (min)	Copolymer (costabilizer: 0.3% Ca-stearate) (min)
Without antioxidant	4	3
2,2′-Methylenebis-(4-methyl-6-tert.butylphenol)	16	14
1,6-Hexandiol bis-3-(3,5-ditert.butyl-4-hydroxyphenyl)-propionate	30	25

Table 17. Oven life of polyacetal at 140 °C. Compression molded plaques, thickness 1 mm, failure criterion: time to 4% weight loss.

Antioxidant	Time to 4% weight loss	
	Homopolymer (costabilizer: 0.5% polyamide) antioxidant: 0.2% (days)	Copolymer (costabilizer: 0.3% Ca-stearate) antioxidant: 0.5% (days)
Without antioxidant	2	2
2,2′-Methylenebis-(4-methyl-6-tert.butylphenol)	16	14
1,6-Hexandiol bis-3-(3,5-ditert.butyl-4-hydroxyphenyl)-propionate	23	31

1.8.18 Polycarbonate (PC)

Thermooxidative damage of polycarbonate manifests itself in yellowing. Because of the transparency of polycarbonate, already very slight yellowing is readily seen. For this reason, stabilization against discoloration is considered to be important. The stabilizers are generally added during the pelletizing step. Exceptionally severe requirements are imposed on non-volatility and thermostability of the additives, because the processing temperatures are extraordinarily high (ca. 320 °C).

1.8.18.1 Stabilization of polycarbonate during processing

Yellowing of polycarbonate during processing is retarded by the addition of phosphites or phosphonites. They are used in concentrations of 0.05 to 0.15%, possibly in combination with an epoxy compound as acid acceptor.

The addition of these stabilizers in the given concentrations results in a decrease of the yellowness index by 5 to 8 points, which is easily detectable visually. Rise of the melt flow index and negative influence on impact strength are inhibited as well [45, 60].

Thermogravimetric and differential thermoanalytical methods have also been proposed to detect the efficacy of phosphites [77].

1.8.18.2 Long term heat stabilization of polycarbonate

The effective processing stabilizers are not suitable for preventing the yellowing of polycarbonate under conditions of long time use simulated in oven aging tests. The successful approach here is the addition of a sterically hindered phenol. As an example, the data in table 18 show the effectiveness of octadecyl 3-(3,5-ditert.butyl-4-hydroxyphenyl)-propionate.

Table 18. Yellowing of polycarbonate on oven aging at 140 °C. Injection molded plaques, thickness 1 mm.

Antioxidant	Yellowness index (ASTM D 1925-70) after specified number of days			
	0	20	40	90
Without antioxidant	10	18	23	32
Octadecyl 3-(3,5-ditert.butyl-4-hydroxyphenyl)-propionate 0.1%	7	10	13	17

1.9 Index of trade names, manufacturers and suppliers of antioxidants

The following index of trade names, manufacturers and suppliers is based on a choice of 54 representative antioxidants for thermoplastics. No claim is made for completeness. Detailed lists are found in [21] and [78].

The letters A to O symbolize chemical classes of compounds with the following meanings (compare also tables 1 and 2):

A Thiobisphenols (molecular weight 300 to 600)
B Alkylidene-bisphenols (molecular weight 300 to 600)
C Alkylidene-bisphenols (molecular weight >600)
D Alkylphenols
E Di-(3-tert.butyl-4-hydroxy-5-methylphenyl)-dicyclopentadiene
F Hydroxybenzyl compounds (molecular weight 300 to 600)
G Hydroxybenzyl compounds (molecular weight >600)
H Acylaminophenols (4-hydroxyanilides)
I Octadecyl 3-(3,5-ditert.butyl-4-hydroxyphenyl)-propionate
J Poly-hydroxyphenylpropionates (molecular weight >600)
K Amines
L Thioethers
M Phosphites and phosphonites
N Zinc dibutyldithiocarbamate
O Commercial blends of primary and secondary antioxidants

Trade name	Chemical class															Manufacturer/supplier
	A	B	C	D	E	F	G	H	I	J	K	L	M	N	O	
Age Rite											×			×		R.T. Vanderbilt Co. Inc., Norwalk, CT 06855, USA
Vanox		×					×			×	×	×	×			
Anox				×					×	×	×					Ente Nazionale Idrocarburi 00144 Rome, Italy
AO									×	×						Song Woun, Ltd., Suweon 25, Korea
CAO	×	×		×												Ashland Chemical Ltd., Columbus OH, USA
Carstab										×	×					Morten Thiokol Inc., Danvers, MA 01923, USA
Cyanox		×					×					×			×	American Cyanamid Co., Wayne, NJ 07470, USA
Ethanox	×	×					×		×							Ethyl Corporation Baton Rouge LA 70801, USA
Good-rite							×			×	×					B.F. Goodrich Chemical Group, Cleveland, OH 44131, USA
Hostanox			×									×	×			Hoechst AG, Frankfurt/M., D; American Hoechst Corp., Somerville, NJ 08876, USA
HPM			×													Société Française d'Organo-Synthese Nevilly-sur-Seine, F
Garbefix	×	×										×	×			
Ionol				×												Shell Nederland Chemie BV, s'Gravenhage, NL
Ionox							×									
Irgafos													×			Ciba-Geigy AG, Basel, CH; Ciba-Geigy Corp., Hawthorne, NY 10532, USA
Irganox	×	×				×	×	×	×	×	×	×			×	

Trade name	Chemical class A	B	C	D	E	F	G	H	I	J	K	L	M	N	O	Manufacturer/ supplier
Isonox		×		×												Schenectady Chemicals Inc., Schenectady NY, USA
Keminox	×	×	×	×			×									Chemipro Kasei Ltd., Chuo-Ka, Kobe, Japan
Lowinox	×	×		×	×				×	×	×	×	×			Chemische Werke Lowi GmbH, Waldkraiburg, D
Mark												×	×		×	Argus Chemical S.A.-N.V., Bruxelles, B; Witco Chem. Co., New York, NY 10017, USA; Adeka-Argus Chem. Co. Ltd., Urawa City, Saitama Pref. 336 Japan
Mixxim		×											×			Fairmount Chem. Co., Inc., Newark, NJ 07105, USA
Naugard		×	×						×	×	×	×	×	×	×	Uniroyal Ltd., Bromsgrove, Worcs., GB; Uniroyal Chem., Naugatuck, CT 06770, USA
Naugawhite		×														
Polygard													×			
Nonox		×		×							×				×	ICI Ltd., London SW, GB; ICI Americas, Wilmington, DE 19897, USA
Topanol		×		×							×					
Negonox			×													
Oxi-chek		×							×							Ferro Corp. Chem. Div., Walton Hills, OH 44146, USA
Perkanox				×												Akzo Chemie, Nederland NV., Amsterdam, NL
Phosctere												×				
Permanax		×		×												Vulnax International, 922/3 Saint Cloud, F

Trade name	Chemicals class															Manufacturer/supplier
	A	B	C	D	E	F	G	H	I	J	K	L	M	N	O	
Samilizer	×	×							×	×		×	×			Sumitomo Chemical Co., Ltd., Osaka 541, Japan
Antigene											×					
Santonox	×															Monsanto Europe SA, Bruxelles, B; Monsanto Co., Akron, OH 30339, USA
Santowhite	×	×		×							×					
Santoflex											×					
Santicizer	×												×			
Flectol											×					
Seenox	×	×					×					×				Shipro Kasei Ltd. Kita-ku, Osaka, Japan
Tominox									×	×						Yoshitomi Pharmaceutical Ind. Ltd., Osaka 541, Japan
Yoshinox	×	×														
Vulkanox		×		×							×	×				Mobay Chem. Corp. Pittsburgh, PA 15205, USA; Bayer AG, 5090 Leverkusen, D
Weston													×		×	Borg-Warner Chemicals, Parkersburg, WV 26101,USA
Ultranox	×	×		×					×		×		×			
Wing Stay	×			×	×						×					Goodyear SA, Rueil-Malmaison, F; Goodyear, Akron, OH 44316, USA
Wytox		×		×							×		×			Olin Corp. Stamford CT 06904, USA

1.10 References

[1] *J.C. Johnson:* Antioxidants, Syntheses and Applications, Noyes Data Corp., Park Ridge/N.J., 1975.

[2] *M.W. Ramney:* Antioxidants, Recent Developments, Noyes Data Corp., Park Ridge/N.J., 1979.

[3] *Edwin S. Gould:* Mechanismus und Struktur in der org. Chemie, German edition, Verlag Chemie, Weinheim/Bergstr., 1962.

[4] *J.L. Bolland, G. Gee:* Trans. Faraday Soc. 42 (1946) pp. 236, 244.

[5] *J. Pospisil* et al.: J. Polymer Sci.: Symposium No. 40 (1973) p. 209,
Angew. Makromol. Chem. 28 (1973) p. 13,
IUPAC International Symposium on Macromolecules, Aberdeen, 1973,
Angew. Makromol. Chem. 38 (1974) p. 191,
Angew. Makromol. Chem. 39 (1974) p. 107,
Angew. Makromol. Chem. 39 (1974) p. 189,
15th IUPAC Microsymposium on Macromolecules, Prague, 1975,
Angew. Makromol. Chem. 52 (1976) p. 31,
Angew. Makromol. Chem. 65 (1977) p. 197,
J. Kovarova, J. Pospisil: European Polym. J. 13 (1977) p. 975.

[6] *J. Koch:* Angew. Makromol. Chem. 20 (1971) pp. 7, 21, 35.

[7] *H.J.M. Bartelink* et al.: Chem. and Ind. (1978) p. 586.

[8] *J. Pospisil:* Adv. in Polym. Sci. 36 (1980) p. 69.

[9] *G. Scott:* Developments in Polymer Stabilization, Vol. 1, chapter 1, Applied Science Publishers Ltd., London, 1979.

[10] *G.K. Cowell:* Plastics Engineering 32 (1976) 10, p. 51.

[11] *G. Scott:* Developments in Polymer Stabilization, Vol. 2, chapter 4, Applied Science Publishers Ltd., London, 1979.

[12] *D.G. Pobedimskii, P.A. Kirpichnikov:* J. of Polymer Sci., Polym. Chem. Ed. 18 (1980) p. 815.

[13] *W.L. Hawkins:* Polymer Stabilization, John Wiley & Sons, Inc., London/New York, 1972.

[14] *C. Armstrong, M.J. Husbands, G. Scott:* Eur. Polym. J. 15 (1979) pp. 241, 249.

[15] *D.W. Chasar:* Polymer Degradation and Stability 3 (1980–1981) p. 121.

[16] *K.U. Ingold:* Chem. Reviews 61 (1961) p. 563, Accounts of Chemical Research 2 (1969) p. 1.

[17] *Houben-Weyl:* Methoden der Organischen Chemie, 4th edition, volume IV/1b, p. 1049. Editor: E. Müller, Georg Thieme Verlag, Stuttgart, 1975.

[18] *G. Scott:* Developments in Polymer Stabilization, Vol. 4, chapters 1, 2, Applied Science Publishers Ltd., London, 1981.

[19] *H. Lind:* Kunststoffe-Plastics 24 (1977) p. 27.

[20] *J. Voigt:* Die Stabilisierung der Kunststoffe gegen Licht und Wärme, Springer Verlag, Berlin, 1966.

[21] *T.J. Henman:* World Index of Polyolefine Stabilizers, Kogan Page Ltd., London, 1982.

[22] *G. Scott, M.F. Yusoff:* Eur. Polym. J. 16 (1980) p. 497.
[23] *K. Figge, J. Klahn:* Angew. Makromol. Chem. 107 (1982) pp. 101, 117.
[24] *J. Klahn, K. Figge, W. Freytag:* Deutsche Lebensmittel-Rundschau 78 (1982) pp. 215, 241.
[25] *G. Scott:* Developments in Polymer Stabilization, Vol. 4, chapter 6, Applied Science Publishers Ltd., London, 1981.
[26] German Patent 2,453,152 (1973) Coalite and Chemical Products.
[27] German Patent 2,509,654 (1974) G. Scott.
[28] U.S. Patent 3,658,769 (1969) The Goodyear Tire and Rubber Co.
[29] *M. Johnson* et al.: J. Appl. Polym. Sci. 19 (1975) pp. 319, 1745; 20 (1976) p. 2533.
[30] *O. Cicchetti* et al.: Eur. Polym. J. 4 (1968) p. 419.
[31] *R.A. Jackson* et al.: J. Appl. Polym. Sci. 12 (1968) p. 1297.
[32] *Ryong-Joon Roe* et al.: J. Appl. Polym. Sci. 18 (1974) p. 843.
[33] *J. Durmis* et al.: Eur. Polym. J. 11 (1975) p. 219; 12 (1976) p. 173.
[34] *J.E. Guillet:* Polymer Letters Edition 14 (1976) p. 257.
[35] *J.Y. Moisan:* Eur. Polym. J. 16 (1980) pp. 979, 989, 997.
[36] *N.C. Billingham* et al.: J. Appl. Polym. Sci. 26 (1981) p. 3543.
[37] *L.H. Buxbaum:* Angew. Chem. 80 (1968) p. 225.
[38] *W.H. Sharkey, W.E. Mochel:* J. Amer. Chem. Soc. 81 (1959) p. 3000.
[39] *G. Scott:* Atmospheric Oxidation and Antioxidants, Elsevier Publishing Company, Amsterdam/Oxford/New York, 1965.
[40] *G. Geuskens:* Degradation and Stabilization of Polymers, Applied Science Publishers, London, 1977.
[41] *H.H.G. Jellinek:* Aspects of Degradation and Stabilization of Polymers, Elsevier, Amsterdam/Oxford/New York, 1978.
[42] *H. Weidner, R. Tilgner:* Kunststoffe 70 (1980) pp. 773, 837.
[43] *T.R. Crompton:* Chemical Analysis of Additives in Plastics, 2nd edition, International Series of Monographs in Analytical Chemistry, Vol. 46, Pergamon Press, 1977.
[44] *G. Scott:* Developments in Polymer Stabilization, Vol. 1, chapter 2, Applied Science Publishers Ltd., London, 1979.
[45] *C. Swasey:* Plastics World 9 (1976) p. 42.
[46] *F. Mitterhofer:* Kunststoffe 67 (1977) p. 151.
[47] *F. Mitterhofer:* Caoutchoucs et Plastiques 568 (1977) p. 65.
[48] *J.P. Forsman:* S.P.E. Tech. Papers 10 (1964) p. 1.
[49] *D.A. Gordon:* Advances in Chemistry Series 85 (1968) p. 224.
[50] *H. Gysling:* Advances in Chemistry Series 85 (1968) p. 239.
[51] *C.H.R.I. de Jonge* et al.: Advances in Chemistry Series 169 (1978) p. 399.
[52] Ciba-Geigy: unpublished results.
[53] Ciba-Geigy: Irganox 1076, product data sheet, 1974.
[54] Ciba-Geigy: Irganox 1010, product data sheet, 1974.
[55] *B.F. Goodrich:* Good-rite 3114 antioxidant, product data sheet, 1977.
[56] Imperial Chemical Industries: Topanol CA, product data sheet, 1970.
[57] Hoechst: Hostanox 03, product data sheet, 1972.
[58] *F. Mitterhofer:* Kunststoffe 67 (1977) p. 684.
[59] *F. Mitterhofer:* Polym. Eng. Sci. 20 (1980) p. 692.

[60] *Nam P. Suh, Nak-ho Sung:* Science and Technology of Polymer Processing, The MIT Press, Cambridge, Mass. and London, England, 1979, p. 809.
[61] *E.C. Rothstein:* J. Appl. Polym. Sci. 12 (1968) p. 1279.
[62] Weston Chemical: Weston 618 Phosphite antioxidant, product data sheet, 1974.
[63] *G. Obieglo:* Kunststoffe 69 (1979) p. 381.
[64] *H. Vogel:* Kunststoffe-Plastics 3 (1975) p. 11.
[65] *C.G. Fratkina, G. Kirillova:* Vysokomol. soyed. A 12, 10 (1970) p. 2199.
[66] *L. Rosik:* Angew. Makromol. Chem. 16/17 (1971) p. 209.
[67] *L. Rosik:* Annual Meeting of French-Czechoslovak Cooperation on Oxidative Degradation and Combustion of Polymers, October 1976.
[68] *L. Rosik* et al.: 8th conference on differential thermal analysis, Czechoslovakia, October 1979, Chem. Abstr. 92 (1980) 77328v.
[69] *A. Casale, O. Salvatore:* Polym. Eng. Sci. 15 (1975) p. 286.
[70] *M. Ghaeny, G. Scott:* Polymer Degradation and Stability 3 (1981) p. 233.
[71] Deutsche Shell Chemie: Cariflex TR-1000, product data sheet RBX/73/8 (G), 1973.
[72] Ciba-Geigy: unpublished results.
[73] *B.M. Walker:* Handbook of Thermoplastic Elastomers, chapter 4, Van Nostrand Reinhold Company, New York, 1979.
[74] *G.K. Hoeschele:* Angew. Makromol. Chem. 58/59 (1977) p. 299.
[75] *G.K. Hoeschele:* Polym. Eng. Sci. 14 (1974) pp. 848, British Patent 1, 450, 910 (1972) Du Pont de Nemours, Co.
[76] *G. Mathur* et al.: Adv. Urethane Sci. Technol. 6 (1978) p. 103, Chem. Abstr. 90 (1979) 152969u.
[77] *I.I. Levantovskaya:* Vysokomol. soyed. A 17, 10 (1975) p. 2331.
[78] Modern Plastics Encyclopedia, Mc Graw Hill Publications Company, New York, 1982.

Suggested reading

[79] *W.O. Lundberg:* Autoxidation and Antioxidants, Interscience Publishers, New York/London, 1962.
[80] *M.B. Neiman:* Aging and Stabilization of Polymers, Consultants Bureau, New York, 1965.
[81] *W.A.G. Platzer:* Stabilization of Polymers and Stabilizer Processes, Advances in Chemistry Series, Vol. 85, American Chemical Society Publications, Washington, D.C., 1968.
[82] *L. Reich, S. Stivala:* Autoxidation of Hydrocarbons and Polyolefins, Marcel Dekker, New York, 1969.
[83] *K. Thinius:* Stabilisierung und Alterung von Plastwerkstoffen, Akademie Verlag, Berlin, 1969.
[84] *L. Reich, S. Stivala:* Elements of Polymer Degradation, Mc Graw Hill, New York, 1971.
[85] *A.S. Kuz'minski:* The Ageing and Stabilization of Polymers, translated and edited by B.N. Leyland, Elsevier Publishing Company, Amsterdam/London/New York, 1971.

[86] *L. Mascia:* The Role of Additives in Plastics, Edward Arnold Publishers, London, 1974.
[87] *B. Dolezel:* Die Beständigkeit von Kunststoffen und Gummi, Carl Hanser Verlag, München/Wien, 1978.
[88] *G. Scott:* Developments in Polymer Stabilization, Vol. 3, Applied Science Publishers Ltd., London, 1980.
[89] *W. Schnabel:* Polymer Degradation, Principles and Practical Applications, Carl Hanser Verlag, München/Wien/New York, 1981.
[90] *B.N. Gorbunor, J.A. Gurvich, I.P. Maslova:* Chemistry and Technology of Polymer Stabilizers, Chimia, Moskau, 1981 (russ.).

2 Metal Deactivators

Dr. Helmut Müller, Ciba-Geigy, Basel/Switzerland

2.1 Introduction

Thermooxidative degradation of polyolefins proceeds by a typical free-radical chain mechanism (see chapter 1). Hydroperoxides are key intermediates in this mechanism because of their thermally induced homolytic decomposition to free radicals, which in turn initiate new oxidation chains.

Since the monomolecular homolytic decomposition of hydroperoxides into free radicals requires relatively high activation energies, this process becomes effective only at temperatures in the range of 120 °C and higher. However, in the presence of catalytic amounts of certain metal ions, hydroperoxides decompose already at room temperature by a redox reaction [1] to radical products:

$$
\begin{array}{lllllll}
\mathrm{ROOH} & + & \mathrm{M}^{n\oplus} & \longrightarrow & \mathrm{ROO}^{\cdot} + \mathrm{M}^{(n-1)\oplus} & + & \mathrm{H}^{\oplus} \\
\mathrm{ROOH} & + & \mathrm{M}^{(n-1)\oplus} & \longrightarrow & \mathrm{RO}^{\cdot} + \mathrm{M}^{n\oplus} & + & \mathrm{OH}^{\ominus} \\
\hline
2\,\mathrm{ROOH} & & & \xrightarrow{\mathrm{M}^{n\oplus}/\mathrm{M}^{(n-1)\oplus}} & \mathrm{ROO}^{\cdot} + \mathrm{RO}^{\cdot} & + & \mathrm{H_2O}
\end{array}
$$

According to *Emanuel* and coworkers [2] the electron transfer is preceded by the formation of a coordination complex metal ion/hydroperoxide.

The most active catalysts of hydroperoxide decomposition are derivatives of those metals, which are easily oxidized or reduced by one-electron transfer and have different oxidation states with comparable stability. This is true of compounds of metals such as Fe, Co, Mn, Cu, Ce, and V, commonly called transition metals.

The presence of the aforementioned metal ions in the autoxidation of a hydrocarbon increases the decomposition rate of hydroperoxides and the overall oxidation rate to such an extent that even in the presence of antioxidants, the induction period of oxygen uptake is drastically shortened or completely eliminated [3].

The same phenomenon applies to the metal-catalyzed thermal oxidation of polyolefins [4]. Even at rather high concentrations, sterically hindered phenols or aromatic amines do not retard the oxidation rate satisfactorily. A much more efficient inhibition is achieved by using metal deactivators, together with antioxidants. The former are also known as copper inhibitors, because, in practice, the copper-catalyzed oxidation of polyolefins is by far of greatest importance. This is mainly due to the fact that polyolefins are the preferred insulation material for communication wire and power cables, which generally contain copper conductors. Although PVC has still the biggest share as insulation for low voltage power cables ($\leqq 1$ kV) at least in the Federal Republic of Germany [5], efforts are being made to improve the permissible thermal load on low voltage power lines. This will lead to increased use of polyethylene or cross-linked polyethylene (see 2.6.1.1).

Communication wire insulated with polyolefins is the most important outlet for metal deactivators. This is especially true for the USA, where the use of metal deactivators in conjunction with a phenolic antioxidant is now specified for telephone wire insulation. The reason for this goes back to premature failure of polyethylene-insulated communication wires in the southern United States. Such wires had been stabilized in the past merely with a phenolic antioxidant [6, 7].

Apart from cable insulation, the application of copper inhibitors is appropriate when polyolefins are in direct contact with copper or copper containing alloys, e.g. metal inserts in technical parts, or when the presence of catalytically active metal ions is likely in other additives such as pigments, fillers and reinforcing agents (talcum, asbestos).

2.2 Mechanisms of inhibition by metal deactivators

As has been postulated previously [8] the function of a metal deactivator is to form an inactive complex with the catalytically active metal species. Chelating agents are specifically suited for this purpose on account of the possibly high thermal stability of the metal complexes formed.

The general feature of chelating agents is their polyfunctionality, this means that they may contain several ligand atoms like N, O, S, P, alone or in combination with functional groups as hydroxyl, carboxyl, or carbamide groups. The ligand N,N′-bis(o-hydroxybenzal)oxalyldihydrazide is a good example for the above-mentioned structural possibilities. This ligand forms a soluble complex (1) with the first mole of a copper salt by binding at the phenolic groups, a second mole of copper salt binds at the amide nitrogens to form an insoluble complex which is most likely polymeric [9].

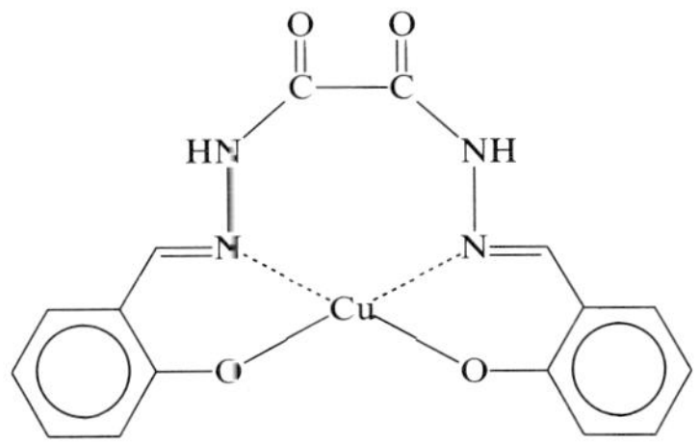

(1)

The efficiency of a specific ligand as a metal deactivator depends on the nature of the metal ion being chelated as *Pedersen* was able to demonstrate [10]. He found that chelated copper can reduce, chelated iron or manganese may even increase, the catalytic activity of the corresponding complexes.

There have been numerous publications on metal-catalyzed oxidation and the determining parameters such as structure of ligands, nature of the central atom, oxidizable substrate etc. Most of this work has been carried out in homogeneous, fluid model systems [1, 3 and references cited therein]. *Allara* and *Chan* [9] were the first investigators to try to obtain a better understanding of the mechanisms on inhibition against copper-catalyzed oxidation of polyethylene by working with structurally defined copper complexes. After synthesis and careful characterization of a number of copper-deactivator complexes based mostly on oxamide derivatives, which as such have been known as efficient copper inhibitors, they examined the catalytic activity of these complexes in low density polyethylene by oxygen uptake tests. They were able to show that deactivators form copper complexes with much less residual catalytic activity than e.g. Cu-I-oxide. Furthermore, their data allow drawing some conclusions on the relationship between reactivity and structure of said copper complexes. There are certainly additional parameters determining the relative deactivator performance as ease of dispersion, solubility and compatibility with the polymeric substrate. Their data show for example that the copper complexes of N,N′-dibenzaloxalyldihydrazide and oxalanilide are of equal reactivity, whereas a comparison of polyethylene-copper samples containing a phenolic antioxidant demonstrates that the oxidative induction period at 120 °C is 2.8 times as long for added N,N′-dibenzaloxalylhydrazide as for added oxalanilide.

2.3 Requirements for metal deactivators

Besides their main function to retard efficiently the metal-catalyzed oxidation of polyolefins, metal deactivators have to possess a number of ancillary properties to be useful in actual service. These are discussed in the following:

- Sufficient solubility or ease of dispersion. Thus, e.g. the commercially available deactivator N,N′-dibenzaloxalyldihydrazide is completely insoluble in polyolefins. For this reason the supplier recommends to treat this additive as a pigment by preparing a masterbatch of the resin that is then repelletized or repowdered and let down into the balance of the polyolefin [11]. Compared to other additives such as antioxidants or light stabilizers, metal deactivators are somewhat peculiar with regard to solubility in the substrates, because the majority of commercially or experimentally offered compounds ought to be completely or predominantly insoluble in polyolefins on account of their structures and melting points [12].
- Good compatibility to avoid exudation connected with loss of performance. That problem can arise only if under actual conditions a metal deactivator is partially or completely molecularly dispersed. Here, there are at least two possibilities: either the additive crystallizes within the polymer or it migrates and forms deposits on the surface, provided it is not removed by vaporization or mechanical wear [13]. In the first instance the crystallized material may serve as a source for the replacement of used up additive.
- High extraction resistance. This property is especially important if petrolatum is being used to fill the interstices between the singles of a telephone cable in order to make it waterproof.
- Low volatility under processing and service conditions. N,N′-diphenyloxamide, the first commercial deactivator is partially volatilized during processing [11].
- No discoloration by interaction with antioxidants or with colorants used for coding wires.
- Sufficient thermal stability under processing and service conditions to avoid the consumption of additive in non-stabilizing reactions and to suppress the formation of decomposition products which may be dangerous to health.
- Handling safety with respect to industrial hygiene, i.e. the deactivators should have low acute toxicity and should not cause irritation of skin and mucous membranes even when carelessly used.

2.4 Structural classes

Within the last 20 years the following chemical structures have been proposed as metal deactivators for polyolefins:

a) Carboxylic acid amides of aromatic mono, and dicarboxylic acids and N-monosubstituted derivatives as e.g. N,N′-diphenyloxamide [4].

b) Cyclic amides such as barbituric acid.

c) Hydrazones [14] (2) and bishydrazones (3) of aliphatic and, above all, aromatic aldehydes as benzaldehyde [15] and salicylaldehyde [16] or of o-hydroxy-arylketones [17]:

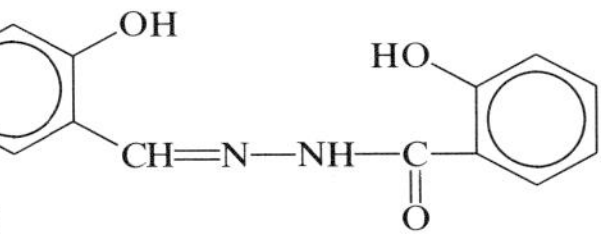
(2)

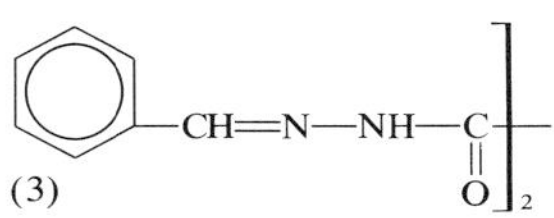
(3)

d) Hydrazides of aliphatic [18] and aromatic [19] mono, and dicarboxylic acids as well as N-acylated derivatives thereof as (4) and (5):

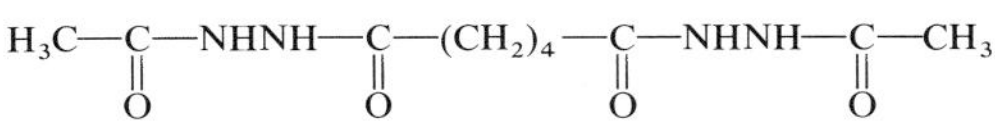
(4)

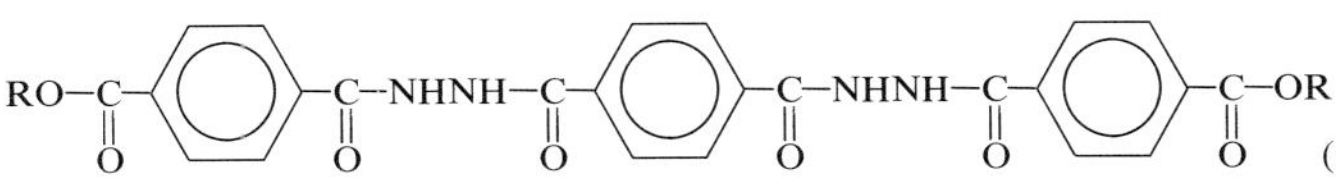
(5)

e) Bis-acylated hydrazine derivatives as (6) [20] and (7) [21]:

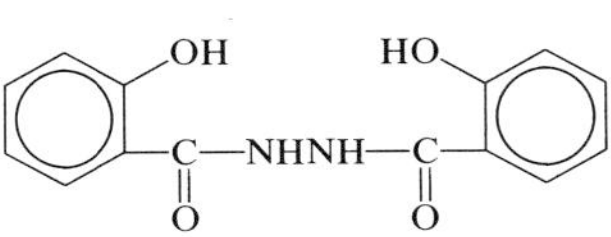
(6)

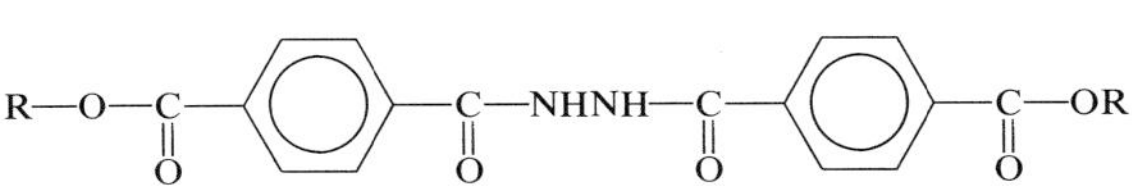
(7)

f) Heterocyclic compounds as melamine, benzotriazoles, 8-oxyquinoline, hydrazones and acylated derivatives of hydrazino-triazines [22] as (8), aminotriazoles and acylated derivatives thereof [23] as (9):

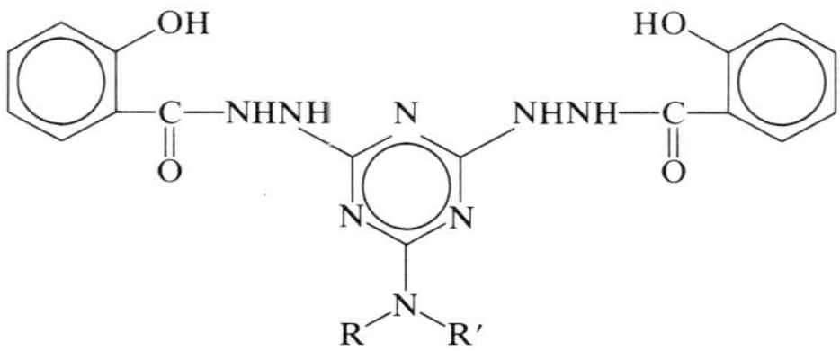

(8)

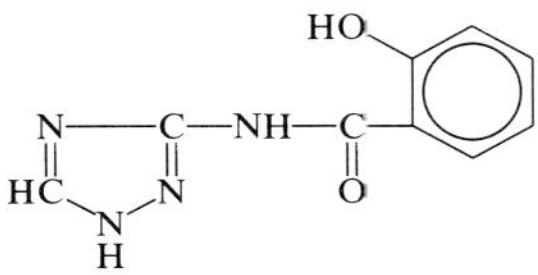

(9)

g) Polyhydrazides [24] as e.g. (10):

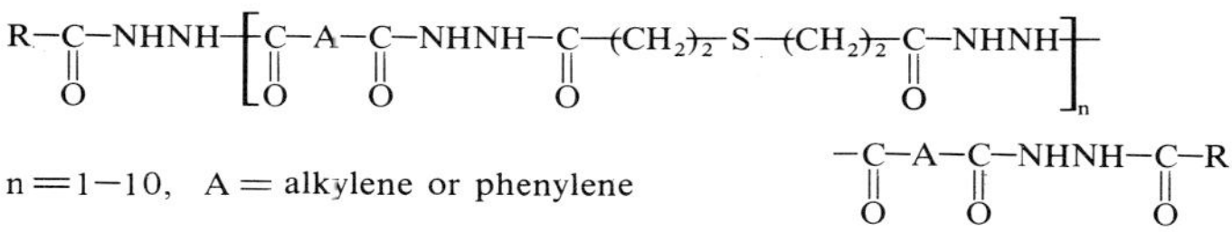

n = 1–10, A = alkylene or phenylene

(10)

h) Molecular combinations of sterically hindered phenols and metal complexing groups as (11) [25], (12) [26], (13) [27] and (14) [28]:

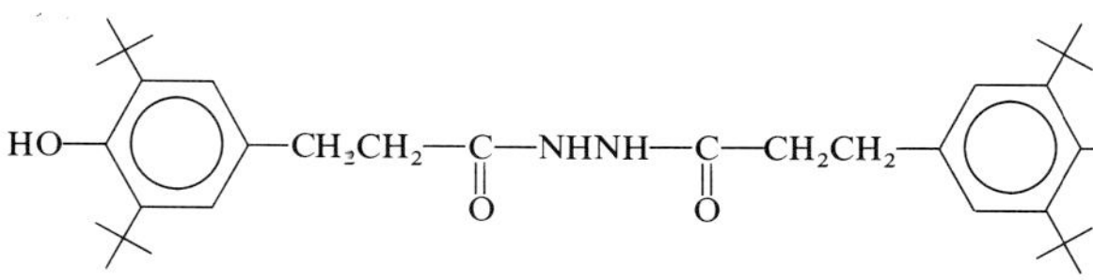

(11)

(+ = tert. butyl group)

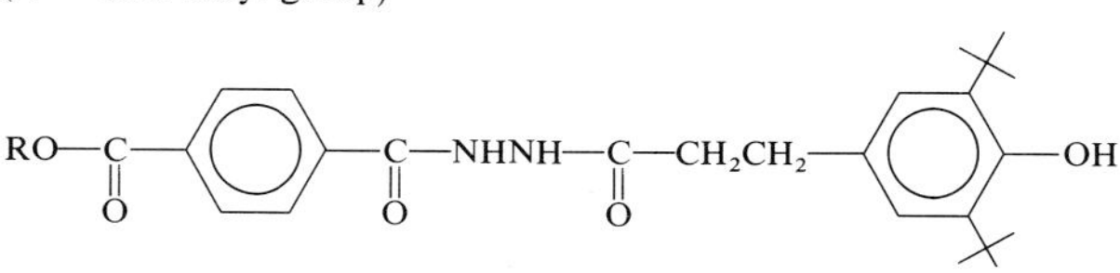

(12)

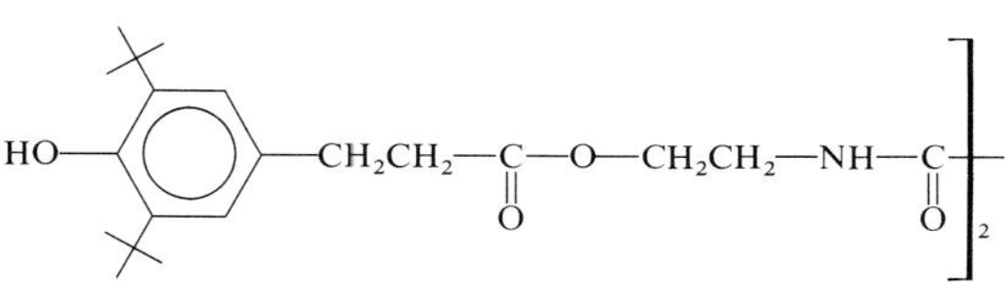

(13)

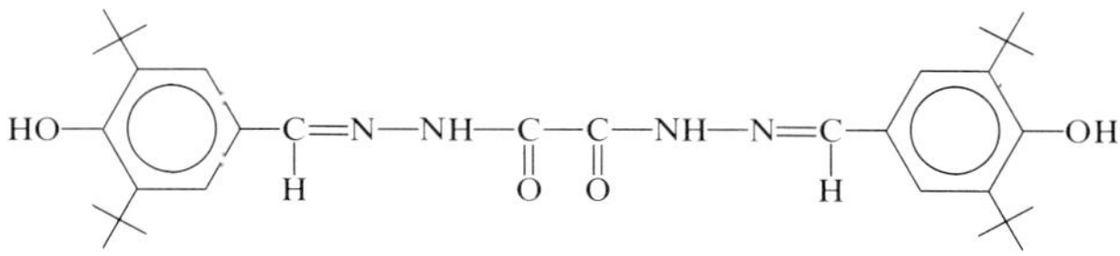

(14)

i) Ni-salts of benzyl phosphonic acids as (15) alone, or in combination with other antioxidants or metal deactivators [29]:

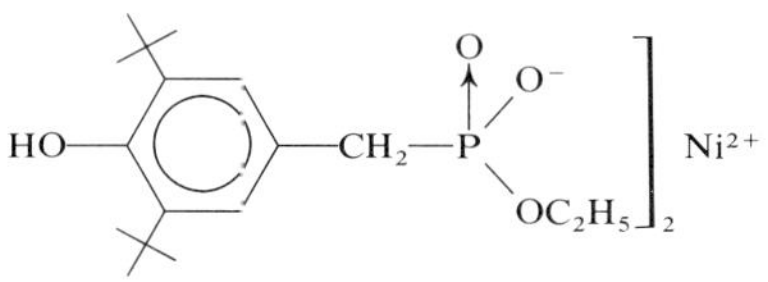

(15)

j) Pyridinethiol-Sn-compounds as (16) [30]:

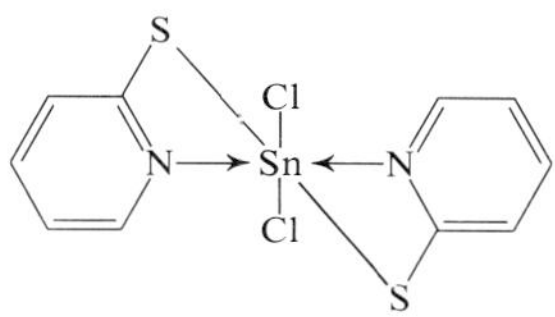

(16)

k) Tert. phosphorous acid ester of a thiobisphenol [31]:

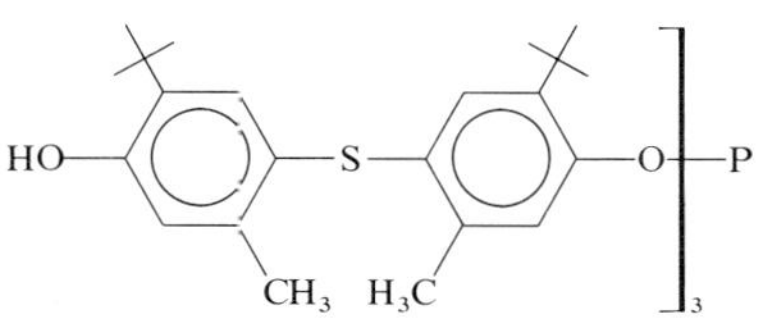

(17)

The patent literature lists numerous additional structural classes as e.g. N,N′-bis-salicylalethylenediimine, salicylaloxime, derivatives of ethylene diamino tetraacetic acid etc. The latter compounds have not been, and probably will not be used in practice as metal deactivators in thermoplastic resins due to low activity, insufficient thermal stability or volatility. According to a recent paper [12] experimentally and commercially supplied metal deactivators belong mainly to the structural class of diacylated hydrazine derivatives (8 out of 13 revealed structures), another 2 compounds are derivatives of oxalic acid dihydrazide. So far only the compounds (2), (3), a modified version of (3) with unknown composition, (6), (9), (11) and (13) have been, or are being, used in practice.

2.5 Testing of metal deactivators

The most common testing methods for evaluating thermal stability of polyolefins in contact with copper are outlined below:

a) Measurement of the oxidative induction time by the well-known oxygen uptake test of polyolefin samples containing either embedded metallic copper (powder, screen or wire) or copper salts such as copper stearate. The stabilization system includes a primary antioxidant along with the metal deactivator, as long as the latter does not represent a molecular combination of an antioxidant and a deactivator, in which case the metal deactivator may serve as the sole stabilizer. Usual temperatures for oxygen uptake are up to 180 °C [32].

b) Thermal analysis (DTA, DSC), in air or oxygen of polyolefin samples that are either in contact with copper (copper pans) or contain copper salts. The more widely used isothermal method involves the measurement of the oxidative induction time at a constant temperature in the range of 140 to 220 °C, whereas the dynamic method consists of heating the sample at a constant rate until thermal oxidation takes place as evidenced by an exotherm [33, 38].

c) Oven aging in static or circulating air at temperatures from 70 to 150 °C of polyolefin wire samples or polyolefin samples containing embedded metallic copper or copper salts. Definition of failure depends on the polymer and the sample form chosen. Appropriate failure criteria are: discoloration, embrittlement, cracking [34] and loss of elongation of the insulation. Another possibility is the determination of "residual thermal stability" as function of oven aging time by isothermal DTA/DSC. The influence of petrolatum, as used in waterproofed communication cables, on thermal stability of the polyolefin insulation is tested by aging wire segments or polyolefin sandwiches with embedded copper screen after dipping in petrolatum or in constant contact with it [29, 34, 36].

Oven aging of wire samples as "pigtails" [34], preferably at temperatures below 100 °C, approaches service conditions better than any other method mentioned. Instead of using wire samples for testing, there are two viable alternatives for preparing test specimens yielding aging data that correlate quite well with those of wire samples. The first method consists of homogenizing the polyolefin resin with the stabilizers in a suitable lab scale kneader (Brabender plastograph), or by milling and adding in the end 1% of a fine copper powder [11] or 0.1% copper stearate. The resulting compound is compression molded into plaques that may be aged as already described. The second method [35, 36] employs polyolefin sandwiches, compression molded from two stabilized sheets after inserting a copper screen.

Life time predictions of polyolefin insulations by extrapolation of high temperature test data (140 to 200 °C) to normal service temperature via an Arrhenius-plot are of questionable value and have to be considered with reservations [7, 33, 37 to 39].

2.6 Stabilization of polyolefins in contact with copper

As mentioned earlier, there are metals other than copper, such as iron, manganese and cobalt which can accelerate thermal oxidation of polyolefins and related polymers such as EPDM. In practice, however, the inhibition of the copper-catalyzed degradation of polyolefins is of paramount importance because of the steadily increasing use of polyolefin insulation over copper conductors. Here it is mandatory to combine a metal deactivator with an antioxidant if the former does not contain moities with radical scavenging function (see compound 11).

The required additive concentrations range from 0.05 to 0.5% depending on polymer, nature of the insulation (solid, cellular) whether the cable is petrolatum filled, and on service conditions.

The most often used antioxidants are the following:

- Pentaerythrityl tetrakis [3(3′,5′-di-tert.butyl-4′-hydroxyphenyl) propionate],
- 1,1,3-tris(5′-tert.butyl-4′-hydroxy-2′-methylphenyl)butane,
- 4,4′-thiobis(3-methyl-6-tert.butylphenol),
- polymeric 2,2,4-trimethyl-1,2-dihydroquinoline.

Metal deactivators in actual use are essentially N,N′-bis[3-(3′,5′-di-tert.butyl-4′-hydroxyphenyl)propionyl]hydrazine and N,N′-dibenzaloxalyldihydrazide. The latter requires predispersion in a masterbatch because of its insolubility in polyolefins [11]. This is not needed for the former compound which at commonly used concentrations is molecularly dispersed in polyolefins after processing.

2.6.1 Polyethylene

Among polyolefins, polyethylene is still the most common primary insulation material for wire and cable. In the USA high density polyethylene and ethylene-propylene copolymers are used in substantial amounts for communication wire insulation. For technical reasons polyethylene resins with densities below 0.92 are being replaced by resins with densities in the range of 0.925 to 0.94 g/cm^3 (MDPE).

Table 1. Thermal stability of medium density polyethylene wire insulation [40].

Test sample: Copper conductor 0.5 mm diameter; insulation: MDPE, $d = 0.927$ g/cm^3; thickness 0.2 mm, 1% TiO_2 (rutile).

Test procedure: Forced air oven aging at 90 and 110 °C, pigtail specimen (10 replicates).

Additives: AO I = pentaerythrityl tetrakis[3(3′,5′-di-tert.butyl-4′-hydroxyphenyl)propionate]
MD I = N,N′-bis[3(3′,5′-di-tert.butyl-4′-hydroxyphenyl)propionyl]hydrazine
MD II = N,N′-dibenzaloxalyldihydrazide

Additives		Oven aging (days)	
AO	MD	90 °C	110 °C
0.05% AO I	–	32	5
–	0.05% MD I	119	29
0.05% AO I	0.1% MD I	410	136
–	0.2% MD I	527	275
0.05% AO I	0.1% MD II	293	96

The data of tables 1 and 2 demonstrate clearly the enhancement of thermal stability of polyethylene in direct contact with the copper conductor achieved by addition of a metal deactivator and a phenolic type antioxidant. Since the structure of MD I combines both the radical scavenger function of a phenolic antioxidant and the complexing ability of a typical deactivator it can be used with quite good results without additional antioxidant (see 2.4, compound 11).

Table 2. Thermal stability of high density polyethylene wire insulation [12].

Test sample: Copper conductor 0.4 mm diameter; insulation: HDPE copolymer, $d = 0.945$ g/cm^3.

Test procedure: Aging in experimental pedestal installation complex (static air) at 100 and 110 °C, days to cracking.

Additives:
AO I = pentaerythrityl tetrakis[3(3′,5′-di-tert.butyl-4′-hydroxyphenyl)propionate]
MD I = N,N′-bis[3(3′,5′-di-tert.butyl-4′-hydroxyphenyl)propionyl]hydrazine
MD III = product of Eastman Chemical Products, structure undisclosed
MD IV = N,N′-bis(3-methoxy-2-naphthoyl)hydrazine

Additives		Oven aging (days)	
AO	MD	100 °C	110 °C
0.1% AO I	0.1% MD I	99	48
0.1% AO I	0.1% MD III	85	37
0.1% AO I	0.1% MD IV	97	49
–	0.2% MD I	92	56

Increasingly, communication cables are filled with petrolatum to improve waterproofness, whereas the insulation may be solid or cellular. Both trends, petrolatum filling and cellular insulation, obviously exert an influence on the oxidative stability of such cables. Table 3 shows a comparison of the stability of solid and cellular polyethylene insulation with and without contact with petrolatum. The data lead to the following conclusions:

- Thermal stability decreases by approx. 35% in changing from solid to cellular insulation.
- Aging in the presence of petrolatum reduces the stability of solid polyethylene by 35% and of cellular polyethylene by 10% to 40%.
- Elevated concentrations of antioxidants or of metal deactivator MD I make it possible to achieve a high level of stability even in the most adverse instance: cellular insulation in contact with petrolatum.

Table 3. Comparison of thermal stability of solid and cellular high density polyethylene wire insulation [41].

Test sample:	Copper conductor 0.65 mm diameter; insulation: 0.35 mm HDPE (solid and cellular); 1% TiO_2 (rutile); blowing agent: 0.5% modif. azodicarbonamide.
Test procedure:	Forced air oven aging at 120 °C, pigtail specimen (5 replicates), days to cracking.
Petrolatum treatment:	Wire samples were dipped for 1 s in petrolatum at 115 °C and preaged at 70 °C for 10 days. Thereafter samples were carefully wiped and aged at 120 °C.
Additives:	see table 1

Additives	Oven aging (days)			
	Without petrolatum treatment		With petrolatum treatment	
	Solid PE	Cellular PE	Solid PE	Cellular PE
0.1% AO I +0.1% MD I	137	79	116	61
0.1% AO I +0.1% MD II	133	85	77	40
0.2% AO I +0.1% MD I	–	105	–	79
0.1% MD I	–	58	–	37
0.2% MD I	158	121	142	71
0.3% MD I	–	150	–	103

The negative influence of foamed insulation is probably not so much caused by the residues of the blowing agent as by the faster oxygen diffusion to the interphase polyethylene/copper [42].

In filled cables migration of additives and rate of oxygen diffusion are important parameters because the insulation is being swollen by the filler to a certain extent.

2.6.1.1 Cross-linked polyethylene

Cross-linked polyethylene insulation is used more and more because of the higher operation temperatures allowable with such cables. Low voltage power cables require a deactivator/antioxidant stabilization, because the insulation is in direct contact with copper and it is not separated by a semi-conducting layer (PE with up to 30% carbon black) or by a polyethylene terephthalate film wrap. In choosing an optimal stabilization system, the cross-linking procedure employed (grafting of silanes, β- or γ-radiation, organic peroxides) has to be taken into account. Stabilizers should not inhibit cross-linking to an appreciable extent and cross-linking should not markedly impair performance. Phenolic antioxidants are being used successfully in peroxide-cured polyethylene. However, in β-radiation induced cross-linking, amine type antioxidants as polymeric 2,2,4-trimethyldihydroquinoline impart higher oxidation stability than phenolic types [43].

In their work on oxidative stability of polyethylene insulation cross-linked by β-radiation, *v. Gentzkow* and *Pieper* [43] have shown that by adequate stabilization – a polymeric amine type antioxidant together

Table 4. Metal deactivator effectiveness in poly(propylene-ethylene) insulated wire [36].

Test sample:	Copper wire 0.65 mm diameter; insulation 0.4 mm.
Test procedure:	Forced air oven aging at 100, 120 and 140 °C (5 replicates).
Petrolatum treatment:	Wire samples were dipped for 1 s into petrolatum stabilized with 0.5% AO I at 115 °C and preaged at 70 °C for 10 days. Thereafter samples were carefully wiped, formed into pigtails and aged until cracking of the insulation.
Additives:	see table 1

Additives 0.2% AO I +0.2% MD	Oven aging (days)					
	Without petrolatum treatment			With petrolatum treatment		
	100 °C	120 °C	140 °C	100 °C	120 °C	140 °C
MD I	589	213	52	416	176	35
MD II	360	152	44	248	112	31

with compound (6) (see page 79) – a limiting temperature of 121 °C according to VDE 0304 can be obtained. This means that after a thermal stress for 25000 h at 121 °C the insulation would still pass the specifications as set forth in VDE 0304. The authors consider this result as sufficient to allow the use of such cables with a conductor temperature of 90 °C in continuous operation.

2.6.2 Polypropylene and copolymers

The inherently higher thermooxidative instability of polypropylene and its copolymers calls for higher additive concentrations than for polyethyl-

Table 5. Comparative antioxidant effectiveness in poly(propylene-ethylene) insulated wire [36].

Test sample:	Copper wire 0.65 mm diameter; insulation 0.4 mm.
Test procedure:	Forced air oven aging at 120 °C (5 replicates).
Petrolatum treatment:	Wire samples were dipped for 1 s into petrolatum (stabilized with 0.5% AO I) at 115 °C and preaged for 10 days at 70 °C. Thereafter samples were carefully wiped, formed into pigtails and aged until cracking of the insulation.
Additives:	AO I = pentaerythrityl tetrakis[3(3′,5′-di-tert.butyl-4′-hydroxyphenyl)]propionate AO III = 3-thiapentanediol-1,5-bis[3(3′,5′-di-tert.butyl-4′-hydroxyphenyl)propionate] AO IV = 4,4′-thiobis(3-methyl-6-tert.butylphenol) AO V = 1,1,3-tris(5′-tert.butyl-4′-hydroxy-2′-methylphenyl)butane AO VI = N,N′-dinaphthylphenylenediamine MD II = N,N′-dibenzaloxalyldihydrazide

Additives 0.2% AO +0.2% MD II	Oven aging (days)	
	Without petrolatum treatment	With petrolatum treatment
AO I	152	112
AO III	116	52
AO IV	21	34
AO V	47	37
AO VI	67	29

ene resins. For telephone cables poly(propylene-ethylene) resins are preferred to propylene homopolymers.

Data on thermal stability of polypropylene insulation with and without petrolatum treatment are given in tables 4 and 5. Again a negative influence of the petrolatum treatment is detectable. The loss in performance ranges from 20 to 50% depending on aging temperature and stabilizer system. In view of the high level of stability that is still achievable in the presence of petrolatum, service performance can be expected to be satisfactory. The data in table 5 indicate quite clearly that the choice of the antioxidant with a given metal deactivator influences oxidative stability very significantly.

2.6.3 Metal deactivators as costabilizers in filled and reinforced polyolefins

During the last decade filled and/or reinforced polyolefins have gained increasing market acceptance on account of the possibility of improving selected mechanical and physical properties (see chapter 9 on fillers and reinforcing agents). At the beginning of this development asbestos was widely used as reinforcing agent. In the meantime emphasis has shifted to glass fibers, talcum and calcium carbonate.

Table 6. Influence of different asbestos types on thermal stability of asbestos-reinforced polypropylene [44].

Test sample: 60/40 PP/asbestos; 1 mm plaques.

Test procedure: Forced air oven aging at 140 °C to embrittlement.

Additives*): 0.1% dodecyl 3,3-bis(4′-hydroxy-3′-tert.butylphenyl)butyrate
0.5% dioctadecyldisulfide
0.5% tris[2-tert.butyl-4-thio(2′-methyl-4′-hydroxy-5′-tert.butyl)phenyl-5-methyl]phenylphosphite = MD

Asbestos type	Oven aging (days to failure)
PP without asbestos	160
Chrysotile	18
Chrysotile, low iron	70
Antophyllite	95
Tremolite	122
Tremolite without MD	56

*) based on PP

It was quickly recognized, that the oxidative stability of resins reinforced with talcum or asbestos is definitely inferior to that of the unfilled base resin. Since these minerals may contain iron and/or manganese in varying amounts, it was obvious that the service life of filled resins had to be improved by adding e.g. metal deactivators.

Of the different asbestos types available, chrysotile is the most common one in polyolefins, antophyllite and tremolite being employed only occasionally. The data in table 6 show convincingly how strongly oxidative stability is influenced by the different asbestos types. Chrysotile, preferred on account of its reinforcing qualities, reduces oxidative stability quite considerably. Tremolite behaves better in that respect. However, it is less used due to its brittleness and reduced tenacity in comparison to chrysotile.

The oven aging data in table 7 for chrysotile-reinforced polypropylene indicate the necessity to optimize the stabilizer system depending on the intended application of the resin. At any rate it is advisable to start optimization trials with rather high additive concentrations.

The influence on the thermal stability of polypropylene caused by two different types of talcum is demonstrated in table 8. Even though both

Table 7. Influence of metal deactivators on thermal stability of asbestos-reinforced polypropylene [45].

Test sample: 70/30 PP/asbestos (chrysotile); 1 mm plaques.

Test procedure: Forced air oven aging at 140 °C, days to embrittlement.

Additives*): 0.4% 1,3,5-trimethyl-2,4,6-tris(3′,5′-di-tert.butyl-4′-hydroxybenzyl)benzene
0.5% dilauryl thiodipropionate
0.5% MD

MD	Oven aging (days to failure)
No MD	12
Benzoic acid hydrazide	19
N,N′-diphenyloxamide	19
Oxalic acid dihydrazide	27
No asbestos	140**)

*) based on PP
**) estimated value

Table 8. Thermal stability of talcum-reinforced polypropylene [40].

Test sample: 60/40 PP/talcum; 1 mm plaques.

Test procedure: Forced air oven aging at 135 and 150 °C, days to embrittlement, data in parentheses: no talcum.

Additives: AO = pentaerythrityl tetrakis[3(3′,5′-di-tert.butyl-4′-hydroxyphenyl)propionate]
MD = N,N′-bis-acetyl-adipic acid dihydrazide
DLTDP = dilauryl thiodipropionate

Additives*)	Oven aging (days to failure)			
	135 °C		150 °C	
	Talcum I	Talcum II	Talcum I	Talcum II
0.5% AO	82 (220)	91	25 (85)	20
0.5% AO +0.5% MD	185 (225)	173	66 (85)	46
0.5% AO +0.5% MD +1.5% DLTDP	191 (255)	172	75 (105)	56

*) based on PP

minerals contained about the same amount of iron and manganese (3500 and 15 ppm respectively), talcum II impaired the oven aging results to a greater extent. Nevertheless, the presence of a metal deactivator raised the oven aging values by a factor of 2 to 3 resulting in a performance loss of only 25 to 30% compared to the unfilled base resin.

Calcium carbonate fillers also influence thermal stability of polypropylene more or less negatively as shown by the data in table 9. The effects are not as spectacular as with the talcum types III and IV. However, it is wortwhile to check that the aging resistance of the formulation chosen is still sufficient for the intended area of application.

The reasons for reduced thermal stability of filled and/or reinforced polyolefins are certainly not due only to the conceivable catalytic effects of metal impurities as found in naturally occurring minerals. The possible sorption of polar additives on the polar surfaces of fillers has to be kept in mind. This phenomenon would result in substantial amounts

Table 9. Influence of a metal deactivator as costabilizer in filled polypropylene [41].

Test sample: 60/40 PP/filler; 3 mm plaques.

Test procedure: Forced air oven aging at 150 °C, days to embrittlement.

Additives: AO = pentaerythrityl tetrakis[3(3′,5′-di-tert.butyl-4′-hydroxyphenyl)propionate]
DSTDP = distearyl thiodipropionate
MD = N-salicylidene-N′-salicoyl-hydrazine

Additives*)	Oven aging (days to failure)**)			
	Talcum III	Talcum IV	$CaCO_3$ I	$CaCO_3$ II
0.1% AO +0.5% DSTDP	14	5	45	24
0.1% AO +0.5% DSTDP +1.0% MD	47	32	58	55

*) based on PP
**) 100 days without filler

of polymer remaining unprotected against the attack of oxygen because at least part of the stabilizers are firmly held by the filler and are hence not available for the intended function as stabilizers.

There are indications in the literature substantiating this hypothesis. It has been shown [46] that the addition of polar organic compounds such as polyols, polyethylene glycols, amides, polyepoxides etc. improved thermal stability of filled polyolefin resins. This has been explained by preferential adsorption of the polar organic compounds on the filler surfaces with the consequence that the active sorption sites of the filler are interdicted to the additives. In recent years "mineral deactivators" based on salts of alkylnaphthalene sulfonic acids are being offered on the market [47, 48]. They are said to enhance thermal stability of e.g. polypropylene compositions which may contain different types of fillers such as calcium carbonate, dolomite, china clay, talcum or wollastonite.

2.7 Index of trade names and suppliers of metal deactivators

No claim is made for completeness of the index of trade names and suppliers.

Trade name	Chemical class	Manufacturer/supplier
Cyanox	Not disclosed	Cyanamid GmbH, D-8190 Wolfratshausen; American Cyanamid Co, Wayne, NJ 07470, USA
Eastman Inhibitor OABH	Dibenzalhydrazone	Eastman Chemical International AG, CH-6301 Zug; Eastman Chemical Products, Kingsport, TN 37662, USA
Hostanox	Phosphorous acid ester	Hoechst AG, Verkauf Kunststoffe, Gruppe Wachse und Kunststoff-Additive, Gersthofen, D-8900 Augsburg
Irganox MD	Diacylhydrazine	Ciba-Geigy AG, Division KA, CH-4002 Basel; Ciba-Geigy Corp., Plastics and Additives Div., Hawthorne, NY 10532
Mark	Not disclosed	Argus Chemical SA, B-1620 Drogenbos; Argus Chemical Corp., Brooklyn, NY 11231
Naugard	Oxalamide derivative	Uniroyal Chemical, Naugatuck, CT 06770 USA

2.8 References

[1] *A.J. Chalk, J.F. Smith:* Trans. Faraday Soc. 53 (1957) pp. 1214/1235.
[2] *N.M. Emanuel, J.K. Maizus, I.P. Skibida:* Angew. Chem. 81 (1969) p. 91.
[3] *L. Reich, S.S. Stivala:* Autoxidation of Hydrocarbons and Polyolefins, Marcel Dekker, Inc., New York, 1969, pp. 247/333.
[4] *R.H. Hansen, C.A. Russell, T. DeBenedictis, W.M. Martin, J.V. Pascale:* J. Polym. Sci. 2A (1964) p. 587.
[5] *G. Sam:* Kunststoffe 68 (1978) p. 656.
[6] *B.B. Pusey, M.T. Chen, G.L. Link:* 20th Int. Wire and Cable Symp., 1971, p. 209.
[7] *J.B. Howard:* 21th Int. Wire and Cable Symp., 1972, p. 329.
[8] *G. Scott:* Atmospheric Oxidation and Antioxidants. Elsevier Publishers, New York, 1965.
[9] *D.L. Allara, M.G. Chan:* J. Polym. Sci. 14 (1976) p. 1857.
[10] *C.J. Pedersen:* Ind. Eng. Chem. 41 (1949) p. 924.
[11] Eastman Chemical Products, Inc.: Product data sheet Nr. D-137 Eastman Inhibitor OABH®, 1974.
[12] *M.G. Chan, H.M. Gilroy, L. Johnson, W.M. Martin:* 27th Int. Wire and Cable Symp., 1978, p. 99.
[13] *R.J. Roe:* Am. Chem. Soc., Div. Org. Coat. Plast. Chem. Pap. 34 (2) (1974) p. 132.
[14] U.S. Patent 3,110,696 (1963) Ciba-Geigy.
[15] U.S. Patent 3,440,210 (1969) Eastman Kodak.
[16] U.S. Patent 3,357,944 (1967) Ciba-Geigy.
[17] U.S. Patent 3,970,637 (1976) Bell Telephone.
[18] U.S. Patent 4,038,247 (1977) Ciba-Geigy.
[19] Japanese Patent 53-24956 (1979) Ube Kosan.
[20] U.S. Patent 3,993,622 (1976) Ciba-Geigy.
[21] U.S. Patent 4,043,976 (1977) Ube Industries.
[22] U.S. Patent 4,007,183 (1977) Ciba-Geigy.
[23] U.S. Patent 4,044,200 (1977) Union Carbide.
[24] U.S. Patent 4,087,405 (1978) Eastman Kodak.
[25] U.S. Patent 3,772,245 (1973) Ciba-Geigy.
[26] Jap. Patent 52-130841 (1977) Ube Industries.
[27] U.S. Patent 4,154,723 (1979) Monsanto.
[28] U.S. Patent 4,073,771 (1978) Bell Telephone.
[29] British Patent 1,444,229 (1976) BICC.
[30] German Patent 2,407,044 (1975) Ube Industries.
[31] Hoechst AG: Product data sheet Antioxidans Hoechst VP OSP 1®, 1974.
[32] *W.L. Hawkins:* J. Appl. Polym. Sci. 1 (1959) p. 37.
[33] *J.B. Howard, H.M. Gilroy:* Polym. Eng. Sci. 15 (1975) p. 268.
[34] Draft of VDE 0209 and FTZ 72 TVI.
[35] *J.M. Farber, W.H. Brown, A. DiBattista, P.P. Klemchuk:* 21th Int. Wire and Cable Symp., 1972, p. 361.

[36] *A. DiBattista, J.M. Farber, M.A. Trainor, P.P. Klemchuk:* 33th SPE Antec, 1975, p. 280.

[37] *E.T. Kokta:* 24th Int. Wire and Cable Symp., 1975, p. 220.

[38] *M.G. Chan, H.M. Gilroy, I.P. Heyward, L. Johnson, W.M. Martin:* 36th SPE Antec, 1978, p. 381.

[39] *G.A. Schmidt:* 26th Int. Wire and Cable Symp., 1977, p. 161.

[40] Ciba-Geigy: Unpublished results.

[41] *D.D. O'Rell, A. Patel:* 24th Int. Wire and Cable Symp., 1975, p. 231.

[42] *F.R. Wright:* J. Cell. Plast. 12 (1976) p. 317.

[43] *W.v. Gentzkow, D. Pieper:* ETG-Fachtagung, 1977, p. 146.

[44] German Patent 1,694,223 (1971) Hoechst.

[45] U.S. Patent 3,649,592 (1972) Naphtachimie.

[46] U.S. Patent 3,553,158 (1971) Shell Oil.

[47] U.S. Patent 4,269,761 (1981) R.T. Vanderbilt.

[48] R.T. Vanderbilt Company, Inc.: Product data sheet Vanox® 3245, 1981.

3 Light Stabilizers for Thermoplastics

Dr. François Gugumus, Ciba-Geigy, Basel/Switzerland

3.1 Introduction

Light and oxygen induce degradation reactions in plastics that may not only modify them visually but also exert a detrimental influence on numerous mechanical and physical properties.

The inhibition of these degradation reactions has considerable technical and economical importance or else, the possible applications of many plastics would be drastically reduced.

Consequently, with the development of synthetic resins it became necessary to look for ways and means to prevent, or at least reduce, the damage caused by the environmental parameters light, air and heat. This can be achieved through addition of special chemicals, light stabilizers or u.v. stabilizers, that have to be adjusted to the nature of the resin and the specific application considered.

Light stabilizers are chemical compounds able to interfere with the physical and chemical processes of light-induced degradation. In spite of the fact that carbon black [1] and other pigments [2] may also protect the plastics from the effects of light, we shall consider in the following only those organic and organo-metallic compounds conferring only slight discoloration or no discoloration at all to the plastics to be stabilized.

The most important light stabilizer classes from a technical point of view are: 2-hydroxybenzophenones, 2-hydroxyphenylbenzotriazoles, hindered amines and organic nickel compounds. In addition, salicylates, cinnamate-derivatives, resorcinol monobenzoates, oxanilides and p-hydroxybenzoates are used. The sterically hindered amines represent the latest development in this field. In numerous synthetic resins they outperform the previously available light stabilizer classes [3 to 5].

Light stabilizers are generally used in concentrations between 0.05 and 2%, the upper limit of this range being employed only exceptionally. Among other factors, the concentrations used will be a function of the inherent light stability of the resin to be stabilized, of the specific application considered, and of the other additives in the formulations, e.g. antioxidants, pigments and fillers.

The data in table 1 show the importance of light stabilization for the practical use of the most important synthetic resins. In 1980 more than 75% of the light stabilizers had been used for polyolefins and more than 30% for polypropylene although the production volume of the latter was significantly below that of polyethylene and polyvinyl chloride.

Table 1. Consumption of light stabilizers in different plastics materials in Western Europe in the years 1970 to 1980.

Plastics material	1970 (t)	1977 (t)	1980 (t)	Proportion (1980) (%)	Mean yearly growth rate 1970 to 1980 (%)
Polypropylene	110	695	765	32	21.4
High density polyethylene	245	560	478	20	6.9
Low density polyethylene	60	400	550	23	24.8
Acrylonitrile/Butadiene/Styrene (ABS)	25	35	72	3	11.1
Polystyrene, impact polystyrene	50	55	96	4	6.7
Polyvinyl chloride	95	270	287	12	4.8
Unsaturated polyester	105	150	118	5	1.2
Total	690	2165	2390	100	13.2

Research on theoretical and practical aspects of light stabilization of the high volume resins is still very impressive despite the fact that considerable advances have been made in recent years [6 to 9]. The main trends identified in this respect concern the development of polymeric light stabilizers for applications requiring very low volatility and high resistance to migration and extraction; the development of "tailor made" systems for specific applications; the preparation of mixtures and concentrates, possibly with pigments and other additives, providing economical advantages; the development of free flowing products for easy addition; and the development of dust-free commercial products to satisfy the increasingly high standards concerning handling safety.

3.2 Photodegradation of synthetic polymers

To influence the degradation processes a thorough knowledge of the photochemical reactions responsible for degradation and of their dependency on the most important environmental factors such as light, oxygen, humidity and temperature is mandatory. On outdoor weathering, the ultraviolet part of sunlight, i.e. the radiation between about 290 and 400 nm, initiates degradation.

3.2.1 Ultraviolet spectrum of sunlight

The radiation from the sun reaching the outer air layers of the earth shows a continuous energy spectrum in the wavelength region between 0.7 and approximately 3000 nm. On passing the atmosphere, part of the long-wavelength radiation is absorbed by water vapor and carbon dioxide and finally only the short-wavelength part of infrared radiation reaches the earth's surface.

The short-wavelength u.v. radiation below 175 nm is absorbed by oxygen in the layers above 100 km whereas the radiation between 175 and 290 nm is absorbed by the ozone layer of the stratosphere which begins at about 15 km above sea level for mean latitudes and shows a maximum in density between 25 and 30 km.

In addition to the absorption of part of the u.v. radiation by ozone, the scattering of sunlight on interaction with air and aerosol particles (water drops, dust) in the atmosphere has to be considered. As a consequence, the radiation responsible for the aging of plastics, i.e. the radiation reaching the surface of the earth, is composed of direct sunlight and scattered light. Light scattering is a function of the inverse of the 4th power of the wavelength (*Rayleigh's* law). This means that on a cloudy day, although direct radiation from the sun is reduced by absorption, the u.v. part of the global radiation may be increased because scattered short-wavelength radiation reaches preferentially the earth's surface.

The intensity and spectral distribution of the global radiation are, in a first approximation, a function of the daily and seasonal position of the sun that determines the depth of the air layer to be passed by light and, therefore, its absorption. Table 2 [10] shows the data valid for vertical incidence. It can be seen that the u.v. part of the radiation responsible for polymer degradation represents only about 6% of the global radiation.

Table 2. Intensity of the global radiation on vertical incidence.

Wavelength (nm)	Intensity ($W \cdot m^{-2}$)		Proportion (%)	
<280	0		0	
280– 320	5		0.5	
320– 360	27		2.4	
360– 400	36	$\Sigma = 68$	3.2	$\Sigma = 6.1$
400– 440	56		5.0	
440– 480	73		6.5	
480– 520	71		6.3	
520– 560	65		5.8	
560– 600	60		5.4	
600– 640	61		5.5	
640– 680	55		4.9	
680– 720	52		4.6	
720– 760	46		4.1	
760– 800	41	$\Sigma = 580$	3.7	$\Sigma = 51.8$
800–1000	156		13.9	
1000–1200	108		9.7	
1200–1400	65	$\Sigma = 329$	5.8	$\Sigma = 29.4$
1400–1600	44		3.9	
1600–1800	29		2.6	
1800–2000	20		1.8	
2000–2500	35		3.1	
2500–3000	15	$\Sigma = 143$	1.3	$\Sigma = 12.7$

The lower wavelength limit (absorption edge) of the global radiation spectrum is essentially determined by the absorption of part of the extra-terrestrial radiation in the ozone layers of the stratosphere. This absorption edge is very steep, i.e. the radiation intensity varies considerably within a narrow wavelength range. Since the ozone content is subject to seasonal and geographic changes, the position of the short-wavelength absorption limit and, therefore, the intensity of the short-wavelength radiation, are strongly dependent on location and on season. The radiation intensity calculated for a plane perpendicular to the direction of direct radiation from the sun at its highest position for the corresponding day and for the wavelength 297.5 nm is shown in figure 1 as a function of latitude and season [11]. At the equator, only small variations of

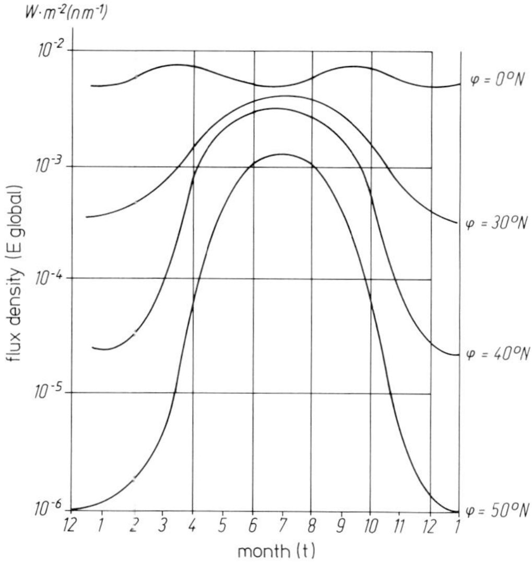

Fig. 1. Flux density $E=f(\varphi, t)$ for $\lambda=297.5$ nm as a function of season and latitude: $\varphi=0$ (~equator), $\varphi=30°$ N (~northern Africa, Florida), $\varphi=40°$ N (~Spain, southern Italy), $\varphi=50°$ N (~central Europe).

the radiation intensity with season are found, whereas for the latitude $\varphi=50°$ North these variations amount to about 3 orders of magnitude.

In addition to u.v. radiation, the influence of temperature and humidity depending on the climatic zone, has to be considered. This is, because for one, with increasing temperature, thermal degradation may overlap photochemical degradation and eventually become dominant and, for another, water may influence the course of degradation by extractive or hydrolytic processes as well as by participating in photochemical reactions involving some pigments, titanium dioxide, for example.

In connection with the optical properties of the atmosphere, various emissions are considered whose photochemical products may lead to the formation of oxidants (smog) [21] or may even disrupt the ozone layer of the stratosphere [13, 14].

3.2.2 Mechanisms of polymer degradation

3.2.2.1 Physico-chemical processes occurring on light absorption

Incident light will either be reflected from the surface or scattered or absorbed in the bulk of the polymer. According to the first law of photochemistry (*Grotthus-Draper*) only that part of the light effectively absorbed leads to photochemical transformations, i.e. to degradation. This has been complemented by the second law of photochemistry (*Stark-Einstein*) which in its updated version states that the absorption of light by a molecule is a one-quantum process, so that the sum of the primary process quantum yields φ must be unity.

The absorption of light by synthetic resins is first of all related to their structure. It is well known that saturated hydrocarbons do not absorb above 250 nm. In the presence of double bonds (chromophores) the absorption is shifted to longer wavelengths. This is particularly pronounced for double bonds between carbon atoms and heteroatoms so that carbonyl compounds absorb already in the wavelength region above 290 nm.

Light of wavelength above 290 nm may also induce the degradation of polymers which do not contain chromophores in their structures to absorb it directly. This is caused by traces of impurities resulting from manufacturing or by structural irregularities often present in technical polymers, e.g. catalyst residues or oxidation products. The latter may show absorption in the u.v. range which can lead to photochemical transformations. Moreover, in semi-crystalline polymers, on account of the scattering of light by the crystalline particles [15, 16], the lightpath is

Table 3. Polymer absorption maxima and most efficient solar wavelength for polymer photodegradation.

Polymers	Absorption maximum (nm)	Most efficient wavelength (nm)
Polyethylene	<150	300
Polypropylene	<200	310
Polyvinyl chloride	<210	310
Polymethyl methacrylate	<240	290–315
Polyvinyl acetate	<250	<280
Polystyrene	<260	318
Polycarbonate	260	295
Polyethylene terephthalate	~ 290	290–320

considerably increased in comparison with amorphous materials. Thus, even at low concentrations of chromophoric groups, appreciable quantities of energy may be absorbed. The solar wavelengths of maximum photodegradation for some polymers are given in table 3. They are derived from so-called activation spectra, i.e. measurements of photodegradation as a function of the wavelength of the incident light [17 to 20].

From the energetical point of view the sensitivity of some polymers to light is not astonishing. In table 4 it can be seen that the energies of the photons found in the ultraviolet region between 293 and 419 kJ·Einstein^{-1} (1 Einstein = 1 mole photons = $6 \cdot 10^{23}$ photons) are significantly higher than the bond energies of some typical bonds found in synthetic resins.

Table 4. Energy of radiation of different wavelengths and bond energy of typical bonds found in polymers [24, 117].

Wave-length (nm)	Energy		Bond type	Bond energy (kJ·mole^{-1})
	(kcal·Einstein^{-1})	(kJ·Einstein^{-1})		
290	100	419	C—H	380–420
300	95	398	C—C	340–350
320	90	375	C—O	320–380
350	81	339	C—Cl	300–340
400	71	297	C—N	320–330

After photon absorption the chromophores are raised to higher energy levels (excited states). This is shown in figure 2. Through spin inversion, singlet states may convert to triplet states (intersystem crossing). The excited states are able to lose the excitation energy by several processes (fluorescence, phosphorescence and radiationless decay). An additional possibility is energy transfer in which the energy of the excited state (singlet, triplet) is transferred to a suitable acceptor molecule. Considering the time scale of all these processes, triplet states are relatively long lived. Therefore, they may become chemically active and thus play an important role in polymer photodegradation.

The ratio of the polymer molecules reacting chemically to the number of photons absorbed is designated as quantum yield. For some plastics the quantum yields for chain scissions found in the literature [21 to

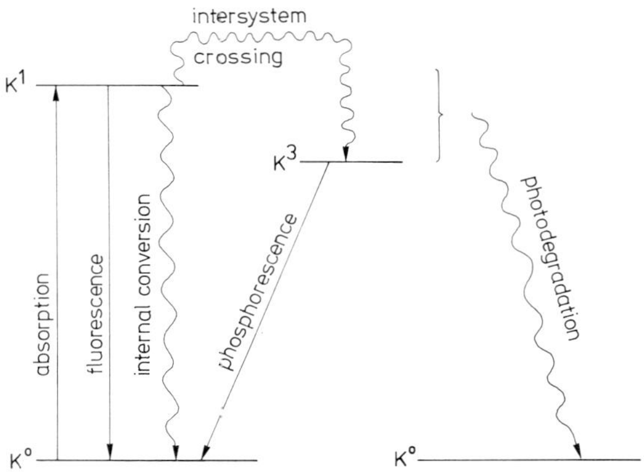

Fig. 2. *Jablonski*-diagram of optical transitions; ⟶ designates transitions involving absorption or emission of light quanta ⇝ indicates radiationless transitions; K^0, K^1 and K^3 mean K (chromophore) in ground state, singlet state and triplet state respectively.

23] are between 10^{-2} and 10^{-5}. This means that among 100 to 100000 molecules that in fact have absorbed light, only one polymer molecule really degrades. These quantum yield values, low in comparison with those observed in the photochemistry of small molecules, are due to the fact that in plastics the energy which is initially locally absorbed can be distributed to nearby domains. As a consequence the probability of chain scission is reduced. Moreover, in solid polymers, the cage effect induces very often the recombination of the broken chain.

3.2.2.2 Photooxidation scheme

Free radicals are formed in polymers exposed to light, as a consequence of the excitation of absorbing functionalities in the polymer. This is a function of the energy of the light and of the structure of the polymer molecule. The recombination of the radicals which escape the primary cage is very slow because they have to find each other through diffusion in the polymer matrix. In the presence of oxygen the plastics will simultaneously oxidize (photooxidation). Then it is often difficult to distinguish the pure photochemical processes from the thermal processes (oxidation) which are superimposed. Some of the fundamental mechanisms involved have been elucidated [24, 25] only recently through systematic studies involving also technical plastics.

The mechanism developed initially for the thermal oxidation of rubber can be applied to other substrates and also to photooxidation. The scheme in figure 3 illustrates a possible reaction sequence.

CHAIN INITIATION

$$\left.\begin{array}{l}\text{Hydroperoxides POOH}\\ \text{Carbonyl compounds } {>}C{=}O\\ \text{Catalyst residues (Ti, \ldots)}\\ \text{Charge-transfer complexes } (PH,O_2)\end{array}\right\} \xrightarrow[Me^{++}/Me^{+++}]{\Delta,\ h\nu} \begin{array}{l}\text{Free radicals}\\ (P^{\cdot},\ PO^{\cdot},\\ HO^{\cdot},\ HO_2^{\cdot},\ldots)\end{array}$$

CHAIN PROPAGATION

$$P^{\cdot} + O_2 \longrightarrow PO_2^{\cdot} \qquad (1)$$

$$PO_2^{\cdot} + PH \longrightarrow PO_2H + P^{\cdot} \qquad (2)$$

CHAIN BRANCHING

$$POOH \xrightarrow[Me^{++}/Me^{+++}]{\Delta \text{ or } h\nu} PO^{\cdot} + {}^{\cdot}OH \qquad (3)$$

$$2\,POOH \xrightarrow[Me^{++}/Me^{+++}]{\Delta \text{ or } h\nu} PO_2^{\cdot} + PO^{\cdot} + H_2O \qquad (4)$$

$$PO^{\cdot} + PH \longrightarrow POH + P^{\cdot} \qquad (5)$$

$${}^{\cdot}OH + PH \longrightarrow H_2O + P^{\cdot} \qquad (6)$$

CHAIN TERMINATION

$$\left.\begin{array}{lr}P^{\cdot} + P^{\cdot} \longrightarrow & (7)\\ P^{\cdot} + PO_2^{\cdot} \longrightarrow & (8)\\ PO_2^{\cdot} + PO_2^{\cdot} \longrightarrow & (9)\end{array}\right\} \text{non radical products}$$

Fig. 3. Oxidation and photooxidation scheme.

Table 5. Spectral data and rate of free radical formation for potential chromophores in commercial polypropylene.

Chromophore	Approximate concentration (mole·l^{-1})	$\varepsilon_{310\,nm}$ ($l \cdot mole^{-1} \cdot cm^{-1}$)	Quantum yield	Rate of free radical formation (outdoor weathering) (mole·s^{-1})
Hydroperoxides				
—CH_2—C(CH_3)CH_2— (C bearing OOH)	$5 \cdot 10^{-3}$ (fibers) 1 to $5 \cdot 10^{-4}$ (films)	0.4	1 (scission)	$\sim 5 \cdot 10^{-8}$
Ketones				
—CH(CH_3)CH_2C(=O)CH_2—	$<1 \cdot 10^{-3}$	10	0.07 (Norrish I)	$<1 \cdot 10^{-8}$
—CH(CH_3)CH_2C(=O)CH_3	$<1 \cdot 10^{-3}$	10	0.01 and 0.08 (Norrish I and II)	$<1 \cdot 10^{-9}$
Catalyst residues				
as $TiOCl_2$, $Ti(OBu)_4$	1 to $10 \cdot 10^{-4}$	40 to 1300	~ 0.02	$\sim 6 \cdot 10^{-8}$
Charge-transfer complexes				
Polymer···O_2	–	–	~ 0.02 (photooxidation)	$\sim 4 \cdot 10^{-11}$

3.2.2.3 Photooxidation of polyolefins

Photodegradation of polyolefins is caused mainly by the presence of catalyst residues, hydroperoxide groups, carbonyl groups, and double bonds introduced during polymer manufacturing, processing or storage [26]. All these species absorb u.v. light above 290 nm to varying degrees and can participate in numerous photochemical reactions. The present knowledge about some polypropylene chromophores is summarized in table 5 [26 to 29].

The high initiating capability of hydroperoxides and probably also of peroxides is due essentially to the high quantum yields (approximately 1) with which they decompose into free radicals (reaction 3).

Carbonyl groups seem to have only a small initiating efficiency [26]. The free radicals formed initially in the *Norrish* type I reaction (reaction 10) can recombine obviously before any further reaction. The products from the *Norrish* type II reaction (reaction 11) are inefficient photoinitiators. According to recent data, however, α, β-unsaturated carbonyl groups may participate in the photodegradation of polypropylene [30]. A relation between fluorescence intensity of these structures and light stability of polypropylene has been found [31].

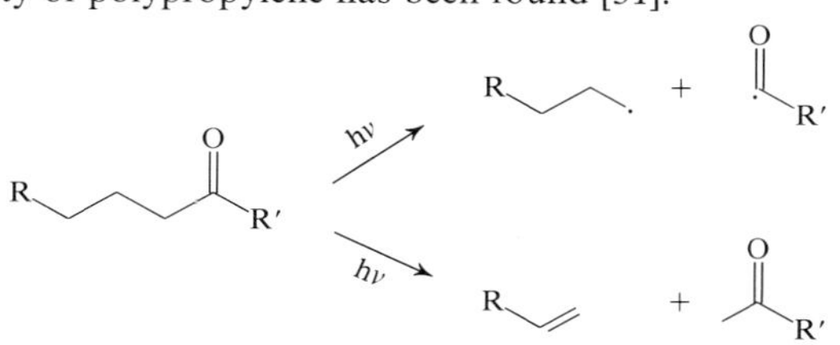

(10)

(11)

Some previous work [32] shows carbonyl compounds to be involved in the formation of singlet oxygen ($^1O_2$) through energy transfer from their excited triplet state to oxygen in the ground state ($^3O_2$) (reaction 12). The lifetime of singlet oxygen in polymers is long enough to allow its reaction with double bonds leading to the formation of allylic hydroperoxides (reaction 13). The contribution of allylic hydroperoxides to the photodegradation of polyolefins has been discussed recently [33, 34].

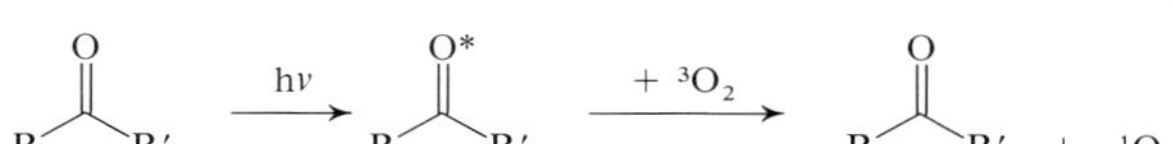

(12)

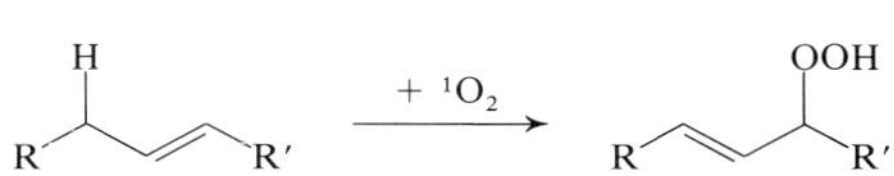

(13)

The formation of singlet oxygen may also be induced by trace amounts of polynuclear aromatics (naphthalene, anthracene) found in polypropylene [35] and by quinones resulting from the destruction of antioxidants and colorants [36].

In connection with the initiation reaction, charge-transfer complexes between oxygen and polypropylene [37], high density polyethylene [38] and poly-4-methylpentene [39] have been considered. These complexes can absorb light of longer wavelength than the polymer as such and thus may become active photochemically (table 5).

Recently it has been shown that after bromination the induction time of polypropylene photooxidation disappears completely [40]. This result led to the conclusion that unsaturated groups in polypropylene, essentially fluorescing α,β-unsaturated carbonyl groups, contribute to the photooxidation induction period, i.e. that they exert a stabilizing influence. This finding has been substantiated by other experiments [41]. As a matter of fact, hexane extraction of the polymer, which removes the unsaturated groups having the structure of α,β-unsaturated carbonyl compounds, leads to a reduction of the induction period of polypropylene photooxidation. On the other hand, the addition of hexane extracts increases the induction time markedly. In the latter experiment, the stabilizing effect of the double bonds dominates the sensitizing effect of the additional atactic material introduced with the hexane extract. This is in contradiction with older results [33] which had pointed towards an extraction of sensitizing polynuclear aromatic compounds. The stabilizing effect of the unsaturated compounds has been related to an isomerization reaction of these compounds [42].

Several new results [40, 41, 43] point to the fact that in the absence of hydroperoxides, charge-transfer complexes between oxygen and polypropylene play a decisive role in the initiation of hydroperoxide formation and, therefore, in the initiation of photooxidation. The corresponding reaction can be represented as follows [44]:

$$PH + O_2 \rightarrow [P\text{—}H \cdots O_2] \xrightarrow{h\nu} [PH^{+} \cdots O_2^{-}] \rightarrow P^{\cdot} + {}^{\cdot}O_2H \quad (14)$$
$$\downarrow$$
$$POOH$$

Recently, the discussion about the relative importance of hydroperoxides and carbonyl compounds in the photooxidation of polyolefins has been renewed [40]. On irradiation with light of wavelength >370 nm which still photolyzes hydroperoxides but not carbonyl compounds, it has been shown that photooxidation is at least considerably reduced if not completely stopped. This means that, despite the photolysis of hydroperoxides accompanied by the formation of carbonyl compounds, photooxidation is inhibited as long as carbonyl compounds cannot react photochemically themselves. This result affords a better understanding of the dual role of carbonyl compounds and hydroperoxides in polypropylene photooxidation. The importance of energy transfer from ketones to hydroperoxide and/or of induced decomposition of the latter by ketones in an excited state has already been stressed previously [45].

The carbon radicals formed in the photochemical primary processes can react very fast with oxygen (reaction 1). The alkylperoxy radicals ($RO_2^{\cdot}$) formed this way have a key role as chain carriers because they abstract mainly hydrogen from the polymer and thus produce the polymer radicals necessary for the propagation of the oxidation (chain propagation, reaction 2). In this respect it has to be considered that the reactivity of a specific substrate is not only a function of its structure, i.e. the ease of hydrogen abstraction, but is also determined by the solubility and the diffusion of oxygen. In semi-crystalline plastics, the density of the crystalline phase is usually much higher than that of the amorphous phase and usually impermeable to oxygen. Hence, oxidation occurs exclusively in the amorphous material. Poly-(4-methyl-1-pentene) (PMP) constitutes an exception in this respect: the densities of the crystalline and amorphous phases are almost identical at 60 °C, and at room temperature the density of the amorphous phase is even higher than that of the crystalline phase [46]. It has been found that isotactic PMP is more susceptible to oxidation than atactic PMP [47, 48]. This has been attributed to the fact that in isotactic PMP crystals oxidation takes place preferentially at the chain foldings because the tension resulting from the chain configuration reduces the oxidation activation energy [47] and also because the crystal dimensions are such that the crystalline regions are permeable to oxygen.

For poly-(1-butene) too, the activation energy of the oxidation reaction has been found to be lower for the isotactic than for the atactic polymer [49].

In polypropylene, as an example, isolated hydroperoxide groups and sequences of hydroperoxide groups are formed in intermolecular (reaction 16) and intramolecular (reactions 17 and 18) oxidation steps [50]. Therefore, after a short exposure period, during which the chromophores initially present (table 5) become photochemically active, appreciable quantities of hydroperoxides may be formed (reactions 15 to 18) so that the following degradation will be determined by the decomposition reactions of the hydroperoxides.

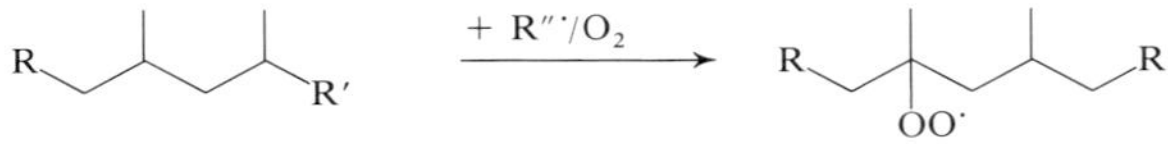

(15)

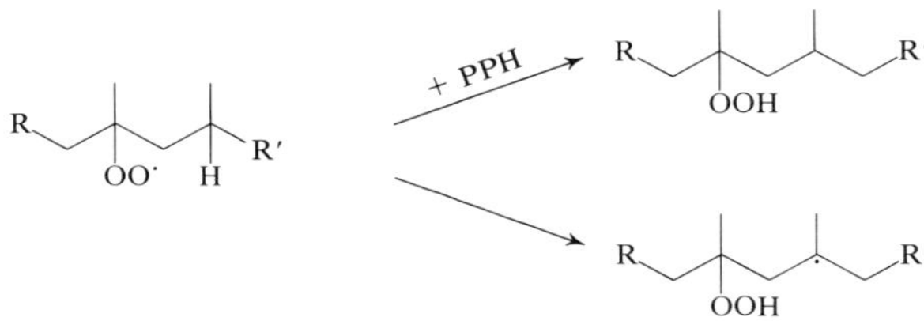

(16)

(17)

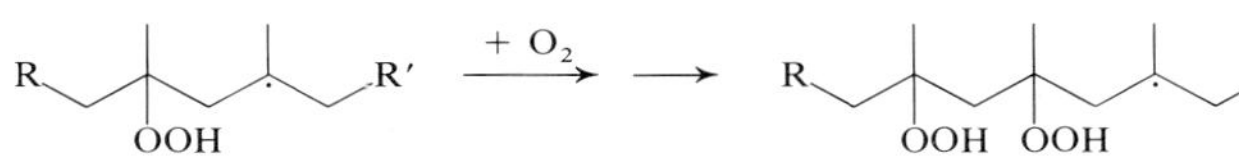

(18)

The alkoxy radicals resulting from the photolysis of hydroperoxides do not only abstract hydrogen (reaction 19) as the alkylperoxy radicals but may also undergo decomposition reactions resulting in polymer chain scission (β-scission, reaction 20). β-scission has been considered as the main reaction responsible for polypropylene and polyethylene degradation [28]. The alkoxy radicals resulting from the decomposition of allylic hydroperoxides in polyethylene may also lead to disproportionation and cross-linking [51].

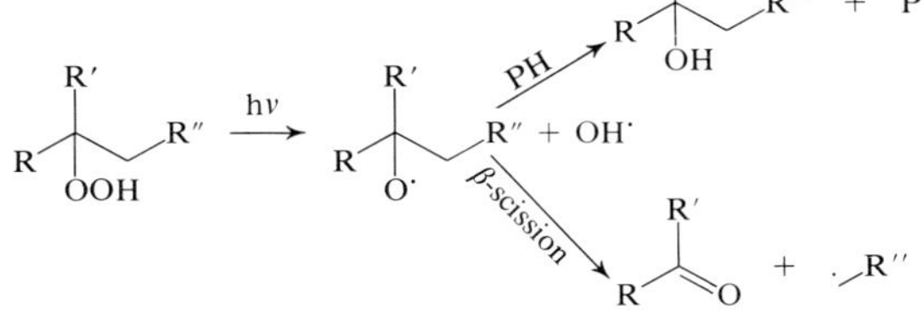

(19)

(20)

The importance of the β-scission in polyolefin degradation has become a matter of discussion recently [52]. It has been shown with ethylene-propylene copolymers that the photolysis of the hydroperoxides at 365 nm does not produce any ketones in the polymer chain. Ketones are considered primary photooxidation products resulting from the bimolecular chain termination between secondary peroxy radicals (*Russell* mechanism). On exposure to 365 nm u.v. light, no ketone formation has been observed and, moreover, the ketones already present are even partially destroyed. This cannot be due to their photolysis because it does not take place at 365 nm. Simultaneously with the decrease in ketone concentration, the formation of carboxylic acids, esters and peroxyesters has been observed. The results have been explained with the following reactions taking into account the photolysis of isolated hydroperoxides as well as the photolysis of hydroperoxides associated through hydrogen bonding [52]:

Isolated hydroperoxides:

The alkoxy radicals and hydroxyl radicals $RO^{\cdot}$ and $OH^{\cdot}$ formed in reaction 3 react with ketones to form carboxylic acids and esters (reactions 21 and 22).

$$^{\cdot}OH + \sim\overset{\overset{\displaystyle O}{\|}}{C}\sim \longrightarrow \sim\overset{\overset{\displaystyle O}{\|}}{C}{-}OH + \cdot\sim \tag{21}$$

$$RO^{\cdot} + \sim\overset{\overset{\displaystyle O}{\|}}{C}\sim \longrightarrow \sim\overset{\overset{\displaystyle O}{\|}}{C}{-}OR + \cdot\sim \tag{22}$$

Associated hydroperoxides:

The alkoxy radicals and peroxy radicals $RO^{\cdot}$ and $RO_2^{\cdot}$ formed in reaction 4 give rise to esters (reaction 22) and peroxyesters (reaction 23).

$$ROO^{\cdot} + \sim\overset{\overset{\displaystyle O}{\|}}{C}\sim \longrightarrow \sim\overset{\overset{\displaystyle O}{\|}}{C}{-}OOR + \cdot\sim \tag{23}$$

Reactions 21 to 23 are associated with chain scission whereas reactions 22 and 23 may also lead to chain branching. Alkoxy radicals and hydroxyl radicals may, of course, also react according to reactions 5 and 6 with formation of alcohols and water. Reactions 21 to 23 obviously occur not only on exposure to 365 nm u.v. light but also with 310 nm radiation for which the *Norrish* type II reaction (reaction 11) is also possible. Reactions 21 to 23 described for an ethylene-propylene copolymer are considered to be valid for polypropylene and polyethylene too [52].

3.2.2.4 Photooxidation of styrene polymers

Double bonds are not regarded as the chromophores responsible for initiation of photooxidation in polystyrene because they absorb below 300 nm. Head to head structural units that have been considered in connection with the "weak links" found on thermal oxidation, have been ruled out as initiators of photodegradation by experiments using anionically prepared polystyrenes [53]. However, peroxide groups in the polymer chain, resulting from the copolymerization of oxygen with styrene, are definitely photolabile. Moreover, oxidation products are formed during processing of styrene polymers at high temperatures. Aromatic ketones of the acetophenone type have been detected even in purified samples by using emission spectroscopy [54 to 56]. The aromatic-ketone-induced initiation may proceed according to different mechanisms. It has been shown by quenching experiments that singlet oxygen produced through energy transfer from the aromatic ketone triplet to oxygen in the ground state plays a negligible role in polystyrene photooxidation [57].

Aromatic ketones (AK) in the triplet state are able to abstract hydrogen from polystyrene (PSH) (reaction 24).

$$\mathrm{AK} \xrightarrow{h\nu} \mathrm{AK^*}$$

$$\mathrm{AK^*} + \mathrm{PSH} \longrightarrow \mathrm{AKH^{\cdot}} + \mathrm{PS^{\cdot}} \qquad (24)$$

This reaction is considered the most important initiation mechanism for polystyrene photooxidation in the presence of aromatic ketones [57]. It has been shown with model compounds that only aromatic ketones whose lowest triplet state is of the n-π^* type can abstract hydrogen and thus initiate photooxidation [58], π-π^* type triplet states do not initiate photodegradation.

Another initiation mechanism becomes more and more important with increasing polymer oxidation level. It involves energy transfer from excited aromatic ketones to hydroperoxides, followed by the decomposition of the latter [57] (reactions 25, 26).

$$\mathrm{AK} \xrightarrow{h\nu} \mathrm{AK^*}$$

$$\mathrm{AK^*} + \mathrm{PSOOH} \longrightarrow [\mathrm{AK} \cdots \mathrm{PSOOH}]^* \qquad (25)$$

$$[\mathrm{AK} \cdots \mathrm{PSOOH}]^* \longrightarrow \mathrm{AK} + \mathrm{PSO^{\cdot}} + \mathrm{^{\cdot}OH} \qquad (26)$$

This sensitized hydroperoxide decomposition will become the dominant initiation reaction already in a relatively early stage of polystyrene photooxidation. This can be explained by a inhomogeneous distribution of the hydroperoxides, i.e. by an accumulation of hydroperoxide groups in the vicinity of aromatic ketones. Recently it has been shown that an exciplex (exciplex = complex existing only in the excited state) is formed between the excited benzophenone triplet and cumyl hydroperoxide [59]. However, no sensitized decomposition of the hydroperoxide has been detected but hydrogen abstraction by the ketone with formation of a peroxyl radical. If this finding is applied to polystyrene, it means that reaction 25 is followed by reaction 27 instead of reaction 26.

$$[\mathrm{AK}\cdots\mathrm{PSOOH}]^* \longrightarrow \mathrm{AKH}^{\cdot} + \mathrm{PSOO}^{\cdot} \qquad (27)$$

For styrene copolymers containing polybutadiene, impact polystyrene for example, the thermally produced hydroperoxides are considered as responsible for photooxidative degradation [60]. The rubber phase (polybutadiene) is more sensitive to oxidation and is thought to sensitize the photooxidation of the polystyrene whereas carbonyl compounds do not seem to play an appreciable role in photoinitiation of the oxidation of these polymers [60]. With styrene-acrylonitrile copolymers (SAN) however, photooxidation is once again initiated by acetophenone groups formed during processing [61].

After initiation, the oxidation of polystyrene proceeds according to the general scheme (figure 3) with formation of hydroperoxides. The decomposition of the latter produces alcohols, ketones of the acetophenone type, and water. These are, with the hydroperoxides, the main products of polystyrene photooxidation. Conjugated polyenes, responsible for polymer yellowing, are formed in smaller yields. Decomposition of the tertiary hydroperoxides (reaction 28) is considered the most probable cause for the observed chain scissions [56, 62]. This mechanism also accounts for the agreement between the quantum yields measured for chain scissions and acetophenone groups. Polyene formation may be

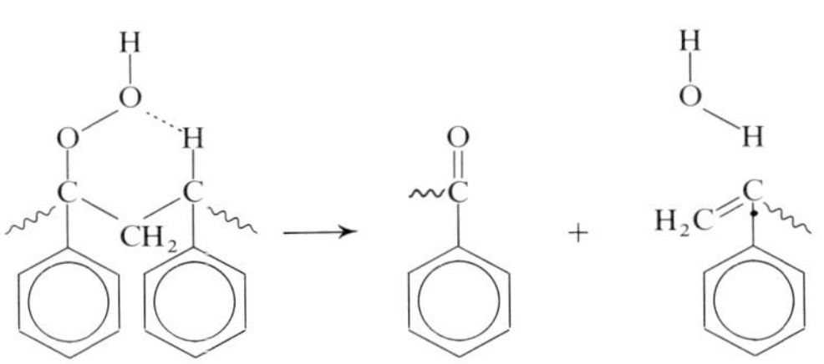

(28)

explained by further reactions of the chain-end double bonds formed on chain scission [56] (reaction 29).

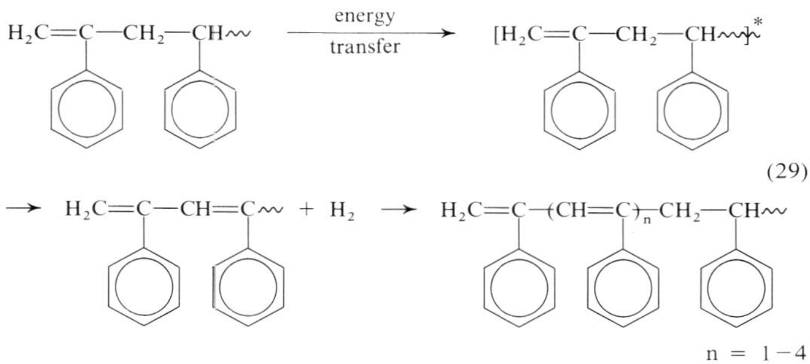

(29)

The homolytic scission of the hydroperoxide group and the subsequent hydrogen abstraction reactions by the scission products is responsible for the formation of tertiary alcohols and additional quantities of water.

In polybutadiene-modified polymers, the allylic carbon-hydrogen bonds are especially prone to oxidation. As a result, the formation of hydroxyl groups and carbonyl groups is very pronounced on exposure of acrylonitrile-butadiene-styrene copolymers and is accompanied by a decrease of the double bond content of the butadiene component [63].

3.2.2.5 Photooxidation of polyvinyl chloride

"Pure" polyvinyl chloride (PVC) does not absorb any light above 220 nm. Therefore, different functional groups which may arise during polymerization and processing have been considered as possible initiating chromophores. They include irregularities in the polymer chain as well as hydroperoxides, carbonyl groups and double bonds. The relative importance of these chromophores for initiation is still a matter of discussion. Experiments performed in solution lead to the conclusion that unsaturated structures of PVC are responsible for photoinitiation in the absence of oxygen [64]. The observation that in the early stages of photooxidation quantum yields remain constant means that, at least during these early stages of degradation, decomposition of photoproducts does not influence oxidation appreciably [64, 65]. This has been attributed to a very pronounced internal filter effect exerted by the growing polyene sequences. This is in contradiction with other work considering the peroxides formed during processing as the primary photoinitiators for PVC [66]. As a matter of fact, the concentrations of carbonyl groups and double bonds contributing to photoinitiation increase with processing temperature and processing time [66]. The attack on PVC by singlet oxygen formed through quenching of excited polyenes and carbonyl groups is also considered as a possibility [67]. In impact-PVC (ABS, MBS or MABS modified) a rapid photooxidation of the elastomer component is observed, leading to an acceleration of PVC degradation [68].

Photooxidation of PVC is accompanied by evolution of hydrochloric acid, discoloration, and modification of the mechanical properties (brittleness), the latter being attributed to chain scission and cross-linking. Polyenes, peroxides, carbonyl and hydroxyl groups are the most important functional groups formed.

As already stated, the initiation reaction is still a matter of discussion. However, the absorption of u.v. light by polyenes and the subsequent reactions 30 and 31 can be considered as the most probable initiation process [65].

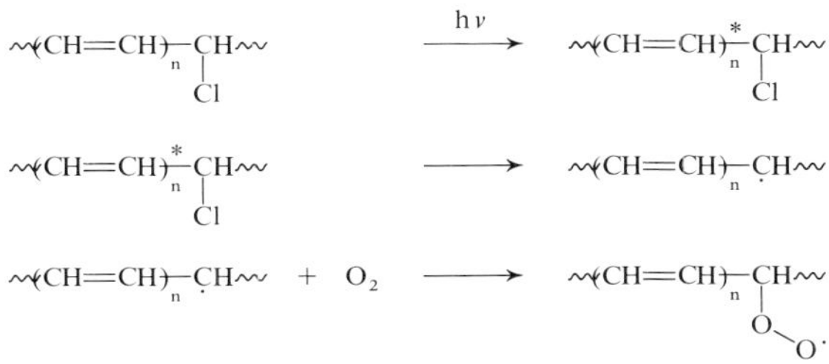

(30)

(31)

(I)

The main oxidation products as well as chain scission and chain branching can be explained through reactions of peroxy radicals. The latter are formed in the initiation reaction (peroxy radical I) as well as on abstraction of a tertiary hydrogen by a peroxy radical (α-chloro peroxy radical II) [64, 65].

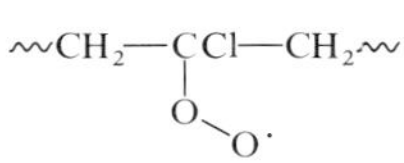

(II)

Hydrogen abstraction by radicals (I) and (II) produces hydroperoxides. The other functional groups observed on photooxidation of PVC can be explained by the termination reactions of radicals (I) and (II) (reactions 32 and 33) and by the reactions of the alkoxy radicals formed in non-terminating bimolecular reactions of tertiary peroxy radicals (reactions 34 to 37) [64, 65].

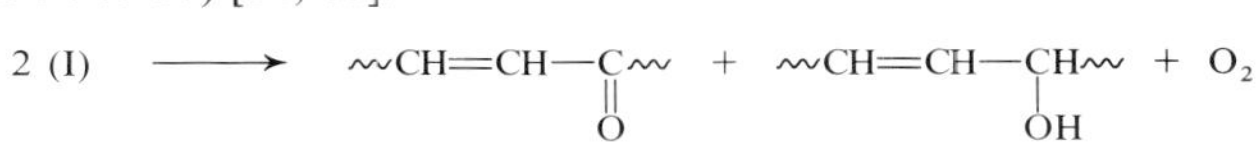

(32)

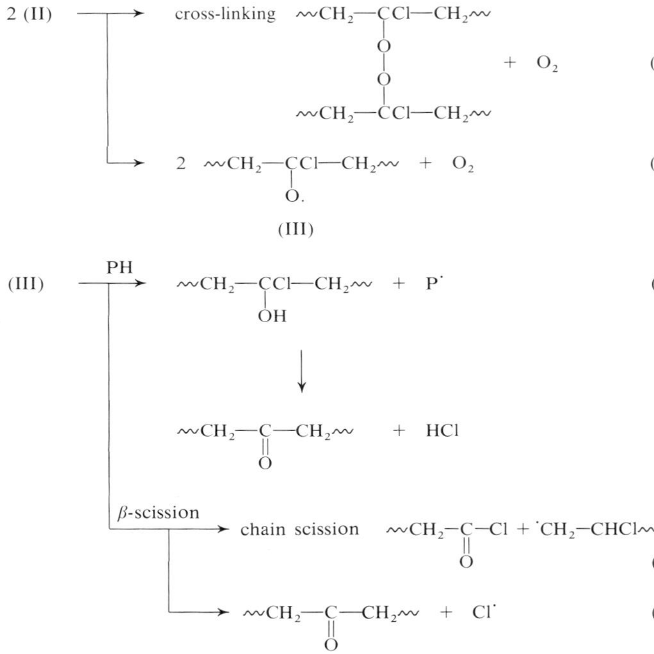

(33)

(34)

(35)

(36)

(37)

Because of polarization effects the abstraction of a secondary hydrogen atom from the methylene groups by a chlorine atom (reaction 38) is much more likely than the abstraction of a tertiary hydrogen [69].

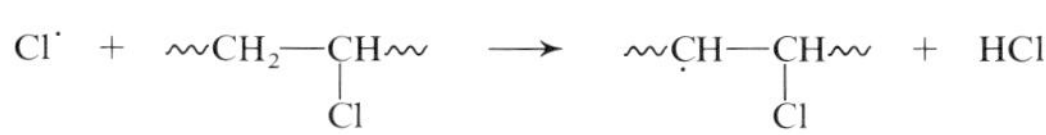

(38)

It is assumed that dehydrochlorination is proceeding zip-like in the polymer chain to form polyene structures, because the β-chloro-alkyl radicals formed in reaction 38 easily lose $Cl^{\cdot}$ with double bond formation [64]. The diminution of hydrochloric acid formation to be expected in the presence of oxygen as a consequence of the reaction of O_2 with the polyene radical (reaction 31) is more than compensated by hydrochloric acid formed in reactions 34 to 38.

Considering the β-scission reactions of the tertiary alkoxy radicals, elimination of chlorine (reaction 37) is favored over chain scission (reaction 36) [70]. It is obvious that photolysis of hydroperoxides may also produce alkoxy radicals susceptible to undergo the reactions cited, in as much as this process gains importance with proceeding oxidation.

3.2.2.6 Photooxidation of polyacrylates

The results of polymethyl acrylate (PMA) photolysis are best explained on the assumption of light absorption by the carbonyl chromophore of the polymer [71]. The excited states formed this way lead to a loss of ester side groups at random with formation of free radicals (reaction 39).

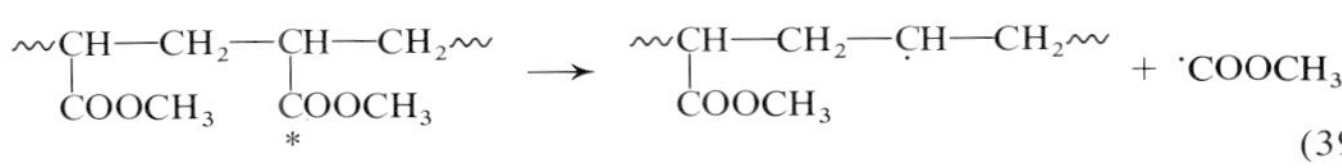

(39)

The macroradicals formed in this reaction may either lead to chain scission directly or, in the presence of oxygen, induce oxidation processes causing also chain scission.

Even "pure" polymethyl methacrylate (PMMA) absorbs in the wavelength region from 290 to 400 nm. The intensity of this absorption increases especially below 300 nm [72]. By means of molecular weight determinations it has been shown that degradation increases steadily at wavelengths below 330 nm. Above 400 nm, on the other hand, only slight degradation occurs and only in a superficial layer not more than 50 μm deep. There are no hints to selective degradation effects at wavelengths between 260 and 300 nm, i.e. no additional, new degradation reactions are induced with increasing energy of the photons, but essentially a larger number of chain scissions.

Recently it has been found that the excited state obtained on exposure to 253.7 nm u.v. light is a dissociative one, causing decomposition with a quantum yield of 1 [73]. This indicates that the outstanding stability of PMMA towards sunlight is due primarily to its weak absorption in the region of the u.v. light reaching the earth.

PMMA degradation may be initiated through loss of side groups as with PMA [74]. It seems that oxygen exerts only a small, rather inhibitory effect [74].

Recently it has been shown that monomeric methyl methacrylate has a distinctly negative effect on the light stability of PMMA [75]. It is assumed that the formation of methyl methacrylate hydroperoxide is the first step in this antagonistic reaction [76]. Reactions 40 to 42 show a possible mechanism of formation of this hydroperoxide [77].

$$\text{MMA} \xrightarrow{h\nu} (\text{MMA})^* \tag{40}$$

$$\text{MMA}^* + \text{O}_2 \longrightarrow \text{MMA} + {}^1\text{O}_2 \tag{41}$$

$$\text{MMA} + {}^1\text{O}_2 \longrightarrow \begin{array}{l} \text{H}_2\text{C}{=}\text{C}{-}\text{CH}_2\text{OOH} \\ \quad\quad\;\; | \\ \;\;\text{O}{=}\text{C}{-}\text{OCH}_3 \end{array} \tag{42}$$

Photolysis of this hydroperoxide generates free radicals initiating the polymer oxidation and leads finally to molecular weight decrease.

3.2.2.7 Photooxidation of polyurethanes

Degradation of polyurethanes (PUR) is observed at wavelengths even up to 410 nm [78]. Though aromatic polyurethanes absorb in the whole u.v. range of sunlight their absorption is especially pronounced below 320 nm [79].

In spite of the technical importance of polyurethanes the primary reactions responsible for their photooxidative degradation are still not established beyond any doubt. According to a hypothesis of *Allen* and *Mc Kellar* [80] photochemical initiation in some aromatic polyurethanes is due to a benzophenone type impurity formed during polymer processing. Experiments with model compounds and triplet quenchers show that this time it is the singlet state that is photochemically reactive [81].

It is generally assumed that photolysis of polyurethanes involves the scission of the N—C and C—O bonds of the urethane group. The photo-*Fries* rearrangement leading to NH_2 groups fixed on the aromatic ring has been confirmed with model compounds [79, 81]. The intramolecular hydrogen abstraction according to the *Norrish* type II reaction can be considered too if there is a hydrogen atom in position γ to the carbonyl group.

Even if the primary reactions are not definitely established, it is well known that the photochemically initiated oxidation of polyurethanes proceeds as that of other polymers, involving hydroperoxides and their thermal and photolytic decomposition products. Considering the structures of the polyurethanes and the ease of the hydrogen abstraction reactions, the hydrogen atoms attached to carbon atoms in position α to the carbamate nitrogen, to the aromatic ring, to the ether group and to the ester group are preferential sites of free radical attack and peroxidation [79, 82]. It is not only the aliphatic part of polyurethanes that is attacked on photooxidation but also the aromatic part [81].

Yellowing of aromatic polyurethanes is generally attributed to further reaction products of the NH_2 and OOH groups formed on aging [81]. Recently, yellowing has been ascribed to the formation of traces of diquinone imide (IV) [82] (reaction 43).

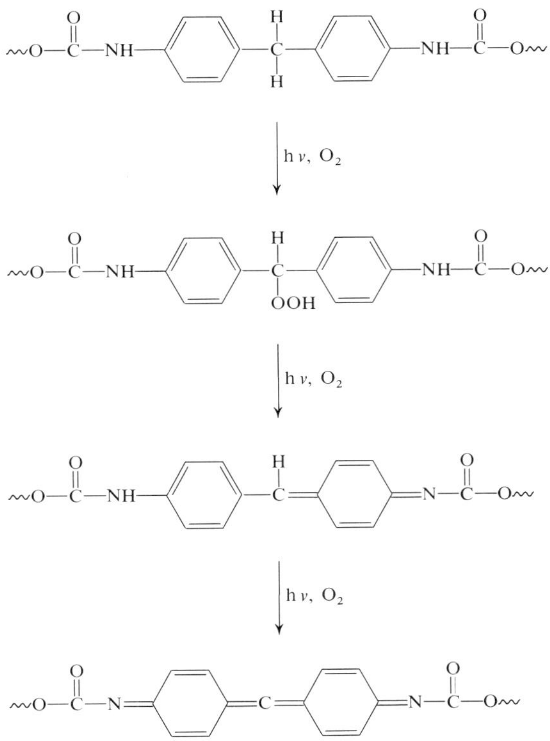

(43)

(IV)

Polyurethanes prepared from aliphatic diisocyanates are also prone to photooxidative degradation [81]. However, since no products absorbing above 400 nm are formed, degradation is not accompanied by discoloration. In a more recent study too, no fundamental difference has been found between aromatic and aliphatic polyurethanes with respect to photooxidation [82]. Polyetherurethanes, however, are significantly more sensitive towards photooxidation than polyesterurethanes because of the presence of the ether group.

It has been shown lately that aromatic urethane groups are able to inhibit the photooxidation of the polyether component [83]. This inhibition is attributed to the formation of complexes with peroxy compounds. It is believed to be a function of the electron donating properties of the urethane groups because a correlation has been found between the stabilizing effect and the ionization potential of the urethane group.

3.2.2.8 Photooxidation of polycarbonate

Bisphenol-A polycarbonate (PC) absorbs u.v. light below 360 nm but its absorption is intense only below 300 nm.

Most studies [84 to 86] attribute the aging, essentially a yellowing, to the photo-*Fries* rearrangement. The latter gives first a phenyl salicylate and after absorption of a second photon and subsequent rearrangement, 2,2′-dihydroxybenzophenone groups (reaction 44).

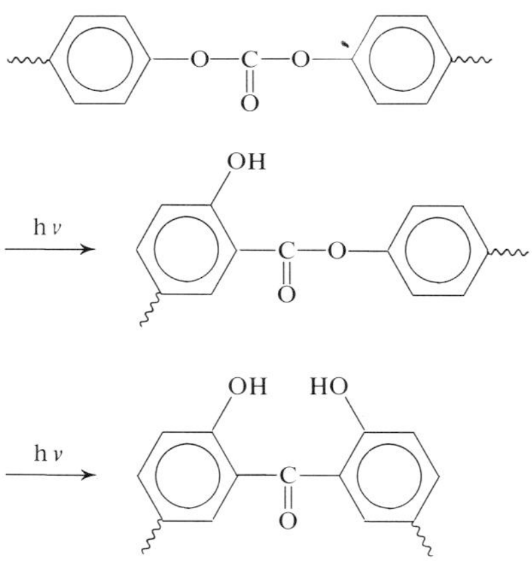

(44)

The absorption of these groups reaches into the visible region and PC yellowing had been essentially attributed to them. The photoproducts thus formed would lead to a reduction of photolysis through u.v. absorption as well as by quenching of excited states [85].

Recently attention has been brought upon the important role of oxygen in PC photoaging [87]. With the help of spectroscopic measurements, product analysis and oxygen absorption determinations, it has been shown that photooxidation is more important than the photo-*Fries* rearrangement for PC photoaging. In addition to the *Fries* reaction, two other primary reactions are considered. The first one is based on the formation of O_2-charge-transfer complexes in PC, similar to those found in polyolefins and leading to the formation of hydroperoxides. Charge-transfer complexes O_2/PC are considered to contribute significantly to photooxidation in the early stages. The second possible primary reaction is a light-induced reduction of metallic impurities (reaction 45).

$$Fe^{3+}OH^{-} \xrightarrow{h\nu} Fe^{2+} + {}^{\cdot}OH \qquad (45)$$

Secondary reactions such as photolytic decomposition of hydroperoxides and ketones as well as free radical attack on methyl groups and the resulting peroxidation have been considered so far. The formation of the photooxidation products already identified i.e. carboxylic acids, esters, anhydrides, alcohols, phenols and ketones can be easily explained this way.

3.2.2.9 Photooxidation of polyacetals

Polyoxymethylene (POM) is markedly unstable towards light, even u.v. radiation of wavelengths as high as 365 nm may initiate its degradation [88]. The species primarily responsible for light absorption have not yet been identified unambiguously, but oxidation products are the prime candidates, especially peroxide and hydroperoxide groups formed on the activated methylene groups. Depolymerization produces formaldehyde which gives formic acid on oxidation by peroxides and the acid thus formed catalyzes the hydrolytic degradation of polyacetals.

3.2.2.10 Photooxidation of polyamides

Aliphatic polyamides

Aliphatic polyamides absorb only slightly in the short-wavelength region of sunlight but, nevertheless, more than corresponding model compounds. This is attributed to the absorption of impurities present in polyamides. Direct chain scission at wavelengths below 300 nm and photosensitized oxidation above 300 nm have already been considered a long time ago as responsible for photooxidation [89]. An aldehyde carbonyl group has been proposed as sensitizing chromophore [90]. However, the identity of the species responsible for initiation of photooxidation of aliphatic polyamides above 290 nm is still a matter of debate [91, 92]. Recently [93], the hypothesis has been put forward that the rate of free radical formation in polyamide is a function of the formation and photochemical transformation of an intermediate of unknown structure.

Anyway, scission of the amide bond seems to be the primary photochemical process [94] (reaction 46).

$$\sim\!\!\underset{\underset{\displaystyle O}{\|}}{C}\!-\!NH\!-\!CH_2\!\sim \xrightarrow{h\nu} \sim\!\!\underset{\underset{\displaystyle O}{\|}}{C^{\cdot}} + {}^{\cdot}NH\!-\!CH_2\!\sim \qquad (46)$$

In the presence of sensitizers or under free radical attack, abstraction of hydrogen in α-position to the NH groups is the main reaction (reaction 47).

$$\sim\!\!\underset{\underset{\displaystyle O}{\|}}{C}\!-\!NH\!-\!\underset{\underset{\displaystyle H}{|}}{C}H\!\sim + X^{\cdot} \longrightarrow \sim\!\!\underset{\underset{\displaystyle O}{\|}}{C}\!-\!NH\!-\!\dot{C}H\!\sim + XH \qquad (47)$$

There is much more agreement on the oxidation reactions following initiation, essentially the addition of oxygen to the primary radicals and the subsequent reactions. The main photooxidation products, peroxides, aldehydes, pyrroles, carboxylic acids and carbon monoxide as well as scission of the polymer chain may be explained this way [94 to 96].

Aromatic polyamides

The u.v. absorption of aromatic polyamides such as poly-(1,3-phenylene-isophthalamide) is generally very high in the wavelength region between 290 and 400 nm [97]. On u.v. exposure, *in the absence of oxygen*, photo-*Fries* rearrangement and cross-linking are the essential reactions observed [97, 98] (reaction 48).

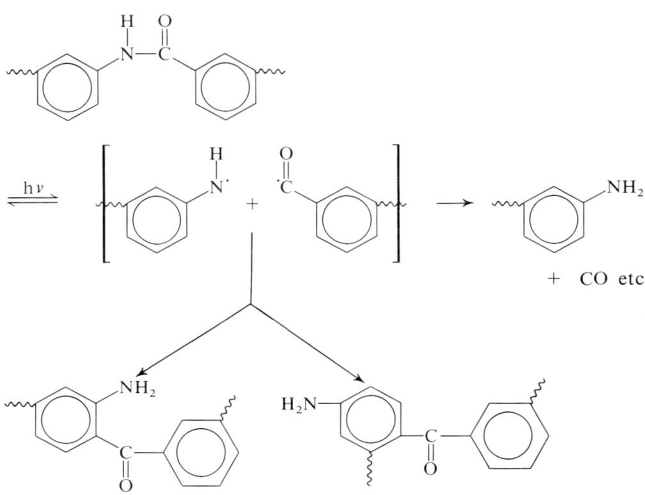

(48)

The formation of a radical pair postulated for the photolytic scission of the amide group explains the formation of CO and the results of the photolysis of model compounds [97].

On exposure of aromatic polyamides *in air* [97, 99] yellowing is less pronounced than in absence of oxygen and no cross-linking is observed. The formation of carboxylic acid end-groups and the absence of *Fries* photoproducts are best explained by assuming a reaction of the radical pair with oxygen (reaction 49).

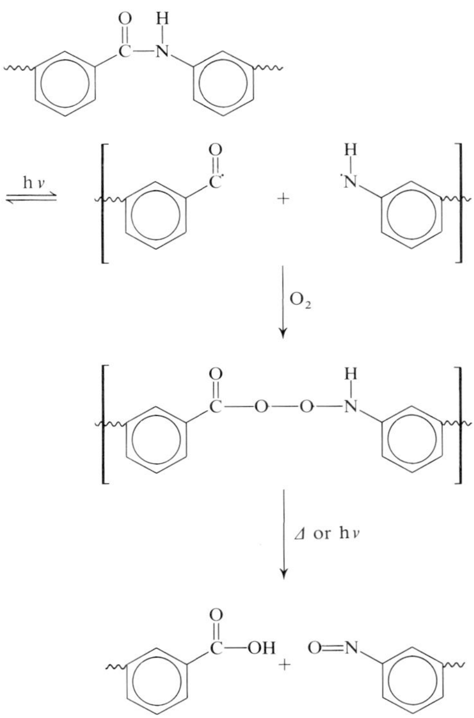

(49)

In corresponding polyamide fibers *Fries* photoproducts are also formed in the presence of oxygen. This is attributed to the fact that the rate of diffusion of oxygen to the places where bonds have been broken is lower in fibers compared to films because of the higher crystallinity of the former.

3.2.2.11 Photooxidation of linear polyesters

The absorption of polyesters made from aromatic dicarboxylic acids such as polyethylene terephthalate (PET) begins already at 360 nm. It increases significantly below 320 nm and becomes very pronounced below 300 nm.

In the presence of oxygen mainly chain scissions are observed, and no chain branching. CO, CO_2 and carboxylic acids are by far the main products of photooxidation. Besides direct photolysis of ester groups, oxidation reactions have to be considered too, through peroxy radicals they may also contribute to CO_2 formation.

The most probable primary photolytic chain scissions are shown in reactions 50 to 52:

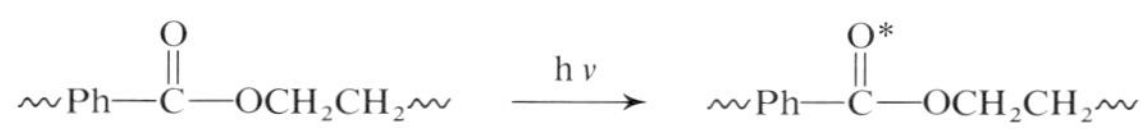

(50)

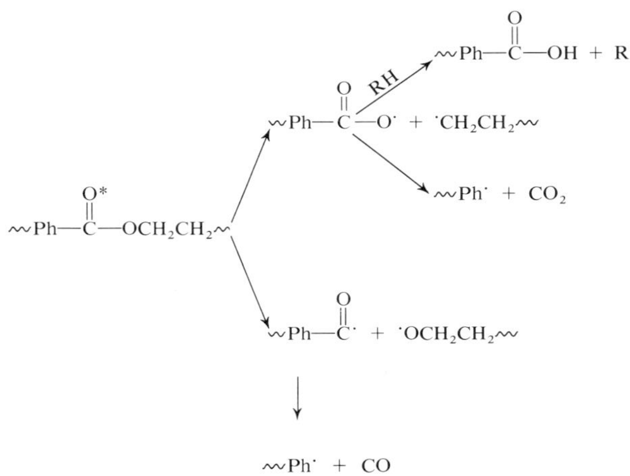

(51a)

(51b)

(52)

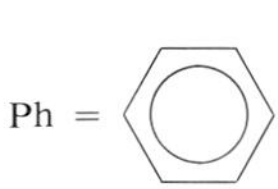

The formation of carbon monoxide is attributed mainly to reaction 52, that of carboxylic acids may be explained primarily by the *Norrish* type II reaction involving intramolecular hydrogen abstraction (reaction 53). This reaction has been clearly proven with model compounds, with aliphatic as well as with aromatic esters. From quenching experiments it was recently concluded that the *Norrish* type II rearrangement involves preferentially a triplet state [100].

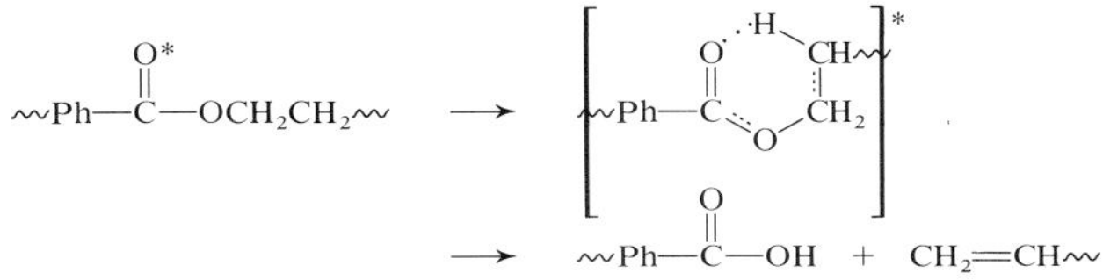

(53)

Photooxidation of PET, according to the general scheme in figure 3, through hydrogen abstraction in α-position to the ester group, gives rise to hydroperoxides and their decomposition products as well as to related chain scissions [101].

Photodegradation of glassy polyesters, aromatic or aliphatic, such as those used for example in lacquers, is similar to that of crystalline PET [102].

3.2.2.12 Photooxidation of polyphenylene oxide

U.V. light absorption by poly-(oxy-2,6-dimethyl-1,4-phenylene) (PPO) begins already at 380 nm [103]. The chromophoric groups are of the quinoid type (V), able to abstract hydrogen after absorption of a photon and thus initiate the formation of free radicals. In fact this is considered to be the most important source of free radicals in the early stages of PPO photooxidation [104] though PPO always contains small amounts of hydroperoxides. Their catalytic decomposition, induced by traces of copper (originating from the polymerization catalyst) or other transition metals, may also contribute to initiation. In the early stages of photooxidation, u.v. light absorbed by the hydroperoxides can probably be neglected in comparison with the overall light absorbed by PPO. With advancing oxidation according to the general autooxidation scheme (figure 3) hydroperoxide concentration is increasing so that its contribution to initiation through direct photolysis or photosensitized decomposition may increase.

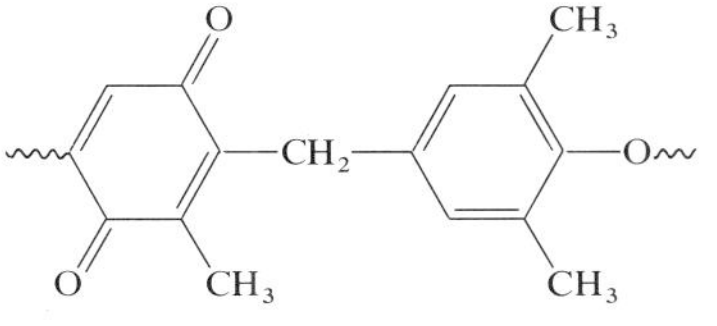

(V)

The main products of PPO photooxidation are alcohols, esters and aldehydes [105]. The latter are oxidized in subsequent reactions to carboxylic acids, peracids and peresters being also formed. The chain scissions observed on photooxidation may be explained by reaction 54 in analogy with thermal oxidation [104].

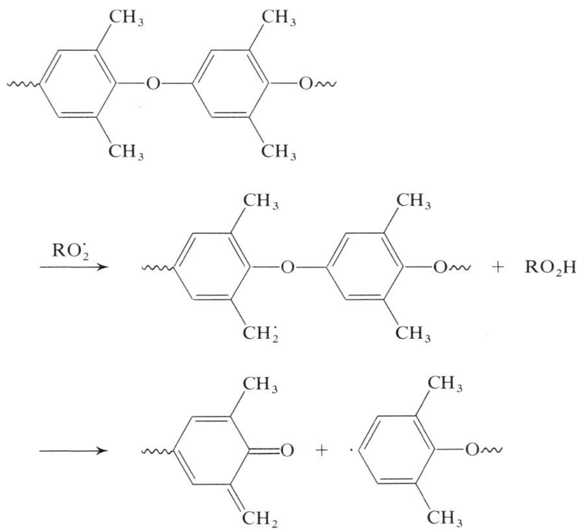

(54)

This interpretation has been challenged lately [213] since it was found that the free radical mechanism of oxidation of the methyl groups is not a major pathway in the photooxidation of PPO. It has been convincingly shown that upon photooxidation the aromatic rings of PPO disappear much faster than the methyl groups. An electron transfer mechanism has been postulated leading finally to the formation of a PPO radical cation and superoxide O_2^-. The combination of these two species gives an unstable, presumably endoperoxidic entity that undergoes further reactions yielding the observed degradation products.

The photoinitiated oxidation depends strongly on the mobility of the polymer segments [106]. In this respect it has been found that oxygen absorption and carbonyl buildup increase strongly in the temperature range between 45 and 60 °C. This temperature range corresponds to the β-relaxation attributed to the oscillation of the aromatic rings around the C—O—C bonds. The oxidation rate increase has been related to variations of the solubility and rate of diffusion of oxygen [106].

3.2.2.13 Photooxidation of polysulfone

Polysulfone, for example the polymer prepared from 2,2′-bis-(4-hydroxyphenyl)-propane and 4,4′-dichlorodiphenyl sulfone (structural unit VI), is degraded by u.v. light of wavelengths below 320 nm [107]. The good correlation between u.v. dose and polymer degradation has been used as an actinometer for measurement of the short-wavelength part of u.v. light [108]. For initiation of polysulfone photooxidation, u.v. absorption by the sulfone group has to be considered rather than impurities in the polymer [109].

Photooxidation of polysulfone becomes visible through broad hydroxyl and carbonyl absorption bands in the infrared [110, 111]. The decrease of the absorption of the methyl groups points to their oxidation. Next to CO and CO_2, low molecular weight sulfonic acids have been also identified.

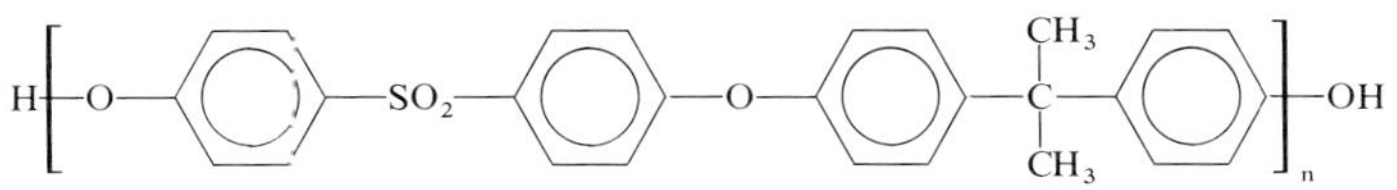

(VI)

3.3 Mechanisms of u.v. stabilization

The overview of plastics' degradation shows that in the absence of appreciable u.v. absorption by a polymer, even small amounts of impurities or chromophores are sufficient to induce photooxidative degradation. Inhibition, or at least retardation, of the reactions responsible for degradation is, of course, a necessity for successful u.v. stabilization. The general principles of u.v. stabilization of plastics are best understood in the light of the main photooxidation reactions shown in figure 3.

The first and most obvious possibility of protection against u.v. light is based on the prevention of u.v. light absorption or the reduction of the amount of light absorbed by the chromophores. The second possibility is given by the reduction of the initiation rate through deactivation of the excited states of the chromophoric groups. The third possibility of intervention in the photooxidative degradation process arises when the chain branching step is considered; then u.v. protection implies the transformation of hydroperoxides into more stable compounds, without generation of free radicals, before the hydroperoxides undergo photolytic cleavage. This, of course, means again a reduction of the initiation rate and is sometimes considered as the most important step of u.v. stabilization. In addition to that, the deactivation or complexing of chromophoric groups other than hydroperoxides may be of some importance. The fourth and last possibility to stop photooxidative degradation consists in the scavenging of the free radicals as soon as possible after their formation, either as alkyl radicals or as peroxy radicals.

The mechanisms described in the literature [112 to 114] correspond essentially to the four possibilities considered above. Even the designation of the light stabilizer classes is based on these mechanisms, that is to say, u.v. absorbers, quenchers of excited states, hydroperoxide decomposers and free radical scavengers. It has to be mentioned at this point that this classification is a simplification and that some compounds or classes of compounds may be active in more than one way and rather often do so.

3.3.1 U.V. absorbers

The protection mechanism of u.v. absorbers is based essentially on absorption of the harmful u.v. radiation and its dissipation in a manner that does not lead to photosensitization, i.e. dissipation as heat. Besides having a very high absorption themselves, these compounds must be very light stable, because otherwise they would be consumed too fast in non-stabilizing reactions. A fundamental disadvantage of u.v. absorbers is the fact that they need a certain absorption depth (sample thickness) for good protection of a plastic. Therefore, the protection of polymer surfaces and of thin articles such as fibers and films is only moderate.

Hydroxybenzophenones and hydroxyphenylbenzotriazoles are the most extensively studied u.v. absorbers. Hydrogen bonding of the o-hydroxyphenyl groups exerts a decisive influence on their spectroscopic and photochemical properties.

The hydroxybenzophenone type u.v. absorbers are essentially derived from 2,4-dihydroxybenzophenone. Less commonly used representatives are derivatives of 2,2′,4-trihydroxy or 2,2′,4,4′-tetrahydroxybenzophe-

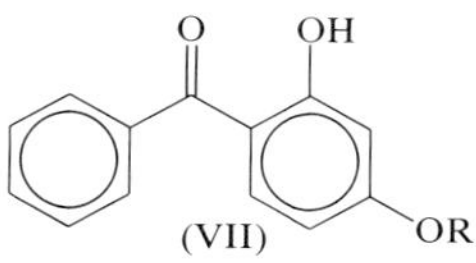

(VII)

R = H, CH_3 to $C_{12}H_{25}$

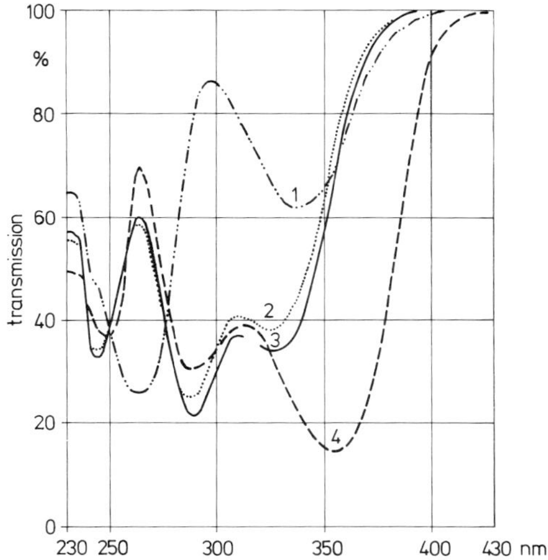

Fig. 4. U.V. transmission spectra of benzophenones in chloroform, $4{,}4 \cdot 10^{-5}$ molar: path length 1 cm; *1:* 2-hydroxybenzophenone, *2:* 2,4-dihydroxybenzophenone, *3:* 2-hydroxy-4-octoxybenzophenone, *4:* 2,2′-dihydroxy-4,4′-dimethoxybenzophenone.

none. The influence of the substituents on the spectral properties of benzophenones is shown in figure 4.

The main absorptions of 2-hydroxybenzophenone are situated in the uninteresting wavelength domain around 260 nm. Electron donating substituents push this absorption towards longer wavelengths, between 300 and 400 nm, and at the same time total absorption in the u.v. is increasing. However, compounds such as 2,2′-dihydroxy-4,4′-dimethoxybenzophenone are already distinctly yellow because of tailing absorptions above 400 nm, so that, depending on the concentration used, discoloration of the substrate may be observed.

Through choice of adequate alkyl groups R in the alkoxy group it is possible to optimize the protective power and the compatibility with the plastics to be stabilized [115, 116].

Concerning the energy transformation in hydroxybenzophenones, it has been shown with 2-hydroxybenzophenone (VIII) [117] that on exposure to light (VIII) is transformed into enol (IX) which turns back into its initial form (VIII) on losing thermal energy to the medium (reaction 55). On the strength of spectroscopic data it is concluded that the energy

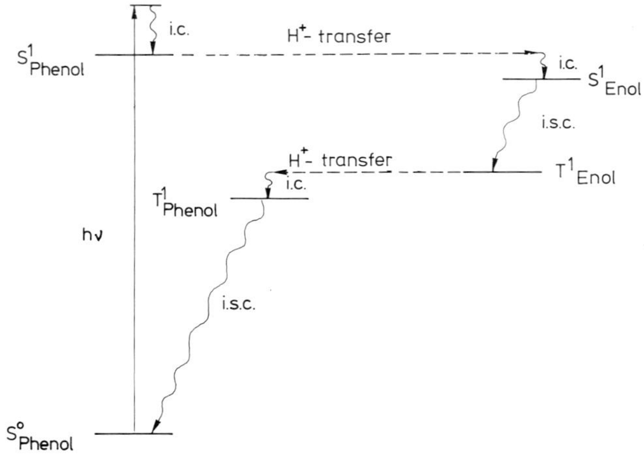

Fig. 5. Intramolecular proton transfer in 2-hydroxybenzophenone (reaction 55); Scheme of main radiationless transitions during energy dissipation (i.c. = internal conversion, i.s.c. = intersystem crossing).

transformation involves exclusively fast radiationless transitions. Recently the scheme shown in figure 5 has been proposed [118]. This scheme shows the important role of proton transfer. In this respect it had already been shown previously with the aid of NMR measurements [119] that the u.v. stabilizing effectiveness of 2-hydroxybenzophenones in polyethylene is increasing with the strength of the hydrogen bond in (VIII). Moreover, it is well known that in highly polar media preventing the formation of the intramolecular hydrogen bonds, 2-hydroxybenzophenones are not stable photochemically and may possibly even act as sensitizers [120].

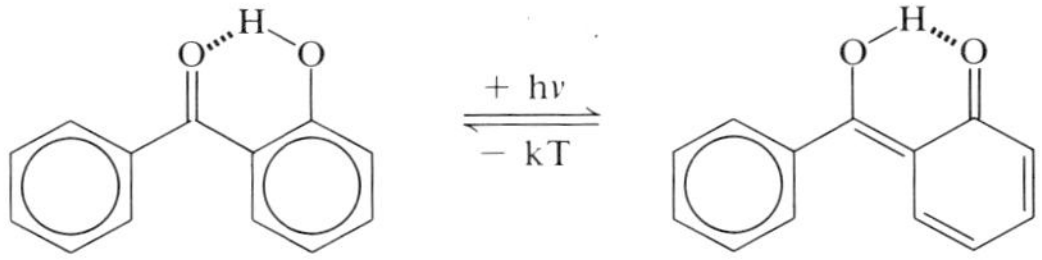

(VIII) (IX) (55)

The transmission spectra of some technically important u.v. absorbers from the 2-(2′-hydroxyphenyl)-benzotriazole class (X) are represented in figure 6. They show the bathochromic shift caused by the chlorine substituent in position 5 and the alkyl substituents in positions 3′ and 5′. In comparison with the 2-hydroxybenzophenones, the 2-(2′-hydroxyphenyl)-benzotriazoles have higher molar extinction coefficients and steeper absorption edges towards 400 nm [121]. The elimination of the excitation energy in the form of harmless heat seems to proceed according to mechanisms similar to those discussed for 2-hydroxybenzophenones [113].

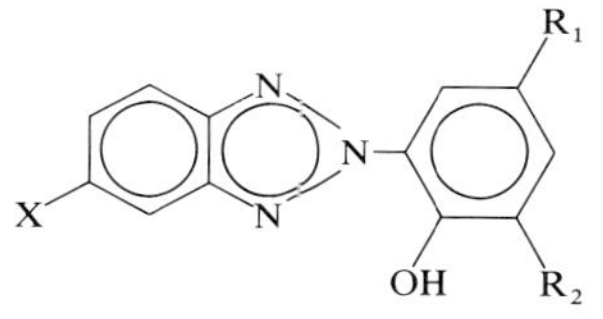

(X)

X = H, Cl
R_1 = CH_3 to C_8H_{17}, linear and branched alkyl
R_2 = H and branched alkyl

The u.v. absorber protection mechanism based solely on a filter effect has become a matter of discussion recently. In this connection, it has

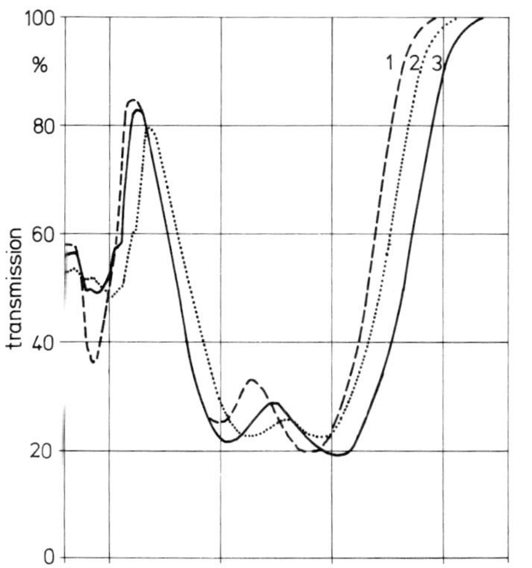

Fig. 6. U.V. transmission spectra of benzotriazoles in chloroform, $4{,}4 \cdot 10^{-5}$ molar; path length 1 cm; *1:* 2-(2′-hydroxy-5′-methylphenyl)-benzotriazole, *2:* 2-(2′-hydroxy-3′,5′-di-tert.butylphenyl)-benzotriazole, *3:* 2-(2′-hydroxy-3′,5′-di-tert.butylphenyl)-5-chlorobenzotriazole.

been stressed that in plastics with a pronounced u.v. absorption, for example in aromatic polyamides, the absorption by the plastics is definitely superior to that of the stabilizers incorporated in technically common concentrations. The fact that in aromatic polyamides 2-(2′-hydroxyphenyl)-benzotriazoles are already effective at concentrations much below the concentration needed theoretically, led to the elaboration of a model involving energy transfer [122]. Similar considerations are valid for polypropylene. For this polymer it was found that the stabilizing effectiveness of some u.v. absorbers goes beyond the pure light filter effect [123, 124]. According to some studies 2-(2′-hydroxyphenyl)-benzotriazoles are acting exclusively as u.v. absorbers whereas 2-hydroxybenzophenones confer additional stability through energy transfer [125].

U.V. absorbers of the cinnamate type correspond to structure (XI). They absorb at much shorter wavelengths than 2-(2′-hydroxyphenyl)-benzotriazoles and 2-hydroxybenzophenones (range of wavelength 290–320 nm, figure 7).

Because of a large number of possibilities of vibration, cinnamates are able to transform the absorbed energy into heat with high efficiency

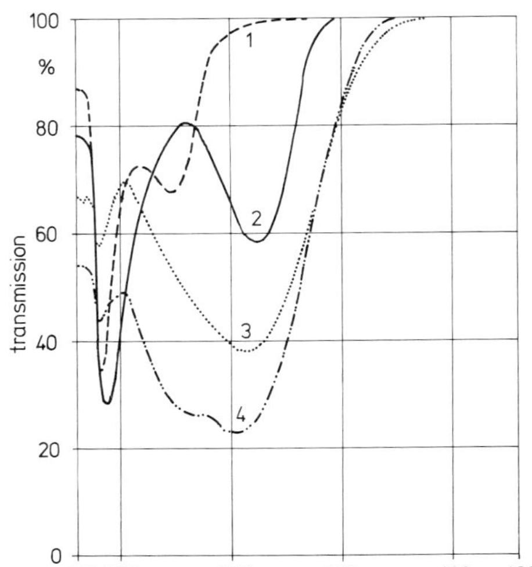

Fig. 7. U.V. transmission spectra of *1:* resorcinol monobenzoate, *2:* tert.octylphenyl salicylate, *3:* 2-cyano-3-phenyl ethyl cinnamate, *4:* N,N′-(2-ethyl-2′-ethoxyphenyl)-oxalamide (chloroform, $4{,}4 \cdot 10^{-5}$ molar; path length 1 cm).

[126]. Photostabilization of plastics with cinnamic acid esters is based on the filter effect on one side, and on the quenching of excited states on the other side. Quenching is attributed to the marked overlap between the absorption bands of the cinnamic acid derivatives and the luminescence bands of e.g. polypropylene and polystyrene [126]. The overall poor efficiency of this type of stabilizers is ascribed to an insufficient compatibility with the polymers to be stabilized [126].

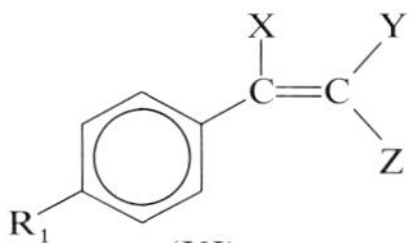

(XI)

R_1 = H, OCH_3
X = H, CH_3, C_6H_5
Y, Z = CN, CO_2 alkyl

The u.v. absorbers of the oxanilide type (XII) (figure 7) available commercially, absorb even at shorter wavelengths than the cinnamates. No details about the protection mechanism of oxanilides have been published so far.

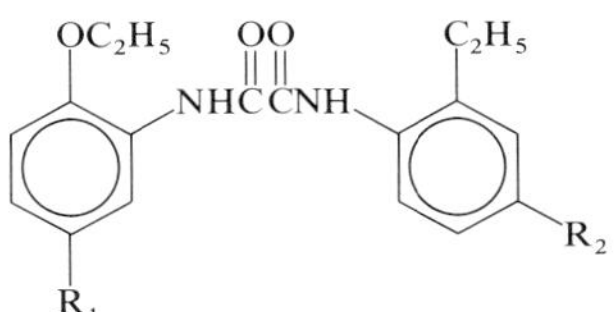

(XII)

$R_1 = R_2$ = H, tert.C_4H_9

Contrary to the typical u.v. absorbers described above, compounds such as (XIII) and (XIV) cannot be considered as absorbers in the same sense (figure 7). It is only on u.v. exposure that these compounds rearrange to give 2-hydroxybenzophenones which may be considered as the active species. The monobenzoate of resorcinol (XIV), for example, rearranges to 2,4-dihydroxybenzophenone (XV) (reaction 56).

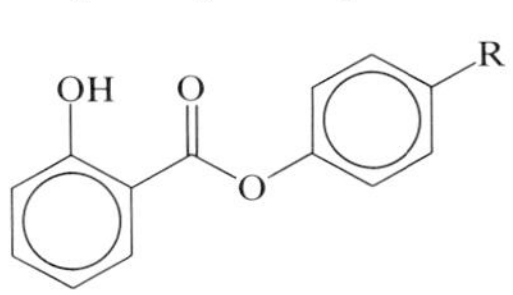

(XIII)

R = tert.C_4H_9, tert.C_8H_{17}

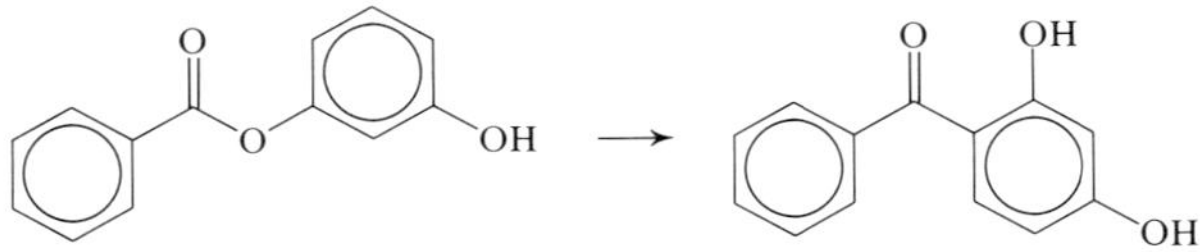
(XIV) (XV) (56)

Because of the necessity of rearrangement prior to effective stabilization, the efficiency of compounds such as (XIII) and (XIV) is rather limited. Additional difficulties arise because the photo-*Fries* rearrangement [127] is not quantitative so that secondary products formed simultaneously may lead to complications such as discoloration.

3.3.2 Quenchers

Quenchers (Q) are light stabilizers able to take over the energy absorbed by the chromophores (K) present in a plastics material and to dispose of it efficiently to prevent degradation (reactions 57 to 60). The energy can be dissipated either as heat (reaction 59) or as fluorescent or phosphorescent radiation (reaction 60).

$$K + h\nu \longrightarrow K^* \quad (57)$$
$$K^* + Q \longrightarrow K + Q^* \quad (58)$$
$$Q^* \longrightarrow Q + \text{heat} \quad (59)$$
$$Q^* \longrightarrow Q + h\nu' \quad (60)$$

For energy transfer to occur from an excited chromophore (donor) to the quencher (acceptor) the latter must have lower energy states than the donor. The transfer can proceed according to two general mechanisms. The first process, the long range energy transfer or *Förster* mechanism, is based on a dipole-dipole interaction and is usually observed in the quenching of excited singlet states. The distance between chromophore and quencher may be as large as 5 or even 10 nm, provided there is a strong overlap between the emission spectrum of the chromophore and the absorption spectrum of the quencher. The *Förster* mechanism has been considered as a possible stabilization mechanism by typical u.v. absorbers with extinction coefficients $\geqq 10000$ [113]. Though quenching of carbonyl compounds through this mechanism has been postulated several times it has not been shown unequivocally. As a matter of fact, with carbonyl compounds this deactivation type can be excluded on theoretical grounds [128].

The second process is the so-called *contact,* or *collisional,* or *exchange energy transfer*. For an efficient transfer to take place the distance between quencher and chromophore must not exceed 1.5 nm. This means that the stabilization that can be achieved will depend on the concentration of the quencher and on the lifetime of the excited donor. Considering the longer lifetimes of excited triplet states (10^{-6} to 10^{-3} s) compared to those of singlet states (10^{-10} to 10^{-6} s), energy transfer from triplet states is more likely [113]. From calculations based on the assumption of random distribution of both stabilizer and sensitizer in the polymer, it is concluded that exchange energy transfer cannot contribute significantly to stabilization [129]. This conclusion will not hold if some kind of association between sensitizer and stabilizer takes place, for example through hydrogen bonding. Another possibility to stabilize through energy transfer, despite random distribution of chromophores and quenchers, is given by exciton migration (exciton = delocalized electronic excited state). The latter is of some importance in polymers with aromatic side groups. It has been verified experimentally for the first time in poly-N-vinyl carbazole [130].

Considering the dominating role of hydroperoxides in polyolefin photooxidation (cf. 3.2.2.3), quenching of excited —OOH groups would contribute significantly to stabilization. However, the —OOH excited state is dissociative, i.e. its lifetime is limited to one vibration of the O—O bond. During this very short time (ca. 10^{-13} s) contact energy transfer to the quencher is highly unlikely if the —OOH group is not already associated with the quencher. From a practical point of view, quenchers are especially interesting because their action is independent of the thickness of the samples to be protected. Thus, they are specifically useful for the stabilization of thin section articles such as films and fibers.

Because energy transfer to paramagnetic transition metal compounds belongs for a long time to the photochemical reactions studied successfully [131, 132], it has been considered relatively early for the elucidation of the mechanisms of nickel-containing light stabilizers. Thus it has been shown that the nickel-phenolate (XVI) inhibits very efficiently the pho-

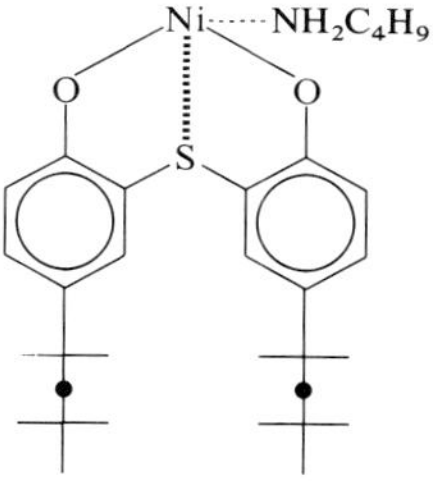

(XVI)

tooxidation of cumene sensitized by diethylketone through quenching of the triplet state of the latter [133]. It has also been shown that there is a correlation between the efficiency of stabilization of polypropylene by a number of diamagnetic nickel compounds such as (XVIII) and their efficiency in quenching excited triplet states in polypropylene [134, 135].

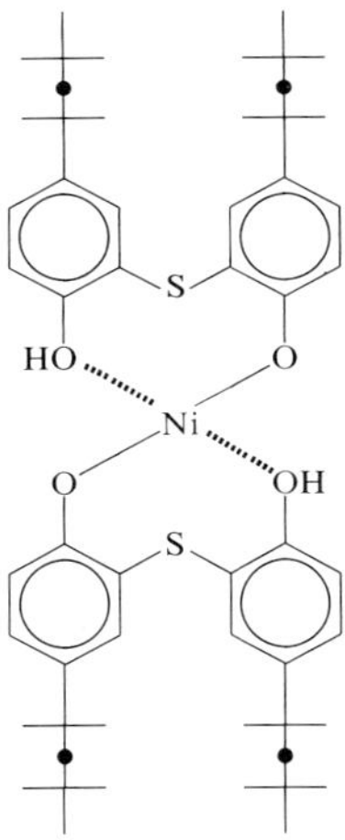

(XVII)

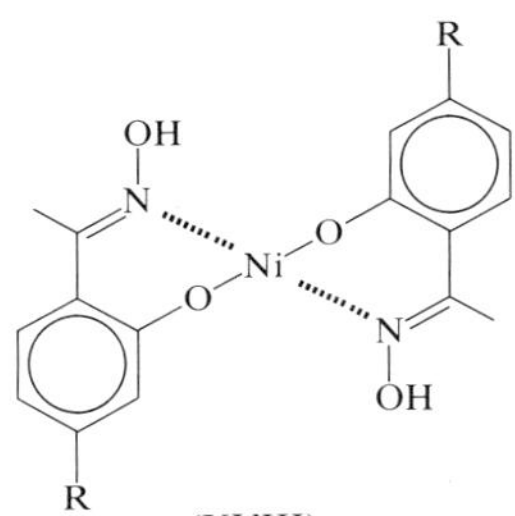

(XVIII)

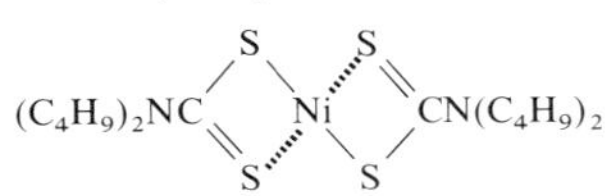

(XIX)

On consideration of the u.v. spectra of nickel-containing light stabilizers, it can be easily understood that their efficiency must not be due exclusively to triplet quenching. In figure 8 it can be seen that (XVI), (XVII) and (XIX) absorb strongly in the u.v. region between 290 and 400 nm and that the absorption of the dithiocarbamate (XIX) is even more pronounced than that of the 2-hydroxybenzophenones and 2-(2′-hydroxyphenyl)-benzotriazoles (figures 4 and 6). Therefore, it is reasonable to assume that the efficiency of such light stabilizers is due, at least in part, to u.v. absorption.

Moreover, it has been shown that some nickel-containing stabilizers may also act as radical scavengers and/or as hydroperoxide decomposers in

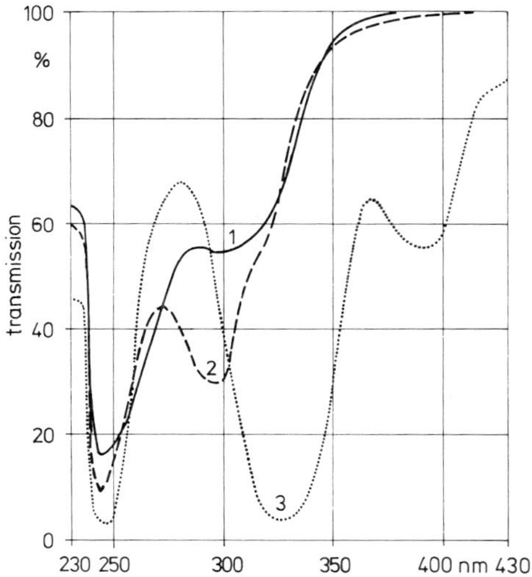

Fig. 8. U.V. transmission spectra of nickel stabilizers in chloroform, $4{,}4 \cdot 10^{-5}$ molar; path length 1 cm; *1:* n-butylamin-nickel-2,2′-thio-bis-(4-tert.octylphenolate), *2:* nickel-bis-[(2,2′-thio-bis-(4-tert.octylphenolate)], *3:* nickel-di-butyldithiocarbamate.

polypropylene [136] and polyethylene [137]. Nickel compounds may even deactivate singlet oxygen [138]. However, there is no correlation between the light stabilizing activity in polypropylene and the singlet oxygen quenching rate constants measured in model systems [114]. This result leads to the conclusion that singlet oxygen is of minor importance in polypropylene degradation.

3.3.3 Hydroperoxide decomposers

Hydroperoxides play a determining role in the photooxidative degradation of polymers. It is of interest to point out that the possibility for typical u.v. stabilizers to protect polymers by peroxide decomposition has been suggested as early as 1965 [139]. However, the decomposition of hydroperoxides in polymer films has been demonstrated unambiguously only in 1974 [140].

Metal complexes of sulfur-containing compounds such as dialkyldithiocarbamates (XX), dialkyldithiophosphates (XXI) and thiobisphenolates (XVI) are very efficient hydroperoxide decomposers even if used in almost catalytic quantities. These metal complexes reduce the hydroperoxide content of pre-oxidized polypropylene films already at room temperature and are at the same time very efficient u.v. stabilizers for this polymer [140].

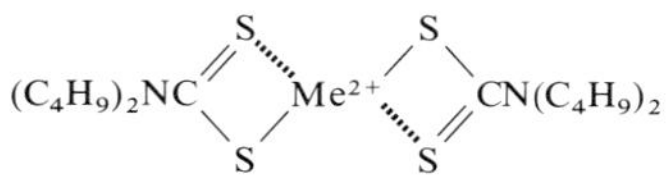

(XX)

Me = Zn, Ni

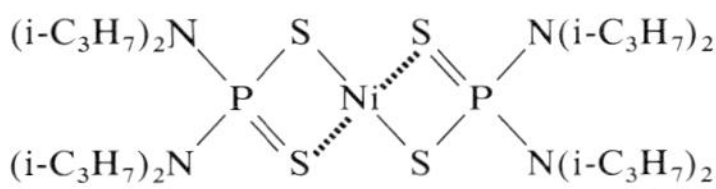

(XXI)

The improvement in u.v. stability observed technically on combining u.v. absorbers with phosphite [141] or nickel compounds [142] can be easily understood with this mechanism.

3.3.4 Free radical scavengers

Apart from the absorption of harmful radiation by u.v. absorbers, the deactivation of excited states by quenchers and the decomposition of hydroperoxides by some compounds containing phosphorus and/or sulfur, the scavenging of free radical intermediates is another possibility for stabilization, analogous to that used in thermal degradation. The importance of free radical termination reactions for u.v. stabilization had been pointed out long ago [29, 139, 143]. However, these reactions were not taken into consideration for a long time, probably because phenolic antioxidants do not inhibit photooxidation efficiently.

It has been shown that compounds (XXII), (XXIII), (XXIV) and (XVI) are effective radical scavengers [140]. On u.v. exposure of hydroperoxide-containing polypropylene films it was found that, depending on the added stabilizer, the quantities of carbonyl groups, hydroxyl groups and water formed, are compatible with the following reactions:

$$\mathrm{PPOOH} \xrightarrow{h\nu} \mathrm{PPO^{\cdot}} + \mathrm{OH^{\cdot}} \tag{61}$$

$$\mathrm{^{\cdot}OH} \xrightarrow{+\ \mathrm{PPH}} \mathrm{H_2O} + \mathrm{PP^{\cdot}} \tag{62}$$

$$\mathrm{^{\cdot}OH} \xrightarrow{+\ \mathrm{InH}} [\mathrm{InH} \cdots \mathrm{^{\cdot}OH}] \tag{63}$$

$$\mathrm{PPO^{\cdot}} \xrightarrow{\beta\text{-scission}} \mathrm{PP{=}O} + \mathrm{PP^{\cdot}} \tag{64}$$

$$\mathrm{PPO^{\cdot}} \xrightarrow{+\ \mathrm{InH}} \mathrm{PPOH} + \mathrm{In^{\cdot}} \tag{65}$$

Whereas stabilizer-free samples, because of homolytic cleavage of hydroperoxide groups, gave almost quantitative yields of water (from hydroxyl radicals, reaction 62) and carbonyl compounds (from alkoxy radicals, reaction 64), the water yield was considerably reduced in presence of n-butylamine-nickel-2,2′-thio-bis-(4-tert.octylphenolate) (XVI) or nickel-bis-(3,5-di-tert.butyl-4-hydroxybenzyl)-phosphonic acid monobutylester (XXII).

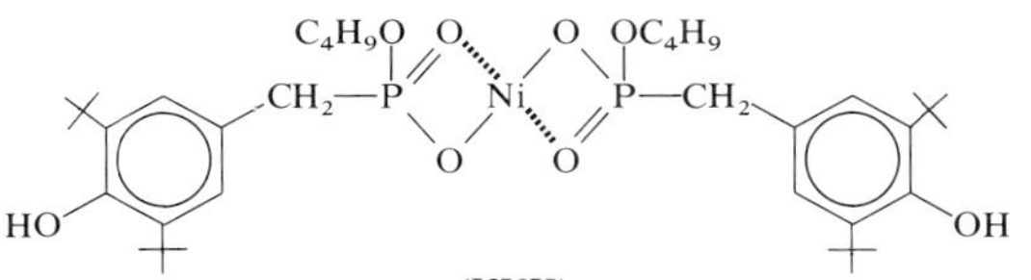

(XXII)

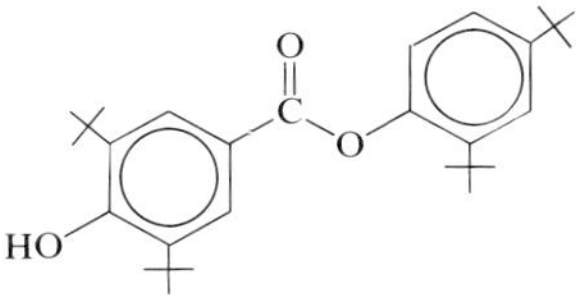

(XXIII)

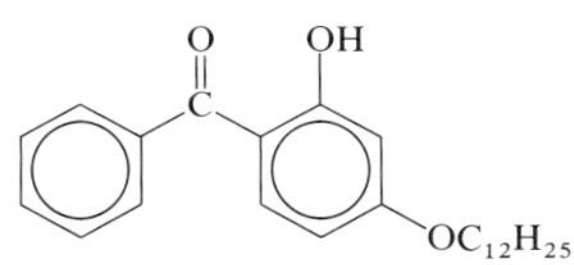

(XXIV)

Similar experiments with (XVI), (XXII), 3,5-di-tert.butyl-4-hydroxybenzoic acid 2,4-di-tert.butylphenylester (XXIII) or 2-hydroxy-4-dodecyloxybenzophenone (XXIV) showed a smaller increase in carbonyl groups than stabilizer-free films. This points to the scavenging of the alkoxy radicals by these stabilizers.

The latest development in this field is represented by hindered amine-type light stabilizers. Bis-(2,2,6,6-tetramethyl-4-piperidyl)-sebacate (XXV) does not absorb any light above 250 nm and thus cannot be considered a u.v. absorber or a quencher of excited states. This has been confirmed in polypropylene through luminescence measurements [135]. The details known so far on the protection mechanisms of sterically hindered amines can be summarized as follows:

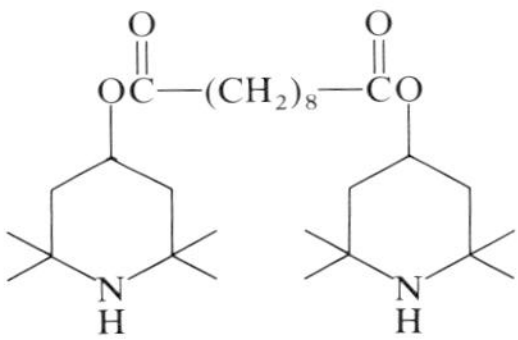

(XXV)

From ESR spectroscopic measurements it is concluded that, under photooxidative conditions, sterically hindered amines are converted, at least in part, to the corresponding nitroxyl radicals (XXVI) [144, 145]. The latter, through reaction 66, are thought to be the true stabilizing species involved [146, 147]. However, although reaction 66 is very fast in solution [148], a rough evaluation for polypropylene [145] shows that no efficient stabilization can be realized through alkyl radical trapping alone.

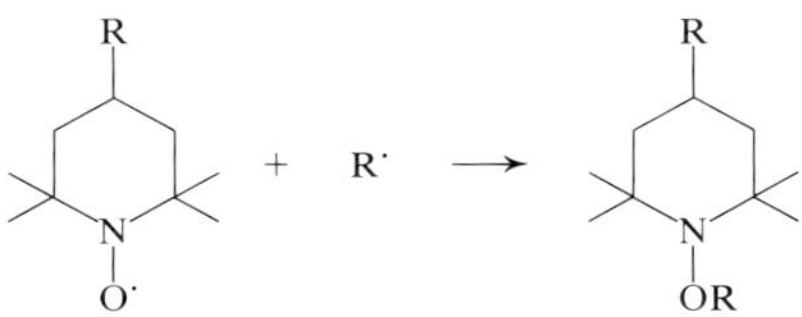

(XXVI) (XXVII) (66)

Therefore, another explanation of the u.v. stabilizing mechanism of sterically hindered amines (HALS) involves the hydroxylamine ethers formed in reaction 66 [149]. There is indirect evidence that, in polypropylene, peroxy radicals can react very quickly with (XXVII) with regeneration of nitroxyl radicals (reaction 67), resulting in an overall slowdown of the usual chain oxidation reactions 1 and 2.

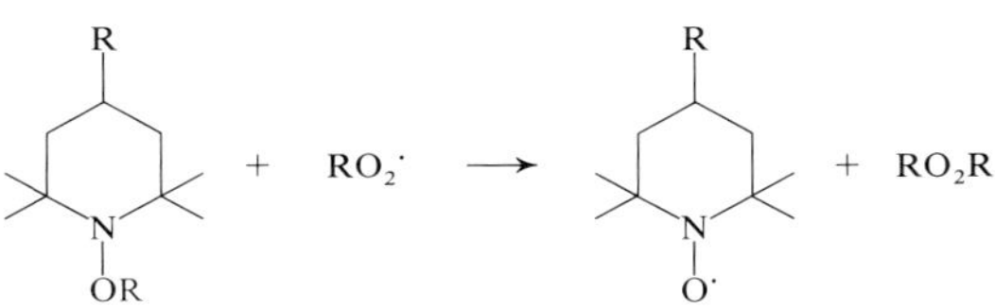

(XXVII) (XXVI) (67)

Consideration of the rate constants of reactions 66 and 67 and of the probable concentrations of the reactive components in comparison with the competitive reactions 1 and 2 lead to the conclusion that even the "*Denisov* cycle" (reactions 66 and 67) cannot explain the outstanding performance of HALS [150, 151]. This is strictly valid only from the point of view of a homogeneous distribution of additives and oxidation products. Nitroxyl radicals, however, may associate, e.g. with hydroperoxides through hydrogen bonding. This has been verified with model systems in the liquid phase [152] as well as in solid polypropylene [153] (reaction 68).

$$[>NO^{\cdot}\cdots PP] + PPOOH \rightleftharpoons [>NO^{\cdot}\cdots HOOPP] + PP \qquad (68)$$

In addition to reactions 66 and 67, the formation of hydroxylamine (reaction 69) is considered together with its reactions with peroxy radicals (reaction 70) and hydroperoxides (reaction 71) [154, 155].

$$>NOP \rightleftharpoons >NO^{\cdot} + P^{\cdot} \longrightarrow >NOH + >C{=}\overset{|}{C}{-}P' \qquad (69)$$

$$>NOH + PO_2^{\cdot} \longrightarrow >NO^{\cdot} + PO_2H \qquad (70)$$

$$>NOH + POOH \longrightarrow >NO^{\cdot} + H_2O + PO^{\cdot} \qquad (71)$$

The fact that polypropylene photooxidation begins to accelerate although high concentrations of PP—ON—R—NO—PP are still present, is attributed to the formation of reaction products which are no longer u.v. stabilizers. One of these products may be of the (XXVIII) type, resulting from the reaction of $>NO^{\cdot}$ with radicals from the ketone photolysis [155, 156].

$$>NO\underset{\underset{O}{\|}}{C}R$$

(XXVIII)

The HALS-induced decomposition of hydroperoxides could not be shown with model compounds such as tert.butyl-hydroperoxide in the liquid phase but has been demonstrated unequivocally with peroxidized polypropylene in the solid state at room temperature and in the absence of light (reaction 72) [154, 155, 157].

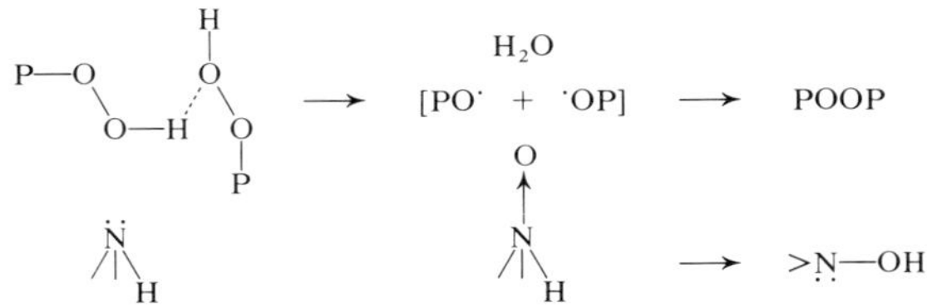

(72)

The formation of associations between sterically hindered amines and hydroperoxides and the reaction of these associations with peroxy radicals represents another possibility of retarding photooxidation (reactions 73 and 74) [156, 158].

$$>NH + HOOP \rightleftharpoons [>NH \cdots HOOP] \tag{73}$$

$$[>NH \cdots HOOP] + PO_2^{\cdot} \longrightarrow POOP + H_2O + >NO^{\cdot} \tag{74}$$

Photolysis of the hydroperoxides associated with sterically hindered amines produces alkoxy and hydroxyl radicals in close proximity to the amines. Then the radicals may abstract a hydrogen atom from the amine and form hydroxylamine and hydroxylamine ether groups (reaction 75) [158].

$$>NH \cdots HOOP \xrightarrow{h\nu} \left[>N{-}H \begin{matrix} ^{\cdot}OH \\ ^{\cdot}OP \end{matrix}\right] \longrightarrow \begin{cases} >NOH + POH \\ >NOP + H_2O \end{cases} \tag{75}$$

Still other mechanisms have been postulated, e.g. the interaction of HALS with α,β-unsaturated carbonyl compounds [159 to 161], the formation of charge-transfer complexes between HALS and peroxy radicals [162] as well as complexing of trace metals [158, 163]. Despite numerous publications in this field we are still away from a complete knowledge of the processes occurring on photooxidation of polymers in the presence of sterically hindered amines. As a matter of fact, many studies were performed with model systems or under special conditions precluding transposition of the results without reservations as to the situation encountered in practice, i.e. outdoor weathering of plastics.

3.4 Light stabilizer testing

The testing of the efficiency of light stabilizers in a given substrate is performed in laboratory trials and/or by outdoor exposure. The latter involves submission of the samples directly to the influence of the weather at a given testing site. Such tests have very often the disadvantage of long testing times that may extend over a period of several years. Moreover, the values obtained in this way are only valid, strictly speaking, for the specific exposure station. In accelerated laboratory tests, exposure devices are used in which the samples are submitted simultaneously or alternately to the influence of light, oxygen, humidity and heat. Under these conditions, the most important environmental parameters may be kept constant relatively easily. Hence, it is possible to generate data with sufficient reproducibility. Moreover, compared to outdoor weathering, artificial weathering offers the advantages of shorter testing time and of independency from site, season and local climate.

In testing the u.v. stability of plastics materials, depending on the nature of the polymer and on the anticipated application (fiber, film, injection molding), different characteristics are considered. These are essentially optical properties such as surface crazes, gloss, chalking, variation of color and transmission as well as mechanical properties such as elongation, tensile strength, impact strength. For more fundamental studies other methods are also considered, that is to say, physical methods such as the determination of the molecular weight and the molecular weight distribution, the viscosity in solution or in the melt and spectroscopic methods, such as u.v., visible and IR spectrometry. Whenever possible, existing standard test methods should be used.

From experience so far available it is found that very often there is a rough correlation between the results of outdoor and accelerated weathering if the test conditions stipulated in several standard methods are observed.

3.4.1 Accelerated weathering

Weathering devices consist usually of a test chamber fitted with air circulation and vertical light sources in the center of the chamber. At a given distance from the light sources the samples to be exposed are mounted on a merry-go-round. Temperature and humidity are kept constant through control and regulation. On selection of suitable programs, water

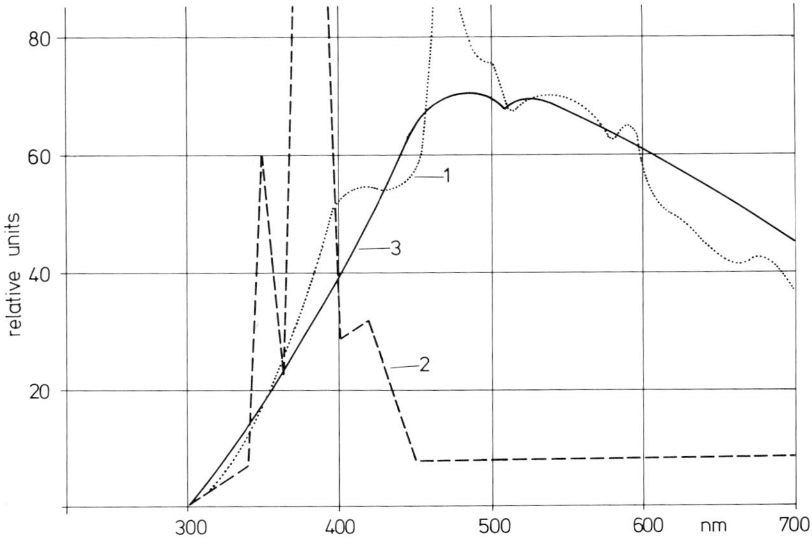

Fig. 9. Spectral energy distribution (schematized): *1:* filtered xenon arc, *2:* closed carbon arc, *3:* global radiation.

spraying of the samples may be achieved as well as a cycle of alternating dark and light periods.

The light sources used mostly are carbon arcs, xenon arcs and combinations of fluorescent lamps. Compared to the energy distribution of terrestrial sunlight the intensity of the carbon arcs in the wavelength region from 290 to 400 nm is too high. The best simulation of terrestrial sunlight is achieved with xenon arcs (figure 9) [6]. The accelerated weathering devices used mostly in Europe are the Xenotests from Original Hanau Quarzlampen G.m.b.H. and the Xenon-Arc-Weather-O-Meter and Fade-O-Meter from Atlas Electrical Devices Co. In these devices the spectra of the light sources are adjusted to that of terrestrial sunlight by use of filter systems.

Standard methods of accelerated weathering have been elaborated nationally and internationally (table 6) to allow rapid but nevertheless correct testing aimed at reasonable correlation with outdoor weathering. The measurement of the changes in properties after weathering (color, tensile strength/elongation, impact strength, bending strength) should

Table 6. ISO, DIN and ASTM standards for accelerated testing of plastics materials with xenon arc radiation.

Main characteristics	ISO 4892–1981	DIN 53387 June 1982	DIN 53389 April 1983	ASTM G 26–77
Energy spectrum	global radiation or filtered through window glass	global radiation	global radiation behind window glass	global radiation or behind window glass
Flux density at sample location	$1000 \pm 200\ W \cdot m^{-2}$	$1000 \pm 200\ W \cdot m^{-2}$	$1000 \pm 200\ W \cdot m^{-2}$	$\leqq 1500\ W \cdot m^{-2}$
Fluctuation of flux density tolerated on samples	±10%	±10%	±10%	±10%
Black panel temperature	45±3 °C 55±3 °C 63±3 °C	45±5 °C	45±5 °C	63±3 °C
Relative humidity	35±5% 50±5% 65±5% 90±5%	60 to 80%	40 to 60%	30±5%
Cycle: Water spraying period / dry period	3/17 min 5/25 min 12/48 min 18/102 min	3/17 min 18/102 min	–	18/102 min
Radiation measuring	lightfastness standards or radiation measuring instrument	radiation measuring instrument	lightfastness standards or radiation measuring instrument	lightfastness standards

be carried out according to the corresponding standards, whenever possible [164 to 166].

Measurement of the amount of radiation is a fundamental problem in relation with accelerated testing. In this connection, it was recognized already early that testing time alone did not give sufficiently reproducible results on light exposure of textiles. The reasons for this are mainly technical limits in lamp fabrication, decrease in output of the lamps with increasing operating time and aging of filters. Such uncertainties may be overcome by expressing the energy received as the product of exposure time and radiation intensity. In a different approach, intensity may be measured at one or several selected wavelengths and adjusted to a constant value.

3.4.2 Outdoor weathering

Outdoor weathering of polymer samples is performed on racks oriented to the south (northern hemisphere) with the exposure surface inclined usually at 45°. The samples may be mounted on suitable backings, preferably stainless steel, or on frames. Outdoor weathering behind glass is another possibility. The standards to be used are DIN 53386 (June 1982), DIN 53388 (provisional standard), the corresponding ISO standards ISO 877–1976 and ISO 4607–1978 or ASTM standard ASTM D 1435–78.

Measurement of incident energy is achieved with pyroheliometers [167] and expressed as Joule$\cdot$cm^{-2} in the SI system (previously Langley, 1 Ly$=$1 cal$\cdot$cm^{-2}$=$4.187 J$\cdot$cm^{-2}). In this way, the global radiation from the ultraviolet to the infrared region is registered but the short-wavelength u.v. radiation mainly responsible for polymer degradation is not measured separately. This is one of the reasons for the variation of weathering results with geographical location [168] and season [169]. For specific measurements in the short-wavelength u.v. region, several chemical actinometers have been proposed [170 to 174]. Through photochemically-induced transformations, they allow the evaluation of the energy received by a given surface during a limited time period.

In outdoor weathering, samples are usually motionless. However, in the so-called EMMA or EMMAQUA devices, the samples are moving with the sun to ensure continuous vertical sunlight. Through additional focusing of the incoming light, "accelerated outdoor weathering" is achieved, used so far mostly in the USA [175 to 177].

The considerations in section 3.2.1 show photooxidative degradation to be heavily dependent on the season. Therefore, the time of the beginning of outdoor weathering may influence the results considerably, especially for short-lived articles, and for this reason it should always be mentioned.

3.4.3 Influence of pigments on light stabilizer performance

The first testing of light stabilizers is usually performed in an unpigmented substrate. Commercial plastics materials, however, are mostly colored. It follows that, on judging weathering stability it should be taken into account that pigments may influence the light stability of the stabilized plastics material in several ways. Pigments may act as sensitizers and thus reduce drastically the light stability of u.v. stabilized samples. This happens especially with some organic yellow and organic red pigments but also with inorganic pigments.

Some other pigments, e.g. phthalo blue and phthalo green, lead to significant increase in u.v. stability of thick sections, especially of stabilized samples. Some pigments may take over the role of the u.v. absorber and prevent the harmful u.v. radiation from reaching the deeper layers of the plastics materials.

Sample temperature too is strongly influenced by pigmentation. Surface temperature of exposed samples, for example, is dependent on the intensity of the radiation absorbed by the sample, on the temperature of the environment and on air circulation. Table 7 [178] shows the results of

Table 7. Surface temperatures of PVC samples containing different pigments. Weather conditions: cloudless sky, outdoor temperature in the sun: 39 °C, black panel temperature: 51 °C, latitude: 51.6° North, time: noon to 2 p.m., August 1973.

Pigmentation	Surface temperature (°C)
white	33
yellow	38
red	40
blue	41
green	43
grey	47
brown	49
black	50

measurements with PVC. It can be seen that surface temperature increases significantly when pigmentation is changed from white to grey or black. Because, as a general rule, oxidation reactions depend heavily on temperature, accelerated oxidation is to be expected in dark pigmented samples.

However, surface temperatures do not depend exclusively on pigmentation. The nature of the plastics material has also a pronounced effect. This can be deduced from the measurements summarized in table 8 [179]. As a matter of fact, the differences in surface temperatures with white and dark pigmented samples are much less pronounced with PMMA than for example with PVC. This has to be attributed to a significantly higher absorption of infrared radiation by PMMA compared to PVC.

Table 8. Surface temperature of different polymers. Outdoor exposure: air temperature 37 °C, black panel temperature 50 °C.

Polymer and pigmentation		Surface temperature (°C)
PVC	white	35.4
PVC	dark brown	46.1
PMMA	white	44.1
PMMA	black	49.1
PE	black	51.7

The values of the surface temperatures shown in tables 7 and 8 do not represent an upper limit. Even in moderate climatic zones, temperatures as high as 60 °C and more can be measured with dark pigmented samples. In tropical climates, surface temperatures up to 80 °C have been measured with dark pigmented samples. In unfavorable circumstances, e.g. with insufficient air circulation, even 90 °C have been found [180].

3.5 Technical aspects of light stabilization

In choosing an additive for a specific application several factors in addition to its protective power, play an important role. In this respect one may cite: physical form (liquid/solid, melting point), particle size, thermal stability, possible interaction with other additives and fillers which may eventually lead to discoloration of the substrate, toxicity (food packaging), volatility and, above all, compatibility with the plastics material considered.

The behavior of an additive with respect to migration after incorporation into a polymer is determined essentially by its solubility and rate of diffusion. Stabilizers are normally dissolved at the processing temperatures usually employed. However, their solubility limit may be exceeded on cooling and may lead to visually observable blooming. These processes are a function of the nature of the substrate considered as well as of its morphology.

Additive diffusion is primarily a consequence of the thermal motion of the polymer chains above the glass transition temperature, Tg, and of the related formation and disappearence of free volumes. If the chains are flexible and move easily, only small quantities of energy are necessary to move the polymer segments. With increasing orientation of the polymer chains or on cross-linking and with increasing crystallinity, the diffusion constant decreases. For this reason, the diffusion of additives is faster in low density polyethylene than in high density polyethylene [181]. Numerous studies confirm that the diffusion behavior of u.v. absorbers depends mainly on polymer structure and morphology and to a minor extent on additive structure [182, 183].

The solubility of additives in suitable model liquids may give already some hints to their probable behavior in plastics materials [184]. For practical use, compatibility should be checked in the substrate to be stabilized. An additive can be considered as compatible if during a long period of time no blooming or turbidity is observed at room temperature and at the elevated temperatures which may be encountered during projected use of the plastics material. However, even if significant blooming should occur, the migrated quantity usually does not amount to more than a few per cent of the additive originally incorporated into the polymer.

The importance of mobility for light stabilizer effectiveness is not established beyond doubt. In this respect, e.g. in polystyrene, no difference in u.v. stabilizing efficiency has been found between free, mobile 2-hydroxy-4-alkoxybenzophenone and copolymerized 2-hydroxy-4-alkoxy benzophenone [185]. Similar results have been found in polypropylene [186].

Among the light stabilizer classes available commercially today only a few may be used in a broad range of plastics. Thus, nickel compounds are used almost exclusively in polyolefins whereas polyvinyl chloride, e.g. is stabilized with u.v. absorbers only. Moreover, nickel-containing additives are also used as "dyesites" because they allow the dying and printing of polypropylene fibers with dyestuffs susceptible to complexation with metals.

3.5.1 Polypropylene

Polypropylene is especially sensitive towards u.v. radiation and this has to be taken into account for every application under consideration. Consequently, not only articles to be used outdoors have to be stabilized against u.v. light but, also items intended for indoor use.

The low stability of unstabilized polypropylene exposed outdoors manifests itself by loss of gloss, formation of surface crazes, chalking and breakdown of its mechanical properties. To be sure, the same phenomena are observed in light stabilized formulations too but with a more or less pronounced time lag.

The light stability of an article will be influenced also, more or less, by parameters other than the effectiveness of the added light stabilizer. Among these parameters the following can be mentioned: sample thickness, degree of orientation and crystallinity [187], pigmentation, presence of fillers and reinforcing agents and of other additives such as phenolic antioxidants and costabilizers [188].

For the light stabilization of polypropylene, representatives of the following light stabilizer classes are mainly used presently: 2-(2′-hydroxyphenyl)-benzotriazoles, 2-hydroxy-4-alkoxybenzophenones, nickel-containing light stabilizers, 3,5-di-tert.butyl-4-hydroxybenzoates as well as sterically hindered amines (HALS). Nickel-containing light stabilizers are used exclusively in thin sections such as tapes and films whereas all other classes may be used in thin and thick sections, however, u.v. absorbers have limited effectiveness in thin sections.

The stabilizers used in the application data relative to polypropylene (figure 10 and tables 9 to 15) will be designated by the following abbreviations:

AO-I = Pentaerythrityl tetrakis-[3-(3′,5′-di-tert.butyl-4′-hydroxyphenyl)-propionate]
AO-II = Calcium (3,5-di-tert.butyl-4-hydroxybenzyl)-monoethylphosphonate
Ph-I = Tris-(2,4-di-tert.butylphenyl)-phosphite
UVA-I = 2-(2′-Hydroxy-3′,5′-di-tert.butylphenyl)-5-chlorobenzotriazole

UVA-II = 2-(2′Hydroxy-3′-tert.butyl-5′-methylphenyl)-5-chloro-benzotriazole
UVA-III = 2-Hydroxy-4-n-octoxybenzophenone
Ni-I = Nickel bis-(3,5-di-tert.butyl-4-hydroxybenzyl)-monoethylphosphonate
Ni-II = n-Butylamine-nickel-2,2′-thio-bis-(4-tert.octylphenolate)
Bz-I = 3,5-Di-tert.butyl-4-hydroxybenzoic acid-2′,4′-di-tert.-butylphenyl-ester
HALS-I = Bis-2,2,6,6-tetramethyl-4-piperidyl sebacate
HALS-II = Poly-(N-β-hydroxyethyl-2,2,6,6-tetramethyl-4-hydroxy-piperidyl succinate)
HALS-III = Poly-{[6-[(1,1,3,3-tetramethylbutyl)-imino]-1,3,5-triazine-2,4-diyl] [2-(2,2,6,6-tetramethylpiperidyl)-amino]-hexamethylene-[4-(2,2,6,6-tetramethylpiperidyl)-imino]}
HALS-IV = Di-(1,2,2,6,6-pentamethyl-4-hydroxypiperidyl)-butyl-(3′,5′-di-tert.butyl-4-hydroxybenzyl)-malonate

The effectiveness of the different light stabilizer classes in polypropylene tapes on accelerated exposure is shown in figure 10. It can be seen clearly that u.v. stability may be increased by factors between 2 and approx. 20 by using identical concentrations of different stabilizers [189, 190]. The outstanding performance of the sterically hindered amine is noteworthy (curve 5). Even at half the concentration it is definitely superior to the other light stabilizers. Similar results are obtained on outdoor exposure in Florida (table 9).

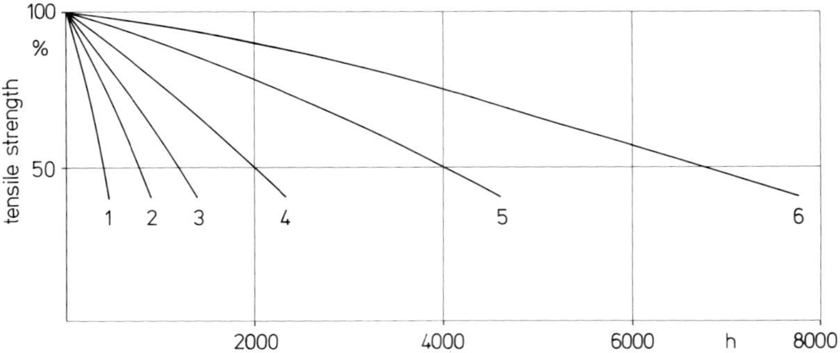

Fig. 10. Efficiency of light stabilizers in 30 μm thick polypropylene tapes in Xenotest 150 [3]; antioxidant: 0.1% AO-I, *1:* without light stabilizer, *2:* 0.5% UVA-I, *3:* 0.5% Ni-I, *4:* 0.5% Bz-I, *5:* 0.25% HALS-I, *6:* 0.5% HALS-I (abbreviations: see p. 150–151).

The u.v. stabilizing efficiency of the sterically hindered amine HALS-I in PP tapes is very pronounced even at rather low concentrations (table 10). This holds for accelerated exposure in the Weather-O-Meter as well as for outdoor weathering in Florida: in the whole concentration range considered, the performance increases significantly with concentration

Table 9. Light stability of polypropylene tapes (30 μm) on weathering in Arizona, 45° South, direct (790 $kJ \cdot cm^{-2} \cdot year^{-1}$), started May 1975 [162].
Base stabilization: 0.1% AO-I.
Abbreviations: see p. 150–151.

Light stabilizers	Energy received ($kJ \cdot cm^{-2}$) till 50% loss of tensile strength
Control	80
0.5% UVA-I	170
0.5% Ni-I	210
0.5% Bz-I	440
0.25% HALS-I	1020

Table 10. Light stability of polypropylene tapes (50 μm).
Florida exposure: 45° South, direct (580 $kJ \cdot cm^{-2} \cdot year^{-1}$), started March 1978 [191].
Base stabilization: 0.1% Ca-stearate + 0.05% AO-I + 0.05% Ph-I.
Abbreviations: see p. 150–151.

Light stabilizer	Time (h) or energy ($kJ \cdot cm^{-2}$) to 50% retained tensile strength			
	Weather-O-Meter exposure		Florida exposure	
	Unpigmented	0.4% TiO_2 (Rutile)	Unpigmented	0.4% TiO_2 (Rutile)
Control	495	750	110	130
0.05% HALS-I	1500	5180	270	730
0.10% HALS-I	2760	8620	690	1270
0.15% HALS-I	4140	–	930	2030
0.30% HALS-I	6720	–	1930	2640
0.60% HALS-I	11300	–	2780	>3140
1.20% HALS-I	16050	–	>3140	>3140

of HALS-I. The light stability observed with the white pigmented tapes is even superior to that of the unpigmented tapes: white pigmented tapes stabilized with 0.6% HALS-I show still more than 50% of their initial tensile strength after more than 5 years outdoor exposure in Florida.

A polymeric sterically hindered amine, HALS-II, developed recently, does not reach completely the performance of the low molecular weight HALS-I. It is nevertheless superior to the other common light stabilizers used at a threefold concentration (table 11). This superiority is again

Table 11. Light stability of polypropylene tapes (50 μm) [191].
Florida exposure: 45° South, direct (580 $kJ \cdot cm^{-2} \cdot year^{-1}$), started November 1978.
Base stabilization: 0.1% Ca-stearate + 0.05% AO-I + 0.05% Ph-I.
Abbreviations: see p. 150–151.

Light stabilizer	Energy received ($kJ \cdot cm^{-2}$) to 50% tensile strength retention	
	Unpigmented	0.4% TiO_2 (Rutile)
Control	180	200
0.3% UVA-II	280	230
0.3% UVA-III	330	200
0.3% Ni-II	300	250
0.1% HALS-II	360	480

more marked in white pigmented tapes (table 11). An additional advantage of HALS-II is its broad food approval (BGA, FDA, etc.).

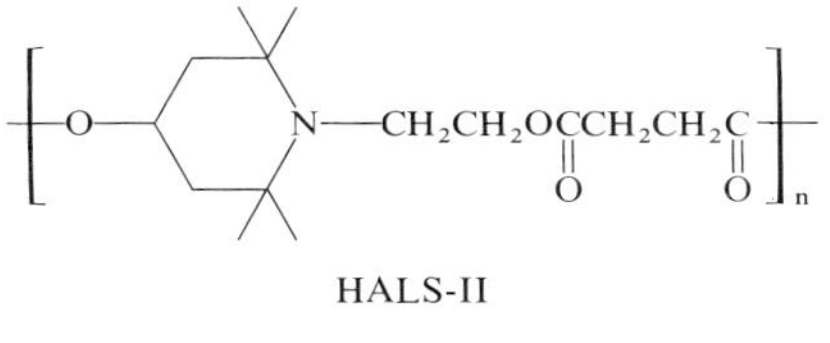

HALS-II

The superior performance of the sterically hindered amines in comparison with other light stabilizers is as remarkable in PP fibers as in PP tapes (table 12). In special applications, for example in PP fine fibers,

Table 12. Light stability of PP multifilaments (120/12) in Weather-O-Meter WRC 600 (without water spraying) [3]. Base stabilization: 0.1% AO-I.
Draw ratio: 1:5.6.
Abbreviations: see p. 150–151.

Light stabilizer	Hours to 50% tensile strength retention
Control	250
0.5% UVA-I	500
0.5% Ni-II	820
0.5% Bz-I	980
0.5% HALS-I	3500

the relative volatility and the limited migration and extraction resistance of low molecular weight HALS may cause a reduction of performance. For such applications it is advantageous to use the polymeric HALS-II and HALS-III. It can be seen in table 13 that the polymeric HALS-II and HALS-III and the low molecular weight HALS-I and HALS-IV give comparable performances in PP fibers so that if low volatility and extraction resistance are not specially requested they may be used interchangeably.

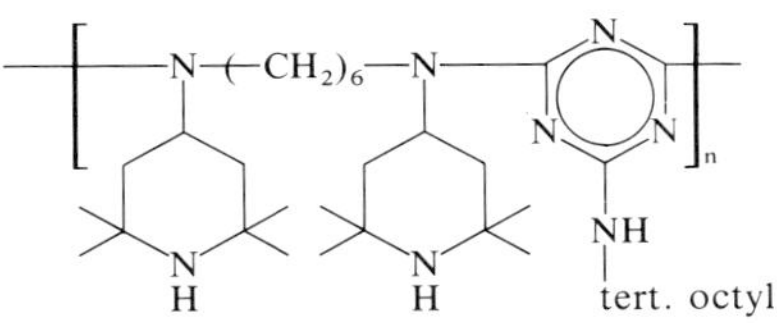

HALS-III

Table 13. Light stability of PP multifilaments (130/37) on Xenotest 1200 exposure (without water spraying) [191].
Base stabilization: 0.1% Ca-stearate + 0.05% AO-II.
Pigment: 0.25% TiO_2 (Rutile). Draw ratio: 1:3.2.
Abbreviations: see p. 150–151.

Light stabilizer concentration	Hours to 50% tensile strength retention				
	Control	HALS-I	HALS-IV	HALS-II	HALS-III
Control	490	–	–	–	–
0.15%	–	3200	2200	2800	2600
0.30%	–	4400	3200	3800	3800
0.60%	–	6600	4400	5100	5200

Table 14. Light stability of injection-molded PP plaques (2 mm) on Florida exposure, 45° South, direct (580 $kJ \cdot cm^{-2} \cdot year^{-1}$), started May 1980 [191].
Base stabilization: 0.1% Ca-stearate + 0.05% AO-I + 0.05% Ph-I.
Impact strength with Dynstat apparatus (DIN 51 230).
Abbreviations: see p. 150–151.

Light stabilizer	Energy received ($kJ \cdot cm^{-2}$) till 50% of the samples break					
	Unpigmented	0.5% TiO_2 (Rutile)	0.5% Chromium yellow	0.5% Cadmium red	0.5% Phthalo green	0.5% Phthalo blue
Control	100	100	100	120	210	210
0.3% UVA-III	190	150	250	250	210	210
0.1% HALS-I	520	940	940	730	>1 570	>1 420
0.05% HALS-II	170	190	250	210	>1 570	>1 570
0.10% HALS-II	210	480	730	480	>1 570	>1 570
0.20% HALS-II	250	730	1 360	1260	>1 570	>1 570
0.05% HALS-II +0.05% UVA-II	250	250	860	290	>1 570	>1 570
0.10% HALS-II +0.10% UVA-II	480	420	1 360	670	>1 570	>1 570

As already noted, pigments may exert a strong influence on the light stability of polymers. This can be seen in table 14 showing the results of outdoor exposure in Florida of injection-molded PP plaques containing various pigments. Large differences in u.v. stability are found on changing from unpigmented to white, yellow, red, green and blue pigmented samples (see 3.4.3). With the u.v. absorber UVA-III (benzophenone type), titanium dioxide has a slightly negative effect whereas chromium yellow and cadmium red show a slight positive contribution. With phthalo green and phthalo blue, the u.v. absorber UVA-III does not show an effect anymore, presumably its role has been overtaken by the pigments. In combination with the low molecular weight HALS-I and the polymeric HALS-II all the pigments used in the study give a significant positive contribution to light stability. The low molecular weight HALS-I shows a better performance than the polymeric HALS-II in the pigmented samples too, but its superority is much less evident than in the unpigmented samples and depends on the pigment. Whereas little influence of HALS-II concentration can be seen with the unpigmented plaques, with titanium dioxide, chromium yellow and cadmium red the

HALS-II concentration-effect is very pronounced. It is too early to draw any conclusions regarding phthalo blue and phthalo green.

The performance of the polymeric HALS-II in unpigmented samples may be improved significantly by combination with a u.v. absorber (table 14). This combination shows also interesting effects in some pigmented samples. As a matter of fact, whereas titanium dioxide does not give any contribution to the stabilizing efficiency of the combination HALS-II/UVA-II, cadmium red gives a slight improvement and chromium yellow an outstanding one. With chromium yellow, the combination HALS-II/UVA-II (1:1) is almost equivalent to the low molecular weight HALS-I used at the same concentration as the combination.

Not only light stabilizers and pigments may influence polymer light stability. The effect of other factors such as, for example, polymer quality and sample preparation should not be underestimated. This can be deduced from a comparison of the results in table 14 with corresponding results of a second test series (table 15) in a different polymer batch: With everything else comparable (sample preparation, exposure, light stabilizer concentrations etc.) the light stability conferred by HALS-I to the unpigmented samples in the second test series is twice that found in the first series. A similar result is observed in the white-pigmented samples.

The influence of different climatic zones may also be seen in table 15. Exposure of the samples stabilized with HALS-I is still going on but the comparison of the u.v. stabilities reached in central Europe and in

Table 15. Light stability of injection-molded PP plaques (2 mm) on exposure in Florida (45° South, direct, 580 $kJ \cdot cm^{-2} \cdot year^{-1}$, started October 1978) and in Basel (45° South, direct, 290 $kJ \cdot cm^{-2} \cdot year^{-1}$, started September 1978) [191].
Base stabilization: 0.1% Ca-stearate + 0.05% AO-I + 0.05% Ph-I.
Impact strength with Dynstat apparatus (DIN 51230).
Abbreviations: see p. 150–151.

Light stabilizer	Energy received ($kJ \cdot cm^{-2}$) till 50% of the samples break			
	Florida		Basel	
	Unpigmented	0.5% TiO_2 (Rutile)	Unpigmented	0.5% TiO_2 (Rutile)
Control	80	100	190	190
0.3% UVA-III	290	170	420	290
0.1% HALS-I	1340	>2170	>1170	>1170

Florida with the controls and the UVA-III stabilized samples shows already clearly that the samples last longer in the temperate climate not only considering the time involved but also with respect to the energy received. This may be due essentially to a smaller proportion of short-wavelength u.v. light and to a lower mean temperature in central Europe.

Already in the early years of HALS development it was found [192 to 194] that sulfur-containing additives such as dilauryl thiodipropionate (DLTDP) or distearyl thiodipropionate (DSTDP) may have a negative influence on the performance of sterically hindered amines. Therefore, it is advisable to avoid using sulfur-containing antioxidants or thiosynergists in combination with HALS or to check the possible negative influence in preliminary tests. The antagonism of sulfur compounds has been attributed to the reaction of sulfenyl radicals with the nitroxyl radicals considered as responsible for the stabilizing activity of HALS, thus producing inactive sulfonamides [195].

3.5.2 Polyethylene

As all polyolefins, the different polyethylene grades are sensitive to u.v. radiation, although less than polypropylene. For outdoor use they need special stabilization against u.v. light. The light stabilizers are in principle the same as for polypropylene.

The stabilizers considered in the application data relative to polyethylene (tables 16 to 26) will be designated by the following abbreviations:

AO-III = n-Octadecyl 3-(3′,5′-di-tert.butyl-4′-hydroxyphenyl)-propionate
AO-IV = 2,6-Di-tert.butyl-4-methylphenol
Ni-II = n-Butylamine-nickel-2,2′-thio-bis-(4-tert.octylphenolate)
UVA-II = 2-(2′-Hydroxy-3′-tert.butyl-5′-methylphenyl)-5-chlorobenzotriazole
UVA-III = 2-Hydroxy-4-n-octoxybenzophenone
UVA-IV = N,N′-(2-ethyl-2′ethoxy-5′-tert.butylphenyl)-oxalamide
HALS-I = Bis-2,2,6,6-tetramethyl-4-piperidyl sebacate
HALS-II = Poly-(N-β-hydroxyethyl-2,2,6,6-tetramethyl-4-hydroxy-piperidyl succinate)
HALS-III = Poly-{[6-[(1,1,3,3-tetramethylbutyl)-imino]-1,3,5-triazine-2,4-diyl] [2-(2,2,6,6-tetramethylpiperidyl)-imino-hexamethylene-[4-(2,2,6,6-tetramethylpiperidyl)-imino]}

3.5.2.1 High density polyethylene

High density polyethylene (HDPE) is used essentially for blow molding, injection molding and extrusion. Table 16 shows the results of Weather-O-Meter and Florida exposure of unpigmented and white-pigmented HDPE tapes. On accelerated weathering, the sterically hindered amines show much better performance in the unpigmented tapes than UVA-III, despite the much higher concentration of the latter. The comparison between the hindered amines is in favor of the polymeric HALS-III. This is also valid for the white-pigmented samples for which the performance data are on a somewhat higher level.

The results of Florida outdoor exposure confirm the data of accelerated exposure. The improvement factors, however, are more modest. Again HALS-III shows a distinct superiority over the other light stabilizers tested and has the same performance when added at a concentration of 0.05% as HALS-I and HALS-II at 0.1%.

As the amount of radiation absorbed by a u.v. absorber is related to its concentration and extinction coefficient as well as to the sample thick-

Table 16. Light stability of *Ziegler* polyethylene tapes (50 μm) on Weather-O-Meter and Florida exposure, 45° South, direct (580 $kJ \cdot cm^{-2} \cdot year^{-1}$), started November 1980 [196, 197].
Base stabilization: 0.1% Ca-stearate + 0.05% AO-III; draw ratio: 1:8.5.
Abbreviations: see p. 157.

Light stabilizer	Exposure time (h) or energy ($kJ \cdot cm^{-2}$) to 50% tensile strength retention			
	Weather-O-Meter (without water spraying)		Florida	
	Unpigmented	0.4% TiO_2 (Rutile)	Unpigmented	0.4% TiO_2 (Rutile)
Control	945	1025	410	390
0.05% HALS-I	4280	4560	540	750
0.10% HALS-I	6500	9250	880	1090
0.05% HALS-II	2920	3770	630	800
0.10% HALS-II	4370	6100	770	940
0.05% HALS-III	5850	7200	920	1030
0.10% HALS-III	10200	10700	1150	1440
0.30% UVA-III	1690	1600	470	420

Table 17. Light stability of injection-molded *Ziegler* polyethylene plaques (2 mm) on Arizona exposure, 45° South, direct (790 $kJ \cdot cm^{-2} \cdot year^{-1}$), started: September 1973 [191].
Base stabilization: 0.03% AO-III + 0.003% AO-IV.
Abbreviations: see p. 157.

Light stabilizer	Energy received ($kJ \cdot cm^{-2}$) to 50% loss of impact strength (DIN 53448)					
	Unpigmented	0.5% TiO_2 (Rutile)	0.5% Cadmium-yellow	0.5% Iron oxide red	0.25% Phthalo-green	0.25% Phthalo-blue
Control	170	210	150	250	400	520
0.3% UVA-III	670	820	> 840	920	750	860
0.3% UVA-IV	650	730	690	750	730	> 750
0.3% HALS-I	940	>2090	>2090	>2090	>2090	>2090

ness by the *Beer-Lambert* law, a better performance can be expected from u.v. absorbers if sample thickness is significantly above that of tapes (50 μm). This is exactly what is found. The results obtained with pigmented, injection-molded plaques of *Ziegler*-type HDPE on outdoor exposure in Arizona are summarized in table 17. It can be seen that

Table 18. Light stability of injection-molded *Phillips* polyethylene plaques (2 mm) on Weather-O-Meter and Florida exposure, 45° South, direct (580 $kJ \cdot cm^{-2} \cdot year^{-1}$), started August 1978 [196, 197].
Base stabilization: 0.06% AO-III; unpigmented (dumbbells cut out after exposure).
Abbreviations: see p. 157.

Light stabilizer	Hours or energy ($kJ \cdot cm^{-2}$) to 50% retained elongation	
	Weather-O-Meter (102′ light exposure, 18′ water spraying)	Florida
Control	1200	230
0.30% UVA-II	6000	2510
0.15% HALS-I	16000	3640*)
0.15% HALS-II	13500	3640*)

*) extrapolated

the pigments used in this study already confer some protection to the samples without light stabilizer. This protection increases significantly on addition of the u.v. absorbers UVA-III and UVA-IV. Another increase in performance is found with the sterically hindered amine HALS-I. After approximately $2^1/_2$ years test duration, most of the pigmented samples still showed the initial tensile impact strength.

In *Phillips*-type HDPE too, outstanding protection efficiency is found (table 18). On Weather-O-Meter exposure the u.v. absorber gives an improvement of the u.v. stability by a factor of 5. The low molecular weight sterically hindered amine HALS-I and the polymeric hindered amine HALS-II, at half the concentration of the u.v. absorber, have again considerable performance advantage. On Florida exposure, the superiority of the UVA-II-containing samples over the control is even more pronounced and reaches a factor of 10. The HALS-stabilized samples are superior once again but the extent of this superiority is not as pronounced as on Weather-O-Meter exposure. In another test series, this time with white-pigmented *Ziegler*-type HDPE injection-molded plaques, significant differences between the HALS can be seen on Florida

Table 19. Light stability of injection-molded *Ziegler* polyethylene plaques (2 mm) on Florida exposure, 45° South, direct (580 $kJ \cdot cm^{-2} \cdot year^{-1}$), started July 1979 [196, 197].
Base stabilization: 0.1% Ca-stearate + 0.025% AO-III.
Pigmented: 0.5% TiO_2 (Rutile).
Abbreviations: see p. 157.

Light stabilizers	Energy received ($kJ \cdot cm^{-2}$) to 50% loss of tensile impact strength (DIN 53448)
Control	250
0.2% UVA-III	290
0.1% HALS-I*)	1460
0.1% HALS-II	1170
0.1% HALS-III	>2510

*) only 0.06% HALS-I found on analysis

exposure (table 19). Whereas the benzophenone-type u.v. absorber UVA-III gives almost no contribution to u.v. stability, the polymeric HALS-II and the low molecular weight HALS-I show improvement factors of approximately 5 and 6 respectively. HALS-III shows already a considerable superiority over HALS-I and HALS-II. Even if we take into account that the concentration of HALS-I found on analysis is considerably inferior to the theoretical one and that, therefore, the superiority of HALS-III over HALS-I may be only apparent, there is no doubt that the efficiency of HALS-III in HDPE thick sections is very pronounced and comparable to that observed in HDPE tapes.

The polymeric HALS-II may not be as effective in unpigmented *Ziegler*-type HDPE as in the presence of pigments when it is used alone [197]. The performance can be significantly improved to a level comparable with pigmented formulations on combining HALS-II with u.v. absorbers.

3.5.2.2 Low density polyethylene

Low density polyethylene (LDPE) is used essentially for film manufacturing, much less than HDPE for molding. The applications of LDPE films in construction (covers) and especially in agriculture (film tunnels, greenhouses) have grown considerably. Among the numerous commercial light stabilizers only a few are suitable for low density polyethylene. This is mainly due to the fact that most light stabilizers are not sufficiently compatible at the concentrations necessary for the required protection, i.e. they bloom more or less rapidly [198].

Initially, u.v. absorbers of the benzophenone and benzotriazole types were used to protect LDPE materials. With the development of nickel-quenchers significant improvement of the light stability of LDPE films has been achieved (table 20). For economical reasons, combinations with u.v. absorbers are mainly used. The service life of such films may be increased by raising the concentrations of the light stabilizers.

Table 20. Light stability of low density polyethylene blown film (200 μm) on Florida exposure, 45° South, direct (580 $kJ \cdot cm^{-2} \cdot year^{-1}$), started April 1977 [191].
Base stabilization: 0.03% AO-III.
Film backing: Plexiglass.
Abbreviations: see p. 157.

Light stabilizer concentration	Energy received ($kJ \cdot cm^{-2}$) to 50% retained elongation			
	Control	UVA-III	Ni-II	Ni-II + UVA-III (1:1)
Control	150	–	–	–
0.15%	–	300	–	–
0.30%	–	510	670	–
0.60%	–	630	940	1150
1.20%	–	–	1420	1460

Table 21. Light stability of low density polyethylene blown film (200 μm) on Florida exposure, 45° South, direct (580 $kJ \cdot cm^{-2} \cdot year^{-1}$), started March 1978 [191].
Base stabilization: 0.03% AO-III.
Film backing: Plexiglass.
Abbreviations: see p. 157.

Light stabilizer concentration	Energy received ($kJ \cdot cm^{-2}$) to 50% retained elongation			
	Control	UVA-III	Ni-II	HALS-II
Control	150	–	–	–
0.15%	–	270	420	630
0.30%	–	500	670	1000
0.60%	–	670	1090	1570

Additive contents close to 2% may be found in greenhouse films thought to last up to 3 years outdoors (3 star films, quality mark from the Centre d'Etudes des Matières Plastiques, Paris).

With the development of sterically hindered amines, a further step forward could be expected in LDPE u.v. stability, similar to that already achieved in PP and HDPE. However, LDPE compatibility of the sterically hindered amines available in the early years was insufficient so that relatively poor performance was achieved, especially on outdoor weathering. It is only with the development of polymeric HALS that these difficulties have been overcome [196, 199].

The performance of the polymeric HALS-II in LDPE blown film is shown in table 21. Within the whole concentration range under consider-

Table 22. Light stability of low density polyethylene blown film (200 μm) on Florida exposure, 45° South, direct (580 $kJ \cdot cm^{-2} \cdot year^{-1}$), started March 1979 [197].
Base stabilization: 0.03% AO-III.
Film backing: Plexiglass.
Abbreviations: see p. 157.

Light stabilizer concentration	Energy received ($kJ \cdot cm^{-2}$) to 50% retained elongation		
	Control	HALS-II	HALS-II + UVA-III (1:1)
Control	180	–	–
0.15%	–	640	–
0.30%	–	880	1170
0.60%	–	1210	1510

ation HALS-II is significantly superior to the u.v. absorber UVA-III and to the Ni-stabilizer Ni-II. The same performance as with Ni-II can be achieved by using half the corresponding amount of HALS-II.

The combinations of the polymeric HALS-II with a u.v. absorber lead to significant improvement of the efficiency in comparison with the HALS used alone at the same concentration as the combination (table 22). This time, the advantages of the combination are not only economical.

A further boost of the performance can be achieved through use of HALS-III. As a matter of fact, the performance of HALS-II can be reached by using HALS-III at half the concentration and even less than that at the higher concentrations (table 23). The superiority of HALS-III becomes even more pronounced if film thickness is below 200 μm (table 24).

Table 23. Light stability of low density polyethylene blown film (200 μm) on Florida exposure, 45° South, direct (580 $kJ \cdot cm^{-2} \cdot year^{-1}$), started March 1979 [196, 197].
Base stabilization: 0.03% AO-III.
Film backing: Plexiglass.
Abbreviations: see p. 157.

Light stabilizer concentration	Energy received ($kJ \cdot cm^{-2}$) to 50% retained elongation				
	Control	UVA-III	Ni-II	HALS-II	HALS-III
Control	180	–	–	–	–
0.075%	–	–	–	480	670
0.15%	–	370	380	640	1260
0.30%	–	640	710	850	2010
0.60%	–	840	920	1210	>2400

The comparison of the data in tables 23 and 24 shows the influence of film backing: u.v. stability of all formulations considered is significantly lower on aluminum than on Plexiglass backing. From numerous data on u.v. stability of LDPE films on aluminum, Plexiglass or in absence of any backing, following relation has been deduced [199]:

$$E_{50\ \text{Plexiglass backing or without backing}} = 1.45\ E_{50\ \text{aluminum backing}}$$

where E_{50} stands for the energy received to 50% retained elongation.

Even if the influence of the backing is taken into account this way, it has to be pointed out that the values given in the tables have to

Table 24. Light stability of low density polyethylene blown film (200 μm); influence of film thickness on Florida exposure, 45° South, direct (580 $kJ \cdot cm^{-2} \cdot year^{-1}$), started May 1980 [196, 197].
Base stabilization: 0.03% AO-III.
Film backing: aluminum.
Abbreviations: see p. 157.

Light stabilizer	Energy received ($kJ \cdot cm^{-2}$) to 50% retained elongation as a function of film thickness		
	50 μm	100 μm	200 μm
Control	90	100	130
0.15% UVA-III	130	180	230
0.15% Ni-II	190	280	360
0.15% HALS-II	210	330	630
0.15% HALS-III	510	720	1090

be considered as upper limits of the stabilizing performance reached under "ideal" conditions. In commercial applications LDPE films used as greenhouse covers, for example, are exposed to additional influences such as film strain, wind pressure and possibly also air pollutants and agricultural chemicals (pesticides, insecticides) that may accelerate photooxidative degradation.

The results obtained with LDPE films may be quite different if they contain also mineral fillers (talcum, chalk, kaolin, etc.). Such fillers are used in LDPE films as opacifiers or infrared barriers. Preliminary results [214] point to a considerable detrimental effect of kaolin on the light stability of such films. The effect of chalk is much less pronounced [214].

A new type of polyethylene has gained importance in recent years, linear low density polyethylene (LLDPE). The results of Weather-O-Meter exposure of compression molded stabilized LLDPE films are summarized in table 25. In this polyethylene type too, the polymeric sterically hindered amines HALS-II and HALS-III show much better performance than the other light stabilizers tested. The blooming observed with the low molecular weight HALS-I is similar to that found with classical LDPE.

In more polar substrates such as e.g. ethylene-vinyl acetate copolymers (EVA) the low molecular weight HALS-I can be used, however, the polymeric HALS-II and HALS-III are significantly superior to the low molecular weight HALS in this substrate too (table 26).

Table 25. Light stability of linear low density polyethylene, compression-molded films (120 μm) [197].
Base stabilization: 0.03% AO-I.
Abbreviations: see p. 157.

Light stabilizer	Hours Weather-O-Meter (without water spraying) to 0.1 carbonyl absorbance
Control	600
0.15% UVA-III	2070
0.15% HALS-I	4000 (blooming)
0.15% HALS-II	5900
0.15% HALS-III	11100

Table 26. Light stability of an ethylene/vinyl acetate copolymer (14% VA); blown film (200 μm), Florida exposure, 45° South, direct (580 $kJ \cdot cm^{-2} \cdot year^{-1}$), started March 1981 [191].
Base stabilization: 0.03% AO-III.
Film backing: aluminum.
Abbreviations: see p. 157.

Light stabilizer concentration	Energy received ($kJ \cdot cm^{-2}$) to 50% retained elongation					
	Control	UVA-III	Ni-II	HALS-I	HALS-II	HALS-III
Control	170	–	–	–	–	–
0.15%	–	230	230	420	630	1420
0.30%	–	290	350	560	790	>1510
0.60%	–	420	690	730	>1510	>1510

3.5.3 Styrenic polymers

Styrenic plastics such as ABS (acrylonitrile/butadiene/styrene graft copolymers) and impact-resistant polystyrenes are very sensitive towards oxidation, essentially because of their butadiene content. Degradation on weathering starts at the surface and results in fast loss of the mechanical properties such as impact strength [200]. Because of the lack of efficient light stabilizers, ABS has not been used outdoors on a large scale.

In the application data relative to styrenic polymers (figures 11 and 12, table 27) the stabilizers considered will be used under the following abbreviations:

AO-III = n-Octadecyl 3-(3,′5′-di-tert.butyl-4′-hydroxyphenyl)-propionate
UVA-V = 2-(2′-Hydroxy-5′-methylphenyl)-benzotriazole
HALS-I = Bis-2,2,6,6-tetramethyl-4-piperidyl sebacate

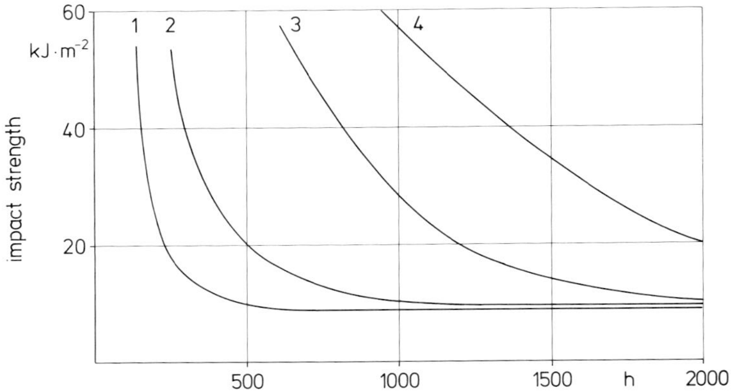

Fig. 11. Variation of impact strength of ABS test samples (unpigmented, small standard test bars according to DIN 53453) on Xenotest 450 exposure; antioxidant: 0.15% AO-III, *1:* without light stabilizer, *2:* 1% UVA-V, *3:* 1% HALS-I, *4:* 0.5% UVA-V + 0.5% HALS-I (abbreviations: see p. 166).

It can be seen in figure 11 that by combining two light stabilizers with different protection mechanisms it is possible to achieve good stabilization even in ABS [201]. Obviously, true synergism is being observed, the u.v. absorber UVA-V protecting the deeper layers whereas the sterically hindered amine HALS-I assures surface protection. This is especially important for ABS. At the same time discoloration of the polymer is reduced significantly. The same holds for impact-modified polystyrenes [201].

Yellowing is especially important for crystal polystyrene and styrene/acrylonitrile copolymers (SAN) in practical use. Hence, light stabilization is necessary for articles for which u.v. exposure can be expected (covers for fluorescent lights).

With u.v. absorbers, it is possible to improve significantly the light stability of crystal polystyrene, especially in comparison with antioxidants used alone. Monophenolic antioxidants have in fact been considered as stabi-

Table 27. Influence of the stabilization on the discoloration of injection-molded polystyrene plaques (2 mm) on exposure in a Weather-O-Meter WRC 600 [191]. Abbreviations: see p. 166.

Stabilization	Color development (yellowness index according to ASTM D 1925–70) after hours of exposure				
	0	1000	1500	2000	3000
Control	−1.4	18.9	24.8	29.4	32.1
0.2% UVA-V	−1.5	3.9	11.5	11.4	14.0
0.2% UVA-V + 0.15% AO-III	0.1	3.2	4.8	8.5	12.1
0.2% HALS-I	0.0	4.1	5.4	7.0	11.6
0.1% HALS-I + 0.1% UVA-V	−0.3	1.1	2.4	2.5	6.0
0.25% HALS-I + 0.25% UVA-V	−0.9	1.1	1.4	1.4	2.7

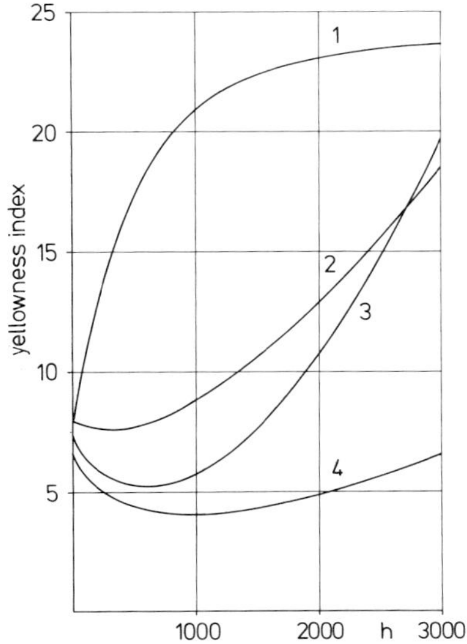

Fig. 12. Yellowing (yellowness index according to ASTM D 1925–70) of SAN copolymer (75:25) on Weather-O-Meter WRC 600 exposure; injection-molded 2 mm plaques; *1:* without light stabilizer, *2:* 0.3% HALS-I, *3:* 0.3% UVA-V, *4:* 0.25% HALS-I + 0.25% UVA-V (abbreviations see p. 166).

lizers retarding yellowing [202]. U.V. absorbers of the 2-(2′-hydroxyphenyl)-benzotriazole type are used almost exclusively in crystal polystyrene because of a reduced effect on initial color. The data in table 27 show that here too, the new system composed of the sterically hindered amine HALS-I and the u.v. absorber UVA-V yields significant performance increase in comparison with the previously used system comprising the phenolic antioxidant AO-III and the u.v. absorber UVA-V.

For light stabilization of SAN the same is valid in principle as for the styrenic polymers already considered, i.e. the best protection is attained with HALS/u.v. absorber combinations (figure 12).

3.5.4 Polyvinyl chloride

The light stability of polyvinyl chloride (PVC), not specially u.v. stabilized, is first a function of the thermal stabilizer used. Ba/Cd salts [203] and organic tin carboxylates, for example, confer already some u.v. stability to PVC on outdoor exposure [204] whereas sulfur-containing organic tin compounds contribute only moderately to light stabilization [205]. For transparent or translucent PVC articles requiring high u.v. stability, the light stability conferred by thermal stabilizers is not sufficient. The addition of light stabilizers is therefore mandatory [178, 206]. So far, u.v. absorbers yield the best results in practical use.

The stabilizers used in the application data relative to PVC (figures 13 to 15, tables 28 and 29) will be designated by the following abbreviations:

UVA-III	=	2-Hydroxy-4-n-octoxybenzophenone
UVA-V	=	2-(2′-Hydroxy-5′-methylphenyl)-benzotriazole
UVA-VI	=	2-(2′-Hydroxy-3′,5′-di-tert.butylphenyl)-benzotriazole
UVA-VII	=	2,2′-Dihydroxybenzophenone
UVA-VIII	=	2-Hydroxy-4-methoxybenzophenone
UVA-IX	=	N,N′-(2-ethyl-2′-ethoxyphenyl)-oxanilide
UVA-X	=	2-Carbomethoxy-4′-methoxymethyl cinnamate
UVA-XI	=	2-Cyano-3-methyl-4′-methoxymethyl cinnamate
HALS-I	=	Bis-2,2,6,6-tetramethyl-4-piperidyl sebacate
Sn/S	=	Thermal stabilization with organic tin/sulfur stabilizer
Ba/Cd	=	Thermal stabilization with barium/cadmium stabilizer
Sn-carboxylate	=	Thermal stabilization with organic tin carboxylate

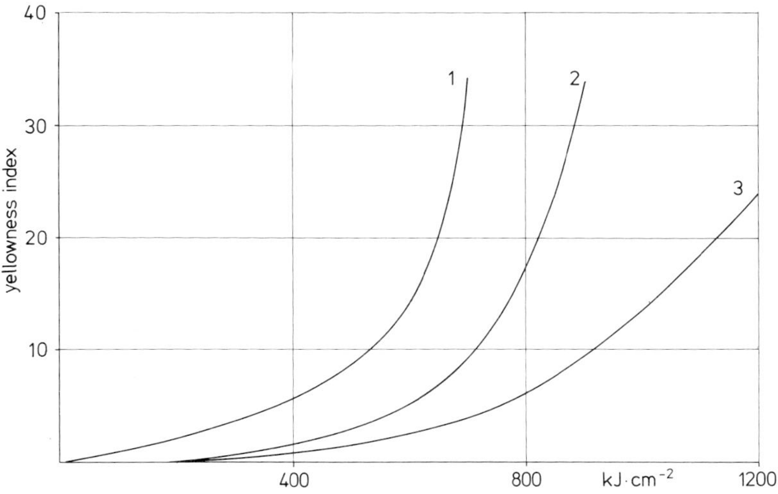

Fig. 13. Yellowing (yellowness index according to ASTM D 1925–70) of 0.5 mm thick rigid PVC sheets on outdoor exposure in Arizona, 45° South, direct (790 kJ · cm^{-2} · $year^{-1}$, started June 1972); formulation: 100 p. S-PVC (K-value 60), 2.5 p. organotin carboxylate, 2 p. epoxidized soy bean oil, 1 p. lubricant. *1:* without light stabilizer, *2:* 0.2% UVA-III or UVA-X or UVA-XI, *3:* 0.2% UVA-VI (abbreviations see p. 168).

The effectiveness of different light stabilizers in transparent rigid PVC observed on Xenotest 1200 exposure is summarized in table 28. The hydroxyphenylbenzotriazoles UVA-V and UVA-VI are superior to the hydroxybenzophenones UVA-VII and UVA-VIII and to the oxanilide UVA-IX. Table 28 shows also the influence of the light stabilizer concentration on their efficiency in reducing color formation. The sterically hindered amine included in the tests, HALS-I, has almost no effect. Similar results are obtained on outdoor weathering (figure 13).

The influence of thermal stabilizers on the performance of a hydroxyphenylbenzotriazole u.v. absorber on Xenotest exposure can be seen in table 29: on changing the thermal stabilization from the system Sn/S to Sn-carboxylate and Ba/Cd u.v. stability increases considerably. A doubling of the u.v. absorber concentration from 0.1 to 0.2% gives the highest relative increase in u.v. stability with the Sn/S stabilization but the improvement is also very pronounced with the Ba/Cd stabilization.

Table 28. Light stability of rigid PVC (1 mm, transparent) on Xenotest 1200 exposure [191].
Influence of light stabilizer class and concentration.
Thermal stabilization: Sn-carboxylate; yellowness index according to ASTM D 1925–70.
Abbreviations: see p. 168.

Light stabilizer	Hours to yellowness index increase of 20 units		
	Without	0.1% LS	0.2% LS
Control	600	–	–
UVA-V	–	3200	4000
UVA-VI	–	2900	4000
UVA-VII	–	2400	3100
UVA-VIII	–	2550	3200
UVA-IX	–	2400	2800
HALS-I	–	600	800

Table 29. Light stability of rigid PVC (1 mm, transparent) on Xenotest 1200 exposure [191].
Influence of thermal stabilization.
Yellowness index according to ASTM D 1925–70.
Abbreviations: see p. 168.

Light stabilizer	Hours to yellowness index increase of 20 units as a function of thermal stabilization		
	Sn/S	Sn-carboxylate	Ba/Cd
Control	560	600	1400
0.1% UVA-V	1880	3200	4850
0.2% UVA-V	3500	4000	6700

The influence of thermal stabilization of PVC is similar on outdoor weathering (figure 14).

For plasticized PVC, whose intrinsic properties may be influenced by the nature and the quantities of plasticizers used [207], the improvement obtained with the most efficient u.v. absorbers reaches a factor of 2 (figure 15).

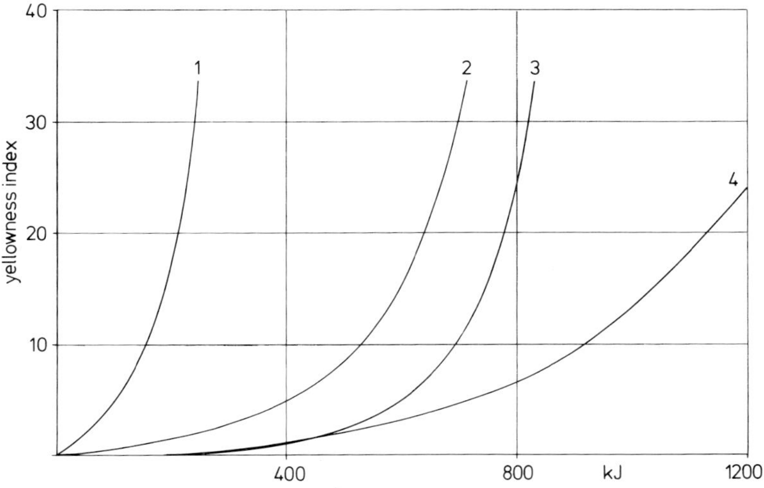

Fig. 14. Yellowing (yellowness index according to ASTM D 1925–70) of rigid PVC sheets on outdoor exposure in Arizona; influence of thermal stabilization; same conditions as in figure 13; *1:* 2.5 p. Sn/S, *2:* 2.5 p. Sn-carboxylate *3:* 2.5 p. Sn/S + 0.2% UVA-VI, *4:* 2.5 p. Sn-carboxylate + 0.2% UVA-VI (abbreviations see p. 168).

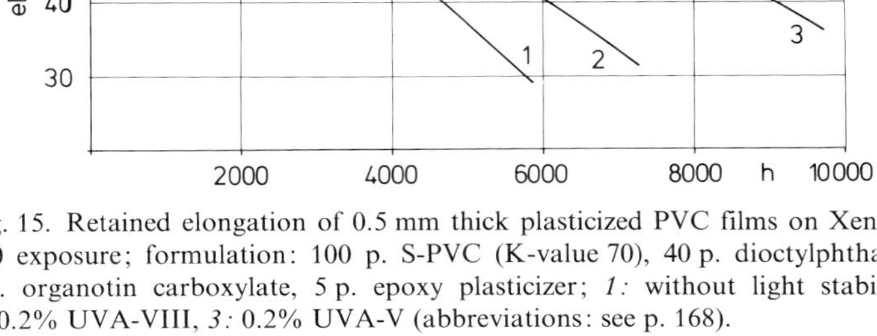

Fig. 15. Retained elongation of 0.5 mm thick plasticized PVC films on Xenotest 150 exposure; formulation: 100 p. S-PVC (K-value 70), 40 p. dioctylphthalate, 1 p. organotin carboxylate, 5 p. epoxy plasticizer; *1:* without light stabilizer, *2:* 0.2% UVA-VIII, *3:* 0.2% UVA-V (abbreviations: see p. 168).

3.5.5 Polyacrylates

Polymethyl methacrylate (PMMA) is highly transparent in the u.v. region and thus much more light stable than other thermoplastics. Light stabilizers may by used, for example, to confer a u.v. filter effect to PMMA articles. Window panes for solar protection containing 0.05 to 0.2% 2-(2′-hydroxy-5′-methylphenyl)-benzotriazole are well known examples of this application. The rear lights of motor cars, electric signs and covers for fluorescent lights, among others, are important applications for which PMMA is u.v. stabilized.

The excellent light stabilizing performance of the sterically hindered amines has been recently confirmed also for PMMA: when PMMA is stabilized with HALS practically no molecular weight decrease is found on outdoor weathering as well as on accelerated testing [77].

In methyl methacrylate/acrylonitrile copolymers yellowing and deterioration of the mechanical properties occur faster than in PMMA because of the unfavorable influence of the acrylonitrile component. Yellowing on outdoor weathering can be reduced considerably by addition of u.v. absorbers [208].

Copolymerization of methyl methacrylate with methyl acrylate gives a material with even better resistance towards weathering than PMMA [77]. This is attributed to a lower residual monomer content in the copolymer compared to the homopolymer.

3.5.6 Polyurethanes

Light stability of polyurethanes (PUR) depends to a large extent on their chemical structure, bearing in mind that both components, i.e. isocyanate and polyol, have an influence. Polyurethanes based on aliphatic isocyanates and polyester diols show the best light stability if yellowing is considered, whereas polyurethanes based on aromatic isocyanates and polyether diols are worst in this respect. It has been seen previously (see 3.2.2.7) that this does not necessarily hold for the changes in mechanical properties.

Light stabilizers are used mainly in the coatings industry (textile coatings, synthetic leather). Loss of elongation and yellowing are important test criteria for polyurethanes based on aromatic isocyanates. Because high concentrations of light stabilizers are necessary to reach a sufficient level of u.v. stability, checking compatibility on testing the corresponding formulations is mandatory if blooming is to be avoided later on.

The stabilizers used with the application data relative to PUR (tables 30 to 34) will be designated by the following abbreviations:

AO-I = Pentaerythrityl tetrakis-[3-(3′,5′-di-tert.butyl-4′-hydroxyphenyl)-propionate]
AO-III = n-Octadecyl 3-(3′,5′-di-tert.butyl-4′-hydroxyphenyl)-propionate
AO-V = 1,6-Hexanediol bis-3-(3′,5′-di-tert.butyl-4′-hydroxy-phenyl)-propionate
AO-VI = 2,2′-Thiodiethanol bis-3-(3′,5′-di-tert.butyl-4′-hydroxyphenyl)-propionate
AO-VII = 30% 2,4-Dinonylphenol + 20% 2,2′-Methylene-bis-(4,6-dinonylphenol) + 50% Di-isodecylphenylphosphite
AO-VIII = 3,6-Oxy-1,8-octane-diyl-bis-[3-(4-hydroxy-3-methyl-5-tert.butyl)-propionate]
UVA-V = 2-(2′Hydroxy-5′-methylphenyl)-benzotriazole
HALS-I = Bis-2,2,6,6-tetramethyl-4-piperidyl sebacate
HALS-II = Poly-(N-β-hydroxyethyl-2,2,6,6-tetramethyl-4-hydroxy-piperidyl succinate)
HALS-III = Poly{[6-[(1,1,3,3-tetramethylbutyl)-imino]-1,3,5-triazine-2,4-diyl][2-(2,2,6,6-tetramethylpiperidyl)-imino]-hexamethylene-[4-(2,2,6,6-tetramethylpiperidyl)-imino]}
HALS-V = Bis-1,2,2,6,6-pentamethyl-4-piperidyl sebacate

In addition to some u.v. absorbers of the 2-(2′-hydroxyphenyl)-benzotriazole type [209], sterically hindered amines used alone or in combination with benzotriazoles are especially effective stabilizers.

Table 30 shows that for retention of mechanical properties of polyetherurethane the system HALS-I/UVA-V is significantly superior to the sys-

Table 30. Light stability of aromatic polyetherurethane films (~50 μm) on Florida exposure behind glass, 45° South (580 kJ·cm^{-2}·year^{-1}) [191].
Abbreviations: see p. 173.

Stabilization	Days to 50% loss of	
	Tensile strength	Elongation
Control	21	21
1% UVA-V + 1% AO-I	14	35
0.5% HALS-I	50	100
1% HALS-I	50	>100 (63%)
0.5% HALS-I + 0.5% UVA-V	83	>100 (86%)

Table 31. Light stability of aromatic polyesterurethane films ($\sim$60 μm) on outdoor weathering in Basel, 45° South (290 $kJ \cdot cm^{-2} \cdot year^{-1}$) started July 1980 (behind glass), July 1981 (direct) and in Florida, 45° South (580 $kJ \cdot cm^{-2} \cdot year^{-1}$) started August 1980 (direct and behind glass) [191].
Test criterion: loss of transmission at 420 nm (yellowing).
Abbreviations: see p. 173.

Stabilization	Energy received ($kJ \cdot cm^{-2}$) to 15 units transmission loss			
	Basel		Florida	
	Direct	Behind glass	Direct	Behind glass
Control	10	20	10	20
0.5% UVA-V +0.5% AO-I	60	100	50	120
0.5% HALS-I	40	250	60	390
0.5% HALS-II	20	80	30	290
0.5% HALS-III	40	180	50	330
0.5% HALS-V	40	230	50	380

tem AO-I/UVA-V used previously. The same holds for polyesterurethane. The data in tables 31 and 32 demonstrate that the superiority of the low molecular weight and polymeric sterically hindered amines is observed principally on exposure behind glass. On *direct exposure* in a central European climate, less so in a tropical climate, sample discoloration (table 31) occurs much faster than loss of elongation (table 32). The results are different on *exposure behind glass*: in the central European climate, discoloration and loss of mechanical properties appear practically simultaneously, whereas in Florida, discoloration of the HALS-stabilized samples occurs significantly later than loss of elongation. The comparison of the *energies received* in Florida and in Basel to 50% loss of elongation or 15% loss of transmission is very interesting: contrary to corresponding results with polypropylene (significantly higher values of the received energy in Basel, see table 15) the values observed with PUR are comparable or significantly higher in Florida. The one exception in this respect concerns the loss of elongation on direct exposure: in this instance the Florida values are significantly below the Basel values. This observation points to the decisive influence of short-wavelength u.v. light on the loss of mechanical properties.

Table 32. Light stability of aromatic polyesterurethane films (~60 μm) on outdoor weathering in Basel, 45° South (290 $kJ \cdot cm^{-2} \cdot year^{-1}$) started July 1980 (behind glass), July 1981 (direct) and in Florida, 45° South ($kJ \cdot cm^{-2} \cdot year^{-1}$) started August 1980 (direct and behind glass) [191].
Test criterion: decrease in elongation at break.
Abbreviations: see p. 173.

Stabilization	Energy received ($kJ \cdot cm^{-2}$) to 50% retained elongation			
	Basel		Florida	
	Direct	Behind glass	Direct	Behind glass
Control	20	30	20	40
0.5% UVA-V +0.5% AO-I	190	100	50	120
0.5% HALS-I	180	210	70	280
0.5% HALS-II	110	100	50	130
0.5% HALS-III	250	190	60	270
0.5% HALS-V	200	210	70	200

Light stabilization of closed-cell polyurethane foams used for example for shoe sole compounds has gained importance recently, mainly for aesthetic reasons. This time the sterically hindered amines used alone are definitely inferior to the combination benzotriazole/phenolic antioxidant

Table 33. Light stability of polyesterurethane foam on Xenotest 150 exposure. Yellowness index according to ASTM D 1925–70, measured at the foam surface [191].
Abbreviations: see p. 173.

Stabilization	Hours to yellowness index increase of 20 units
Control	5
0.5% AO-I + 0.5% UVA-V	60
0.5% HALS-I	6
0.5% HALS-I + 0.2% AO-I	18
0.5% HALS-I + 0.2% UVA-V	15
0.5% HALS-I + 0.2% UVA-V + 0.2% AO-I	63
0.5% HALS-I + 0.5% UVA-V + 0.25% AO-I	155
0.5% HALS-I + 0.5% UVA-V + 0.5% AO-I	180

(table 33). The combination of HALS-I with the phenolic antioxidant AO-I or with the u.v. absorber UVA-V yields an increase in performance. However, it is only with the three-component system HALS/phenolic antioxidant/u.v. absorber that the effectiveness of the combination phenolic antioxidant/u.v. absorber can be matched. The best results finally are obtained with increasing concentration of the u.v. absorber in the three-component system. This is quite different from the results observed with PUR films: for the latter the best performance is achieved with HALS used alone or in combination with benzotriazole u.v. absorbers. For optimum light stability of PUR foams, the three-component stabilization system with an additional phenolic antioxidant is mandatory.

Table 34. Light stability of polyesterurethane foam on Xenotest 150 exposure; influence of phenolic antioxidants [191].
Yellowness index according to ASTM D 1925–70, measured at the foam surface.
Light stabilization: 0.5% HALS-V + 0.5% UVA-V.
Abbreviations: see p. 173.

Antioxidants 0.25%	Hours to yellowness index increase by 20 units
AO-VII	80
AO-VI	130
AO-III	140
AO-V	140
AO-I	180
AO-VIII	220

The results summarized in table 34 show that the nature of the phenolic antioxidant used for PUR foam stabilization also has a significant influence: u.v. stabilization remaining the same, variation in protection against yellowing can reach a factor of 3 depending on the phenolic antioxidant considered.

3.5.7 Polycarbonate

Insufficient light stability of polycarbonate (PC) on outdoor use is manifested by rapidly increasing yellowing. Until now, only u.v. absorbers are in use for practical stabilization. In choosing a u.v. absorber, intrinsic performance, adequate thermal stability at the elevated processing temperatures and effect on initial color of the polycarbonate articles should be considered.

The u.v. stabilizers used in conjunction with the polycarbonate application data (figure 16, table 35) will be designated by the following abbreviations:

Ph-II = Tris-nonylphenyl-phosphite
UVA-V = 2-(2′-Hydroxy-5′-methylphenyl)-benzotriazole
UVA-IX = N,N′-(2-ethyl-2′-ethoxyphenyl)-oxanilide
UVA-X = 2-Carbomethoxy-4′-methoxymethyl cinnamate
UVA-XII = 2-(2′-Hydroxy-5′-tert.octylphenyl)-benzotriazole
UVA-XIII = 2-(2′-Hydroxy-3′-sec.butyl-5′-tert.butylphenyl)-benzotriazole
UVA-XIV = 2-(2′-Hydroxy-3′,5′-di-isopropylphenyl-phenyl)-benzotriazole

Xenotest exposure of stabilized samples leads first of all to a decrease in discoloration. The time to reach again the original yellow shade (yellowing induction time) is a function of the nature and the concentration of the additive (figure 16). The best performance is found with 2-(2′-hydroxyphenyl)-benzotriazoles (figure 16). Table 35 shows the effectiveness

Table 35. Light stability of injection-molded polycarbonate plaques (2 mm) [191].
Base stabilization: 0.07% Ph-II.
Yellowness index according to ASTM D 1925–70.
Abbreviations: see p. 177.

Light stabilizer	Yellowing induction time hours Xenotest 1200 (without water spraying)
Control	0
0.3% UVA-V	3200
0.3% UVA-XII	3100
0.3% UVA-XIII	2900
0.3% UVA-XIV	2400

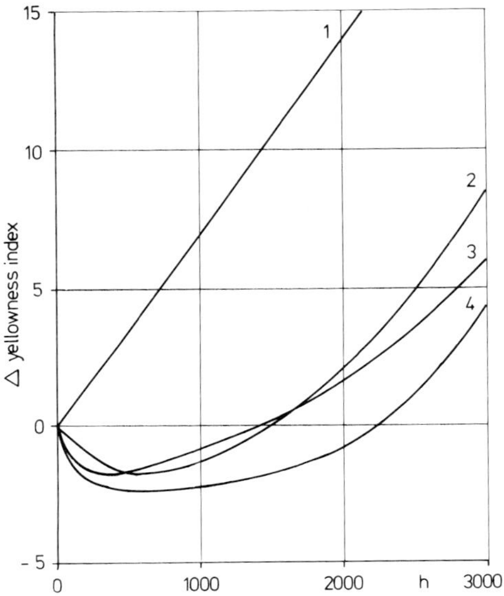

Fig. 16. Yellowing (yellowness index according to ASTM D 1925–70) of 2 mm thick injection-molded polycarbonate plaques on Xenotest 1200 exposure; *1:* without light stabilizer, *2:* 0.25% UVA-X, *3:* 0.25% UVA-IX, *4:* 0.25% UVA-V (abbreviations see p. 177).

of some benzotriazoles. Only minor differences exist in this respect. Therefore, the choice among the benzotriazoles available is often dictated by the application under consideration. For extrusion of plates and profiles, for example, low volatility is a main requirement to minimize deposits of light stabilizers on chill rolls and gaging rolls.

3.5.8 Polyacetal

Polyacetal (POM) cannot be used outdoors if it does not contain any light stabilizers. After a relatively short weathering period, surface crazes and pronounced chalking are observed.

The stabilizers considered in the application data relative to POM (tables 36 to 39) are designated by following abbreviations:

AO-V = 1,6-Hexanediol bis-3-(3′,5′-di-tert.butyl-4′-hydroxyphenyl)-propionate
AO-VIII = 3,6-Oxa-1,8-octane-diyl-bis-[3-(4-hydroxy-3-methyl-5-tert.butyl)-propionate]
AO-X = 2,2′-Methylene-bis-(4-methyl-6-tert.butylphenol)
HALS-II = Poly-(N-β-hydroxyethyl-2,2,6,6-tetramethyl-4-hydroxypiperidyl succinate)
HALS-V = Bis-1,2,2,6,6-pentamethyl-4-piperidyl sebacate
UVA-V = 2-(2′-Hydroxy-5′-methylphenyl)-benzotriazole
UVA-XIII = 2-(2′-Hydroxy-3′-sec.butyl-5′-tert.butylphenyl)-benzotriazole
UVA-XV = 2-(2′-Hydroxy-3′,5′-di-tert.amylphenyl)-benzotriazole

Carbon black (0.5 to 3%) is a good stabilizer for polyacetal when sample color is not important. Other possibilities for stabilization are the use of 2-hydroxybenzophenone and especially 2-hydroxyphenylbenzotriazole type u.v. absorbers (table 36) [210]. Sterically hindered amines impart

Table 36. Efficiency of light stabilizers in polyacetal copolymer [210].
Standard bars according to DIN 53455.
Xenotest 450, cycle: 29 min exposure to light followed by 1 min water spraying with light on. Outdoor weathering: central Europe, 45° South, direct.

Light stabilizer type	Time to 60% retained elongation	
	Hours Xenotest 450	Months outdoor weathering
Control	~ 600	~ 3
1.5% 2-Hydroxybenzophenone	~ 850	~ 4
1.0% 2-Hydroxyphenylbenzotriazole	>1000 (~65%)	~ 5.5
0.5% Carbon black	>1000 (~85%)	~36

to polyacetal a stability comparable to that observed with u.v. absorbers (table 37). With all types of stabilizers, doubling the concentration produces only a slight improvement in performance. Even the combination of HALS with benzotriazoles does not yield more than a modest increase in light stability (table 37). This is valid only on exposure to light without water spraying (table 38). On weathering (light exposure and water spray-

Table 37. Light stability of polyacetal copolymer on Xenotest 1200 exposure [191] (without water spraying); 1 mm compression-molded plaques; base stabilization: 0.3% Ca-stearate + 0.3% AO-V.
Abbreviations: see p. 179.

Light stabilizer	Hours to chalking
Control	1400
0.25% UVA-XIII	3800
0.25% UVA-V	4000
0.50% UVA-V	5000
0.25% HALS-II	3200
0.50% HALS-II	3200
0.25% HALS-V	3600
0.50% HALS-V	4000
0.25% HALS-II + 0.25% UVA-V	5000
0.25% HALS-V + 0.25% UVA-V	5000

Table 38. Light stability of polyacetal copolymer [191]; 1 mm compression-molded plaques; base stabilization: 0.3% Ca-stearate + 0.3% AO-V.
Abbreviations: see p. 179.

Light stabilizer	Hours to chalking	
	In Xenotest 1200 (without water spraying)	In Weather-O-Meter WRC 600 (102′ light exposure + 18′ water spraying)
0.25% UVA-XV	3800	1000
0.25% UVA-XV +0.25% HALS-II	4800	2400
0.25% UVA-XV +0.25% HALS-V	4800	3200

ing), the samples stabilized with combinations HALS/u.v. absorber are definitely superior to those containing only a u.v. absorber.

The data in table 39 show the influence of the phenolic antioxidant on weathering.

Table 39. Light stability of polyacetal copolymer [191]; 1 mm compression-molded plaques; 0.3% Ca-stearate.
Light stabilization: 0.25% UVA-XV + 0.25% HALS-II.
Abbreviations: see p. 179.

Antioxidant	Hours to chalking	
	In Xenotest 1200 (without water spraying)	In Weather-O-Meter WRC 600 (102′ light exposure + 18′ water spraying)
0.3% AO-V	4800	2400
0.3% AO-VIII	4800	2600
0.3% AO-X	4800	2000

3.5.9 Polyamides

There is extensive literature on the light stabilization of polyamide (PA) fibers whereas not much has been published on the behavior of polyamide injection moldings in outdoor use. The complexity of polyamide photooxidation processes has already been discussed in detail (see section 3.2.2.10). Other factors likely to influence polyamide aging are particularly humidity [95] and corrosive atmospheric impurities such as ozone and nitrogen dioxide [211]. Thus, it is not astonishing that aging data vary significantly with the location of the weathering site and the considered test criteria.

For the polyamide application data (tables 40 to 42) the abbreviations used for the stabilizers are the following:

AO-IX = N,N′-Hexamethylene-bis-[3-(3′,5′-di-tert.butyl-4′-hydroxyphenyl)-propionamide]
Ph-I = Tris-(2,4-di-tert.butylphenyl)-phosphite
HALS-I = Bis-2,2,6,6-tetramethyl-4-piperidyl sebacate
HALS-II = Poly-(N-β-hydroxyethyl-2,2,6,6-tetramethyl-4-hydroxy-piperidyl succinate)
HALS-IV = Di-(1,2,2,6,6-pentamethyl-4-hydroxypiperidyl)-butyl-(3′,5′-di-tert.butyl-4-hydroxybenzyl)-malonate
UVA-I = 2-(2′-Hydroxy-3′-5′-di-tert.butylphenyl)-5-chloro-benzotriazole

Table 40. Light stability of injection-molded polyamide-6 dumbbells (1 mm). Outdoor exposure in Arizona, 45° South, direct (790 $kJ \cdot cm^{-2} \cdot year^{-1}$), started February 1980; Basel 45° South, direct (290 $kJ \cdot cm^{-2} \cdot year^{-1}$), started November 1979 [191].
Abbreviations: see p. 181.

Stabilization	Time (hours) or energy received ($kJ \cdot cm^{-2}$) to 70% tensile strength		
	Xenotest 1200 (without water spraying)	Arizona (direct)	Basel (direct)
Control	750	290	110
95 ppm $Cu^{2\oplus}$ + 1830 ppm $J^{\ominus}$	9750	370	120
0.5% AO-IX	5000	350	150
0.5% HALS-I	10250	>420	120
0.5% AO-IX + 0.5% HALS-I	10000	>420	160
0.5% AO-IX + 0.5% UVA-I	4850	390	210

The data in table 40 obtained with PA-6 on accelerated exposure and outdoor weathering are rather puzzling. On Xenotest 1200 exposure to 70% retained tensile strength all stabilized samples show a remarkable superiority over the control. This holds particularly for the samples stabilized with copper or with HALS which exhibit approximately twice the light stability of the samples stabilized with a phenolic antioxidant or

Table 41. Light stability of injection-molded polyamide-6.6 dumbells (1 mm). Outdoor weathering in Basel, 45° South, direct (290 $kJ \cdot cm^{-2} \cdot year^{-1}$), started May 1979 [191].
Abbreviations: see p. 181.

Stabilization	Time (hours) or energy received ($kJ \cdot cm^{-2}$) to 50% retained elongation	
	Xenotest 450	Basel
Control	1650	150
0.2% AO-IX	2800	230
0.4% AO-IX	5250	270
0.2% AO-IX + 0.2% Ph-I	9200	>420
0.2% or 0.4% HALS-I	7650	220

the combination of the latter with a u.v. absorber. On outdoor weathering in Florida and Basel this superiority vanishes almost completely.

U.V. stabilization of PA-6.6 presents some pecularities compared to that of PA-6. The data in table 41 show that the phenolic antioxidant AO-IX is already conferring good light stability on Xenotest 450 exposure. The sterically hindered amine HALS-I is significantly superior but no increase in u.v. stability is observed when its concentration is raised from 0.2 to 0.4%. The combination of the phenolic antioxidant AO-IX with the phosphite Ph-I gives the best performance in this test. This is comfirmed on outdoor exposure. The efficiencies of AO-IX and of HALS-I used without coadditive are, however, considerably reduced outdoors and are practically undistinguishable.

Table 42. Light stability of injection-molded polyamide-12 dumbbells (1 mm) [191]. Abbreviations: see p. 181.

Stabilization	Hours Xenotest 450 to	
	50% retained tensile strength	50% retained elongation
Control	3400	2800
0.2% AO-IX	5600	4600
0.1% AO-IX + 0.1% Ph-I	5300	4500
0.2% HALS-IV	>17500	>18000
0.2% HALS-II	>17500	14800
0.2% or 0.4% HALS-I	>17500	17800

The behavior of PA-12 is much different. The data in table 42 indicate that the phenolic antioxidant AO-IX and its combination with the phosphite Ph-I result in about the same light stability on exposure in a Xenotest 450. U.V. stability of the samples stabilized with sterically hindered amines reaches a multiple of it. An increase in concentration from 0.2 to 0.4% brings no change in stability. After about $1\,^1/_2$ years outdoor weathering in Basel (direct exposure) only the unstabilized control had shown loss of mechanical properties, the stabilized samples all retained their initial elongation and tensile strength.

3.5.10 Polyesters

Polyethylene terephthalate and polybutylene terephthalate are intrinsically more light stable than polyurethanes or polyamides. However, after prolonged weathering, yellowing and brittleness of the material may be observed. Because surface deterioration predominates, the deeper polymer layers are more or less unattacked. U.V. absorbers are efficient stabilizers for polyethylene terephthalate films on outdoor exposure [212].

3.6 List of trade names and light stabilizer manufacturers

The list of trade name and manufacturers does not claim to be complete. More data can be found in: *T.J. Henman*, World Index of Polyolefine Stabilizers, Kogan Page Ltd., London, 1982, and in Modern Plastics Encyclopedia, McGraw-Hill Inc., New York, 1982.

The following key has been used to relate trade names to light stabilizer classes:

Benzophenones (A), benzotriazoles (B), nickel compounds (C), salicylates (D), cyanocinnamates (E), benzylidene malonates (F), benzoates (G), oxanilides (H), sterically hindered amines (I) and polymeric sterically hindered amines (K).

Trade name	Compound type	Manufacturer
Aduvex	A	Ward Blenkinsop and Co. Ltd., Wembley (Middx.), GB
Anti-UV	A, D	Société Française D'Organo Synthèse, Neuilly-sur-Seine, F
Carstab	A	Cincinnati Milacron Chemicals Inc., New Brunswick (N.Y.), USA
Chimassorb	A, C, K	Chimosa S.p.A, 40044 Pontecchio Marconi (Bologna), I
Cyasorb UV	A, B, C, E, F, I, K	American Cyanamid Co., Wayne (N.J.), USA
Eastman Inhibitor	A	Eastman Chemical Products Inc., Kingsport (Tn.), USA

Trade name	Compound type	Manufacturer
UV-Chek AM	A, C, G	Ferro Corp., Cleveland (Oh.), USA
Goodrite UV	I	B.F. Goodrich, Chemical Group, Cleveland (Oh.), USA
Hostavin	A,I,K	Hoechst AG, Frankfurt/M., D
Interstab	G	Interstab Ltd., Liverpool, GB
Irgastab	C	Ciba-Geigy AG, Basel, CH
Mark	A, I	Argus Chemical Corp., Brooklyn (N.Y.), USA
Mixxim	I, K	Fairmount Chemical Co., Newark (N.J.) USA
Rylex	A, C	E.I. Du Pont de Nemours and Co., Wilmington (De.), USA
Salol	A, D	The Dow Chemical Co., Midland (Mi.), USA
Sanol LS	I	Sankyo Co. Ltd., Tokyo 140, J
Seesorb	A, B, C, D, E, G	Sun Chemical Corp., (Towa Building), Tokyo, J
Sumisorb	A, D	Sumitomo Chemical Co. Ltd., Osaka 541, J
Tinuvin	B, H, I, K	Ciba-Geigy AG, Basel, CH
UV-Absorber Bayer	A, E	Bayer AG, Leverkusen, D
Uvinul	A, E	BASF AG, Ludwigshafen, D
Viosorb	A, B, D	Kyodo Chemical Co. Ltd., Tokyo, J

3.7 References

[1] *V.T. Walldner, W.J. Clarke, J.B. Decoste, J.B. Howard:* Ind. Eng. Chem. 42 (1950) p. 2320.

[2] *A. King:* Plast. Polym. 36 (1968) p. 195.

[3] *F. Gugumus:* 4e Conférence Européene des plastiques et des caoutchoucs. Paris, 1974, Kunstst. Plast. 22 (1975) p. 11, Caout. Plast. 558 (1976) p. 67.

[4] Plast. Tech. 22 (1976) p. 57.

[5] Plast. Eng. 34 (1978) 5, p. 46.

[6] *J. Voigt:* Die Stabilisierung der Kunststoffe gegen Licht und Wärme. Springer-Verlag, Berlin, 1966.
[7] *K. Thinius:* Stabilisierung und Alterung von Plastwerkstoffen, chapter 1 and 2, Akademie Verlag, Berlin, 1966.
[8] *W.L. Hawkins:* Polymer Stabilisation. Wiley, New York, 1972.
[9] *B. Dolezel:* Die Beständigkeit von Kunststoffen und Gummi. Carl Hanser-Verlag, München, 1978.
[10] Publication by Commission Internationale de l'Eclairage (CIE) 20 (1972).
[11] *D. Kockott:* 10th colloquium of Danubian countries for natural and artificial aging of polymers, Vienna, 1977, Farbe Lack 85 (1979) p. 102.
[12] *J.N. Pitts,* Jr., *B.J. Finlayson:* Angew. Chem. 87 (1975) p. 18.
[13] *J.P. Jesson:* Angew. Chem. 89 (1977) p. 507.
[14] *M.J. Molina, F.S. Rowland:* Nature 249 (1974) p. 810.
[15] *S.L. Fitton, R.N. Haward, G.R. Williamson:* Br. Polym. J. 2 (1970) p. 217.
[16] *K. Tsuji, T. Seiki:* J. Polym. Sci. A 1, 9 (1971) p. 3063.
[17] *R.C. Hirt, N.J. Searle:* Appl. Polym. Symp. 4 (1967) p. 61.
[18] *A.R. Burgess:* Nat. Bur. Stand. Circ. 525 (1953) p. 149.
[19] *P.A. Mullen, N.J. Searle:* J. Appl. Polym. Sci. 44 (1970) p. 765.
[20] *F. Kügler, P. Frank:* Kunststoffe, Fortschr. Berichte 2 (1976) p. 71.
[21] *R.B. Fox:* Progr. Polym. Sci. 1 (1967) p. 45.
[22] *H.H.G. Jellinek:* Pure Appl. Chem. 4 (1962) p. 419.
[23] *N. Grassie, N.A. Weir:* J. Appl. Polym. Sci. 9 (1965) p. 975/987.
[24] *B. Ranby, J.F. Rabek:* Photodegradation, Photooxidation and Photostabilisation of Polymers. Wiley Interscience, London, 1975.
[25] *W. Schnabel, J. Kiwi:* Photodegradation, in *H.H.G. Jellinek:* Aspects of Degradation and Stabilisation of Polymers. Elsevier, New York, 1978, p. 195.
[26] *D.J. Carlsson, D.M. Wiles:* J. Macromol. Sci., Rev. Macromol. Chem. C 14 (1976) p. 65.
[27] *J.C.W. Chien:* J. Phys. Chem. 69 (1965) p. 4317.
[28] *D.J. Carlsson, D.M. Wiles:* Macromolecules 2 (1969) p. 587.
[29] *O. Cicchetti:* Adv. Polym. Sci. 7 (1970) p. 70.
[30] *N.S. Allen, R.B. Cundall, M.W. Jones, J.F. McKellar:* Chem. Ind. 28 (1976) p. 110.
[31] *N.S. Allen, J. Homer, J.F. McKellar:* Chem. Ind. 28 (1976) p. 692.
[32] *A.M. Trozzolo, F.H. Winslow:* Macromolecules 1 (1968) p. 98.
[33] *D.J. Carlsson, D.M. Wiles:* J. Polym. Sci., Polym. Lett. Ed. 11 (1973) p. 759.
[34] *K.B. Chakraborty, G. Scott:* Polymer 18 (1977) p. 98.
[35] *J. Aspler, D.J. Carlsson, D.M. Wiles:* Macromolecules 9 (1976) p. 691.
[36] *J. Griffiths, C. Hawkins:* J. Soc. Dyers Colourists 89 (1973) p. 173.
[37] *K. Tsuji, T. Seiki:* J. Polym. Sci. B. 8 (1970) p. 817.
[38] *K. Tsuji, T. Seiki:* Polym. J. 2 (1971) p. 606.
[39] *D.G.M. Wood, T.M. Kollman:* Chem. Ind. 24 (1972) p. 423.
[40] *N.S. Allen, K.O. Fatinikun, T.J. Henman:* Polym. Degr. Stab. 4 (1982) p. 59.
[41] *N.S. Allen, K.O. Fatinikun, J.-L. Gardette, J. Lemaire:* Polym. Degr. Stab. 4 (1982) p. 95.

[42] *N.S. Allen, J.F. McKellar, G.O. Phillips,* in *D.L. Allara, W.L. Hawkins* (editors): Stabilisation and Degradation of Polymers, Am. Chem. Soc. Adv. Chem. Ser., No. 169 (1978) p. 68.
[43] *N.S. Allen:* Polym. Degr. Stab. 2 (1980) p. 155.
[44] *A. Garton, D.J. Carlsson, D.M. Wiles,* in *N.S. Allen* (editor): Developments in polymer photochemistry-1. Applied Science Publishers Ltd., London, 1980, p. 93.
[45] *J. Guillet:* Adv. Chem. Ser. 169 (1978) p. 1.
[46] *J.C.W. Chien, D.S.T. Wang:* Macromolecules 8 (1975) p. 920.
[47] *D.C. Bassett:* Polymer 5 (1964) p. 457.
[48] *S.M. Gabbay, S.S. Stivala:* Polymer 17 (1976) p. 137.
[49] *L. Reich, S.S. Stivala:* Autoxidation of Hydrocarbons and Polyolefins. Marcel Dekker, New York, 1969.
[50] *J.C.W. Chien,* in *G. Geuskens:* Degradation and Stabilisation of Polymers. Applied Science Publishers Ltd., London, 1975, p. 95.
[51] *G. Scott:* Mechanisms of Photodegradation and Stabilisation of Polyolefins, in *S.S. Labana:* Ultraviolet Light Induced Reactions in Polymers, ACS Symposium Series 25, Am. Chem. Soc., Washington, 1976, p. 340.
[52] *G. Geuskens, M.S. Kabamba:* Polym. Degr. Stab, 4 (1982) p. 69.
[53] *J.B. Lawrence, N.A. Weir:* J. Polym. Sci. 11 (1973) p. 105.
[54] *W. Klöpffer:* Eur. Polym. J. 11 (1975) p. 203.
[55] *G.A. George, D.K.C. Hodgeman:* Eur. Polym. J. 12 (1976) p. 63.
[56] *G. Geuskens, D. Baeyens-Volant, G. Delaunois, Q. Lu-Vinh, W. Piret, C. David:* Eur. Polym. J. 14 (1978) p. 291, p. 299.
[57] *G. Geuskens, C. David:* Pure Appl. Chem. 51 (1979) p. 2385.
[58] *G. Geuskens, G. Delaunois, Q. Lu-Vinh, W. Piret, C. David:* Eur. Polym. J. 18 (1982) p. 387.
[59] *G. Geuskens, Q. Lu-Vinh:* Eur. Polym. J. 18 (1982) p. 307.
[60] *A. Ghaffar, A. Scott, G. Scott:* Eur. Polym. J. 12 (1976) p. 615.
[61] *G. Geuskens, P. Bastin:* Polym. Degr. Stab. 4 (1982) p. 111.
[62] *G. Geuskens, D. David,* in *G. Geuskens:* Degradation and Stabilisation of Polymers, Applied Science Publishers Ltd., London, 1975, p. 113.
[63] *E. Priebe, P. Simak, K. Stange:* Kunststoffe 62 (1972) p. 105.
[64] *M. Balandier, C. Decker:* Eur. Polym. J. 14 (1978) p. 995.
[65] *C. Decker:* International Conference on Advances in the Stabilization and Controlled Degradation of Polymers, Luzern, Switzerland, 1980.
[66] *B.B. Cooray, G. Scott:* Polym. Degr. Stab. 3 (1980/81) p. 127.
[67] *D. Braun, S. Kull:* Angew. Makromol. Chem. 86 (1980) p. 171.
[68] *G. Scott, M. Tahan:* Eur. Polym. J. 13 (1977) p. 989.
[69] *J.M. Fedder:* Quart. Rev. 14 (1960) p. 336.
[70] *C. Decker:* J. Appl. Polym. Sci. 20 (1976) p. 3321.
[71] *N. Grassie, T.I. Davis:* Makromol. Chem. 175 (1974) p. 2657.
[72] *J. Hennig:* 13th colloquium of Danubian countries for natural and artificial aging of polymers, Budapest, 1980.
[73] *A. Gupta, R. Liang, F.D. Tsay, J. Moacanin:* Macromolecules 13 (1980) p. 1696.
[74] *R.B. Fox, L.G. Isaacs, S. Stokes:* J. Polym. Sci. Part A, 1 (1963) p. 1079.

[75] *V.Ya. Shlyapintokh, V.I. Goldenberg:* Eur. Polym. J. 10 (1974) p. 679.
[76] *M.J. Bowden, E.A. Chandross, I.P. Kaminow:* Polym. Eng. Sci. 14 (1974) p. 494.
[77] *D. Panke, W. Wunderlich, J. Hennig:* International Conference on Advances in the Stabilization and Controlled Degradation of Polymers, Luzern, 1981.
[78] *C.S. Schollenberger, K. Dinbergs:* SPE Trans. 1 (1961) 1, p. 31.
[79] *H.C. Beachel, I.L. Lang:* J. Polym. Sci. A 1, 10 (1972) p. 503.
[80] *N.S. Allen, J.F. McKellar:* J. Appl. Polym. Sci. 20 (1976) p. 1441.
[81] *H. Schultze:* Makromol. Chem. 172 (1973) p. 57.
[82] *J.-L. Gardette, J. Lemaire:* Makromol. Chem. 182 (1981) p. 2723.
[83] *V.A. Kosobutskü, M.N. Kurganova, O.G. Tarakanov, V.K. Belyakov:* Polym. Sci. USSR 22 (1980) p. 1386.
[84] *D. Bellus, P. Hrdlovic, Z. Manasek:* J. Polym. Sci. B 4 (1966) p. 1.
[85] *A. Gupta, A. Rembaum, J. Moacanin:* Macromolecules 11 (1978) p. 1285.
[86] *A. Gupta, R. Liang, J. Moacanin, R. Goldbeck, D. Kliger:* Macromolecules 13 (1980) p. 262.
[87] *A. Factor, M.L. Chu:* Polym. Degrad. Stab. 2 (1980) p. 203.
[88] *N. Grassie, R.S. Roche:* Makromol. Chem. 112 (1968) p. 34.
[89] *R.F. Moore:* Polymer 4 (1963) p. 493.
[90] *N.S. Allen, J.F. McKellar, G.O. Phillips:* J. Polym. Sci., Polym. Chem. Ed. 12 (1974) p. 1233.
[91] *H.A. Taylor, W.C. Tincher, W.F. Hammer:* J. Appl. Polym. Sci. 14 (1970) p. 141.
[92] *A.L. Margolin, I.A. Kabanova, L.M. Postnikov, V.Ya. Shylapintokh:* Polym. Sci. USSR 18 (1976) p. 1258.
[93] *T.S. Popravko, G.G. Makarov, Yu.A. Mikheyev, G.B. Parüskii, L.M. Postnikov, D.Ya. Toptygin:* Polym. Sci. USSR 19 (1978) p. 3108.
[94] *A.L. Margolin, L.M. Postnikov, V.Ya. Shlyapintokh:* Polym. Sci. USSR 19 (1977) p. 2236.
[95] *A.L. Margolin, L.M. Postnikov:* Russ. Chem. Rev. 49 (1980) p. 1106.
[96] *L. Tang, J. Lemaire, D. Sallet, J.-M. Mery:* Makromol. Chem. 182 (1981) p. 3467.
[97] *D.J. Carlsson, L.H. Gan, D.M. Wiles:* J. Polym. Sci., Polym. Chem. Ed. 16 (1978) p. 2353, p. 2365.
[98] *D.J. Carlsson, R.D. Farnell, D.M. Wiles:* J. Polym. Sci., Polym. Lett. Ed. 11 (1973) p. 149.
[99] *D.J. Carlsson, L.H. Gan, R.D. Parnell, D.M. Wiles:* J. Polym. Sci., Polym. Lett. Ed. 11 (1973) p. 683.
[100] *J.A. Dellinger, C.W. Roberts:* J. Appl. Polym. Sci. 26 (1981) p. 321.
[101] *M. Day, D.M. Wiles:* J. Appl. Polym. Sci. 16 (1972) p. 203.
[102] *S. Husband, K.E. Russell, J.A. Stone, J.K.S. Wan:* J. Appl. Polym. Sci. 21 (1977) p. 611.
[103] *Z. Sláma, J. Majer:* Plaste Kautsch. 24 (1977) p. 98.
[104] *J. Petrûj, Z. Sláma:* Makromol. Chem. 181 (1980) p. 2461.
[105] *Z. Sláma, E. Švejdová, J. Majer:* Makromol. Chem. 181 (1980) p. 2449.
[106] *B. Wandelt:* Polym. Bull. 4 (1981) p. 199.
[107] *A. Davis, G.H.W. Deane, B.L. Diffey:* Nature 261 (1976) p. 169.

[108] *A. Davis, B.W. Howes, K.J. Ledbury, P.J. Pearce:* Polym. Degrad. Stab. 1 (1979) p. 121.
[109] *N.S. Allen, J.F. McKellar:* J. Appl. Polym. Sci. 21 (1977) p. 1129.
[110] *B.D. Gesner, P.G. Kelleher:* J. Appl. Polym. Sci. 12 (1968) p. 1199.
[111] *B.D. Gesner, P.G. Kelleher:* J. Appl. Polym. Sci. 13 (1969) p. 2183.
[112] *H.J. Heller, H.R. Blattmann:* Pure Appl. Chem. 30 (1972) p. 145.
[113] *H.J. Heller, H.R. Blattmann:* Pure Appl. Chem. 36 (1973) p. 141.
[114] *D.J. Carlsson, D.M. Wiles:* J. Macromol. Sci., Rev., Macromol. Chem. C 14 (1976) p. 155.
[115] *J. Mikovic, P. Ambrovic, Z. Manasek, M. Karvas:* Chem. Zvesti 27 (1973) p. 255.
[116] *O. Cicchetti, M. Dubini, P. Parrini, G.P. Vicario, E. Bua:* Eur. Polym. J. 4 (1968) p. 419.
[117] *J.G. Calvert, J.N. Pitts,* Jr.: Photochemistry. John Wiley & Sons, Inc., New York, 1967, p. 534.
[118] *W. Klöpffer:* J. Polym. Sci. Polym. Symp. 57 (1976) p. 205.
[119] *J.H. Chaudet, G.C. Newland, H.W. Patton, J.W. Tamblyn:* SPE Trans. 1 (1961) p. 26.
[120] *A.A. Lamola, L.J. Sharp:* J. Phys. Chem. 70 (1966) p. 2634.
[121] *H. Gysling, H.J. Heller:* Kunststoffe 51 (1961) p. 13.
[122] *B. Küster, H. Herlinger:* Angew. Makromol. Chem. 40/41 (1974) p. 265.
[123] *J.P. Guillory, C.F. Cook:* J. Polym. Sci. A 1, 9 (1971) p. 1529.
[124] *J.S. Zannucci, G.R. Lappin:* Macromolecules 7 (1974) p. 393.
[125] *D.J. Carlsson, T. Suprunchuk, D.M. Wiles:* J. Appl. Polym. Sci. 16 (1972) p. 615.
[126] *V.S. Sivokhin:* Polym. Sci. USSR 21 (1980) p. 1207.
[127] *D. Bellus:* Adv. Photochem. 8 (1971) p. 109.
[128] *W. Klöpffer:* 16th International Meeting on Synthetic Fibres, Dornbirn, Austria, 1977.
[129] *B. Felder, R. Schumacher:* Makromol. Chem. 31 (1973) p. 35.
[130] *W. Klöpffer:* J. Chem. Phys. 50 (1969) p. 2337.
[131] *R.P. Foss, D.O. Cowan, G.S. Hammond:* J. Phys. Chem. 58 (1964) p. 3747.
[132] *A.J. Try, R.S.H. Lin, G.S. Hammond:* J. Am. Chem. Soc. 88 (1966) p. 4781.
[133] *J.C.W. Chien, W.P. Connor:* J. Am. Chem. Soc. 90 (1968) p. 1001.
[134] *D.J. Harper, J.F. McKellar, P.H. Turner:* J. Appl. Polym. Sci. 18 (1974) p. 2805.
[135] *N.S. Allen, J. Homer, J.F. McKellar:* Macromol. Chem. 179 (1978) p. 1575.
[136] *D.J. Carlsson, D.M. Wiles:* Macromolecules 7 (1974) p. 259.
[137] *R.P.R. Ramawara, G. Scott:* Eur. Polym. J. 12 (1976) p. 591.
[138] *D.J. Carlsson, G.D. Mendenhall, T. Suprunchuk, D.M. Wiles:* J. Am. Chem. Soc. 94 (1972) p. 8960.
[139] *G. Scott:* Atmospheric Oxidation and Antioxidants. Elsevier, London, 1965.
[140] *D.J. Carlsson, D.M. Wiles:* J. Polym. Sci., Polym. Chem. Ed. 12 (1974) p. 2217.
[141] *J.E. Bonkowski:* Textile Res. J. 3 (1969) p. 243.
[142] *R.S. Halles:* Plast. Eng. 5 (1976) p. 17.

[143] *J.H. Chaudet, J. Tamblyn:* J. Appl. Polym. Sci. 8 (1964) p. 1949.

[144] *W. Iwanow, W. Shlyapintokh, A. Shapiro, O. Chwostatsch, E. Rosanzew:* Iswestija Akad. Wiss. UdSSR, Serie Chimija 8 (1974) p. 1916.

[145] *D.J. Carlsson, D.W. Gratton, T. Suprunchuk, D.M. Wiles:* J. Appl. Polym. Sci. 22 (1978) p. 2217.

[146] *M.V. Sudnik, M.F. Romantsev, A.G. Shapiro, E.G. Rozantsev:* Izv. Akad. Nauk. S.S.S.R., Ser. Khim. (1975) p. 2813.

[147] *D.B. Chakraborty, G. Scott:* Chem. Ind. 30 (1978) p. 237.

[148] *K. Murayama, S. Marimura, T. Yashioka:* Bull. Chem. Soc. Jpn. 42 (1969) p. 1640.

[149] *E.T. Denisov:* Int. Symp. on Degradation and Stabilisation of Polymers. Bruxelles, 1974, p. 137.

[150] *D.J. Carlsson, A. Garton, D.M. Wiles,* in *G. Scott* (editor): Developments in Polymer Stabilisation-1. Appl. Sci. Publishers. Barking, 1979, p. 219.

[151] *D.W. Grattan, D.J. Carlsson, D.M. Wiles:* Polym. Degr. Stab. 1 (1979) p. 69.

[152] *D.W. Grattan, A.H. Reddoch, D.J. Carlsson, D.M. Wiles:* J. Polym. Sci., Polym. Lett. Ed. 16 (1978) p. 143.

[153] *K.H. Chan, D.J. Carlsson, D.M. Wiles:* J. Polym. Sci., Polym. Lett. Ed. 18 (1980) p. 607.

[154] *D.J. Carlsson, K.H. Chan, D.M. Wiles, J. Durmis:* Preprints, Organic Coatings and Applied Polymer Science Proceedings 46 (1982) p. 457.

[155] *D.M. Wiles:* International Conference on Advances in the Stabilization and Controlled Degradation of Polymers, Luzern, Switzerland, 1982.

[156] *B. Felder, R. Schumacher, F. Sitek,* in *S.P. Pappas, F.H. Winslow* (editor): Photodegradation and Photostabilization of Coatings. ACS Symp., Ser. 151, Washington, 1981, p. 65.

[157] *D.J. Carlsson, K.H. Chan, J. Durmis, D.M. Wiles:* J. Polym. Sci., Polym. Chem. Ed. 20 (1982) p. 575.

[158] *J. Sedlar:* 3rd International Conference on Advances in the Stabilization and Controlled Degradation of Polymers, Luzern, Switzerland, 1981.

[159] *N.S. Allen, J.F. McKellar:* Chem. Ind. 29 (1977) p. 537.

[160] *N.S. Allen, J. Homer, J.F. McKellar:* Makromol. Chem. 179 (1978) p. 1575.

[161] *N.S. Allen, J.F. McKellar:* Eur. Polym. J. 16 (1980) p. 553.

[162] *F. Gugumus,* in *G. Scott* (editor): Developments in Polymer Stabilization-1. Appl. Sci. Publishers, Barking, 1979, p. 261.

[163] *J. Sedlar, J. Marchal, J. Petrûj:* Polym. Photochem. 2 (1982) p. 175.

[164] Deutscher Normenausschuß, Prüfnormen für Kunststoffe, Taschenbuch 48, Beuth-Vertrieb GmbH, Frankfurt, 1974.

[165] Annual Book of ASTM Standards 1978, Part 35, Plastics-General Test Methods, ASTM 1916 Race St., Philadelphia, PA 19103.

[166] Annual Book of ASTM Standards 1978, Part 36, Plastics Materials, Film, Reinforced and Cellular Plastics: High Modulus Fibers and Their Composites, ASTM 1916 Race St., Philadelphia, PA 19103.

[167] *C.R. Caryl:* ASTM Bull. 243 (1960) 55, p. 479.

[168] *K. Berger:* 7th colloquium of Danubian countries for natural and artificial aging of polymers, Moscow, 1974.

[169] *K. Berger:* 9th colloquium of Danubian countries for natural and artificial aging of polymers, Gottwaldow, CSSR, 1976.
[170] *G.C. Newland, J.W. Tamblyn:* Appl. Polym. Symp. 4 (1967) p. 119.
[171] *W. Bergk, P. Pluhar, K.H. Reichert:* Farbe Lack 77 (1971) p. 1056.
[172] *A. Davis, G.H.W. Deane, D. Gordon, G.V. Howell, K.J. Ledbury:* J. Appl. Polym. Sci. 20 (1976) p. 1165.
[173] *A. Davis. G.H.W. Deane, B.L. Diffey:* Nature 261 (1976) p. 169.
[174] *A. Davis, B.V. Howes, K.J. Ledbury, P.J. Pearce:* Polym. Deg. Stab. 1 (1979) p. 121.
[175] *C.R. Caryl:* Farbe Lack 77 (1971) p. 524.
[176] *R.J. Martinowitch, G.R. Hill:* Appl. Polym. Symp. 4 (1967) p. 141.
[177] *P.J. Papillo:* J. Paint Tech. 40 (1968) 524, p. 359.
[178] *G. Menzel:* Plastverarbeiter 25 (1974) p. 137.
[179] *L. Glück, G. Poschet:* Kunststoffe 72 (1982) p. 353.
[180] *W. Brockmann, E. Roeder:* Gummi Asbest Kunstst. 35 (1982) p. 265.
[181] *P.N. Lowell, N.G. McCrum:* J. Polym. Sci. A 2, 9 (1971) p. 1935.
[182] *R.G. Hauserman, J.F. Westlake:* J. Appl. Polym. Sci 20 (1976) p. 2533.
[183] *M. Johnson, R.G. Hauserman:* J. Appl. Polym. Sci. 21 (1977) p. 3457.
[184] *R.J. Roe, H.E. Bair, C. Gienicwski:* J. Appl. Polym. Sci. 18 (1974) p. 843.
[185] *Z. Osawa, K. Matsui, Y. Ogiwara:* J. Macromol. Sci., Chem. A 1 (1967) p. 581.
[186] *Y. Mizvtani, K. Kusumoto:* J. Appl. Polym. Sci. 19 (1975) p. 713.
[187] *F.H. McTigue, M. Blumberg:* Appl. Polym. Symp. 4 (1967) p. 175.
[188] *J. Masek, P. Caucik, J. Holcik, M. Karvas, J. Durmis:* Plaste Kautsch. 24 (1977) p. 404.
[189] *K.W. Leu, F. Gugumus, H. Linhart, A. Wieber:* Plaste Kautsch. 24 (1977) p. 408.
[190] *K.W. Leu:* Plast. News (1975) p. 2.
[191] Ciba-Geigy: unpublished results.
[192] *A.B. Shapiro, L.P. Lebedeva, V.I. Suskina, G.N. Antipina, L.N. Smirnov, P.I. Levin, E.G. Rozantsev:* Polym. Sci. USSR 15 (1973) p. 3034.
[193] Ciba-Geigy: Preliminary data sheet: TINUVIN 770, publ.-No. 37327/d, September 1975/15.
[194] *V.N. Levin, G.N. Antipina, E.V. Merkuryeva, A.A. Skripko, A.B. Shapiro:* Preprints, II. International symposium on man-made fibres, Kalinin, USSR, 1977.
[195] *H. Lind, H. Loeliger:* Preprint M 44 in Mechanisms of Degradation and Stabilization of Hydrocarbon Polymers, 19th Prague Microsymposium on Macromolecules, Prague, 1979.
[196] *F. Gugumus:* Preprints, Organic Coatings and Applied Polymer Science Proceedings 46 (1982) p. 449.
[197] *F. Gugumus:* 5èmes Journées d'Etudes sur le Vieillissement des Polymères, Bandol, France, 1982.
[198] *G.C. Newland:* Solar Energy 12 (1968) p. 65.
[199] *F. Gugumus:* International Conference on Advances in the Stabilization and Controlled Degradation of Polymers, Luzern, Switzerland, 1981.
[200] *E. Priebe, K. Stange, K.W. Steiner:* Kunststoffe 60 (1970) p. 617.

[201] *B. Gilg:* 5th European Plastics and Rubber Conference, Paris, 1978, 1, C-13.
[202] *Ch. Savides, J.A. Stretanski, L. Costello:* Adv. Chem. Ser. 85 (1968) p. 287.
[203] *N.L. Perry:* Mod. Plast. 40 (1963) p. 156.
[204] *L.B. Weissfeld, G.A. Thacker, L.I. Nass:* SPE J. 21 (1965) p. 649.
[205] *G.A. Baum:* Appl. Polym. Symp. 4 (1967) p. 189.
[206] *J. Büssing:* Kunstst. Plast. 4 (1975) p. 43.
[207] *B. Dolezel:* Chemicky prumysl 13 (1963) p. 160.
[208] *G. Schreyer:* Angew. Makromol. Chem. 11 (1970) p. 159.
[209] *C.S. Schollenberger* u. *J.D. Stewart:* J. Elast. Plast. 8 (1976) p. 11.
[210] *H. Schmidt, E. Wolters:* Kunststoffe 61 (1971) p. 261.
[211] *H.H.G. Jellinek, A.K. Chaudhuri:* J. Polym. Sci. A 1, 10 (1972) p. 1773.
[212] *B.M.C. Dorset:* Textile Manufacturer 88 (1962) p. 254.
[213] *J.E. Pickett:* International Conference on Advances in the Stabilization and Controlled Degradation of Polymers, Zürich, Switzerland, 1983.
[214] *F. Gugumus:* Kunststoffe 74 (1984) p. 620.

4 PVC Stabilizers

Dr. Holger Andreas, Ciba-Geigy Marienberg GmbH,
Bensheim/Federal Republic of Germany

4.1 The economical and technical importance of PVC stabilizers

Polyvinyl chloride (PVC) is the oldest thermoplastic and has developed to one of the most important bulk plastics in the course of its 50 years' history. The consumption in different geographic areas and the expected demand through 1990, based on our estimates, are shown in figure 1.

In the future, too, PVC will maintain its importance among the thermoplastics, because it is compatible with a large number of other products

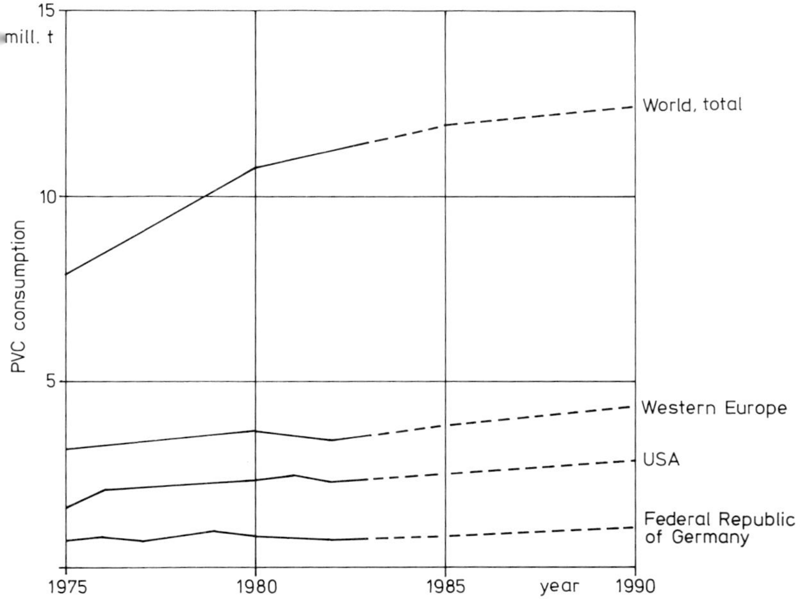

Fig. 1. PVC consumption from 1975 to 1990.

(e.g. plasticizers, impact modifiers) in contrast to other plastics. By this means, the mechanical properties can be adjusted over a wide range from rigid to flexible. Therefore, there are a great number of different kinds of processing methods and applications for PVC. The toxicological problems which arose by manufacture and processing of PVC have now been solved satisfactorily [1, 2].

At the beginning of PVC technology, flexible PVC was dominant, but in the meantime the proportion of rigid PVC has increased constantly and has reached about 60% of total consumption in many countries [2].

The low thermal stability of PVC is very well known. Despite that, processing at elevated temperatures is possible without damage by the addition of specific heat stabilizers. This is one of the chief reasons PVC has developed to become a major bulk plastic. The development and production of suitable heat stabilizers was connected with the production of PVC from the very beginning, and will also be the precondition for its processing and application in the future. In 1980 the worldwide consumption of PVC heat stabilizers was about 200,000 tons of which the largest portion were lead compounds and metal soaps. The total turnover of these additives is assessed at around $ 500 million in 1980.

4.2 Thermal degradation and stabilization of PVC

4.2.1 Degradation by thermal stress

At elevated temperatures, which are necessary for thermoplastic processing, PVC is damaged by dehydrochlorination (release of hydrogen chloride), autooxidation and mechanochemical chain scission. This degradation is caused by the simultaneous sequence of these reactions and is not clarified in all details. Many comprehensive reviews on this subject have been published in recent years [3 to 12].

Dehydrochlorination, the prevailing reaction during processing, leads to increasing discoloration. In the course of the proceeding degradation the physical properties are also changed in the direction of increasing embrittlement.

PVC of ideal constitution should be thermally stable [12, 13], which was concluded from investigations with model substances. Therefore it has to be assumed that the damage, particularly the dehydrochlorination, starts from sites of the macromolecule with labile chlorine-carbon bonds [5, 7, 14 to 23]. As initiation sites the following structures have to be considered:

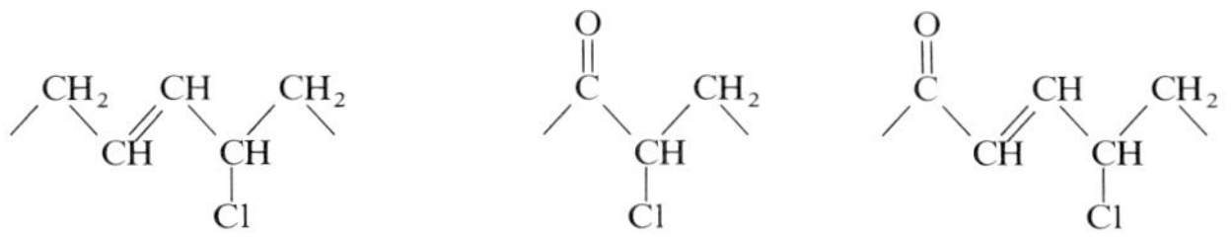

They are formed during the manufacturing process of PVC by autooxidation [18, 24], chain termination [17, 19], still unknown side reactions [15, 16, 19], or copolymerization with traces of oxygen [25], but also during processing by autooxidation [26, 27] and mechanochemical effects [6, 28, 29]. The influence of a branched PVC chain with tertiary C—Cl structure is the subject of controversial publications [5, 12, 16, 17, 23].

It is assumed that unprocessed PVC contains one initiation site within 10^3 to 10^4 repeating units [12, 30], depending on the production process. Thus it can be understood that emulsion PVC differs from suspension or mass PVC with regard to thermal stability[1]). The thermal evolution of hydrogen chloride is initiated at these labile sites. By this reaction conjugated double bonds (polyene sequences) are formed and the allyl activation is maintained. The reaction progresses "zipper-like" and leads to the formation of long polyene sequences, which from seven double bonds on are colored.

As for the mechanism of dehydrochlorination, there are different theories. An ionic as well as a radical reaction course are discussed [12, 27, 31 to 34]. According to *Shapiro* [34], the radical reaction course is said to occur to a larger extent with increasing temperature, being dominant at high temperatures.

Due to the basic character of extended polyene sequences they are able to bind hydrogen chloride with formation of carbenium salts (reaction (1)). This results in further intensification of color (halochromism) [35, 36].

---

[1]) It should also be pointed out that impurities in the resin, which are imparted by the polymerization, may have an influence on heat stability. This can be observed with emulsion PVC (E-PVC) in particular. E-PVC is generally spray-dried immediately after polymerization, and, therefore, separation from the manufacturing aids and its reaction products is not possible. In some cases, alkali salts are added before spray-drying in order to capture hydrogen chloride formed by thermal stress. Since the production aids are proprietary, E-PVC types differ to a large extent in heat stability.

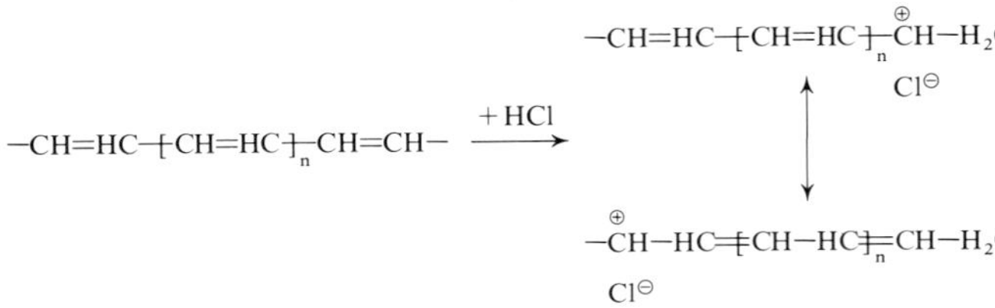

(1)

The autocatalytic influence of hydrogen chloride on dehydrochlorination in the presence of air, first postulated by *Druesedow* [37], has been investigated extensively [12, 38 to 41]. Without any doubt, liberated hydrogen chloride has – at least in the presence of oxygen – an autocatalytic effect on dehydrochlorination [41]. A possible explanation for this fact is the scission of hydrogen peroxides catalyzed by protons (*Hock* scission), which causes new initiation sites (reaction (4)).

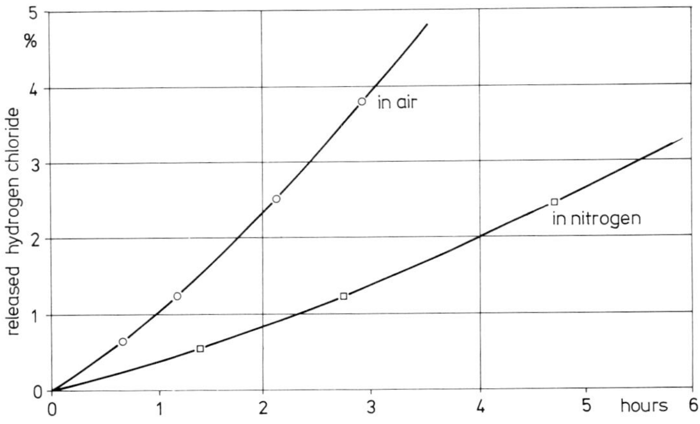

Fig. 2. Thermal dehydrochlorination of PVC at 180 °C in air and in nitrogen.

In addition to dehydrochlorination, autooxidation plays an important part in the thermal degradation of PVC [26, 27, 42 to 46][1]). Thus the thermal dehydrochlorination of PVC is considerably faster in the presence of oxygen than in inert atmosphere (figure 2).

[1]) Regarding the necessity to add an antioxidant to the stabilizing system, refer to pages 201, 212, 216, 227.

Several sites are susceptible to attack by oxygen. Attack on methylene groups leads to the formation of hydroperoxides in a radical chain reaction (reaction (2)):

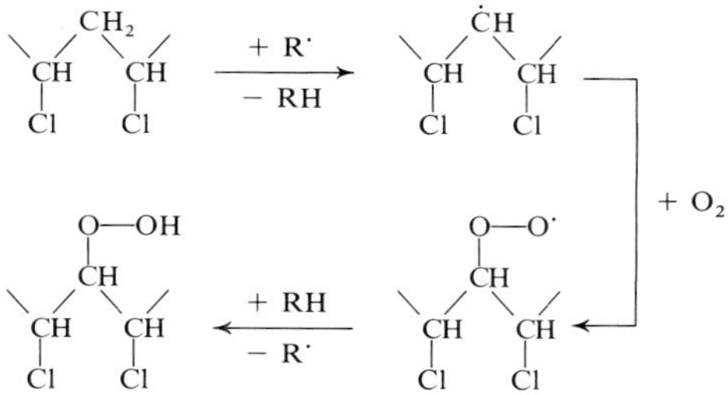

(2)

The decomposition of hydroperoxides by ionic (reaction (4)) as well as by radical (reaction (3)) reaction sequences was proposed [41, 49].

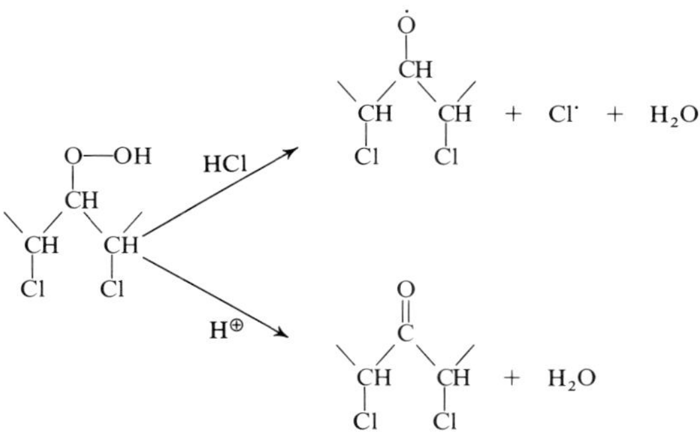

(3)

(4)

Oxygen may also react with polyene sequences of the PVC molecule caused by dehydrochlorination. *Berlin* et al. [42] postulated that hydroperoxides are formed with short polyene sequences and cyclic peroxides (reaction (5)) with longer polyene sequences:

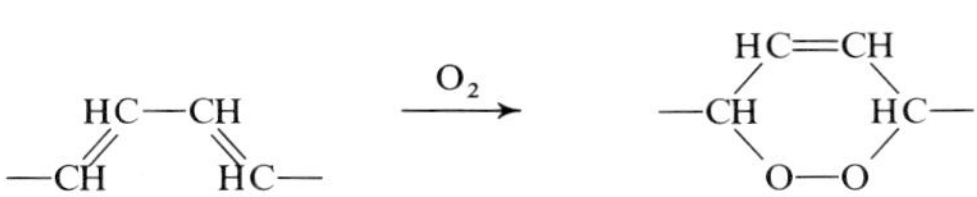

(5)

Oxygen may also combine with carbon free radicals, which are formed by mechanochemical fragmentation [29], to produce peroxy radicals. All oxygen-containing structures (peroxy radicals, peroxides, hydroperoxides) are less stable under processing conditions and undergo further transformation. As a result, new structures are formed which accelerate dehydrochlorination [26]. Dehydrochlorination starting in α-position to a carbonyl group yields keto polyene sequences which are according to reaction (6) able to form strongly colored carbenium salts with hydrogen chloride [47]:

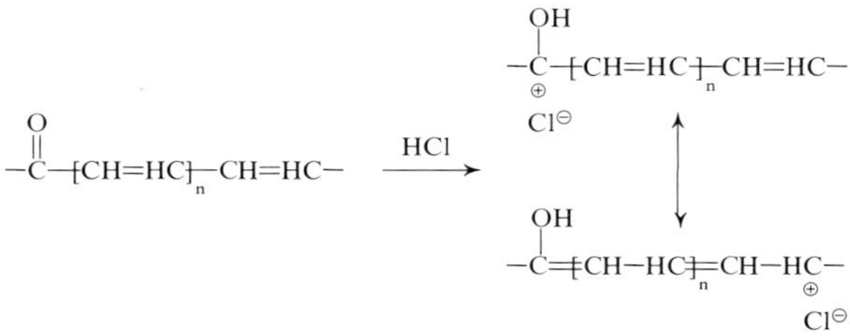

(6)

Fragmentation by shearing forces during processing deserves special mention. In recent investigations by *Scott* [29] this phenomenon in the thermooxidative degradation of PVC has been studied intensively. It has already been mentioned that radicals produced during mechanochemical processes may form peroxy radicals with oxygen which are subjected to further transformation (see reaction (2)). The more labile chlorine produced near the scission site can induce further dehydrochlorination.

Without any doubt, dehydrochlorination is the dominant reaction during thermal stressing. But autooxidation and chain scission should not be underestimated because they contribute to initiation sites [45].

Besides degradation during processing, PVC is also aged by the impact of light and weathering. Details are reported in the chapter "light stabilizers" of this book.

The processing of PVC and the maintenance of properties in service are made feasible by the addition of specific heat stabilizers which retard dehydrochlorination and autoxidation (preventive stabilizing function), but also cure existing damage (curative stabilizing function).

4.2.2 Mode of action of heat stabilizers[1])

The following requirements of a heat stabilizer may be derived from the explanation of the thermal damage of PVC:

---

[1]) Comprehensive papers on the thermostabilization of PVC cf. [3, 6, 7, 11, 12, 48 to 51].

- A heat stabilizer must be capable of preventing the dehydrochlorination reaction or at least must be able to retard this reaction (preventive function) [50].
- Additionally the stabilizer should shorten polyene sequences and destroy carbenium salts (curative function) [50].

One may differentiate between the following stabilizing functions:

preventive	*curative*
absorption of hydrogen chloride	addition to polyene sequences
elimination of initiation sites	destruction of carbenium salts
prevention of autooxidation	

4.2.2.1 Absorption of hydrogen chloride

A common property of all heat stabilizers is the capability to bind hydrogen chloride. This may be illustrated by two examples with metal carboxylates and organotin mercaptides:

$$\mathrm{Pb{-}(O{-}CO{-}C_{17}H_{35})_2 + HCl} \longrightarrow \mathrm{Cl{-}Pb{-}O{-}CO{-}C_{17}H_{35} + HO{-}CO{-}C_{17}H_{35}} \qquad (7a)$$

$$\mathrm{Cl{-}Pb{-}O{-}CO{-}C_{17}H_{35} + HCl} \longrightarrow \mathrm{PbCl_2 + HO{-}CO{-}C_{17}H_{35}} \qquad (7b)$$

$$\mathrm{(C_4H_9)_2Sn{-}(S{-}CH_2{-}CO{-}O{-}iso\text{-}C_8H_{17})_2 + HCl} \longrightarrow \mathrm{(C_4H_9)_2Sn(Cl){-}S{-}CH_2{-}CO{-}O{-}iso\text{-}C_8H_{17} + HS{-}CH_2{-}CO{-}O{-}iso\text{-}C_8H_{17}} \qquad (8a)$$

$$\mathrm{(C_4H_9)_2Sn(Cl){-}S{-}CH_2{-}CO{-}O{-}iso\text{-}C_8H_{17} + HCl} \longrightarrow \mathrm{(C_4H_9)_2SnCl_2 + HS{-}CH_2{-}CO{-}O{-}iso\text{-}C_8H_{17}} \qquad (8b)$$

Binding of hydrogen chloride can also take place by addition to the oxirane ring in e.g. epoxidized fatty acid esters:

$$\mathrm{H_3C{-}(CH_2)_7{-}\underset{\diagdown O \diagup}{CH{-}HC}{-}(CH_2)_7{-}CO{-}O{-}C_8H_{17} + HCl} \longrightarrow \mathrm{H_3C{-}(CH_2)_7{-}\underset{OH}{CH}{-}\underset{Cl}{HC}{-}(CH_2)_7{-}CO{-}O{-}C_8H_{17}} \qquad (9)$$

4.2.2.2 Elimination of initiation sites

One of the most important stabilizing reactions is the elimination of labile chlorine atoms, e.g. by substitution according to the following example:

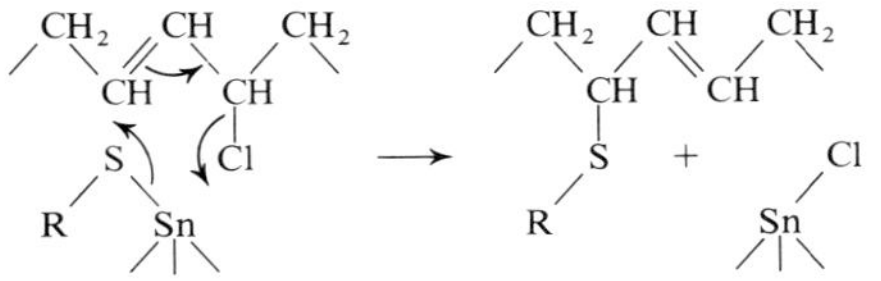

(10)

The specific complex chemistry of the most effective metal compounds favors the substitution reaction over the undesirable elimination of HCl which contributes conjugated double bonds. These effects can be found with organotin, lead, cadmium and zinc compounds, although there are tremendous differences with respect to secondary effects. In this connection special reference is made to the works of *Frye* [52], who postulated already in 1960 the exchange of labile chlorine atoms in PVC on the basis of IR-investigations [53, 54]. In model reactions it also could be proved that metal-free stabilizers, e.g. β-aminocrotonates, are capable of substitution reactions, especially in the presence of $ZnCl_2$ as a catalyst [55].

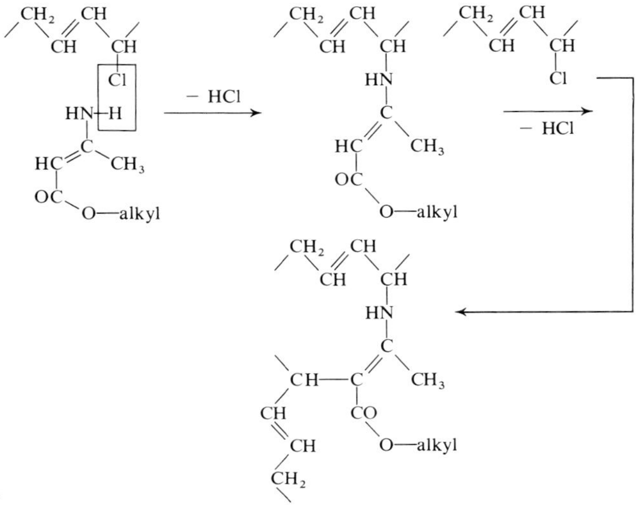

(11)

Besides the chemical elimination of labile chlorine atoms by substitution with groups, which are less susceptible to elimination, the removal of these potential initiation sites by isomerization (allyl transformation) has to be considered (reaction (12)). This isomerization may be the reason for the retardation of the dehydrochlorination in the presence of dialkyltin dichlorides [56]:

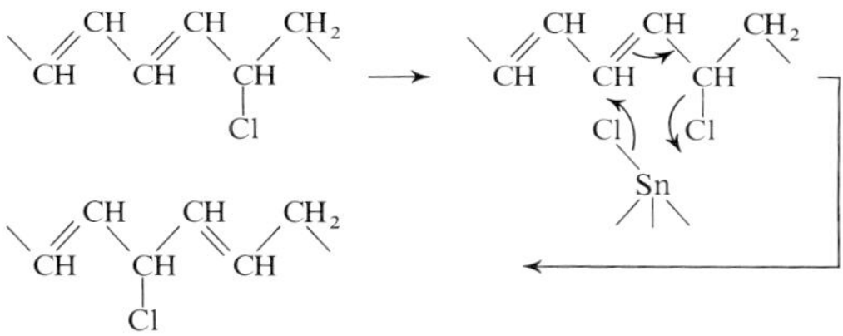

(12)

Now the chlorine is no longer in α-position to a proton which may be eliminated.

4.2.2.3 Prevention of autooxidation

The retardation of the dehydrochlorination by the addition of a phenolic antioxidant is evidence that antioxidants contribute to thermal stabilization (figure 3). Therefore, it is desirable that a good thermostabilizer must also possess antioxidative properties. In the case of organotin mer-

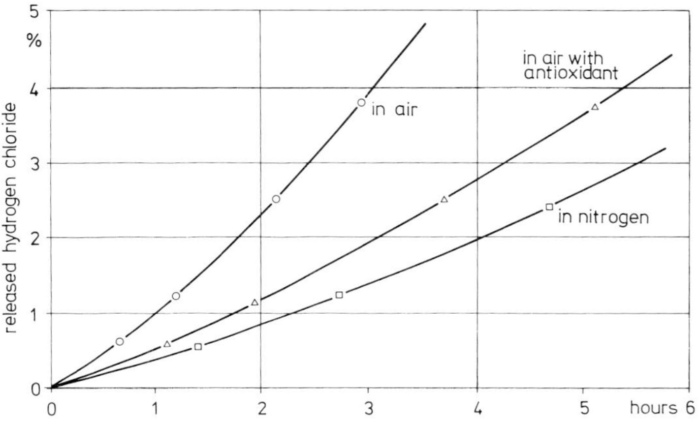

Fig. 3. Thermal dehydrochlorination of PVC at 180 °C in air and nitrogen in the presence of a phenolic antioxidant.

captides a secondary antioxidant effect, i.e. the decomposition of hydroperoxides could be proved in a model reaction (13) with tert.butylhydroperoxide [50, 57]:

$$(C_4H_9)_2Sn{-}(S{-}CH_2{-}CO{-}O{-}iso\text{-}C_8H_{17})_2 + HO{-}O{-}C(CH_3)_3 \quad (13)$$

$$\longrightarrow (C_4H_9)_2SnO + (iso\text{-}C_8H_{17}{-}O{-}CO{-}CH_2{-}S{-})_2 + HO{-}C(CH_3)_3$$

Even if the thermal stabilizer has peroxide-decomposing properties, the addition of a phenolic antioxidant may contribute to the stabilization by providing radical-trapping activity. The addition of a phosphite may be helpful with thermal stabilizers without peroxide-decomposing activity.

4.2.2.4 Addition to polyene sequences

The mercapto compounds which are released in the course of the reaction of the organotin mercaptide with hydrogen chloride are able to add to double bonds corresponding to the reaction (14) [58 to 60]:

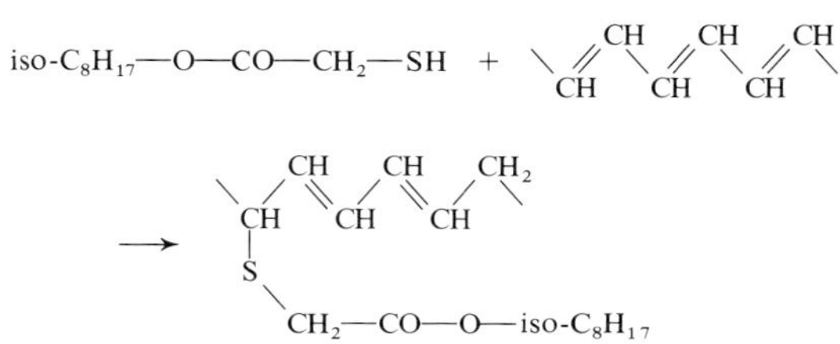

(14)

This reaction is an explanation for the fact that thermally discolored PVC is brightened by treatment with iso-octylmercaptoacetate [51]. This is the consequence of a reduction of the length of the polyene sequences. For the dialkyltin maleates the addition of maleic acid to the polymeric molecule by *Diels-Alder* reaction has also been proved [61]:

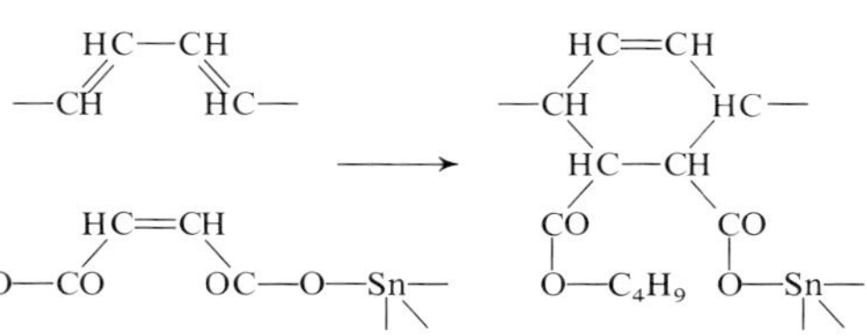

(15)

4.2.2.5 Destruction of carbenium salts

The deep discoloration of thermally damaged PVC is mainly due to carbenium salts. The destruction of carbenium salts by removing hydrogen chloride according to reaction (16) causes a brightening as it has been proved already in 1967 by *Schlimper* [35]:

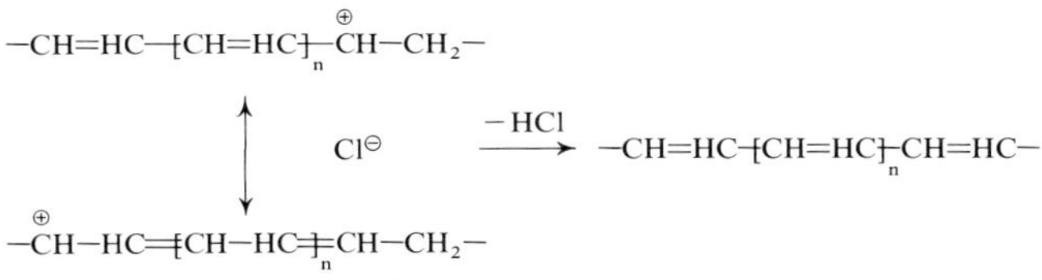

(16)

In principle all stabilizers or stabilizer systems which are able to bind hydrochloric acid, are also capable of this reaction.

It can be summarized that according to the current state of knowledge, PVC stabilizers may have three preventive and two curative functions. The organotin mercaptide stabilizers combine all these functions in one molecule. Other stabilizer systems show these functions only by the common action of different compounds. From this it can be concluded that the efficiency of the different heat stabilizers may be very different.

When selecting PVC stabilizers for the optimum solution of a processing problem, not only the thermostabilizing effects have to be considered, but also quite a number of other factors. It must be mentioned that PVC stabilizers may have e.g. a different influence on the processability (rheology) of a PVC compound, which will be demonstrated in the following chapters.

Furthermore the choice of the stabilizers depends on the formulation of the PVC compound. Because of the lower processing temperature, plasticized compounds usually suffer less from thermal damage during processing in comparison with rigid PVC.

Additionally, the kind of processing plays an important role, especially in the processing of rigid PVC. In the injection molding process, high processing temperatures must be applied in order to lower the melt viscosity. Since the temperatures in the fluidized bed coating process are even higher, very efficient thermostabilizers have to be used. The thermal stress put on PVC in fast rotating calenders and in the extrusion of thick profiles or of profiles with large surfaces is very high too.

In the following chapter recommendations are given for the proper selection of stabilizers. Because of the great number of formulating possibilities, the different types of processing, the different requirements for the finished product, any recommendation has to be rather general. Details are published in special data sheets and in brochures of the stabilizer manufacturers.

4.3 Product groups and their specific chemical and technical performance

4.3.1 Organotin stabilizers

Already in 1936 *V. Yngve* recommended not only tetraalkyltin but also alkyltin carboxylates as PVC stabilizers [62]. In 1950 Firestone filed patent applications for organotin mercaptides which became extremely important for the further development of PVC technology [63]. Organotin compounds with at least one tin-sulfur bond are generally called organotin mercaptides. Stabilizers on the basis of these compounds are also called sulfur-containing tin stabilizers or thiotins.

Organotin compounds, which are salts of alkyltinoxides with carboxylic acids – mainly maleic acid or half esters of maleic acid – are called usually organotin carboxylates, and the corresponding stabilizers are called sulfur-free tin stabilizers.

4.3.1.1 Organotin mercaptides and organotin sulfides

Sulfur-containing organotin compounds are up to now the most efficient and most universal heat stabilizers. The important products which are on the market have the following structures:

R^1	R^2
H_3C—	—CH_2—CO—O—alkyl
n-C_4H_9—	—CH_2—CH_2—CO—O—alkyl
n-C_8H_{17}—	—CH_2—CH_2—O—CO—alkyl
n-$C_{12}H_{25}$—	—alkyl
alkyl—O—CO—CH_2—CH_2—	

Very efficient stabilizers are derived from β-mercaptopropionic acid:

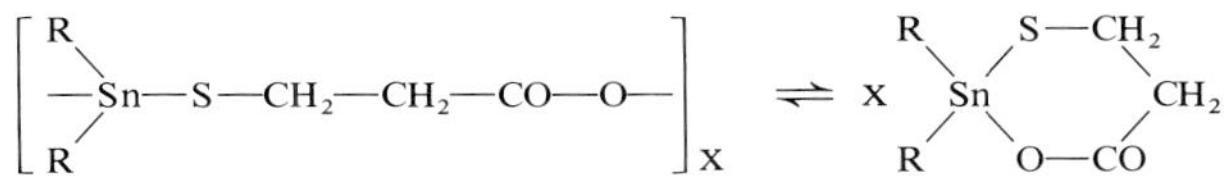

Moreover sulfides of mono- and diorganotin are of importance, e.g. cyclic dibutyltinsulfide and butylthiostannoic acid of adamantane structure [66]:

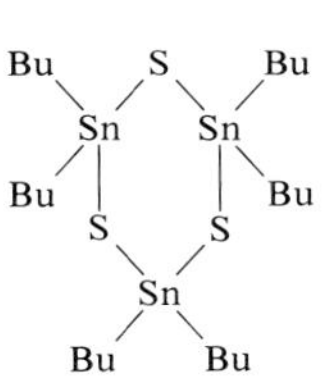

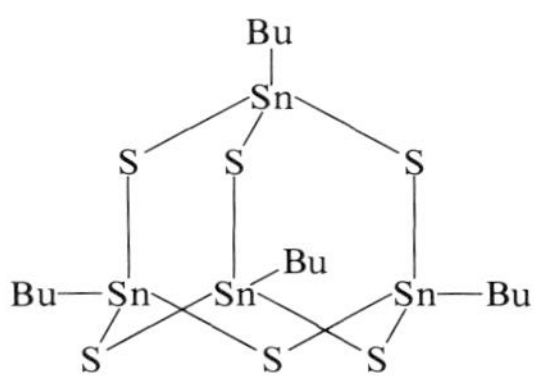

Mono- and diorganotin mercaptides are often used in combination, because these mixtures improve initial color as well as long-term stability of PVC [64, 65]. This is shown in figure 4. In recent years mixtures

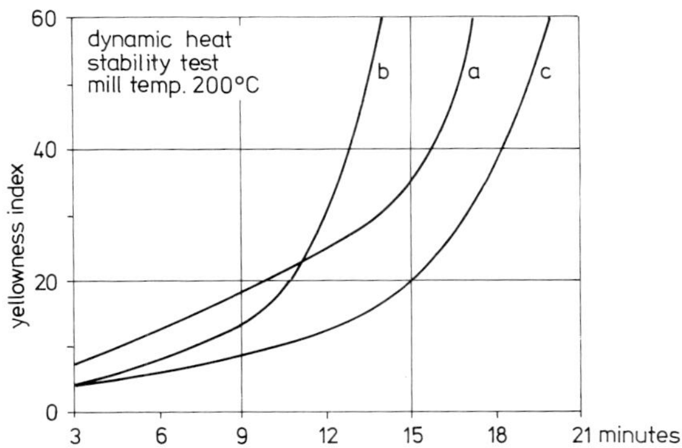

Fig. 4. Synergism of mono- and dioctyltin-isooctylthioglycolates in terms of Yellowness Index curves (YI) [122].

a: dioctyltin-bis(isooctylthioglycolate),
b: monooctyltin-tris(isooctylthioglycolate),
c: 80% dioctyltin-bis(isooctylthioglycolate) and
20% monooctyltin-tris(isooctylthioglycolate).

with a high content of monoalkyltin compounds have found increasing market penetration.

The heat stabilizing effect of these organotin stabilizers is due to the mercapto or sulfide group. These groups are involved in the stabilizing reaction [50, 52], as opposed to the alkyl or esteralkyl groups which only influence processibility, volatility and also toxicity of the tin stabilizers.

The presence of all these stabilizing functions is the reason for the excellent heat stabilizing efficiency of this product group, which has not been surpassed by any other kind of stabilizer. Thus it is logical that sulfur-containing tin stabilizers are used in all types of PVC processing where very high thermostability is required. This is the case in the processing of rigid PVC. Temperatures of 200 to 230 °C can be reached in extruders and on calenders. In the fluidized bed coating process and in injection molding, temperatures peaks can be reached which are still higher. On the other hand, it is an advantage of very efficient stabilizers that they can be used in lower concentrations if the requirements are less demanding.

Sulfur-containing tin stabilizers can be used universally. They are suitable for the stabilization of all homopolymers like emulsion, suspension and mass PVC, as well as for copolymers of vinyl chloride, grafted polymers and polyblends.

One of the most appreciated properties – not only of the sulfur-containing tin stabilizers but also of the whole group of organotin stabilizers – is absolute crystal clarity, which can be achieved by means of a proper formulation. Clarity is required for bottles, containers, corrugated sheets, light panels, all kinds of rigid packaging films and also for hoses, profiles, films, swinging doors and transparent top coats of floor or wall coverings made from plasticized PVC.

Sulfur-containing tin stabilizers do not exhibit sufficient self-lubrication. Therefore, the high processing temperatures which are necessary for optimal transparency may cause sticking of the hot melt to the metal of the processing machine, if this is not prevented by the addition of suitable lubricants. Products are now on the market with improved self-lubrication, but, even with these products, additional lubricants are often needed. Although tin stabilizers influence the melt viscosity of rigid PVC favorably, e.g. decrease the viscosity, the addition of processing aids is advantageous in order to impart better flow properties and to improve the surface quality of the finished articles, e.g. in the case of calendering of films, extrusion of profiles and sheets, blowing of bottles and in the injection molding of fittings.

Formulations which contain both organotin mercaptides and cadmium- or lead-containing stabilizers or pigments may discolor on account of the formation of cadmium or lead sulfide ("sulfur staining"). Organotin mercaptides have a typical odor, which is noticeable during the processing of rigid PVC but not in the finished product. With flexible PVC, the odor of stabilizer in finished articles is not observed in general, because usually only low concentrations are needed.

Organotin stabilizers migrate from rigid PVC only very slightly [67 to 86]. This fact, together with favorable toxicological properties is the basis for the worldwide approval of di-n-octyltin-bis(isooctylthioglycolate) for use in food packaging and potable water pipes [87]. Therefore, di-n-octyltin-bis(isooctylthioglycolate) is used to a large extent as stabilizer for films and bottles from rigid PVC. Recently, new types of monooctyltin mercaptides specifically designed for calendering have found approval by the Health Authorities in some countries [88].

Methyltin stabilizers, monobutylthiostannoic acid and estertin stabilizers are also approved in some countries for the aforementioned applications [88, 89]. Recently, dodecyltin mercaptides have found approval by the West German Health Authorities (BGA).

Methyltin stabilizers have a considerably higher vapor pressure compared with the analogous butyl- or octyltin compounds. Because of the volatility of methyltin stabilizers the monitoring and control of the threshold limit value (TLV) for tin (MAK in Germany 0.1 mg/m^3) has to be enforced strictly. Therefore, methyltin stabilizers when used in extrusion and in calendering require special, expensive security measures. In the USA, mainly monomethyltin mercaptides are used for the production of pipes of all kinds. In Europe, dimethyltin stabilizers are used only in profile extrusions and to some extent for bottles, fittings and fluidized bed coatings.

Butyltin mercaptides are utilized as universal stabilizers for the production of technical articles of all kinds, e.g. films, sheets, profiles, injection-molded articles, floor tiles and wall coverings. In the USA, they also are used for pipe extrusion.

Transparent articles for outdoor use are stabilized with sulfur-containing tin stabilizers in combination with suitable u.v. absorbers. In general, the addition of u.v. absorbers is not necessary for the production of profiles with a high content of titanium dioxide (10 to 15%) like siding, because the pigment provides a sufficient light stabilizing effect.

A special field of application for organotin stabilizers, both sulfur-containing and sulfur-free, is the production of rigid foamed PVC profiles.

By adding a chemical blowing agent, e.g. azodicarbonamide, nitrogen is released at the processing temperature, which expands the PVC after leaving the die of the extruder resulting in a very fine cell structure. Thus articles with low density and special properties can be made by this process. Tin stabilizers are activators (kickers) of the decomposition of the chemical blowing agents and influence the cell structure in a favorable way.

Recommended formulations

Transparent sheets for exterior application

PVC	100.0 parts
butyltin mercaptide	2.0 to 2.5 parts
fatty alcohol	0.5 to 0.8 parts
fatty acid ester	0.5 to 0.8 parts
polyethylene wax	0.1 to 0.2 parts
processing aid	1.0 to 2.0 parts
u.v. absorber (benzotriazole type)	0.3 to 0.5 parts

Bottles for food packaging

PVC	100.0 parts
processing aid	0.5 to 1.0 parts
methyl- or octyltin mercaptide	1.0 to 1.5 parts
lubricants	0.5 to 1.5 parts
impact modifier	5.0 to 10.0 parts

Glass-clear food packaging film

PVC	100.0 parts
octyltin mercaptide	1.0 to 1.5 parts
glycerol ester	0.5 to 1.0 parts
ester wax	0.1 to 0.4 parts

Potable water pipes

PVC	100.0 parts
methyltin mercaptide*)	0.3 to 0.4 parts
or octyltin mercaptide*)	
or butyltin mercaptide*)	
calcium stearate	0.5 to 0.8 parts
solid paraffin wax	0.6 to 0.8 parts
polyethylene wax	0.1 to 0.2 parts

*) butyltin mercaptide only for the US market according to NSF regulation, octyltin and methyltin mercaptides in compliance with specific national regulations.

4.3.1.2 Organotin carboxylates

Only dialkyltin derivatives of the following structures are of technical importance:

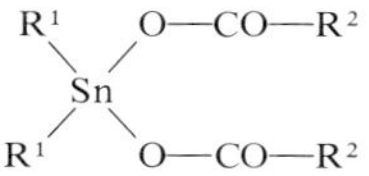

R^1	R^2
n-C_4H_9—	—alkyl
n-C_8H_{17}—	—CH=CH—CO—O—alkyl

Also dialkyltin maleate is of some importance:

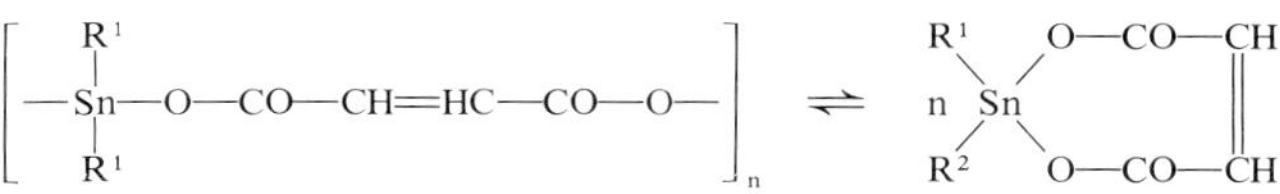

As already mentioned, organotin derivatives of maleic acid have an additional stabilizer function by reacting with polyene sequences according to *Diels-Alder* [61, 90].

The performance of sulfur-free tin stabilizers is good in all types of E-, S-, and M-PVC, copolymers, grafted polymers and polyblends. High transparency can be achieved by correct formulations. In contrast to sulfur-containing stabilizers, tin carboxylates impart excellent light stability. Because of these properties they are destined for rigid and flexible PVC articles in outdoor exposure, e.g. siding, window profiles, bench and fence profiles, films for agriculture, for swimming pools and roofs, tents etc. [91].

The efficiency of sulfur-free tin stabilizers can be improved by the addition of antioxidants. Low-volatile, sterically hindered phenols are most suitable. Certain difficulties may occur in rigid PVC processing by sticking of the melt to the metal of the processing machine. This can be avoided by the correct selection of lubricants or by use of modified tin maleates [92]. Vapors formed during processing may irritate mucous membrances and have to removed by exhaustors.

Recommended formulations

Window profiles

PVC	100.0 parts
impact modifier	6.0 to 12.0 parts
modified tin maleate stabilizer	2.0 to 2.5 parts
paraffin wax	0.6 to 1.2 parts
oxidized polyethylene wax	0.6 to 1.2 parts
antioxidant	0.1 parts
u.v. absorber	0.2 to 0.4 parts
titanium dioxide	2.0 to 4.0 parts
chalk	10.0 parts

Technical flexible film for outdoor application

PVC	100.0 parts
plasticizer	40.0 to 80.0 parts
epoxy plasticizer	2.0 to 3.0 parts
tin carboxylate stabilizer	1.0 to 2.5 parts
u.v. absorber	0.2 to 0.5 parts
oxidized polyethylene wax	0.2 to 0.4 parts

4.3.2 Barium/cadmium stabilizers

These are stabilizer systems based on various ratios of barium and cadmium salts primarily of carboxylic acids but sometimes of phenols. Straight chain natural fatty acids and branched synthetic acids with a chain of 8 to 20 C-atoms as well as alkylphenols and benzoic acid derivatives are used.

With the combination of barium and cadmium compounds a synergistic effect in thermostabilizing efficiency is achieved. Cadmium carboxylates give good early color but bad long-term stability, whereas barium compounds show the opposite behavior. The combination of these two compounds yields a good long-term stability (figure 5).

Many authors have tried to explain this synergism [48, 93 to 98]. Three fundamental reactions can be distinguished in the mode of action of barium and cadmium compounds (figure 6):

- Cadmium carboxylate is able to substitute labile chlorine atoms due to the complexing properties (reaction (a)) of the cadmium ion.

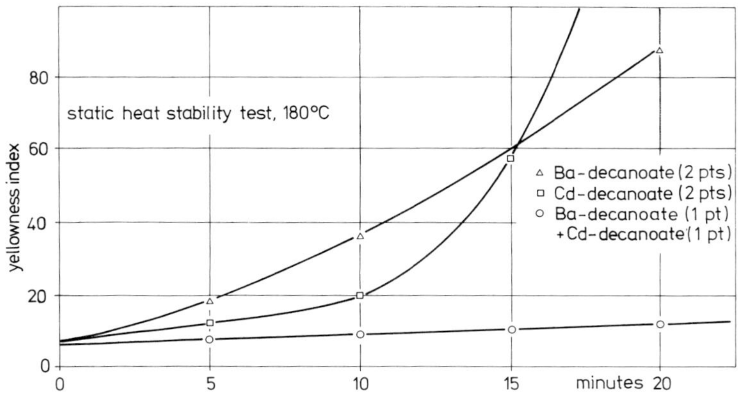

Fig. 5. Synergism of barium and cadmium carboxylate.

Formulation:	S-PVC (K-value 71)	100.0 parts
	dioctylphthalate	50.0 parts
	barium and/or cadmium decanoate (total)	2.0 parts

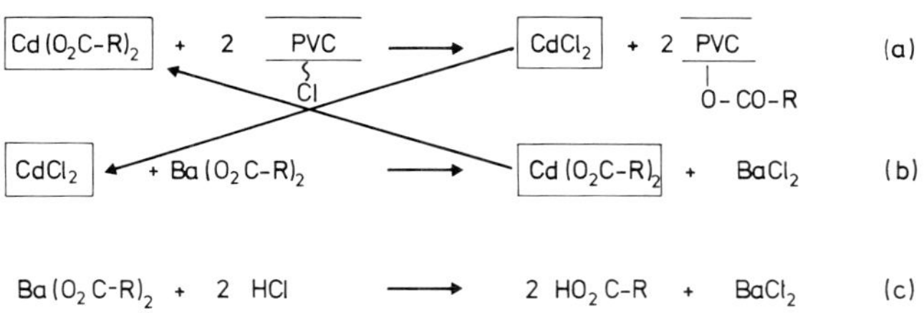

Fig. 6. The stabilizing reactions of barium and cadmium carboxylates in combination [50]

- Barium carboxylate reacts with hydrogen chloride preferably (reaction (c)).
- Cadmium chloride which is formed according to reaction (a) is transformed according to reaction (b) to cadmium carboxylate which may again substitute labile chlorine atoms. By this reaction the destabilizing effect of cadmium chloride, a *Lewis*-acid is blocked (at least temporarily).

Cadmium stabilizers differ from organotin stabilizers in that cadmium chloride formed from the cadmium carboxylate has a destabilizing effect (figure 7). Therefore, it is necessary to add deactivating components which do not contribute directly to the stabilization. In this connection polyols have to be mentioned (cf. 4.3.7). In general the addition of phenolic antioxidants increases the efficiency, especially in the absence of phosphites which have some inherent antioxidant properties.

Because of the complex composition there is a dependence on other components of the formulation (PVC type, plasticizer, fillers), with the consequence that the components of an optimum barium/cadmium stabilizer have to be varied regarding type and amount. This is also necessary with respect to the processing conditions as e.g. calendering, extrusion, injection molding and plastisol processing. Furthermore the requirements for the finished articles, e.g. high transparency, resistance to sulfur staining, printing and welding properties have to be considered. This is the reason why so many different barium/cadmium stabilizers are on the market, and why none of them is hardly identical with any other.

From a practical standpoint barium/cadmium stabilizers are separated into *powders* and *liquids*. Powders are based on metal salts of fatty acids with a chain of 8 to 18 C-atoms and contain also polyols and phenolic antioxidants. For practical applications they have to be combined with liquid phosphites and liquid epoxy plasticizers, which is to some extent

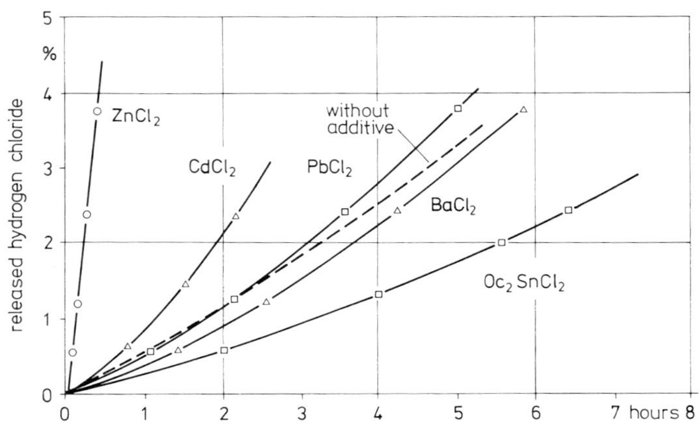

Fig. 7. Thermal dehydrochlorination of PVC at 175 °C in the presence of different metal chlorides, which may be regarded as secondary products of the corresponding stabilizers [50]

a little bit complicated. Barium/cadmium powders with a cadmium content above 10% yield a very good heat stability and are used mainly for the stabilization of rigid PVC. Their main application is in the extrusion of pigmented profiles and sheets in the building sector because of their contribution to light stability. However, highly transparent articles of rigid PVC cannot be produced with this type of stabilizer.

Barium/cadmium powders with a cadmium content below 10% are adequate for flexible PVC because of the lower demands. They are specifically used in this sector for the production of semirigid and flexible articles, as e.g. shoes, hoses, profiles etc. Powdery barium/cadmium stabilizers have generally sufficient self-lubrication. However, sometimes because of the processing conditions the amount of stabilizer has to be increased and so-called "over-lubrication" may occur. Often a combination of powdery and liquid stabilizers may be the solution to this problem. Liquid barium/cadmium stabilizers in general do not exhibit self-lubrication. But they have the advantage that they are easier to handle because the stabilizers already contain a phosphite. Preferably they are premixed with the plasticizer, which guarantees a good distribution. Because of their significantly lower metal content, they are sufficient in heat stability only for flexible PVC.

Very often liquid barium/cadmium stabilizers also contain zinc compounds, which provide improved early color and transparency. Furthermore, the sensitivity to sulfur staining is reduced by this addition. The correct choice of liquid barium/cadmium stabilizers yields very good transparency in flexible PVC. In recent years new liquid barium/cadmium stabilizers have been developed with sufficient self-lubrication for use in flexible PVC.

Nearly all barium/cadmium stabilizers impart to PVC a certain sensitivity to water, which is noticeable in a humid atmosphere by reversible cloudiness of the finished article. *Plate-out* is another big problem. This separation of ingredients of the formulation on hot or cooled metal parts may result in a surface defect of the finished article. Besides the addition of abrasives to reduce this tendency, there are now some new stabilizers on the market, which can be applied without any plate-out. These products have been developed specifically for calendering of flexible PVC. In general, liquid barium/cadmium stabilizers can be used universally for the processing of flexible PVC. By the calendering process all kinds of films are manufactured e.g. for decoration, furniture and agriculture, etc. Flexible profiles and hoses are produced by extrusion, and shoes, sandals and shoesoles by injection molding. In the processing of PVC pastes, liquid barium/cadmium stabilizers are also used to a large extent.

Powders are not suitable for plastisols, as they increase viscosity and may cause problems during incorporation.

In this connection it should be mentioned that the cadmium compound in a barium/cadmium stabilizer has an activating effect on the decomposition of the blowing agent azodicarbonamide. This effect is utilized for the production of foamed artificial leather. The temperature of decomposition of the blowing agent can be varied within certain limits by the addition of cadmium-containing stabilizers, also called "kickers". Zinc, tin and lead compounds are also effective as kickers.

Due to the content of free fatty acids, phenols and certain solvents, many barium/cadmium stabilizers have a typical odor, which is also noticeable in the finished product. In recent years, odorless liquid barium/cadmium stabilizers have been developed and are on the market.

As already mentioned, all barium/cadmium stabilizers in combination with phosphites and epoxy plasticizers impart very good light stability, which can be further improved by the addition of u.v. absorbers. Formulations based on barium/cadmium are therefore very often used in outdoor applications.

Recommended formulations

White outdoor profile

PVC	100.0 parts
impact modifier	6.0 to 12.0 parts
solid Ba/Cd stabilizer	2.5 to 3.0 parts
phosphite	0.5 to 0.7 parts
epoxidized soybean oil	1.0 to 1.5 parts
12-hydroxy stearic acid	0.4 to 0.6 parts
ester wax	0.4 to 0.6 parts
antioxidant	0.1 to 0.2 parts
titanium dioxide	2.0 to 4.0 parts
chalk	0.0 to 10.0 parts

Furniture film

PVC	100.0 parts
plasticizer	15.0 to 25.0 parts
epoxidized soybean oil	2.0 to 3.0 parts
liquid Ba/Cd stabilizer	1.5 to 2.0 parts
u.v. absorber	0.2 to 0.4 parts
bis-amide wax	0.2 to 0.4 parts

processing aid	1.0 parts
pigments	0.5 to 10.0 parts

Calendered floor tiles

PVC	100.0 parts
plasticizer	30.0 to 70.0 parts
epoxidized soybean oil	2.0 to 5.0 parts
solid Ba/Cd stabilizer	1.5 to 2.5 parts
stearic acid	0.1 to 0.4 parts
chalk	50.0 to 100.0 parts

Shoes

PVC	100.0 parts
plasticizer	50.0 to 80.0 parts
epoxidized soybean oil	2.0 to 3.0 parts
liquid Ba/Cd stabilizer	1.5 to 2.0 parts
stearic acid	0.2 to 0.4 parts

4.3.3 Barium/zinc and calcium/zinc stabilizers

This group of stabilizers is very similar in chemical composition to the aforementioned barium/cadmium stabilizers. Here the combination of zinc carboxylate with barium or calcium carboxylate results in a synergistic effect.

The reaction scheme outlined for barium/cadmium carboxylates (figure 6) can be applied for this combination, in which zinc carboxylates possess the ability to substitute labile chlorine atoms.

The destabilizing effect of zinc chloride is much more pronounced than that of cadmium chloride, as can be seen in figure 7. This is due to zinc chloride being the stronger *Lewis*-acid.

The enormous steep rise of the dehydrochlorination curve in the presence of zinc chloride explains the sudden black discoloration of PVC after a certain thermal stress when using e.g. zinc stearate (so-called zinc burning). In order to eliminate or at least retard this destabilizing effect, the inclusion of suitable components is necessary for zinc-containing stabilizers. Barium and calcium carboxylates suppress the detrimental effects of zinc chloride for a certain time, according to the reaction described in figure 6. In addition to that, complex compounds, formed from the chlorides of barium or calcium with zinc chloride, cannot be excluded. Polyols, epoxidized fatty acid esters and phosphites are also suitable to slow down the acceleration of the dehydrochlorination by zinc chloride [99].

Interest in cadmium-free stabilizers has grown in the past years with respect to occupational health and environmental protection. Therefore, *barium/zinc stabilizers* have been developed, which are very similar in their composition to barium/cadmium stabilizers. There are powdery and liquid barium/zinc stabilizers, which contain additional additives, in order to achieve optimum performance. They are used mainly in calendering of plasticized PVC and in plastisol processing. There are liquid barium/zinc stabilizers available, which are comparable with liquid barium/cadmium stabilizers in respect to performance, compatibility with the common plasticizers and transparency of the stabilized PVC. Recently, there are also products on the market which are odorless and allow processing of plasticized PVC without any plate-out and which exhibit excellent self-lubrication. With regard to light stability a slight change for the worse is observed with barium/zinc stabilizers but this can be compensated by a higher level of epoxy plasticizer and by the incorporation of a u.v. absorber. Restrictions have to be made for zinc sensitive PVC types as e.g. mass PVC, certain S-PVC grades and copolymers.

Calcium/zinc stabilizers would not have gained too much importance if the two non-toxic components, calcium and zinc stearate, had not received worldwide approval for the stabilization of food packaging, toys and blood transfusion tubings. By variation of the calcium and zinc ratio the required properties can be adjusted: an increase of calcium results in poorer early color but better long-term stability and an increase in transparency. The combination has good self-lubrication which is sufficient for the processing of flexible PVC. The viscosity of PVC pastes, however, is very much increased by calcium/zinc stabilizers. As far as the regulations permit, combinations of calcium and zinc salts, which are soluble in organic solvents, can be combined with costabilizers, which have only little influence on viscosity. This is of special importance for coatings and rotational molding of plasticized PVC.

The performance of calcium and zinc stearate can be improved by the addition of epoxy plasticizers, polyols, phosphites and antioxidants. The stabilizers on the market consist of such mixtures. Furthermore, there are calcium/zinc combinations for technical applications, which are not approved for food packaging. They fulfill the requirements of plastisol processing but cannot compete with the efficiency of barium/cadmium stabilizers.

Recommended formulations

Non-toxic flexible film

PVC	100.0 parts
plasticizer	30.0 to 70.0 parts
epoxidized soybean oil	2.0 to 1.0 parts
calcium/zinc stabilizer	2.0 to 3.0 parts
stearic acid	0.1 to 0.2 parts

Artificial leathercloth

Base coat

PVC	100.0 parts
plasticizer	50.0 to 100.0 parts
epoxidized soybean oil	2.0 to 3.0 parts
barium/zinc stabilizer	1.5 to 2.0 parts
filler	0 to 20.0 parts

Top coat

PVC	100.0 parts
plasticizer	40.0 to 60.0 parts
epoxidized soybean oil	3.0 to 5.0 parts
barium/zinc stabilizer	1.5 to 2.0 parts
u.v. absorber	0.2 to 0.3 parts

4.3.4 Lead stabilizers

Lead compounds belong to one of the oldest group of heat stabilizers [100, 101]. Bivalent lead possesses the ability for complex formation, which is necessary for the substitution of labile chlorine atoms. Another advantage of lead stabilizers is the fact that lead chloride, formed during the stabilizing process, has no destabilizing effect on PVC (see figure 7). Consequently, costabilizers like polyols, phosphites, etc. do not increase the efficiency of lead stabilizers to a noticeable extent. This is in contrast to metal soaps containing cadmium or zinc.

The most important compounds are:

- tribasic lead sulfate: $3\,PbO{\cdot}PbSO_4{\cdot}H_2O$
- dibasic lead phosphite: $2\,PbO{\cdot}PbHPO_3{\cdot}{}^1/_2\,H_2O$
- dibasic lead phthalate: $2\,PbO{\cdot}Pb(OOC)_2C_6H_4$
- dibasic lead stearate: $2\,PbO{\cdot}Pb(OOCC_{17}H_{35})_2$
- normal lead stearate: $Pb(OOCC_{17}H_{35})_2$
- dibasic lead carbonate: $2\,PbO{\cdot}PbCO_3$

The basic compounds react with hydrogen chloride with the formation of lead chloride. Of the aforementioned products, lead stearate imparts a remarkable lubricating effect. Therefore, this compound has a double function during PVC processing. In general, lead stearate is combined with basic lead salts, a very efficient mixture which is suitable for all PVC types. The basicity of lead stabilizers has to be considered carefully in alkaline-prestabilized emulsion PVC.

Powdery lead stabilizers have a high specific gravity, which may lead to separation in PVC dry blends. Due to their small particle size they tend to be dusting, which may cause occupational health problems. This disadvantage has been overcome in the past years by mixing lead stabilizers, lead stearates and lubricants in so-called one-packs. These one-packs are often packaged in polyester bags, weight-adjusted for the mixing batch, or they are supplied as granules, flakes or dustless, free-flowing pellets.

Because lead stabilizers, as well as their reaction products in PVC, do not increase the conductivity of PVC, the application for cable insulation is one of the preferred domains of this class of stabilizers. Only with lead stabilizers can the required good electrical properties be guaranteed as well as the necessary heat stability required under service conditions of insulating materials.

Transparent PVC articles cannot be produced with lead stabilizers because these stabilizers act as pigments. Furthermore, these stabilizers are sensitive to sulfur-containing compounds. This may lead to discoloration if sufficient dispersion is not achieved. For most of the articles stabilized with lead compounds, this is not of importance, but one has to select suitable pigments in order to avoid color changes due to the reaction with lead salts. In general, one may say that the right combination of basic lead stabilizers with lead stearates and suitable lubricants yields very efficient stabilizing systems.

For this reason and because of their good processibility, lead stabilizers have a broad field of applications. Furthermore, they allow the production of articles with a high Vicat softening point and, therefore, good mechanical properties, e.g. pressure pipes, sewage pipes, gas pipes, electrical conduits, gutters, profiles for in- and outdoor applications.

Recommended formulations

Wire and cable insulating material

PVC	100.0 parts
plasticizer	30.0 to 60.0 parts

tribasic lead sulfate	2.0 to 4.0 parts
normal lead stearate	0.5 to 1.5 parts
chalk	20.0 to 40.0 parts
stearic acid and/or paraffin wax	0.0 to 0.2 parts

Pressure pipes

PVC	100.0 parts
tribasic lead sulfate	0.5 to 1.0 parts
dibasic lead stearate	0.5 to 1.0 parts
calcium stearate	0.2 to 0.4 parts
stearic acid and/or paraffin wax	0.2 to 0.5 parts

4.3.5 Metal-free stabilizers

In 1940, urea derivatives, e.g. diphenylthiourea, had been recommended for the stabilization of PVC [102]. Nearly all metal-free heat stabilizers are nitrogen-containing compounds, characterized by low basicity. In general, they are only efficient in E-PVC, however β-aminocrotonates can also be used for S-PVC.

β-Aminocrotonates

The most important compounds have the following structures:

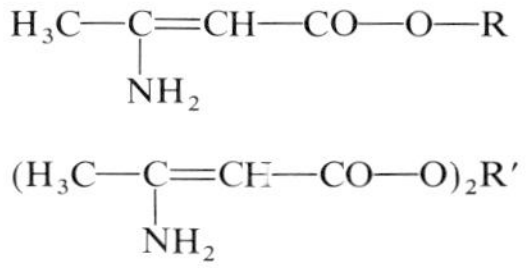

R = higher alkyl group

R′ = $-(CH_2)_4-$, $-CH_2-CH_2-S-CH_2-CH_2-$

Aminocrotonates show a significant induction period in the dehydrochlorination curve, which depends on concentration (figure 8). The existence of an induction period may lead to the conclusion that this type of stabilizer can also bind hydrogen chloride.

From investigations by *Guyot* et al. [55] with model substances it has been concluded that aminocrotonates may also substitute labile chlorine atoms (see reaction (11)).

Aminocrotonates of the above structures are approved for the manufacture of food packaging in the Federal Republic of Germany [87] and in many other countries at levels up to 3%.

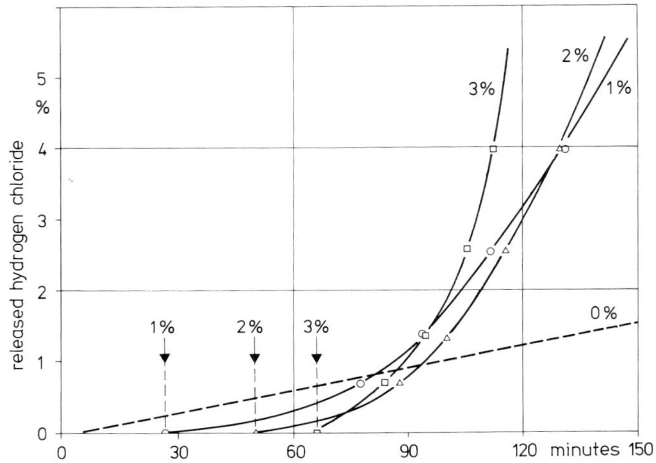

Fig. 8. Thermal dehydrochlorination of PVC at 175 °C in the presence of different levels of thiodiethyleneglycol-bis-β-aminocrotonate [50].

Therefore, aminocrotonates are used despite their moderate heat stabilizing effect (especially in view of the poor early color) for the production of blown films. These blown films are used to produce containers by thermoforming for margarine, cakes and pralines. Waste from this process is recycled immediately. Aminocrotonates are also common for the production of packaging for non-food articles and for the production of records, because of the low background noise achieved with these metal-free stabilizers.

As already mentioned, aminocrotonates are efficient in both suspension and mass PVC as opposed to other metal-free thermostabilizers. Aminocrotonates are used mainly in combination with calcium/zinc stabilizers and epoxy plasticizers, because they considerably improve the heat stabilizing efficiency of calcium/zinc stabilizers (figure 9).

Aminocrotonates cannot be used for the production of mineral water bottles, because of the evolution of ammonia during processing, which must not be detectable in potable water according to legislation.

2-Phenylindole

The prototype of this stabilizer group:

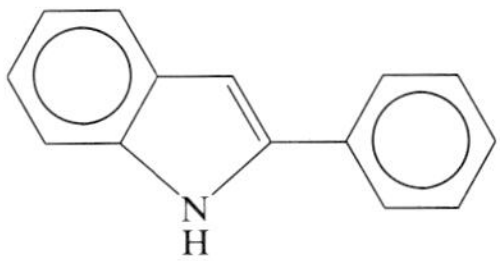

is a long-known, metal-free heat stabilizer [103]. It could be proved with 4-chlorhexene-2 as a model substance that 2-phenylindole reacts with labile chlorine atoms [104]:

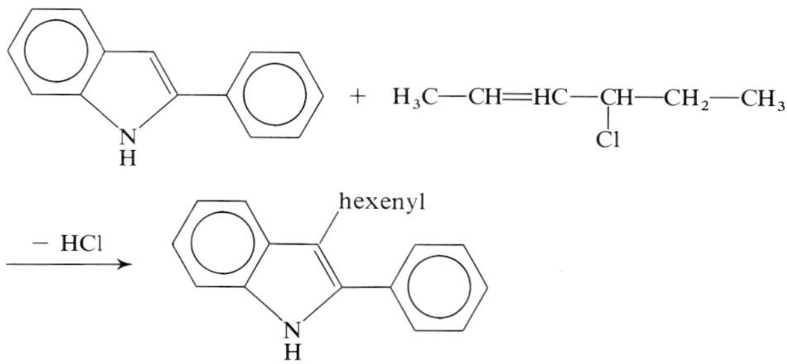

(17)

It is remarkable that the substitution does not occur at the nitrogen but in the 3-position of the indole ring. The binding of hydrogen chloride may be also possible with 2-phenylindole, but in any case a significant improvement of efficiency can be achieved by the combination with typical acceptors for hydrogen chloride.

2-Phenylindole without any additional additive is not a very good heat stabilizer and is only effective in alkaline-prestabilized emulsion PVC. Initial color is especially not satisfactory.

In S-PVC, 2-phenylindole is used exclusively in combination with calcium/zinc compounds, because it imparts significant increases in the efficiency of this system (figure 9). In contrast to aminocrotonates no ammonia is emitted during processing. Therefore, 2-phenylindole was common for many years in France for the production of PVC mineral water bottles. Because of the high volatility of 2-phenylindole and the attendant problems during processing, less volatile, substituted 2-phenylindole derivatives have been developed.

Recently 2-phenylindole and the aforementioned derivatives have been removed from the list of approved stabilizers in France (July 1978).

Two new metal-free compounds are used in France to an increasing extent for the improvement of calcium/zinc stabilizers for mineral water bottles:

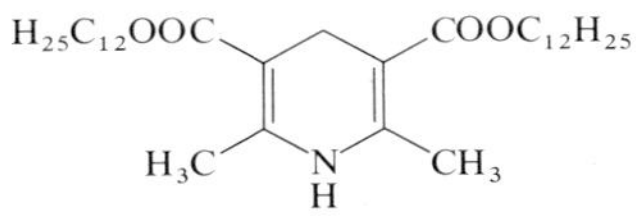

1,4-Dihydro-2,6-dimethyl-3,5-dicarbododecycloxypyridine [105, 106, 108]

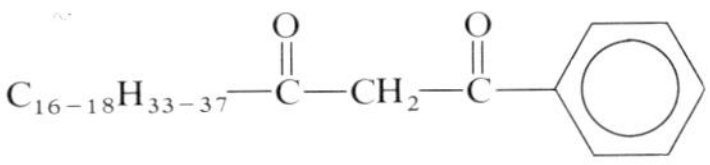

(Palmitoyl)stearoylbenzoylmethane [107 to 109]

These products impart excellent early color in combination with calcium/zinc systems at low concentrations of ca. 0.2%. Therefore, they are gaining increasing importance for this application.

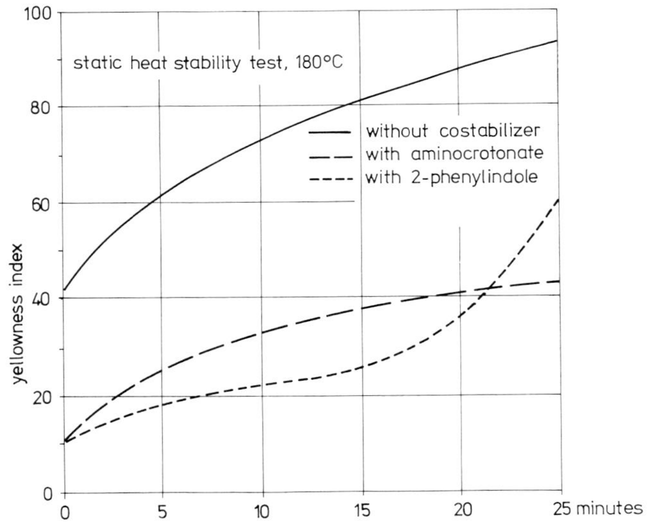

Fig. 9. Costabilizing efficiency of 2-phenylindole and thiodiethyleneglycol-bis-β-aminocrotonate.

Formulation:	
S-PVC (K-value 58)	100.0 parts
epoxidized soybean oil	2.0 parts
calcium stearate	0.9 parts
zinc stearate	0.6 parts
tris(nonylphenyl)phosphite	0.5 parts
costabilizer	0.4 parts

Urea derivatives

No investigations are known on the mode of action of the two urea derivatives, phenylurea and diphenylthiourea:

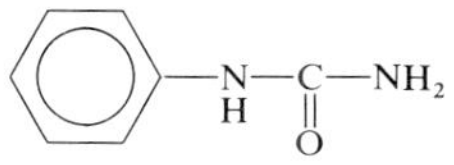

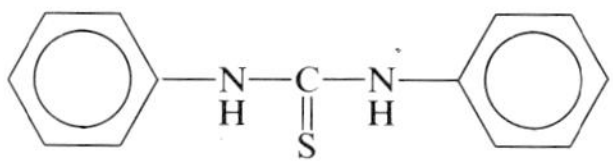

Diphenylthiourea is approved for the manufacture of food packaging in many countries. It is used in the Luvitherm process for sheets made from alkaline-prestabilized emulsion PVC.

4.3.6 Other heat stabilizers

Although there are many metals recommended as components for PVC stabilizers in the patent literature, other metal compounds, besides the ones already mentioned, have gained only modest importance.

Antimony-tris-mercaptides impart in low amounts sufficient heat stability to rigid PVC, especially in combination with calcium compounds. This stabilizing system was put to use for records, recently also for pipes in the USA. A disadvantage of all antimony-containing stabilizers up to now is their insufficient light stability. Recently, improved antimony stabilizers have been introduced into the US market.

4.3.7 Costabilizers

Costabilizers are compounds which do not possess a thermostabilizing effect as such, but improve the efficiency of other stabilizer systems. To this group belong esters of phosphorous acid (phosphites), epoxy compounds (epoxy plasticizers), polyols and phenolic antioxidants. Also stearoylbenzoylmethane has to be regarded as costabilizer, because this substance has no thermostabilizing effect as such.

Organic phosphites

Organic phosphites are important costabilizers, especially in combination with barium/cadmium, barium/zinc and calcium/zinc systems. Of special importance are the alkyl/aryl phosphites:

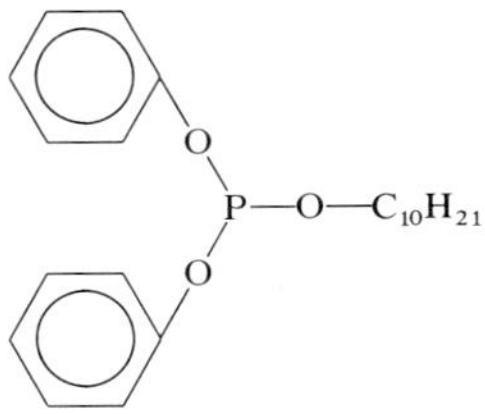

Diphenyl-decylphosphite

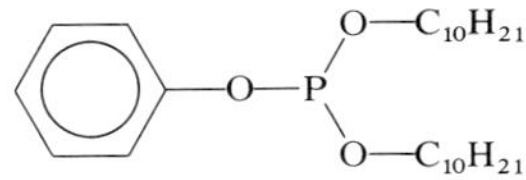

Phenyl-didecylphosphite

Their costabilizing efficiency in barium/cadmium stabilizers is remarkable (figure 10).

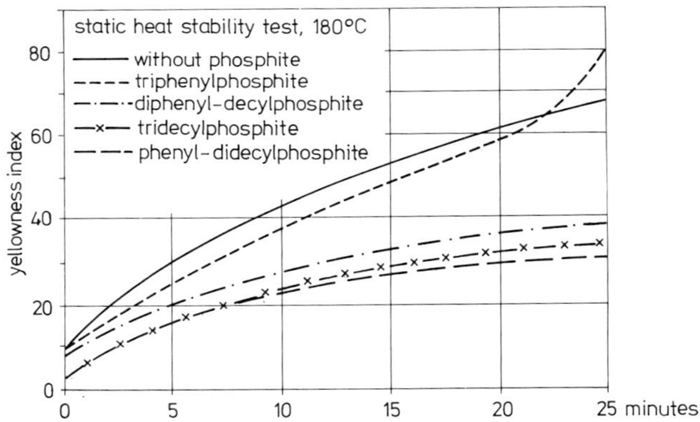

Fig. 10. Costabilizing efficiency of different phosphites in combination with barium/cadmium stearate.

Formulation:	
S-PVC (K-value 64)	100.0 parts
Ba/Cd stearate	1.2 parts
phosphite	0.4 parts

A number of experts have tried to elucidate the mode of action of phosphites [110 to 117]. From these investigations the following stabilizing reactions may be assumed as proved:

- binding of hydrogen chloride,
- addition to polyene sequences,
- antioxidative effect.

In unplasticized PVC, early color, transparency and light stability are improved by addition of aryl/alkyl phosphites. In plasticized PVC, the heat stability also can be improved. Furthermore, the addition of liquid phosphites may decrease the melt viscosity and may so contribute to an easier and smoother processing. The level of phosphites ranges from 1:3 to 1:7 in relation to the heat stabilizer.

Epoxy compounds

The most common costabilizers on the basis of epoxy compounds are epoxidized soybean oil and epoxidized esters of oleic acid and tall oil fatty acids. Epoxidized castor oil, linseed oil and sunflower oil also are used occasionally, as well as epoxy resins. The most important representatives have the following structure:

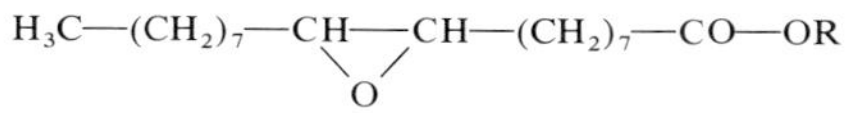

The mode of action in combination with metal carboxylates has been studied extensively [118, 119]. According to these investigations such compounds are not only able to bind hydrogen chloride according to reaction (18):

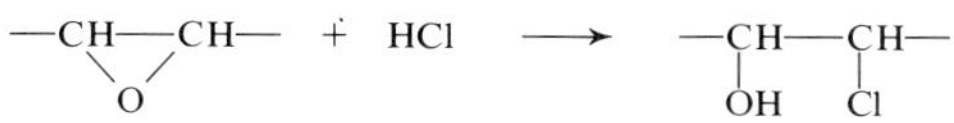

(18)

but they are also able to substitute labile chlorine atoms in PVC under the catalytic influence of cadmium or zinc ions:

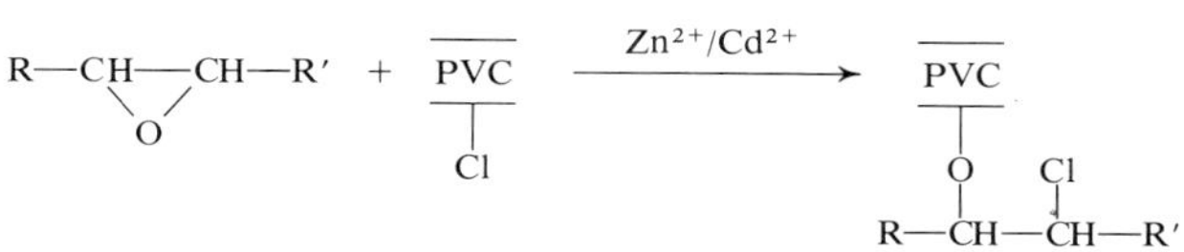

(19)

Epoxidized oils and oleates act also as plasticizers. Therefore, the amount of primary plasticizer can be reduced to a certain extent in flexible PVC formulations. The increase in efficiency by epoxy stabilizers depends on the stabilizing system and has different effects on thermal and light stability. The most significant effects in thermal stability are observed with calcium/zinc, barium/zinc and metal-free stabilizers. Still noticeable effects can be seen with barium/cadmium and lead stabilizers, whereas with organotin stabilizers an improvement is often not noticeable. With regard to light stability the addition of an epoxide has the strongest influence on barium/zinc and barium/cadmium stabilizers.

Epoxy plasticizers have to be considered as secondary plasticizers, i.e. their sole use in larger amounts may lead to exudation. Epoxidized oleates have lower viscosity and confer good flexibility at lower temperatures. On the other hand the volatility of epoxidized oils (triglycerides) is lower and resistance to extraction is increased. The effect of binding hydrogen chloride is especially useful for the stabilization of chloroparaffin-containing mixtures. Figure 11 shows the costabilizing effect of epoxidized soybean oil and phenyl-didecylphosphite and their mixture with barium/cadmium stearate.

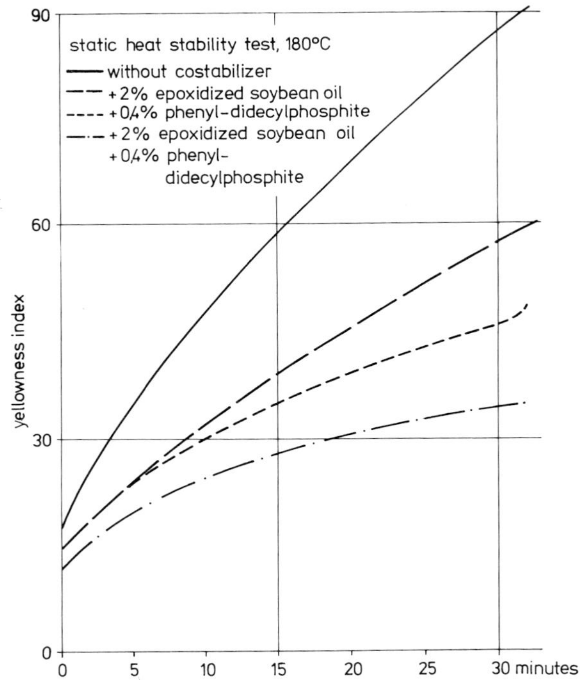

Fig. 11. Costabilizing efficiency of epoxidized soybean oil and phenyl-didecylphosphite as well as improvement by combinations of both costabilizers.

Formulation:

S-PVC (K-value 64)	100.0 parts
Ba/Cd stearate	1.2 parts

Polyols

These are polyvalent alcohols such as the following:

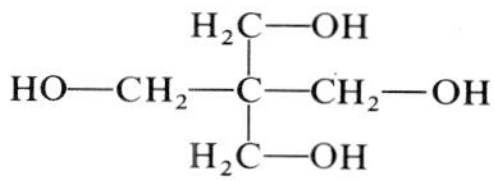

Pentaerythritol

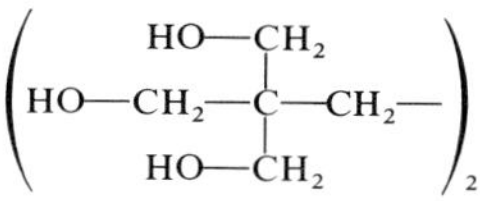

Dipentaerythritol

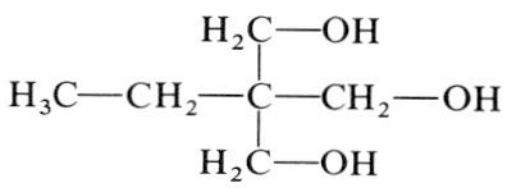

Trismethylolpropane

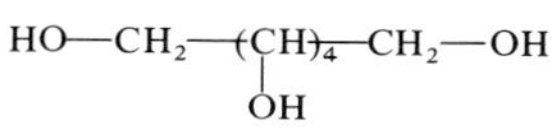

Sorbitol

The function of these polyols is complex formation and, therefore, deactivation of zinc or cadmium chloride in the corresponding stabilizer systems [50]. They are ingredients in many of the common stabilizing systems.

Antioxidants

A heat stabilizer should also have antioxidative effectiveness. This property is only inherent in sulfur-containing organotin stabilizers. The combination of heat stabilizers with phosphites leads in general to an antioxidative effect of some extent. Otherwise, the addition of phenolic antioxidants causes a considerable increase in efficiency. Most of the commercial stabilizers already contain an antioxidant. The addition of antioxidants is essential to PVC, which contains also other polymers which may be degraded by autooxidation.

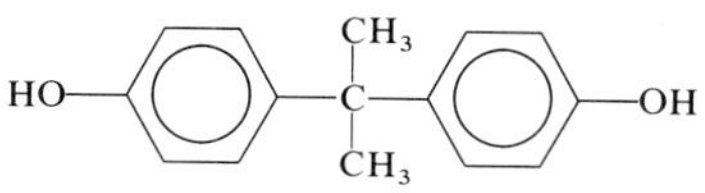

Bis-phenol A

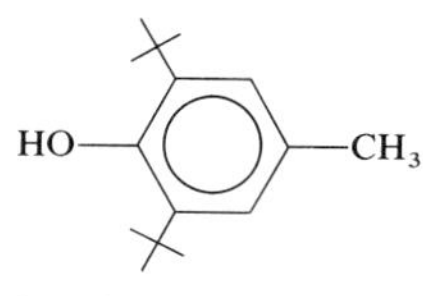

2,6-Di-tert.butyl-
4-methylphenol (BHT)

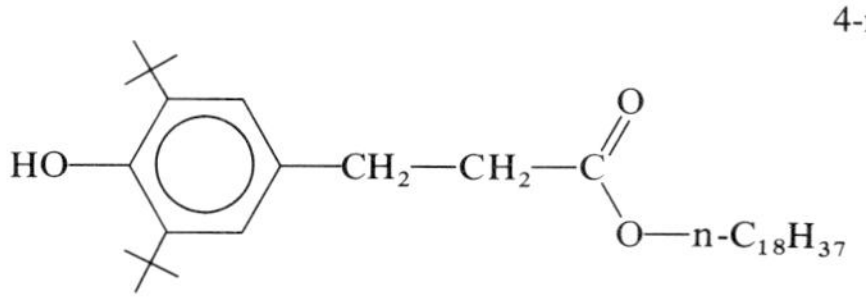

Octadecyl 3(3,5-di-tert.butyl-4-hydroxyphenyl)propionate

Table 1. Fields of application of heat stabilizers in rigid PVC

Article		Processing technique*)	Stabilizers**)							
			MeSnS	OcSnS	BuSnS	BuSnO	Ba/Cd powder	Ca/Zn	Pb	metal-free stab.
Films	transparent or pigmented	cal., extr.								
	for non-food application				×		×			
	for food application			×				×		×
Bottles	transparent or pigmented	blow molding								
	for non-food application				×					
	for food application		×	×				×		×
Pipes, fittings	for potable water	extr.	×	×					×	
	sewage	i.m.							×	
	cable conduit				×					
Sheets, profiles	transparent, indoor	extr.	×		×					
	outdoor				×	×				
	pigmented, indoor	extr.	×		×		×		×	
	outdoor					×	×		×	
Profiles	foamed	extr.			×		×		×	
Records		press			×				×	×

*) *Manufacturing processes:*
cal.: calendering
extr.: extrusion
i.m.: injection molding
press: compression molding

**) *Stabilizers:*
MeSnS: S-containing methyltin compounds
OcSnS: S-containing octyltin compounds
BuSnS: S-containing butyltin compounds
BuSnO: S-free butyltin compounds

Table 2. Fields of application of heat stabilizers in flexible PVC

Article		Processing technique*)	Stabilizers**)						
			BuSnS	BuSnO	Ba/Cd powder	Ba/Cd liquid	Ba/Zn	Ca/Zn	Pb
Films	transparent or pigmented	cal.							
	for non-food application		×	×	×	×	×		
	for food application							×	
Profiles, flex. tubes	transparent or pigmented	extr.							
	for non-food application		×	×	×	×			
	for food application							×	
Cable insulation and jacketing		extr.							×
Shoes, sandals, soles		i.m.	×	×	×	×			×
Coatings, artificial leather		coating	×	×		×	×	×	
Dippings				×		×	×	×	

*) *Manufacturing processes:*

cal.: calendering
extr.: extrusion
i.m.: injection molding

**) *Stabilizers:*

BuSnS: S-containing butyltin compounds
BuSnO: S-free butyltin compounds

4.3.8 Summary of recommendations for application

The most important guidelines for the use of heat stabilizers in relation to the PVC processing methods are given in tables 1 and 2.

4.3.9 Test methods

As the requirements for a heat stabilizer may be varied and, as heat stabilizers may have a great influence on the properties of a PVC compound, they are subjected to a versatile test program. The test methods, which have been developed over many years, are partly subject of national and international standardization [120]. Laboratory-scale test methods do not always permit conclusions for practical use in a short time and at low expense. However, valuable knowledge of the PVC processing and the properties of the finished article under service conditions can be generated from these laboratory results, if they are evaluated in the right manner. The influence of the processing technique, of the production machinery and of the design parameters cannot always be calculated. Here the test results have to be confirmed by test runs on production equipment.

In the following chapters the most important test methods are outlined as they have been introduced for testing and assessment of heat stabilizers in application laboratories.

4.3.9.1 Heat stability

Static heat tests

A very well-known procedure for testing the heat stabilizing efficiency of stabilizers is to assess the change in appearance of test sheets at constant elevated temperatures as a function of time.

Test sheets are prepared according to standard procedures and formulations, in which all ingredients except the stabilizer system are unchanged. Commonly, the preparation of the test sheets is performed on a two-roll mill. It is essential that the thermal stress remains constant. Pieces of the sheet of equal size are aged in a test oven at constant temperatures, e.g. 180 °C. For this purpose special test ovens have been designed, which guarantee constant temperature. It is advantageous to use an oven fitted with a rotating cylinder with self-levelling shelves on which the samples are laid, in order to equalize heat differences in the oven. At set intervals, e.g. every 10 or 15 minutes, a sample is taken in order to assess the change of color. The test is continued until a significant change to darker colors has occurred (ASTM D 2115, ISO 305–1976). The optical comparison can be done visually as well as by measuring the Yellowness Index according to ASTM D 1925–70 [121].

Visual comparison of the aged PVC samples assesses, which stabilizer imparts better early color or long-term stability. Procedures are available for visual color matching by experts, but in practice standardized conditions are often missing. This is the reason for errors due to subjective factors. A far better assessment of stabilizer performance via color changes is possible by measuring the Yellowness Index. This method permits the exact detection of slight yellowing in the very early stage of discoloration. Spectrophotometers or three-filter photometers are suitable instruments for this purpose. Results may be plotted easily in the form of graphs or tabulations and allow precise documentation and production of copies.

In addition to the already explained heat test, there are some other static test methods. There are test ovens on the market with the possibility of continuous withdrawal of strip-shaped samples on their support. By this method an assessment of continuous discoloration is possible.

Thermal stressing of sheets in a multiplaten press where oxygen is excluded permits evaluating the effects of thermal stress during extrusion or injection molding.

Infrequently-used static heat tests include immersion of samples into inert heat transfer oils or the procedure according to DIN 53381, part 2. In this test, specimen are placed on specially shaped pieces of aluminum fitting into test tubes, which are heated by a constant temperature oil bath.

With plastisols, best results can be obtained by coating them on a support, e.g. release paper, glass plates or textiles followed by gelation. The samples are subsequently oven aged (DIN 54800).

Dynamic heat tests

In the dynamic heat test the sample is not only thermally stressed but also subjected to shearing forces. The following dynamic heat tests are used:

- continuous stress on a two-roll mill (continuous rolling test),
- plastograph test,
- multiple extrusion.

For the continuous rolling test, the PVC compound is first fused on a two-roll mill at constant temperature and subjected to continuous stress (180 to 200 °C) on the same mill. At set intervals samples are taken from the center of the sheet and used for the assessment of color change.

The dynamic test may also be performed in the kneading chamber of a plastograph, e.g. Brabender at elevated temperatures. After plasticization, samples are taken at constant time intervals and pressed to plaques of equal thickness. A further possibility for a dynamic heat test is multiple extrusion. The test material is fed to an extruder and the extruded material is recycled. The change in color is assessed after each cycle.

Determination of dehydrochlorination

Thermal damage of PVC is detectable not only by discoloration but also by dehydrochlorination. In this test the amount of hydrogen chloride emitted at constant temperature can be used for the assessment of heat stability (DIN 53381, part 3, ISO R 182–1970, ASTM D 793).

But there is no definite correlation between the discoloration and the amount of dehydrochlorination. For this reason, the determination of dehydrochlorination is more of scientific interest. So this method is used for the screening of new stabilizers, because it allows more insight into the chemical reactions. The time until hydrogen chloride is released (induction period) and the rate of the subsequent hydrogen chloride evolution are measured (see figure 3). The equipment has a device to use either PVC powder mixtures or cut PVC particles. This device is heated at constant temperature and the evolved hydrogen chloride is transferred by an air or nitrogen stream to an absorption flask, where the HCl concentration can be measured either using the conductivity method or according to the pH value method.

Residual heat stability

The assessment of the residual heat stability of PVC articles is performed according to the static heat test by storing samples in a test oven or according to the dehydrochlorination method. A simplified method, very common in the cable industry, is the Congo Red Test (DIN 53381, part 1, ISO R 182–1970). For this test the sample is cut into small pieces and heated in a test tube. Hydrogen chloride released from the sample causes a color change of wet Congo Red or Universal Indicator Paper on top of the test tube. The time is measured from the insertion of the sample in the heating medium until the color change of the indicator paper. From the results one may conclude deficiency in stabilization or over-stabilization. If there are comparative data available, a certain conclusion may be drawn regarding light stability, too.

4.3.9.2 Influence of heat stabilizers on processing properties

The assessment of processing properties of rigid and plasticized PVC in the laboratory is based mainly on the gelation properties and the rheology of the PVC melt at processing temperature. For this purpose there are a number of test devices with which other processing properties can be checked too, e.g. plate-out, sticking to metals, defects on the surface, melt elasticity, etc. The test on a two-roll mill is very common, whereby the milling properties and the material in the roller gap are visually assessed. Two-roll mills for laboratories can also be equipped with devices for the evaluation of pressure and torque on the rolls. They also permit the adjustment of a definite friction.

Torque rheometers have found approval as universal test devices. They have a closed kneading chamber, which can be heated to different temperatures. After charging the PVC mixture, which quantity may vary within certain limits, and after setting the speed of rotation and the temperature, a torque curve is plotted against time. From this the course and the speed of plasticization, as well as the resulting melt viscosity can be read. Comparing the results with a known formulation, the processing properties can be anticipated. This result can be checked and improved by test runs on a laboratory extruder.

Laboratory extruders equipped with suitable instruments (extrusiometer) allow predictions not only for extrusion, but also e.g. for calendering and injection molding, always in comparison with known formulations. Conclusions regarding the processing behavior can be drawn by evaluating pressure and temperature build-up of the mass in the barrel, by varying the speed of the screw and the heating temperature. Furthermore, torque, output and quality of the extrudate are determined.

The processing behavior of plastisols in the presence of different heat stabilizers can be checked in the laboratory by measuring their viscosity at room temperature via rotational viscosimeter or via a laboratory coating machine.

4.3.9.3 Electrical properties

The measurement of the electrical resistance is performed according to the procedures for testing insulating materials (DIN 53482, IEC 93, ASTM D 275) or according to analogous VDE methods (Federal Republic of Germany).

4.3.9.4 Weather and light stability

Reliable testing of weather and light stability can only be achieved by outdoor exposure in the corresponding climate. This test can be carried out according to DIN 53386 or ASTM D 1435–75. Under natural conditions it takes indeed a long time to get results. Therefore, methods have been developed to simulate outdoor weathering conditions in specific equipment in order to shorten the exposure period. There are several different apparatuses on the market with a variety of artificial light sources including xenon arc lamps, carbon arc lamps, mercury lamps, fluorescent sunlamps. Xenotest machines with high or low pressure xenon lamps are very popular in Europe. By selecting the right filter systems the spectrum of this lamp is the best approximation of global radiation. By applying different cycles of light exposure and water spraying it is possible to improve the simulation of the different natural climates (cf. 3.4.1, 3.4.2). Nevertheless, care has to be taken when results from artificial light exposure have to be correlated with outdoor weathering results, especially if for a stabilizing system the correlation between the exposure in the respective equipment and results of weathering in the respective part of the world is not known [122].

The assessment of weathering results may also differ depending on the parameter used, i.e. whether discoloration, residual stability, or mechanical properties are taken as a measure of the degree of photooxidation.

4.4 Outlook and technological trend

In times of an economy which is characterized by slow growth rates or even stagnation, the increase of costs, caused for instance by higher cost of labor and raw materials, can be absorbed only with more and more difficulty. Simultaneously, it is realized to a growing extent that great efforts have to be made for the protection of our environment. This also leads inevitably to higher costs.

Therefore, the possibility to improve the efficiency of the PVC industry is of great importance for the future. Greater output at equal or even higher product quality will be required; the automation of production deserves high priority. The achievement of these goals should contribute to significant cost reductions.

Additive packages – also called one-packs – may fulfill these demands. Besides heat stabilizers they contain lubricants, costabilizers, antioxidants, processing aids, sometimes also pigments and fillers. For many years solid lead stabilizers have been on the market which are combined with other additives to form dust-free grades, e.g. granulates, flakes or prepacks. These non-dusting products allow safe handling in respect to occupational health and environment.

A rationalization effect can be achieved by these additive packages, because of their simple storage and fewer weighing steps. This means lower costs, fewer errors and very often also the possibility of automatic dosage and a computer-controlled process for the total production. The demand for simpler and easier handling is also achieved by pre-weighed mixtures packed in bags and adjusted to a certain charge of PVC compound.

This tendency is also noticeable with other groups of stabilizers [123]. The development of liquid barium/cadmium stabilizers with self-lubrication was another step in this direction. The complicated process of using powdered barium/cadmium stabilizers together with liquid costabilizers for calendering of flexible PVC could be simplified by these liquid self-lubricating systems. In the future, solid barium/cadmium stabilizers will be on the market only in a dust-free grade.

Some time ago, there was a similar tendency towards additive packages based on tin stabilizers. They had been developed mainly for reasons of rationalization and had to be of a real advantage for the processor. Additive packages, e.g. the mixed metals in the USA, which did not really reduce costs were not successful. Additive packages are of advantage especially when production is constant over a long period of time, e.g. in pipe extrusion. The processor who has to switch formulations rather often, may experience disadvantages with additive packages.

Another tendency is the desire for even more efficient heat stabilizers. In this way the required amounts will be kept as low as possible in order to avoid disadvantageous effects of the stabilizers relative to mechanical properties of the finished product. This can be achieved by the right use of synergistic combinations as explained in the above paragraphs.

From the standpoint of occupational health, stabilizers with low volatility will be appreciated more and more. In this connection, stabilizers with low odor during processing and in the finished product are required.

During the past years the question of the impact of heat stabilizers on processibility was increasingly in the limelight. Careful process control is a prerequisite for optimum production and constant high quality of the finished products at high output. This can be achieved with precise knowledge of the rheological processes during processing which are strongly influenced by heat stabilizers. Therefore, it can be expected in the future that a heat stabilizer will not only provide excellent heat stability but also contribute significantly to the processing properties of the PVC compound.

I want to thank *Dr. J. Büssing* and *Dr. H.O. Wirth* for their fruitful cooperation.

4.5 Index of trade names and suppliers of PVC stabilizers

This list of trade names and suppliers makes no claim for completeness.
Abbreviations used:

MeSnS	methyltin mercaptide
BuSnS	butyltin mercaptide
BuSnO	butyltin carboxylate
OcSnS	octyltin mercaptide
OcSnO	octyltin carboxylate
EstSnS	estertin mercaptide
BaCd p	barium/cadmium stabilizer, powder
BaCd l	barium/cadmium stabilizer, liquid
CaZn	calcium/zinc stabilizer
BaZn	barium/zinc stabilizer
Pb	lead stabilizer

Supplier	Trade name	MeSnS	BuSnS	BuSnO	OcSnS	OcSnO	EstSnS	BaCd p	BaCd l	CaZn	BaZn	Pb	Other metal-containing stab.	Metal-free stab., costab.	Stab. packages
Federal Republic of Germany															
Akzo Chemie GmbH 5160 Düren	Estabex Stanclere	—	×	×	×	×	×	×	×	×	×	×	×	×	×
BASF Farben + Fasern AG Unternehmensbereich Siegle 7000 Stuttgart 30	Sicostab Nuostab	—	×	×	×	×	—	×	×	×	×	×	×	×	×
Bayer AG Bayerwerk 5090 Leverkusen	Stabilisator	—	—	—	—	—	—	—	—	—	—	—	—	×	—
Chemische Werke München Otto Bärlocher AG 8000 München 50	Bärostab Okstan	—	×	×	×	—	—	×	×	×	×	×	×	×	×
Ciba-Geigy Marienberg GmbH 6140 Bensheim	Irgastab Reoplast	—	×	×	×	×	×	×	×	×	×	×	—	×	×
Henkel KGaA 4000 Düsseldorf 1	Edenol Stabilox	—	—	—	—	—	—	—	—	×	—	—	—	×	×

Supplier	Trade name	MeSnS	BuSnS	BuSnO	OcSnS	OcSnO	EstSnS	BaCd p	BaCd l	CaZn	BaZn	Pb	Other metal-containing stab.	Metal-free stab., costab.	Stab. packages
Hoechst AG Verkauf Kunststoffe 8900 Augsburg	Hostastab	—	×	×	×	—	—	—	—	—	—	—	—	×	×
Chemson 6000 Frankfurt/Main 1	Naftovin Listab Naftomix	—	×	—	×	—	—	×	×	×	×	×	×	×	×
GDR															
VEB Greiz-Dölau 66 Greiz-Dölau	Stabilisator	—	×	×	×	—	—	×	×	×	×	×	×	×	×
Austria															
Bleiberger Bergwerks Union 9010 Klagenfurt	Austrostab	—	×	×	×	×	—	×	×	×	×	×	×	×	×
Switzerland															
Acima 9470 Buchs/SG	Metastab	—	×	×	×	×	—	—	—	×	—	—	—	—	—
Meister AG 4313 Möhlin	Meister	—	×	×	×	×	—	×	×	×	×	—	×	×	×

Belgium															
Thiokol/Carstab Corporation 3680 Maaseik	Carstab	×	—	—	—	—	—	—	—	—	—	—	—	—	—
France															
M&T Chimie 92062 Paris Cédex 54	Ecepox	—	×	×	×	—	—	×	×	×	×	×	×	×	×
L'Air Liquide 75321 Paris Cedex 07	Plastipon	—	—	—	—	—	—	—	—	—	—	—	—	×	—
Polytitan Paris 6e, Danton 11–12 Senlis	Polyfix Stannofix	—	×	×	×	×	—	×	×	×	×	—	×	×	×
Rhône-Poulenc Ind. 75360 Paris Cedex 08	Rhodiastab	—	—	—	—	—	—	—	—	—	—	—	—	×	—
United Kingdom															
Lancro Chemicals Ltd. Eccles, Manchester M30 0BH	Lankromark	—	×	×	×	—	—	×	×	×	×	—	×	×	×
Durham Raw Materials London EX 3R 5AB	Durostabe Nuostabe	—	—	—	—	—	—	×	×	×	×	×	×	—	×
Italy															
Commer Vias. Colombano, 47 20075 Lodi (MI)	Prosper	—	×	×	×	×	—	×	×	×	×	×	×	×	

Supplier	Trade name	MeSnS	BuSnS	BuSnO	OcSnS	OcSnO	EstSnS	BaCd p	BaCd l	CaZn	BaZn	Pb	Other metal-containing stab.	Metal-free stab., costab.	Stab. packages
Reagens 40016 S. Giorgio Di Plano (Bologna)	Reagens	—	—	—	—	—	—	×	×	×	×	×	×	×	×
The Netherlands															
Akzo Chemie BV P.O. Bocks 247 Ammersfoort	Estabex Interstab Stanclere Phosclere	—	×	×	×	×	×	×	×	×	×	×	×	×	×
Dr. Haagen Chemie BV Molenweg 10 Roermond	Haro Chem	—	—	—	—	—	—	×	×	×	×	×	×	×	×
Spain															
Asua S.A. Carretera Sangroniz, 20 Sondica-Bilbao	Asua	—	×	×	×	×	—	×	×	×	×	×	×	×	×
Hebron Av. Estacion, 61 La Llagosta (Barcelona)	Vinstab	—	×	×	×	×	—	×	×	×	×	×	×	×	×
Kemichrom S.A. Caracas, 15 Barcelona 16	Kemistab	—	×	×	×	—	—	×	×	×	—	×	×	×	×

Quimica Iberica Plaza da Salamanca Madrid	Mark Drapex	—	—	—	—	—	—	×	×	×	×	—	×	×	—
Traquisa Sta. Maria de Barbara Barcelona	—	—	—	—	—	—	—	—	—	—	—	—	—	×	—
Sweden															
Swedstab AB Fack 28400 Perstorp	Swedstab	—	×	×	×	×	—	—	—	—	×	×	×	×	×
Japan															
Adeka Argus Chemical Co., Ltd. Urawa-city, Saitama Pref.	Mark	×	×	×	×	×	×	×	×	×	—		—	×	—
Akishima Chemical Co., Ltd. Nihonbashi, Chuo-ku, Tokyo	TS/FD/LT	—	×	×	×	×	—	—	×	×	—		—	—	×
Katsuta Kako Co., Ltd. Koto-ku, Tokyo	Advastab	×	×	×	×	×	×	×	×	×	×		—	×	×
Kyodo Yakuhin Co., Ltd. Setagaya-ku, Tokyo	KS/KF/KP	—	×	×	×	×	—	×	×	×	—		—	—	×
Sakai Chemical Ind. Co., Ltd. Sakai-city, Osaka Pref.	TN/LCZ/OW	—	×	×	×	×	—	—	×	×	×		—	—	×

Supplier	Trade name	MeSnS	BuSnS	BuSnO	OcSnS	OcSnO	EstSnS	BaCd p	BaCd l	CaZn	BaZn	Pb	Other metal-containing stab.	Metal-free stab., costab.	Stab. packages
Sankyo Organic Chemical Co. Kawasaki-city, Kanagawa Pref.	Stann	—	×	×	×	×	—	—	—	—	—		×	—	—
Taikyo Kasei Ind. Co., Ltd. Adachi-ku, Tokyo	LX/SX	—	×	×	—	—	—	—	×	×	—		—	—	×
Tokyo Fine Chemical Co., Ltd. Chiyoda-ku, Tokyo	Embilizer	—	×	×	×	×	—	—	×	×	—		—	—	×
Nissan Ferro Organic Chemical Nihonbashihoncho, Chuo-ku, Tokyo	Therm-Chek	—	—	—	—	—	×	×	×	×	—		—	—	×
Nitto Kasei Co., Ltd. Higashi-ku, Osaka	TVS	—	×	×	×	×	—	—	×	—	—		—	—	×
Mizusawa Chemical Ind. Co. Higashi-ku, Osaka	Stabinex	—	—	—	—	—	—	—	—	—	×		—	—	×
Kyowa Gas Chemical Ind. Co. Chuo-ku, Tokyo	Alkamezer	—	—	—	—	—	—	—	—	—	—	—	×	—	—
Dainippon Ink Chem. Ind. Co. Chuo-ku, Tokyo	Stanclere	—	×	×	×	×	—	—	×	×	—		—	—	×

Kikuchi Colour & Chem. Corporation, 5-1 Toshima-ku, Tokyo	TS/NS/ BS/CS	–	–	–	–	–	–	–	×	×	×		–	–	×
Shin Chuo Chemical Co., Ltd. Higashi-ku, Osaka	NK	–	×	×	–	–	–	–	×	×	×		–	–	×
Shinagawa Chemical Ind. Co., Ltd. Chuo-ku, Tokyo	Sinacalead	–	–	–	–	–	–	–	–	–	×		–	–	×
USA															
R.T. Vanderbilt Norwalk, CT 06855	Vanstay	–	–	–	–	–	–	×	×	×	×	×	×	×	–
American Cyanamid Bound Brook, NJ 08805	Cyastab	–	–	–	–	–	–	–	–	–	–	×	–	–	–
Halstab Corporation Hammand, IN 46325	Halstab	–	–	–	–	–	–	–	–	–	–	×	–	–	–
Mobay Chemical Corp. Union, NJ 07083	Vestablit	–	–	–	×	–	–	–	×	×	–	×	×	×	–
Nuodex Piscataway, NJ 08854	Nuostabe	–	×	×	–	–	–	×	×	×	×	×	×	×	–
Associated Lead Philadelphia, PA 19125	Lectro	–	–	–	–	–	–	–	–	–	–	×	–	–	×

Supplier	Trade name	MeSnS	BuSnS	BuSnO	OcSnS	OcSnO	EstSnS	BaCd p	BaCd l	CaZn	BaZn	Pb	Other metal-containing stab.	Metal-free stab., costab.	Stab. packages
Cardinal Chemical Columbia, SC 29202	Cardinal	—	×	×	×	—	—	—	—	—	—	—	—	—	×
M-R-S Chemicals Maryland Heights, MO 63043	MiRaStab	—	—	—	—	—	—	—	—	×	×	×	×	×	×
Borg-Warner Parkersburg, WVA 26101	Weston	—	—	—	—	—	—	—	—	—	—	—	—	×	—
American Hoechst Somerville, NJ 08876	Hostastab	—	×	×	×	×	—	—	—	×	—	×	—	—	—
M & T Chemicals Rahway, NJ 07065	Thermolite	—	×	×	×	×	—	—	—	—	—	—	×	×	×
Carstab Corp. Cincinnati, Ohio 45215	Carstab	×	—	—	—	—	—	—	—	—	—	—	—	—	×
Synthetic Products Cleveland, Ohio 44112	Synpron	—	×	×	—	—	—	×	×	×	×	—	×	×	—
Interstab Chem. New Brunswick, NJ 08903	Stanclere Estabex	—	×	×	—	—	×	×	×	×	×	×	×	×	—
Ferro Chemicals Bedford, OH 44146	Therm-Chek Plas-Chek	—	×	×	—	—	—	×	×	×	×	×	×	×	×

Argus Chemical Brooklyn, NY 11231	Mark Drapex	×	×	×	×	×	—	×	×	×	×	×	×	×	×
Santech Toronto, Ont. Canada	—	—	—	—	—	—	—	—	—	—	—	×	—	—	×
Eagle Picher Cincinnati OH 45201	Epistatic	—	—	—	—	—	—	—	—	—	—	×	—	—	×

4.6 References

[1] Verband Kunststofferzeugende Industrie e.V.: VC/PVC: Beispiel einer Problemlösung, 1976.

[2] *K.H. Michl:* Kunststoffe 70 (1980) p. 591.

[3] *G. Ayrey, B.C. Head, R.C. Poller:* J. Polym. Sci., part D, Macromol. Rev. 8 (1974) p. 1.

[4] *C. David:* Compr. Chem. Kinet. 14 (1975) p. 78.

[5] *D. Braun:* Degradation and Stabilization of Polymers, *G. Geuskens,* Ed., Wiley, New York, 1975, pp. 23/41.

[6] *H.H.G. Jellinek:* Aspects of Degradation and Stabilization of Polymers, Elsevier Publishers Ltd., London, 1977.

[7] *W.H. Starnes, Jr.:* Amer. Chem. Soc. Div. Polym. Chem., Polym. Prepr. 18 (1977) p. 493.

[8] *J.R. Dunn:* Rubber Chem. Technol. 51 (1978) p. 686.

[9] *W. Starnes, Jr.:* Adv. Chemistry Ser. 169 (1978) p. 301.

[10] *F.H. Winslow:* Makromol. Chem. 2 (1979) Suppl., p. 27.

[11] *L.I. Nass* in *Kirk Othmer:* Encyclopedia of Chemical Technology, 3rd. Ed., Vol. 12, John Wiley & Sons, New York, 1980, p. 249.

[12] *D. Braun:* Pure Appl. Chem. 53 (1981) p. 549.

[13] *Z. Mayer:* J. Macromol. Sci., part C, Rev. Macromol. Chem. 10 (1974) p. 263.

[14] *A. Guyot, M. Bert, P. Burille, M.F. Llauro, A. Michel:* J. Pure Appl. Chem. 53 (1981) p. 401.

[15] *K. Mitani, T. Ogata, H. Awaya, Y. Tomari:* J. Polym. Sci., Polym. Chem. Ed. 13 (1975) p. 2813.

[16] *L. Valko, I. Tvaroska, P. Kovarik:* Eur. Polym. J. 11 (1975) p. 411.

[17] *K.B. Abbas, E.M. Sörvik:* J. Appl. Polym. Sci. 20 (1976) p. 2395.

[18] *K.S. Minsker, A.A. Berlin, V.V. Lisicky, T.S. Kolesov:* Vysokomol. Soedin. Ser. A 19 (1977) p. 32.

[19] *A. Caraculacu, E.D. Buruiana, G. Robila:* J. Polym. Sci., Polym. Chem. Ed. 16 (1978) p. 2741.

[20] *D. Braun, M. Wolf:* Angew. Makromol. Chem. 70 (1978) p. 71.

[21] *J. Svetly, R. Lukas, M. Kolinsky:* Makromol. Chem. 180 (1979) p. 1363.

[22] *J. Svetly, R. Lukas, J. Michalcova, M. Kolinsky:* Makromol. Chem., Rapid Commun. 1 (1980) p. 247.

[23] *E.C. Buruiana, A. Airinei, G. Robila, A. Cararulacu:* Polym. Bull. 3 (1980) p. 267.

[24] *A. Garton, M.H. George:* J. Polym. Sci., Polym. Chem. Ed. 12 (1974) p. 2779.

[25] *J. Bauer, A. Sabel:* Angew. Makromol. Chem. 47 (1975) p. 15.

[26] *T.T. Nagy, B. Turcsanyi, T. Kelen, F. Tüdös:* React. Kinet. Catal. Lett. 8 (1978) p. 7.

[27] *V.P. Gupta, L.E. St. Pierre:* J. Polym. Sci., Polym. Chem. Ed. 17 (1979) p. 797.

[28] *K.B. Andersson, E.M. Sörvik:* J. Polym. Sci., part C, Polym. Symp. 33 (1971) p. 247.

[29] *G. Scott, M. Tahan, J. Vyvoda:* Eur. Polym. J. 14 (1978) p. 377.
[30] *G. Lindenschmidt:* Angew. Makromol. Chem. 47 (1975) p. 79.
[31] *W.H. Starnes, Jr., R.C. Haddon, D.C. Hische, I.M. Plitz, C.L. Schosser, C.F. Schilling, D.J. Freed:* Polymer Preprints 21 (1980) p. 138.
[32] *L.G. Close, R.D. Gilbert, R.E. Fornes:* Polym. Plast. Technol. Eng. 8 (1977) p. 177.
[33] *R.A. Varbanskaya, V.S. Pudov:* Vysokomol. Soed. B 8 (1979) p. 583.
[34] *K.P. Nolan, J.S. Shapiro:* J. Polym. Sci. Polym. Symp. 55 (1976) p. 201.
[35] *R. Schlimper:* Plaste Kautsch. 14 (1967) p. 657.
[36] *E.D. Owen, I. Pasha:* J. Polym. Sci., Polym. Lett. Ed. 16 (1978) p. 429.
[37] *D. Druesedow, C.F. Gibbs:* Nat. Bur. Stand. Circ. 525 (1953) p. 69.
[38] *T. Hjertberg, E.M. Sörvik:* J. Appl. Polym. Sci. 22 (1978) p. 2415.
[39] *B.B. Cooray, G. Scott:* Chem. Ind. (1979) p. 741.
[40] *T.T. Nagy, T. Kelen, B. Turcsanyi, F. Tüdös:* Polym. Bull. 2 (1980) p. 77.
[41] *B.B. Cooray, G. Scott:* Eur. Polym. J. 16 (1980) p. 169.
[42] *R.M. Aseeva, Y.G. Aseev, A.A. Berlin, V.I. Kasatochkin:* Zurn. Strukturnoy Chim. 6 (1965) p. 47.
[43] *T.T. Nagy, T. Kelen, B. Turcsanyi, F. Tüdös:* Plaste Kautsch. 23 (1976) p. 894.
[44] *T.T. Nagy, T. Kelen, B. Turcsanyi, F. Tüdös:* Magy. Kem. Poly. 83 (1977) p. 486.
[45] *T.T. Nagy, B. Turcsanyi, T. Kelen, F. Tüdös:* Müanyages Gumi 14 (1977) p. 349.
[46] *R.A. Varbanskaya, V.S. Pudov:* Vysokomol. Soedin B 23 (1981) p. 211.
[47] *P. Pfeiffer:* Liebigs Ann. Chem. 412 (1916) p. 253.
[48] *M. Onozuka, M. Asahina:* J. Macromol. Sci., part C, Rev. Macromol. Chem. 3 (1969) p. 235.
[49] *K. Figge, W. Findeiß:* Angew. Makromol. Chem. 47 (1975) p. 141.
[50] *H.O. Wirth, H. Andreas:* Pure Appl. Chem. 49 (1977) p. 627, cf. *H.O. Wirth:* J. Macromol. Sci., part A, Chem. 12 (1978) p. 345.
[51] Ciba-Geigy Marienberg GmbH: Handbuch der PVC-Additive, 1971.
[52] *A.H. Frye, R.W. Horst, M.A. Paliobagis:* J. Polym. Sci., part A 2 (1964) pp. 1765, 1785, 1801.
[53] *A.H. Frye, R.W. Horst:* J. Polym. Sci. 45 (1960) p. 1.
[54] *A.H. Frye, R.W. Horst:* J. Polym. Sci. 40 (1959) p. 419.
[55] *A. Michel, T.V. Hoang, A. Guyot:* J. Macromol. Sci., part A, Chem. 12 (1978) p. 411.
[56] *H.O. Wirth, H.A. Müller, W. Wehner:* J. Vinyl Technol. 1 (1979) p. 53.
[57] *B.B. Cooray, G. Scott:* Eur. Polym. J. 17 (1981) p. 233.
[58] *F. Alari-Moghadam, G. Ayrey, R.C. Poller:* Eur. Polym. J. 11 (1975) p. 649.
[59] *B.W. Rockett, M. Hadlington, W.R. Poyner:* Polym. Lett. 9 (1971) p. 371.
[60] *W. Jasching:* Kunststoffe 52 (1962) p. 458.
[61] *L.S. Troitskaya, B.B. Troitski:* Plast Massy (1968) p. 12.
[62] U.S. Patent 2219463 (1936) Carbide & Carbon Chem. Corp.
[63] U.S. Patent 2731484 (1950) Firestone.
[64] *G. Abeler, J. Büssing:* Kunststoffe 71 (1981) p. 315.
[65] *J.W. Burley, R.E. Hutton:* Polym. Degrad. Stabil. 3 (1980/81) p. 285.

[66] *C. Dörfelt, A. Janeck, D. Kobelt, E.F. Paulus, H. Scherer:* J. Organometal. Chem. 14 (1968) p. 22.

[67] *H. Woggon, D. Jehle, W.J. Uhde:* Nahrung 13 (1969) p. 343.

[68] *H. Wieczorek:* Dt. Lebensmitt. Rdsch. 66 (1970) p. 92.

[69] *H. Woggon, W.J. Uhde, H. Säuberlich:* Ernährungsforsch. 16 (1971) p. 645.

[70] *H. Seidler, M. Härting, H. Woggon, W.J. Uhde:* Nahrung 15 (1971) p. 187.

[71] *J. Koch, K. Figge:* Z. Lebensmitt. Unters. Forsch. 147 (1971) p. 8.

[72] *W. Pfab:* Dt. Lebensmitt. Rdsch. 68 (1972) p. 350.

[73] *K. Figge, H. Piater:* Dt. Lebensmitt. Rdsch. 68 (1972) p. 313.

[74] *K. Figge, A. Zeman:* Kunststoffe 63 (1973) p. 543.

[75] *H. Woggon, W.J. Uhde, K. Romminger, H. Hoppe:* Nahrung 18 (1974) p. 671.

[76] *K. Figge, W.-D. Bieber, J. Koch:* Verpack. Rundsch. 26 (1975) p. 59.

[77] *G. Neubert, H.O. Wirth:* Z. Anal. Chem. 273 (1975) p. 19.

[78] *J. Koch, L. Robinson, K. Figge:* Fette, Seifen, Anstrichm. 78 (1976) p. 371.

[79] *K. Figge, D. Cmelka, J. Koch:* Food Cosmetics Tox. 16 (1976) p. 165.

[80] *W.J. Uhde, H. Woggon:* Plaste Kautsch. 24 (1977) p. 389.

[81] *W.J. Uhde, H. Woggon:* Nahrung 21 (1977) p. 377.

[82] *K. Figge:* Kunstst. Plast. 24 (1977) 5, p. 54.

[83] *C.G. vom Bruck, W. Hammerschmidt:* Verpack. Rundsch. 28 (1977) p. 1.

[84] *C.G. vom Bruck, W.R. Eckert, F.B. Rudolph, J. Koch, K. Figge:* Fette, Seifen, Anstrichm. 80 (1978) p. 72.

[85] *K. Figge:* Acta Pharm. Technol. 24 (1978) p. 47.

[86] *C.G. vom Bruck, W.R. Eckert, K. Figge, J. Koch, F.B. Rudolph:* Mitt. Lebensmittelunters. Hyg. 69 (1978) p. 20.

[87] *R. Franck, H. Mühlschlegel:* Kunststoffe im Lebensmittelverkehr, Carl Heymann Verlag, 1977.

[88] Bundesgesundheitsblatt 23 (1980) p. 50.

[89] Eur. Chem. News 32 (1978) p. 10.

[90] *B.B. Troitski, L.S. Troitskaya, V.N. Dinisova, M.A. Novikova, Z.B. Luzinova:* Eur. Polym. J. 13 (1977) p. 1033.

[91] *J.G. Blok:* Tin and its Uses 123 (1980) p. 13.

[92] *J. Büssing:* Kunstst. Plast. 22 (1975) 4, p. 43.

[93] *W.I. Bengough, M. Onozuka:* Polym. 6 (1965) p. 625.

[94] *T. Shimura, A. Sakai:* Bull. Chem. Soc. Japan 40 (1967) p. 995.

[95] *D.F. Anderson, D.A. McKenzie:* Amer. Chem. Soc., Div. Org. Coat., Plast. Chem. Prepr. 31 (1971) p. 740.

[96] *D. Braun, D. Hepp:* Angew. Makromol. Chem. 44 (1975) p. 131.

[97] *T.V. Hoang, A. Michel, A. Guyot:* Eur. Polym. J. 12 (1976) p. 337.

[98] *D. Braun, D. Hepp:* Angew. Makromol. Chem. 66 (1976) p. 23.

[99] *G. Briggs, N.F. Wood:* J. Appl. Polym. Sci. 15 (1971) p. 25.

[100] U.S. Patent 2075543 (1934) Union Carbide.

[101] U.S. Patent 2141126 (1934) Union Carbide.

[102] German Patent 746081 (1940) I.G. Farben Ind. AG.

[103] German Patent 862512 (1942) Bayer AG.

[104] *T.V. Hoang, A. Michel, A. Guyot:* Eur. Polym. J. 12 (1976) p. 357.

[105] French Patent 2239496 (1973) Société Chimique des Charbonnages.

[106] French Patent 2433034; French Patent 2405974, French Patent 2429806 (1977) Labaz.
[107] French Patent 2297227, French Patent 2324681 (1975) Rhône-Poulenc.
[108] Communication from Service de la Répression de Fraudes, July 20, 1979.
[109] Bundesgesundheitsblatt 23 (1980) p. 335.
[110] *E.N. Zilberman, S.B. Meiman, A.E. Kulikova:* Vysokomol. Soyed. A 9 (1967) p. 1554.
[111] *D. Ryšavy, Z. Slama:* Angew. Makromol. Chem. 9 (1969) p. 129.
[112] *L.V. Verizhikov, P.A. Kirpičhnikov, N.S. Kolyubakina, N.A. Mukmeneva, S.I. Karpova:* Vysokomol. Soyed. A 13 (1971) p. 714.
[113] *G.A. Rasuvaev, L.S. Troitskaya, B.B. Troitski:* J. Polym. Sci. A-1, 9 (1971) p. 2673.
[114] *R.D. Deanin, R.M. Foss, P.G. Gilbert, R.F. Guerard, E.A. Muccio:* Polym. Eng. Sci. 13 (1973) p. 96.
[115] *T.V. Hoang, A. Michel, Q.T. Pham, A. Guyot:* Eur. Polym. J. 11 (1975) p. 475.
[116] *K.S. Minsker, N.A. Mukmeneva, A.A. Berlin, D.V. Kazačenko, M.J. Janberdina, S.I. Agadžanjan, P.A. Kirpičnikov:* Dokl. Akad. SSSR 226 (1976) p. 1088.
[117] *N.A. Mukmeneva, S.I. Agadžanjan, P.A. Kirpičnikov, K.S. Minsker:* Dokl. Akad. SSSR 233 (1977) p. 375.
[118] *D.F. Anderson, D.A. McKenzie:* J. Polym. Sci. 8 (1970) p. 2905.
[119] *T.V. Hoang, A. Michel, A. Guyot:* Eur. Polym. J. 12 (1976) p. 347.
[120] DIN-Normenblattverzeichnis, Beuth-Verlag, Berlin, 1977.
[121] *G. Abeler, W. Schramm:* Kunststoffe 66 (1976) p. 369.
[122] *G. Menzel:* Angew. Makromol. Chem. 47 (1975) p. 181, 202.
[123] *E. Weinberg:* Plast. Eng. 34 (1978) p. 35.

5 Plasticizers

Ing. Werner Sommer, Hoechst AG, Frankfurt/Main,
Federal Republic of Germany

5.1 Historical survey

The plasticizing effects of oils, waxes and balsams in brittle natural resins were known in antiquity and have long been used in "plastic" resinous compounds and varnishes. In the search for new materials, the first successful discoveries were made over 100 years ago. At that time, it was already possible to nitrate cotton, polymerize some vinyl compounds and obtain resinous substances by condensation of phenol and formaldehyde. However, the translation of these discoveries into large-scale industrial manufacture had to wait until the 20th century. Then, cellulose derivatives, condensation resins and in particular vinyl polymers met with an enthusiastic response which heralded an undreamt-of boom in their production.

In the early days, the situation with regard to plasticizers – required as processing aids – was far from ideal and manufacturers of celluloid or cellulose lacquers had to make do with natural camphor and castor oil at first. A turning point, the implications of which were not immediately recognized, came with the discovery in 1912 of triphenyl phosphate, later to be sold as a camphor substitute. This product launched the era of ester plasticizers. The most important product in this series was tricresyl phosphate which we still happily fall back on today. For cellulose derivatives, tributyl phosphate was popular for a time but diminished in importance owing to its replacement by less volatile products. The same thing happened to glycerol acetates. From 1920 onwards, the first phthalic esters enlarged the range of available plasticizers which at that time was still small. Within this group, dibutyl phthalate attained a dominant position which it held for many years and continues to hold in the case of polyvinyl acetate dispersions.

The spontaneous rise in the fortunes of dioctyl phthalate which had been known since 1930 started two decades later. The market development of this plasticizer was closely linked with the expansion of polyvinyl chloride. This top consumer of plasticizers finally determined that phthalic esters should have worldwide domination in terms of number and volume.

The development of speciality plasticizers was given impetus by increasingly high quality requirements, the desire for products with distinct characteristics, and compatibility problems relating to particular products. Food legislation, industrial safety and commercial reasons also played a not inconsiderable role. Over the last 40 years, these factors have led to the development of today's wide and varied range of plasticizers.

To begin with only a few fatty acid esters, benzoates, tartrates and chlorinated hydrocarbons were available. These were soon augmented by esters of adipic, azelaic and sebacic acid which rapidly attracted interest because of their outstanding low-temperature resistance. Numerous polymeric plasticizers followed with high resistance to extraction and migration. Epoxidized fatty acid esters attained great popularity because of their synergistic heat stabilizing effect. Increasingly high safety requirements in the automobile industry were taken into account in the development of ethylbutyric esters and pelargonic esters for polyvinyl butyral film which is used in the production of laminated safety glass. The need of the cable industry, in particular, for low-volatility plasticizers with relatively long-term heat stability was first met with di-iso-tridecyl phthalate and later with several trimellitic esters. The poor light stability of the triaryl phosphates was successfully overcome by using alkylated phenols as esterifiers. Linear or predominantly linear C_6-C_{11} alcohols made available by new technologies have given further powerful momentum in recent years.

Through systematic esterification of proven aliphatic or aromatic carboxylic acids, a large number of high quality plasticizers have been obtained. Naturally, this survey can only illustrate the general trend and give some limited examples of problem solving.

5.2 Definition of terms

Plasticizers are substances added to plastics compounds to improve their flexibility, extensibility and processability. It is characteristic of plasticizing substances that they lower the melting temperature, elastic modulus and second order transition temperature of polymers [1] but do not alter the chemical nature of the macromolecules. The great advantage of this so-called *external* plasticization lies in its scope for variation which enables the degree of flexibility of the plasticized material to be regulated according to the type and amount of plasticizer added. At low plasticizer concentrations, some degree of embrittlement can occur and it is therefore essential not to add less than the minimum quantity appropriate for the particular plasticizer used [2]. A disadvantage in some cases is the undesirable extraction of plasticizer from the plasticized material.

The principle of *internal* plasticization is that monomers yielding homopolymers with a high second order transition temperature are copolymerized with monomers whose homopolymers have a lower second order transition temperature.

The advantage of internally plasticized products lies in the stable chemical linking of rigid and flexible segments which cannot be separated by extraction. There are limitations to this method in the sense that each copolymer can only be matched to certain specific requirements. Since technology and cost considerations narrow the scope for internal plasticization, it is used far less than external plasticization.

One extremely versatile and promising technique is the modification of rigid thermoplastics with elastomers or other flexible polymers. This method originated in the USA and gave rise to new terms such as "polyblends", "elastomer blends", "plastics alloys" and "impact modifiers". The most advanced development in this field is in the compounding of PVC. Chlorinated polyethylene, ethylene/vinyl acetate copolymers, polyacrylates and a number of synthetic rubbers have been used very successfully for quite some time now. In principle what happens is that the flexible component is embedded in the matrix of the base polymer, forming a mixture similar to an alloy (cf. chapter 8). To minimize any reduction in the heat distortion resistance of the base product, the maximum acceptable concentration for the particular type of modifier should not be exceeded.

Of course, internally plasticized polymers can also be externally plasticized. Even polyblends are perfectly accessible to external plasticization. Combined systems such as this make it possible to achieve maximum effect with minimum plasticizer content.

The classification of plasticizing organic substances for PVC into primary plasticizers, secondary plasticizers and extenders will serve as an illustration. *Primary plasticizers* are required to gel the polymer rapidly in the normal processing temperature range, should be suitable for use on their own and should not exude from the plasticized material. *Secondary plasticizers* on the other hand have lower gelation capacity and limited compatibility with the polymer so that combination with a primary plasticizer is advisable. *Extenders* have very poor gelation properties in PVC, tend to exude heavily when used on their own and are only suitable for use in limited quantities as reducing agents for primary plasticizers.

5.3 Quality requirements

Quality requirements can relate to purity criteria, the processing behavior of the plasticizers or to the properties a plasticizer is intended to impart to the plasticized material [3].

Commercial plasticizers are normally supplied in the form of low to high viscosity liquids and more rarely as solid products. Generally speaking, high purity, absence of color, neutral reaction, freedom from taste or odor and consistent quality are required. In the case of esters derived from alcohol fractions, care should be taken to ensure that the product conforms to specifications and contains no highly volatile first runnings.

Since many thermoplastics require relatively high processing temperatures, plasticizers must have good heat resistance to prevent discoloration or decomposition phenomena. Epoxy plasticizers are even required to have a costabilizing effect.

Low volatility is an advantage from several points of view. It eliminates noxious plasticizer vapors during processing and also increases the ageing resistance of the plasticized material, particularly when this is exposed to short-term high temperature stresses during service. Phthalates prepared from $C_{10}-C_{13}$ aliphatic alcohols are very suitable in this respect. Similar esters of trimellitic acid are favored by cable manufacturers because of their low volatility and good electrical properties.

For articles which have to withstand the complex effects of weathering, not only low volatility but also properties such as resistance to light, water extraction and fungal attack and low-temperature resistance come into play. It is not always easy to bring all of these requirements down to a common denominator. Even with combined plasticizer systems, compromises still have to be made.

Migration problems can occur under unfavorable circumstances in multilayer artificial leathers and films and can also be triggered off by contact between plasticizer-containing and plasticizer-free polymers. Nitrocellulose films and polystyrene articles are particularly sensitive in this respect. They readily take up a number of low molecular plasticizers thus becoming tacky. If necessary, polymeric plasticizers can be used. Among this group, too, products with low solubility in oils, fats and blowing agents can be found.

Flame retardancy requirements are met by phosphoric esters and chlorinated paraffins with a high chlorine content. Trichloroethyl phosphate even reduces the flammability of nitrocellulose and also enjoys considerable popularity for polyurethane foams.

Great care has to be taken in plasticizing consumer articles for food contact or medical applications. Statutory regulations or official recommendations should be strictly observed [4, 5]. Of course, it is always important to ensure that the plasticizer used is perfectly compatible with the polymer being processed. Caution is advisable when strong gelling agents are combined with extenders because if compatibility limits are exceeded this inevitably leads to exudation.

5.4 Market situation

Over 500 plasticizers are marketed worldwide. Confusing trade names often hide the fact that some of these products have the same or very similar structures and overlaps are difficult to recognize. The number can be drastically condensed to about 100 different plasticizers of industrial importance with a corresponding market share.

The reasons for this discrepancy can best be explained with the aid of a few examples. If we look closely at chlorinated paraffins, we can see that most manufacturers' products contain between 30 and 70% chlorine. In the popular medium range there are however seemingly incomprehensibly fine gradations such as 40, 42, 45 or 46% Cl. Depending on the type of base paraffins (chain length, chain length distribution, melting points), such differences are necessary to attain optimum properties and viscosity values. The same applies to the phthalates. Di-n-butyl phthalate cannot just simply be replaced by di-isobutyl phthalate because among other things there are volatility differences at high temperatures. According to how narrowly cut the alcohol fraction used as esterifier is, the end products will again differ in their properties. The ester resulting from a defined n-alcohol gives plasticized material better low temperature resistance than a comparable ester with a high degree of branching. The intrinsic viscosity and viscosity/temperature graphs confirm this. A halfway position in this respect is occupied by mixtures of linear and branched alcohols known as ®Linevols. By esterifying these with different acids, plasticizers with a wide variety of properties can be obtained. An additional complication with polymeric plasticizers is that they are not only produced from different diols with an aliphatic or aromatic dicarboxylic acid but also from variable mixtures and can have a relative molecular mass (molecular weight) ranging from 1000 to 8000. On the raw material side, availability and price among other factors contribute to the diversity.

Table 1. Production capacity for plasticizers (as of 1980).

Region	Production capacity (1000 t/yr)
Western Europe	1820
Eastern Europe	300
North America	1150
Central America	5
South America	130
Asia	715
Australia	30
Africa	50
World	4200

Table 2. Development of plasticizer production in the USA.

Year	Production (t)
1939	13530
1950	>100000
1956	>200000
1965	>400000
1970	>800000
1979	967491

World production of plasticizers in 1977 amounted to more than 3 million tons and in 1980 passed the 4 million ton mark. Expansion of existing production plants is taking place slowly because planned spare capacity is still covering demand (table 1).

Useful information on the growth (table 2) and structure (table 3) of the US plasticizer range is supplied by the annual reports of the US International Trade Commission.

It is very difficult to forecast growth rates over the next few years in the present economic climate. Some countries can be expected to increase their production capacity to cover their own demand or to take advantage of their geographical location to export plasticizers into neighboring regions.

Table 3. Breakdown of plasticizer production in the USA 1979.

Type of plasticizer	Quantity (t)	Share (%)
Phthalic esters	585583	60.53
Trimellitic esters	9311	0.96
Triaryl and alkylaryl plasticizers	47939	4.95
Trialkyl phosphates	8824	0.91
Adipic esters	34232	3.54
Sebacic esters	442	0.05
Polyesters, polymeric plasticizers	25526	2.64
Fatty acid esters (oleates, palmitates, stearates etc.)	18077	1.87
Fatty acid esters, epoxidized	54164	5.60
All other cyclic plasticizers	123959	12.81
All other acyclic plasticizers	59434	6.14
Total	967491	100.00

The dominant position of phthalic esters is unmistakable. Esters from C_8 aliphatic alcohols account for 23% of these. Together with C_6, C_{10}, C_{11}, and C_{13} phthalates, they make up almost 50% of the group. PVC and VC copolymers consume up to 90% of these plasticizers.

The US and European product ranges are comparable since they are mainly determined by the plasticizer requirement of PVC. In Western Europe, speciality plasticizers have also been developed for PVC and other polymers. Generally speaking, however, standard plasticizers are preferred because of limited availability of storage tanks, simplicity of formulation, suitability for automatic conveying and metering units or for commercial reasons. If finished articles with distinct properties are required, there is usually still scope for blending with speciality plasticizers. Such requirements are taken into account by the wide spread of the plasticizer range.

5.5 Methods of incorporation

In the production of paints and adhesives the plasticizer is added to the normal dissolving tanks together with the solvent. Because the solvent is always geared to the binder, it almost invariably has good plasticizer dissolving power as well. Solvent and plasticizer may even activate each other, as processors of alcohol-moistened collodion cotton have long known.

A certain amount of experience is required to incorporate plasticizers into aqueous polymer dispersions. Because of the shear sensitivity of many dispersions, it is advisable to use slow running mixers, introduce the dispersion first and then stir in the plasticizer gradually in a thin jet. In many cases, the addition of 0.5% methanol, ethanol or glycol to the plasticizer assists incorporation. Coarse dispersions usually have better plasticizer absorption properties than fine dispersions. With the latter, it may be necessary first to add protective colloids such as concentrated aqueous solutions of cellulose ethers, alkaline solutions of polyacrylic acids or alkaline solutions of vinyl acetate/crotonic acid copolymers. Of course, it is also possible to add an emulsifier to the plasticizer, providing it does not impair or even destroy the effect of the dispersing agent already contained in the dispersion, for example antagonism such as between anionic and cationic emulsifiers must be avoided. Only in the case of very sensitive dispersions must the plasticizer be emulsified before being stirred in.

For the preparation of plastisols, planetary mixers, internal mixers and dissolvers are suitable. It is an advantage to use mixers with vacuum deaeration. To prevent powder clusters in the forming plastisol, it is advisable with less intensive mixers to work from thick to thin, i.e. to mix only part of the plasticizer with the paste-making PVC to begin with and to add the rest later.

In a dissolver, considerable frictional heat is generated which can lead to a sharp increase in viscosity when temperatures exceeding 30 °C are reached. Unless this thickening is deliberate, it can be countered by suitable cooling or a reduction in speed.

Free-flowing plasticized PVC compounds can be prepared in fluid mixers by dry blending. For this process, PVC suspension and mass polymers with high plasticizer absorption capacity are suitable. PVC emulsion polymers with good particle cohesion can be used alone up to a certain plasticizer content but otherwise it is better to blend them with PVC suspension polymers. When the polymer has been charged into a hot mixer, the material is heated up to 80–140 °C, depending on the type of compound, and the plasticizer sprayed in. Through the sight glass, the point of optimum plasticizer absorption can be watched for and further controlled via the increase in power input. When this stage has been reached, usually after a few minutes, the material is discharged into the cold mixer below and cooled with further stirring. The dryblends so obtained are suitable for charging extruders, roll mills and injection molding machines and as sintering powder.

In heated internal batch mixers, practically all thermoplastics and elastomers can be mixed with plasticizers and other compounding ingredients.

For dispersing pigments on a roll mill, a standard plasticizer such as DOP is normally used. By combining with fatty acid esters of epoxy plasticizers, color concentrates, or color masterbatches as they are called, can be obtained.

5.6 Technological functions

External plasticizers fulfill several functions. They act as process aids, enter into physical interaction with the polymer and enable materials to be "tailor-made", or very nearly so, to specific property requirements. Since polyvinyl chloride is the main consumer of plasticizers, this thermoplastic will best serve to illustrate the functions of a plasticizer.

5.6.1 Solvency and gelation properties

To adopt a simple interpretation, plasticizers may be regarded as solvents for high molecular materials and classified accordingly. It has always been customary to classify chemical compounds which have a plasticizing effect on cellulose esters, particularly nitrocellulose, as either solvents or non-solvents. The following classification criteria are recommended by *K. Thinius* [6]:

Group A: Plasticizers which peptize dry collodion wool to completely clear, fiber-free solutions within 3 hours.

Group B: Plasticizers which only slightly dissolve dry collodion wool within 3 hours but form a clear solution within 24 hours.

Group C: Plasticizers which do not dissolve or only incompletely dissolve dry collodion wool after 24 hours.

The author specifies *critical solubility temperatures* T_{kL} of plasticizers for PVC. These are determined by uniformly heating a 24 percent suspension of an emulsion polymer. Some values are shown in table 4. A standard method for determining the solubility temperature of PVC in plasticizers is described in DIN 53408 (ISO 4574–1978).

Table 4. Critical solubility temperature T_{kL} of plasticizers for PVC.

Plasticizer	T_{kL} (°C)
Tributyl phosphate	58
Dibutyl phthalate	90– 95
Butyl benzyl phthalate	96–100
Tricresyl phosphate	101–105
Trioctyl phosphate	111–115
Dioctyl phthalate	116–120
Dioctyl adipate	121–125
Dioctyl sebacate	161–165

A low critical solubility temperature indicates that the plasticizer has a high gelation capacity and this in turn has technological implications. Plasticizers with a strong gelling action permit a gradual reduction in processing temperatures or shorter mixing times for plastic compounds; in PVC plastisols on the other hand they cause certain viscosity problems. For example, the very low viscosity tributyl phosphate initially forms low viscosity plastisols with PVC but these thicken within a short time. If the frictional heat generated in the mixing process is not removed, there is a risk of gel-like agglomerates being formed. Although dibutyl phthalate and butyl benzyl phthalate have a much weaker solvating action on polymer particles, both plasticizers are normally used only in blends with other plasticizers for plastisol production.

The gelation characteristics of plasticizers in polymer dispersions can be studied by mixing typical compounds at normal processing temperature in a Brabender Plastograph. The small mixing head of this instrument is run at constant speed while the instrument draws a graph of torque versus time which is used as a measure of progressive gelation [7, 8]. The graph in figure 1 shows the effect of the type of plasticizer on the gelation characteristics of PVC at 170 °C. This method enables products to be compared and compound formulations and temperature settings to be worked out.

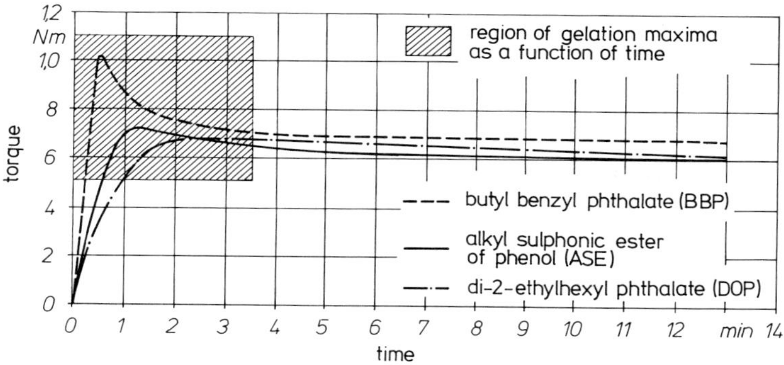

Fig. 1. Torque measured in the Brabender Plastograph as a measure of the gelation characteristics of plasticized PVC compounds, temperature 170 °C, 30 g material, speed 40 rpm, material composition: 40 pts. by wt. plasticizer, 60 pts. by wt. PVC, 1 pt. by wt. stabilizer (according to [9]).

High pressure capillary viscometers give valuable information on the flow properties of high molecular materials at normal processing temperatures.

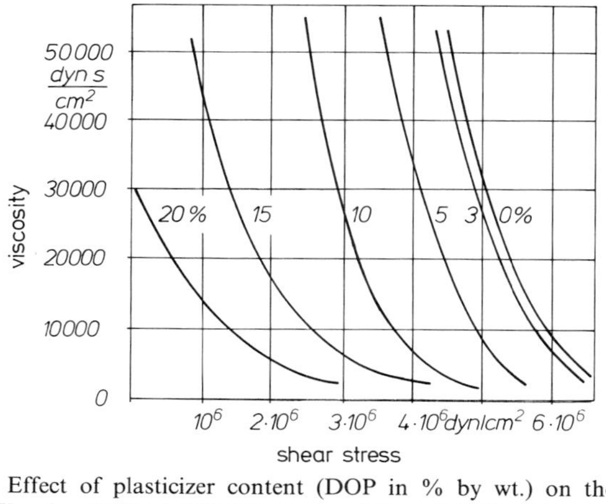

Fig. 2. Effect of plasticizer content (DOP in % by wt.) on the flowability of PVC (K-value 65.8) at 180 °C, measured in a Göttfert high pressure capillary viscometer, nozzle diameter 1 mm, nozzle length 20, 30 and 40 mm, shear stresses and shear rate corrected according to *Bagley*, *Weissenberg* and *Rabinowitsch* [10].

As figure 2 shows, the melt viscosity of PVC compounds is steadily lowered and their flowability improved by increasing the amount of plasticizer. Even low plasticizer concentrations in the brittle range are clearly effective here.

Optimum flow properties facilitate molding of plastic compounds and subsequent processing operations. This has a beneficial effect on productivity and on such processes as welding, thermoforming and embossing.

5.6.2 Effects on hardness

External plasticization reduces the hardness of high molecular materials. The efficiency of plasticizing substances is specific to the type of plasticizer and also depends on the particular product, the quantity used and the temperature. Useful information on the sensitivity of rigid or semi-rigid plastics to small-point pressure loading can be obtained from tests

Table 5. Shore hardness A and D of plasticized PVC with increasing plasticizer concentration, test temperature 20 °C.

Plasticizer[1]	Shore A/Shore D hardness			
	25%	30%	35%	40%
DBP	90/54	84/40	76/29	68/24
DOP	95/59	91/46	82/35	77/30
DINP	98/65	94/54	86/41	79/32
DOA	91/50	86/38	77/28	70/23

Table 6. Effect of temperature on the Shore hardness A and D of plasticized PVC with a plasticizer content of 30%.

Plasticizer[1]	Shore A/Shore D hardness								
	−10	0	+10	+20	+30	+40	+50	+60	+70
DBP	98/75	97/69	94/58	85/−	78/−	71/−	66/−	62/−	60/−
DOP	98/76	97/71	94/65	90/47	85/−	79/−	72/−	66/−	62/−
DINP	98/82	98/76	97/70	94/54	90/45	82/−	73/−	67/−	63/−
DOA	97/62	95/56	91/48	86/−	82/−	77/−	70/−	64/−	59/−

[1] Abbreviations cf. 5.8.

of ball indentation hardness (DIN 53456), Vickers hardness (DIN 50133), Rockwell hardness (ASTM D 785–65), Shore D hardness (DIN 53505, ISO 868–1978, ASTM D 2240). For flexible, rubber-like plasticized materials, Shore A hardness values determined on 6 mm plaques (standards cf. Shore D hardness) are normally used. Since the various test methods are based on different principles of measurement, the values resulting from them cannot be interconverted.

The effect of the type and quantity of standard plasticizer on the Shore hardness of PVC suspension polymer (K-value 70) can be seen from table 5. The effect of temperature on hardness is shown in table 6 for plasticized materials with the same plasticizer content.

5.6.3 Effects on tensile strength and elongation

Once a certain product-dependent minimum concentration is exceeded, plasticizers increase the extensibility of macromolecular materials but reduce their tensile strength. In assessing mechanical properties, the effect of temperature should generally be considered. In order not to impair tensile strength excessively, the amount of plasticizer added is kept as small as possible consistent with imparting just sufficient flexibility to the polymer for the intended service temperature range. Extreme requirements necessitate compromises because one property is inevitably favored at the expense of another.

The tensile test (ISO R 527–1966, DIN 53455, ASTM D 638) has been used successfully for many years to test plastics materials and is therefore incorporated into numerous finished product specifications for plastics

Table 7. Tensile strength and elongation at break of PVC with increasing plasticizer content. Test temperature 20 °C.

Plasticizer[1])	Tensile strength (N/mm^2) and elongation at break (%) at plasticizer contents				
	25%	30%	35%	40%	45%
DBP	24/250	21/300	18/350	15/375	11/400
DOP	27/250	24/300	21/350	18/400	14/425
DINP	29/225	26/275	23/325	19/375	15/400
DOA	26/275	23/325	20/375	16/400	12/425

[1]) Abbreviations cf. 5.8

products as an obligatory acceptance test. In tensile testing, the relation of stress of strain at a constant rate of extension is recorded on test specimens to breaking point. Systematic studies have shown the value of this method in providing a meaningful prediction of the effect of plasticizers on polymers. Taking PVC as an example, table 7 shows the effect of increasing amounts of various standard plasticizers on tensile strength and elongation at break at 20 °C. This is supplemented by table 8 which indicates the effect of temperature on mechanical properties [11].

Table 8. Tensile strength and elongation at break of PVC with 30% plasticizer content as a function of temperature.

Temperature (° C)	Tensile strength (N/mm^2) and elongation at break (%) plasticizer			
	DBP	DOP	DINP	DOA
+70	8/675	9/675	11/650	8/700
+60	10/600	12/600	14/575	11/625
+50	12/525	15/525	17/500	14/550
+40	15/450	18/450	20/425	17/475
+30	18/375	21/375	23/350	20/400
+20	21/300	24/300	26/275	23/325
+10	24/250	27/225	28/200	26/275
0	27/200	20/150	30/125	29/225
−10	31/100	33/100	32/50	32/175

Tensile tests can also determine what quantity of a plasticizer or plasticizer blend is required to obtain plasticized materials capable of 100% extension under a given stress (e.g. 10 N/mm^2). Plasticizer requirement clearly characterizes the relative effectiveness of substances with different chemical structures but provides only minimal differentiation between closely related compounds. Although tensile strength and elongation can supply important technical information, the quality of a plasticizer may not be judged entirely on these criteria.

5.6.4 Effects on low-temperature resistance

Below their characteristic second order or glass transition temperature T_g, thermoplastics are in a relatively brittle state. Plasticizers shift down the second order transition temperature considerably, relegating brittleness to a lower temperature region. The second order transition temperature of rigid PVC, which lies between 75 and 80 °C, can be lowered by 140 °C to −55 to −60 °C through external plasticization. The effect of plasticizer content is shown in figure 3, using DOP as an example.

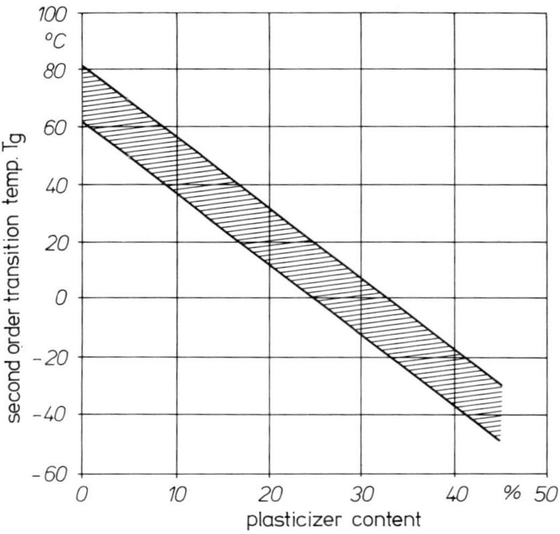

Fig. 3. Second order transition temperature region of plasticized PVC as a function of plasticizer content (DOP) (according to [12]).

Because of their chemical structure, esters of aliphatic dicarboxylic acids, some fatty acid esters and trialkyl phosphates perform particularly well at low temperature. Among the phthalates, esters of linear C_6 to C_{10} alcohols head the list but DOP may still be rated as perfectly good. Triaryl phosphates and chlorinated paraffins with a high chlorine content are less effective.

It is customary to describe plastics products as low-temperature resistant in terms of serviceability when they can withstand the mechanical stresses occurring during their normal prescribed use down to a rather low temperature limit. The falling dart test (DIN 53372, ASTM D 746, D 1790) offers the possibility of carrying out impact tests on annealed film loops but only gives real guidance as to their behavior under shock loading. Table 9 presents a survey of these values.

Table 9. Low-temperature impact test on PVC. Weight of falling dart: 200 g, drop height: 200 mm.

Plasticizer[1])	Temperature (°) at which falling dart test was passed at plasticizer contents of				
	25%	30%	35%	40%	45%
DBP	−15	−25	−35	−45	−50
DOP	− 5	−15	−25	−35	−40
DINP	0	− 5	−10	−15	−20
DOA	−25	−40	−45	−50	−55

The relatively mild bar bending test can clarify where the lower temperature limit is at which flexible planar structures can be handled without risk of fracture. Flex cracking tests at normal temperature and at low temperature are always advisable for artificial leather or other articles which are repeatedly crumpled, folded or rolled up during service for quite long periods. Torsional vibration tests according to DIN 53445 (ISO 537–1980, ASTM D 2236) provide information on the thermomechanical properties of polymers. The advantage of this non-destructive method is that it gives a precise indication of the low-temperature elasticity of externally plasticized products.

The effect of plasticizer concentration on the shear modulus of plasticized PVC test specimens is shown in figure 4a for DIDP and figure 4b for DIDA. The parameter is temperature.

5.6.5 Effects on electrical properties

Materials used to insulate electrical conductors must have certain mechanical, thermal and dielectric properties. PVC has properties which render it very suitable for this application but nevertheless owes its use

[1]) Abbreviations cf. 5.8.

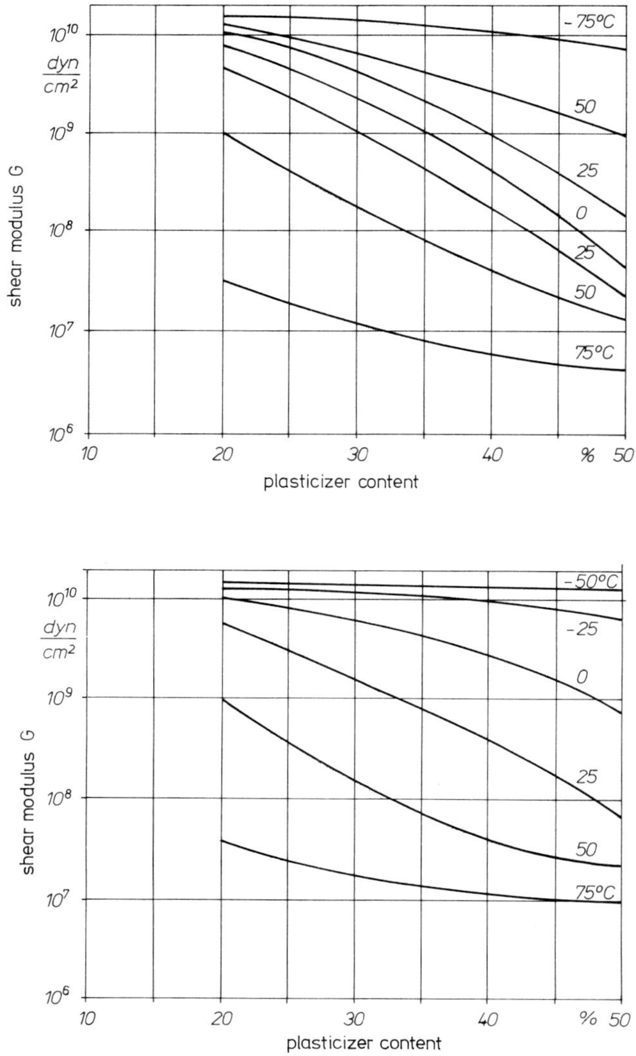

Fig. 4. Shear modulus G of plasticized PVC according to DIN 53445 as a function of plasticizer content (according to [10]), *a*: DIDP, *b*: DIDA.

in the cable industry to the plasticizers specially developed for it. To demonstrate the economic importance of this industry for PVC, it should be noted that the PVC requirement of West German cable factories in 1978 amounted to some 91 000 t.

Plasticized PVC insulating and jacketing compounds are normally permitted for use up to a maximum operating temperature of 70 °C. When there is higher thermal stress over a relatively long period of time, a shorter service life must be expected. The recently agreed specifications are laid down in DIN 57 207, parts 4 and 5. For the cable sector, only plasticizers with high purity, low volatility and high thermal stability can be considered, e.g. phthalic esters of branched or unbranched C_8 to C_{12} aliphatic alcohols. Similar esters of trimellitic acid have a lower plasticizing action in PVC but are distinguished by their low volatility and have considerably less effect on electrical properties.

If it is necessary to improve the low-temperature elasticity of a cable compound, the less moisture-sensitive n-alkyl phthalates should be used instead of the usual aliphatic dicarboxylic acid esters.

The volume resistivity of plasticized PVC decreases with increasing plasticizer content as figure 5 shows for DOP. Temperature increases have the same effect.

In the medium voltage range, plasticized PVC may be used for insulated cables with ratings from 6 to 10 kV so that in addition to insulation

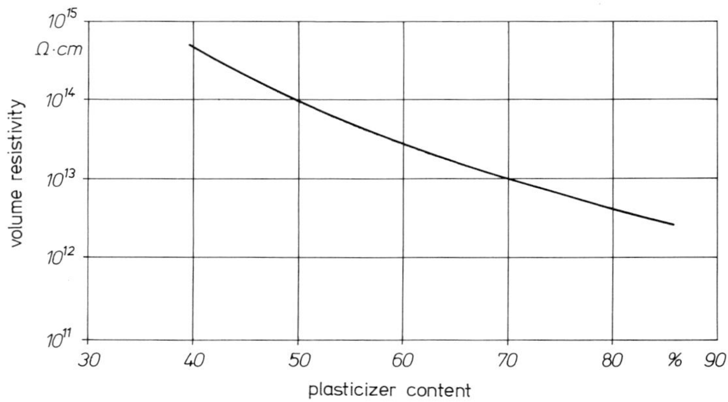

Fig. 5. Volume resistivity of plasticized PVC at 23 °C as a function of plasticizer content (DOP).

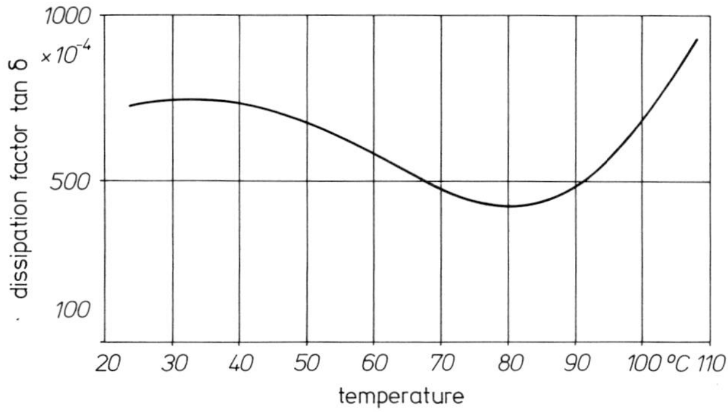

Fig. 6. Effect of temperature on the dissipation factor tan δ of plasticized PVC, voltage 600 V, frequency 50 Hz.

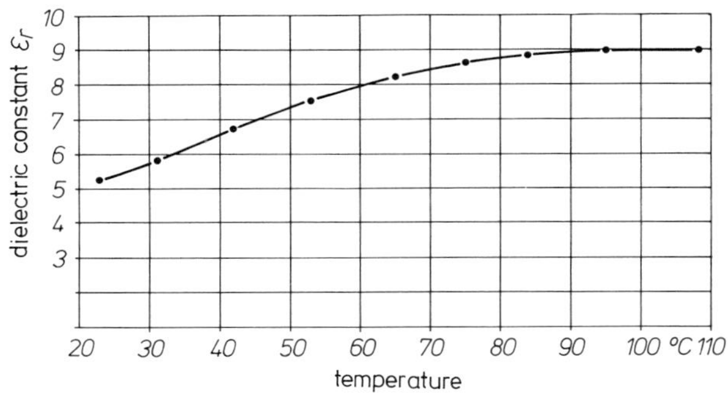

Fig. 7. Effect of temperature on the dielectric constant ε_r of plasticized PVC, voltage 600 V, frequency 50 Hz.

resistance the dielectric properties of the PVC/plasticizer system are very important. The dissipation factor tan δ is determined by the polymer and by the type and quantity of plasticizer, filler or other process aid. Valuable information on the serviceability of the insulation in the critical temperature range from 20 to 120 °C is obtained by plotting the curve for the dissipation factor. At the highest permissible loading, the curve should, if possible, still show a downward movement in order to keep dissipation low. Figure 6 shows an experimentally determined curve for a typical insulating compound composed of:

62 pts. by wt. PVC,
24 pts. by wt. DIDP,
10 pts. by wt. hard kaolin,
3 pts. by wt. tribasic lead sulphate,
1 pt. by wt. lead stearate (28% Pb).

The dielectric constant ε_r of the same insulating compound is plotted in figure 7 as a function of temperature. The dielectric constant increases with increase in plasticizer concentration. The type of plasticizer has considerably less influence.

5.7 The plasticizer range

Today's range of plasticizers is dominated by the phthalic esters. These largely cover the demand of the plastics processing industry which comes mainly from PVC and vinyl chloride copolymers. The phthalates class also contains suitable products for thermoplastic compounds based on cellulose esters which are the next highest consumers of plasticizers, though a long way behind. The same applies to the phosphoric esters although it should be remembered that these make up only about 5% of the total quantity of plasticizers used while the phthalates account for around 70 to 75%. The remainder is reserved for a large number of specialities which again are strongly oriented towards the requirements of PVC, e.g. esters of aliphatic dicarboxylic acids, alkyl sulphonic esters of phenols and cresols, polymeric plasticizers and epoxidized fatty acid esters. Polyvinyl butyral, polyamides, polyvinyl alcohol and cellophane complete the list of high molecular materials processed with plasticizers in the plastics industry. The range of suitable chemical compounds for these plastics is still small.

5.7.1 Phthalic esters

The normal commercial plasticizers in this class are esters of orthophthalic acid, isophthalates behave less favorably. Terephthalates have lower volatility at elevated temperature but customers have not been willing to pay for their extra cost. Economic considerations are among the paramount factors in selecting a plasticizer. They have even played a major part in the rapid growth of the orthophthalates. Because of their rather high volatility, the use of phthalates derived from short chain aliphatic alcohols is restricted to cellulose esters plastics. Even dibutyl and dihexyl phthalate are too volatile to be used on their own for processing. They can only be used in combination with low volatility plasticizers. In this way, their good gelation properties can be exploited. Dioctyl phthalates, particularly those produced from 2-ethylhexanol, are distinguished as PVC plasticizers by their good all-round properties which include high gelation capacity, good compatibility, light stability, low volatility, water resistance and good electrical properties. The C_6-C_{10} and C_8-C_{10} n-alkyl phthalates are notable for their low-temperature resistance which regresses far less on heat aging than with DOA. C_7-C_9 and C_8-C_{11} alkyl phthalates from predominantly linear alcohols behave in a very similar way. Gelation ability decreases with increasing chain length. It is so low with di-iso-tridecyl phthalate that higher processing temperatures or additions of strong gelling agents are required. DINP and DITP confer long-term heat resistance up to temperatures of 70 to 90 °C. BBP has proved very popular as a strong gelling agent for PVC, particularly in chemical foaming of plasticized PVC.

Table 10 gives a general survey of the most commonly used phthalic esters.

5.7.2 Phosphoric esters

The esters of orthophosphoric acid have not profited from the increase in the demand for plasticizers but have lost market share. The extremely low volatility, flame retardant triaryl phosphates on the other hand continue to maintain their position, especially for highly stressed industrial parts based on PVC such as conveyor belts. Solid triphenyl phosphate is still an indispensible speciality product for plastics compounds based on cellulose derivatives. Increasing interest is being shown in phosphates of alkylated phenols as flame retardant plasticizers with improved light stability.

Mixed esters with one alkyl and two aromatic groups almost attain the level of DOP with respect to low temperature resistance. The use of triaryl phosphates – which have particularly good low temperature resistance and light stability – is somewhat restricted by their high volatility. Trichloroethyl phosphate is a good, light-stable plasticizer for cellulose derivatives. Although it has very low compatibility with PVC, small quantities may be added to improve the high frequency welding properties of PVC. This product imparts considerable flame resistance to PUR foams. Phosphoric esters produced from glycols or glycol ethers are not generally popular, despite their good plasticizing action, because they

Table 10. Properties of phthalic esters (Abbreviations used cf. 5.8).

Plasticizer	Standard abbreviation	Molecular weight	Viscosity at 20 °C (mPa·s)
Dimethyl phthalate $o\text{-}C_6H_4(COOCH_3)_2$	DMP	194	12–15
Diethyl phthalate $o\text{-}C_6H_4(COOC_2H_5)_2$	DEP	222	10–12
Dibutyl phthalate $o\text{-}C_6H_4(COOC_4H_9)_2$	DBP	278	19–20
Dihexyl phthalate $o\text{-}C_6H_4(COOC_6H_{13})_2$	DHXP	334	40–50
Di-2-ethylhexyl phthalate $o\text{-}C_6H_4[COOCH_2 \cdot CH(C_2H_5)C_4H_9]_2$	DOP	390	74–76
Di-n-octyl phthalate $o\text{-}C_6H_4(COOC_8H_{17})_2$		390	40–45
Di-iso-octyl phthalate (mixture of phthalic esters)	DIOP	390	155–175
Di-iso-nonyl phthalate (mixture of phthalic esters) or di-3,5,5-trimethylhexyl ester	DINP	418	160–165
Di-iso-decyl phthalate (mixture of phthalic esters)	DIDP	446	115–130

Plastics in which used	Characterization
CA, CAB, CAP, CN, CP	High dissolving capacity for CN. Good compatibility with cellulose esters. Used in celluloid and plastic compounds or films made from cellulose esters. Light stable, highly volatile.
CA, CAB, CAP, CN, CP	Properties similar to dimethyl phthalate. Slightly less volatile.
CN, CAB, CAP, CA (limited), PVC, PVCA	High dissolving capacity for CN. Good gelling agent for PVC and PVCA. Fast to light. Good low-temperature resistance. Causes thickening in PVC plastisols. Is rather volatile. Is only used as a secondary plasticizer in PVC.
CN, CAB, PVC, PVCA	Moderate dissolving capacity for CN. Good gelling agent for PVC and PVCA. Is still rather volatile. Is normally used only as a secondary plasticizer in PVC.
CN, CAB, PVC, PVCA	Still has good compatibility with CN. Standard plasticizer for PVC and PVCA characterized by low volatility, heat resistance, low-temperature resistance, water resistance, high gelation capacity and good electrical properties.
CAB, PVC, PVCA	Slightly inferior gelation properties to DOP, much better low-temperature resistance, good plasticizing action, otherwise similar to DOP.
CAB, PVC, PVCA	Similar to DOP. Slightly poorer plasticizing action, better electrical properties.
CAB, PVC, PVCA	Less volatile than DOP, poorer plasticizing action. Has poorer low-temperature resistance. Good electrical properties. Otherwise similar to DOP.
CAB, PVC	Very low volatility, heat resistant, very good electrical properties. Particularly suitable for cable compounds. Considerably poorer plasticizing action than DOP. Requires higher gelation temperature. Addition of Bisphenol A necessary in applications involving high thermal stress.

Table 10 (continued)

Plasticizer	Standard abbrevi-ation	Molec-ular weight	Viscosity at 20 °C (mPa · s)
Di-iso-tridecyl phthalate $o\text{-}C_6H_4(COOC_{13}H_{27})_2$	DITP (DITDP)	530	290–295
Dicyclohexyl phthalate $o\text{-}C_6H_4(COOC_6H_{11})_2$	DCHP	330	crystalline substance
Dimethylcyclohexyl phthalate $o\text{-}C_6H_4(COOC_6H_{10}CH_3)_2$		358	145–155
Dimethylglycol phthalate $o\text{-}C_6H_4(COOC_2H_4OCH_3)_2$		282	50–60
Dibutylglycol phthalate $o\text{-}C_6H_4(COOC_2H_4OC_4H_9)_2$		366	42
Benzyl butyl phthalate $o\text{-}C_6H_4$ $\langle$ $COOC_4H_9$, $COOCH_2C_6H_5$	BBP	312	50
Diphenyl phthalate $o\text{-}C_6H_4(COOC_6H_5)_2$		318	crystalline substance
C_7-C_9 alkyl phthalates from predominantly linear alcohols		–	~50
C_9-C_{11} alkyl phthalates from predominantly linear alcohols		–	~75

Plastics in which used	Characterization
PVC	Poor gelation. Requires high processing temperature. Has low plasticizing action. Gives plasticized PVC high fatigue strength. Good electrical properties. Particularly suitable for cable compounds.
CN, CAB, CAP, PVC, PVCA	Has low plasticizing action in cellulose esters which is desirable for heat sealing. Limited use for PVC, facilitates processing of rigid PVC. Resistance to extraction by gasoline.
CN, CAB, CAP, PVC, PVCA	Has low plasticizing action in cellulose esters which is of interest for films. Can be used as an aid in processing of rigid PVC. Good light stability. Low volatility.
CA, CAB, CAP, PVC	Very suitable for production of plastic compounds or films from cellulose esters. Less volatile than DBP but more water resistant. Resistant to petroleum hydrocarbons. Of interest for PVC in special cases.
CN, PVC	Behaves similarly to DBP but is less volatile. No other advantages.
PVC, PVCA	Is an excellent gelling agent for PVC. Produces lower flexibility than DOP. Has lower volatility than DBP. Is highly valued for use in expandable plasticized PVC.
CN, PVC, PVCA	Can be used in the manufacture of nitrocellulose films. Produces only slight flexibility. Used as an aid in production of rigid PVC foams.
PVC	Plasticizing action greater than that of DOP. Lower volatility, better low-temperature resistance. Electrical properties slightly poorer. High water resistance.
PVC	Plasticizing action lower than that of DOP. Good low-temperature resistance, high water resistance, remarkable aging resistance. Electrical properties somewhat less favorable.

Table 10 (continued)

Plasticizer	Standard abbreviation	Molecular weight	Viscosity at 20 °C (mPa·s)
C_6-C_{10} n-alkyl phthalates		–	45
C_8-C_{10} n-alkyl phthalates		–	48

are relatively sensitive to water. Particularly careful heat stabilization is necessary with all phosphoric esters.

The properties of the phosphoric esters are summarized in table 11.

5.7.3 Esters of adipic, azelaic and sebacic acid

The esters of aliphatic dicarboxylic acids belonging to this class are used mainly as plasticizers for PVC and PVCA in the plastics industry. They give the polymer outstanding low-temperature resistance and are distinguished by their high light stability. Another characteristic is their low viscosity which is valuable in the manufacture of PVC plastisols. The weak solvating action of these esters on PVC at room temperature has a favorable effect on the initial viscosity and storage stability of plastisols even when, for example, plasticizer mixtures of DOP and an aliphatic dicarboxylic ester with DOP contents of up to 80% are used. Such combinations help to improve gelation.

Even at low concentration the strong plasticizing action of the adipates, azelates and sebacates is evident. Within this group, the low-temperature resistance of comparable esters increases from the adipate to the sebacate. Their sensitivity to water decreases in the same direction. With respect to heat aging, DOA comes off relatively badly, DOZ and DOS have much lower volatility.

To meet higher requirements, isononyl or isodecyl alcohols produced by the oxo process may be used as esterifiers. The resulting esters lead

Plastics in which used	Characterization
PVC	Gives the same flexibility as DOP. Has considerably better low-temperature resistance. Comes out better in heat aging. Gives PVC plastisols low viscosity and storage stability.
PVC	Differs from C_6-C_8 esters in its better low-temperature resistance, particularly after aging. Otherwise very similar properties.

to somewhat more rigid plasticized materials with good low-temperature elasticity and 'drier' surfaces. It is common practice for economic reasons to blend these extremely low-temperature resistant plasticizers with DOP which itself has considerable low-temperature resistance. Esters of aliphatic dicarboxylic acids used in combination with an n C_8-C_{10} phthalate should bring optimum results from both technical and economic viewpoints.

A survey of the properties of adipic, azelaic and sebacic esters is given in table 12.

5.7.4 Fatty acid esters

The esters of higher fatty acids (pelargonates, laurates, myristates, palmitates, stearates and ricinoleates) are included in plasticizer statistics. Strictly speaking, however, they are a series of speciality products acting as extenders, lubricants or secondary plasticizers.

As a plasticizer for nitrocellulose, acetylricinoleic butyl ester has proved successful because it gives films high elasticity, light stability and low-temperature resistance. Similar properties are offered by tetrahydrofurfuryl oleate.

This class includes important lubricants for use in processing PVC. On the other hand, esters of higher fatty acids are unsuitable as plasticizers for PVC or PVCA because of their poor compatibility.

Table 11. Properties of phosphoric esters.

Plasticizer	Standard abbreviation	Molecular weight	Viscosity at 20 °C (mPa·s)
Tributyl phosphate $(C_4H_9O)_3PO$	–	266	4
Tri-2-ethylbutyl phosphate $[C_2H_5CH(C_2H_5)CH_2O]_3PO$	–	350	8
Tri-2-ethylhexyl phosphate $[C_4H_9CH(C_2H_5)CH_2O]_3PO$	TOP	435	14
Trichloroethyl phosphate $(ClC_2H_4O)_3PO$	–	286	41
2-ethylhexyl-di-phenylphosphate $(C_6H_5O)_2$ PO $C_4H_9CH(C_2H_5)CH_2O$	–	362	21
Cresyldiphenyl phosphate $CH_3C_6H_4O$ PO $(C_6H_5O)_2$	–	340	48–50
Triphenyl phosphate $(C_6H_5O)_3PO$	TPP	326	crystalline substance
Tricresyl phosphate $(CH_3C_6H_4O)_3PO$	TCP	368	85–95
Trixylenyl phosphate $[(CH_3)_2C_6H_3O]_3PO$	–	413	230–240

Plastics in which used	Characterization
CN, CAB	Dissolves CN extremely well. Can be blended with up to six times its amount of castor oil. Is exceptionally volatile. Is only used in the plastics sector for plastics compounds based on CN.
CN, PVCA	Gels CN and PVCA. Is less volatile than DBP. Gives CN films flame retardancy. Good light stability. Used only in plastics compounds based on CN.
CN, PVCA, PVC	Gels CN, PVCA and PVC. Is light stable and flame retardant. Has markedly higher volatility than DOP. Gives PVC plastisols low viscosity.
CN, CA, CAB, CAP	Good plasticizer for cellulose esters. Actively flame retarding product. High petrol resistance. Normally used for safety films made from CA. Poor compatibility with PVC. Flame retardant for PUR foam. Lightfast.
PVC, PVCA	Good gelling agent for PVC. Plasticizing action similar to DOP. More light sensitive than this. Water resistant. Good flame retarding effect. Good resistance to extraction by water and petroleum hydrocarbons.
CN, CAB, PVCA, PVC	Suitable for plastics compounds based on cellulose esters, except for CA. Is used in PVC in the same way as TCP. Gives plasticized materials with good mechanical properties. Is flame retardant but sensitive to light. Very low volatility.
CN, CAB, CAP, CA limited	Only of interest for plastics compounds based on cellulose esters. Virtually non-combustible product. Insoluble in petrol.
CN, CAB, PVCA, PVC	Is a good flame retardant plasticizer for technical PVC articles subjected to high mechanical stress. Gives lower flexibility than DOP. Poor low-temperature resistance. Excellent permanence in the plasticized material.
PVC	Exceptionally volatile, exceeding TCP in this respect. Gels PVC somewhat less well than TCP and produces lower flexibility. Preferred for long-lasting PVC articles with low flammability and good mechanical properties. Good resistance to extraction by petroleum hydrocarbons. Not light stable.

Table 12. Properties of adipic, azelaic and sebacic esters.

Plasticizer	Standard abbreviation	Molecular weight	Viscosity at 20 °C (mPa·s)
Di-2-ethylhexyl adipate $COOCH_2CH(C_2H_5)C_4H_9$ $(CH_2)_4$ $COOCH_2CH(C_2H_5)C_4H_9$	DOA	370	14–15
Di-iso-octyl adipate (mixture of adipic esters)	DIOA	370	17–20
Di-iso-nonyl adipate (mixture of adipic esters)	DINP	398	22–25
Di-iso-decyl adipate (mixture of adipic esters)	DIDA	426	30–35
Benzyl butyl adipate $COOCH_2C_6H_5$ $(CH_2)_4$ $COOC_4H_9$	–	292	12–15
Benzyl octyl adipate $COOCH_2C_6H_5$ $(CH_2)_4$ $COOC_8H_{17}$	–	348	16–17
Di-2-ethylhexyl azelate $COOCH_2CH(C_2H_5)C_4H_9$ $(CH_2)_7$ $COOCH_2CH(C_2H_5)C_4H_9$	DOZ	411	16–18
Di-2-ethylhexyl sebacate $COOCH_2CH(C_2H_5)C_4H_9$ $(CH_2)_8$ $COOCH_2CH(C_2H_5)C_4H_9$	DOS	426	21
Di-iso-decyl sebacate (mixture of sebacic esters)	–	474	38–42

Plastics in which used	Characterization
PVC, PVCA	Gels PVC less well than DOP. Good plasticizing action. Excellent low-temperature resistance. Very stable to light. More volatile and sensitive to water than DOP.
PVC, PVCA	Behaves practically the same as DOA.
PVC, PVCA	Gels PVC less well than DOA. Has poorer low-temperature resistance but is considerably less volatile. Gives notably lower flexibility.
PVC, PVCA	Has a high PVC solution temperature. Does not attain the low-temperature resistance of DOA. Is the least volatile of the adipates listed.
PVC, PVCA	Gels PVC better than DOA. Is rather highly volatile. Does not attain the low-temperature resistance of DOA.
PVC, PVCA	Gels PVC better than DOA. Volatility similar to DOA. Does not quite attain the low-temperature resistance of DOA. Plasticizing action similar to DOA.
PVC, PVCA	This plasticizer comes very close to DOS. Is somewhat more volatile and not quite so resistant to low temperature. Less sensitive to water than adipates.
PVC, PVCA	Exceeds the low-temperature resistance of all other products in the group. Has the least sensitivity to water. Gels PVC satisfactorily. Has a relatively low volatility.
PVC, PVCA	Gels PVC less well than DOS. Is less resistant to low temperature but has lower volatility. Gives lower flexibility. Low water sensitivity.

5.7.5 Epoxidized fatty acid esters

Through the action of peracids on esters of unsaturated fatty acids, oxygen adds to the double bond so forming an epoxide group. Since the epoxide group reacts with acids, epoxidized fatty acid esters have become very popular as plasticizers with a stabilizing effect.

The main applications for these plasticizers in plastics are in the processing of rigid and plasticized PVC. They are also suitable for VC copolymers. In combination with stabilizers – even the less effective Ca, Zn compounds – valuable synergistic stabilizing effects result.

Butyl and octyl epoxy stearate are plasticizers with good low-temperature resistance equal to that of DOA. The epoxidized oleic acid octyl ester is less volatile than DOA and is characterized by better water resistance. Cyclohexyl epoxy stearate performs favorably in the gelation of PVC.

Epoxidized soybean oil is added to PVC compounds in amounts up to 10% mainly because of its good stabilizing effect. It has low migration tendency and wide-ranging resistance to extraction by detergents. Epoxidized linseed oil is best for improving the thermal stability of PVC because it has an epoxide oxygen content of 8% compared with 6% for epoxidized soybean oil. The iodine number usually quoted by plasticizer manufacturers for their products is a measure of the content of unconverted ester.

5.7.6 Citric esters

It may be said of triethyl citrate and triethyl acetylcitrate that they are among the few plasticizers to have good solvency for cellulose acetate. In comparison with diethyl phthalates, with which they are in competition, there are slight but not serious differences in volatility and water sensitivity. In these critical points, the citrates leave something to be desired.

Blocking the hydroxyl group by acetylation makes a noticeable improvement but does not yield results superior to those which can be achieved with diethyl phthalate. The step to the less volatile and less water sensitive tributyl acetylcitrate has not been taken because of compatibility problems. The properties of the last-mentioned plasticizer approach those of DOP and meet PVC processing requirements. However only tri-2-ethylhexyl acetylcitrate is on a par with DOP over the full range of properties.

The interest in citric esters is due to a favorable assessment of their physiological properties. In the production of consumer articles which come under food legislation, statutory regulations or official recommendations must be carefully taken into consideration.

5.7.7 Esters of acetic, propionic and butyric acid

All products in this class are hampered by the disadvantages of high volatility and water sensitivity. Since the usual di- or triglycerides of monocarboxylic acids are solvent plasticizers for cellulose acetate, it is not easy to do without them. There are compatibility limitations in the use of glyceryl tripropionate and tributyrate. These monocarboxylic acid esters are unsuitable for vinyl polymers.

5.7.8 Esters of ethylbutyric and ethylhexanoic acid

In the production of polyvinyl butyral films for safety glass, triethylene glycol di(2-ethylbutyrate) is an extremely valuable plasticizer which has a proven record of success over many years. It has the required light stability, a plasticizing effect tailored to the requirements of polyvinyl butyral and gives the films suitable adhesion to glass ensuring good splinter adhesion over the temperature range from −40 °C to +70 °C. Adequate flexibility at low temperature can be seen from the impact test results. Producers of polyvinyl butyral films prefer this plasticizer.

The ethylhexanoic esters of triethylene glycol and polyethylene glycol are good plasticizers for CAB. They have only limited compatibility with polyvinyl butyral. Other possible applications are in the production of PVC plastisols. Small additions can be used to regulate viscosity.

5.7.9 Glycolic esters

Glycolic esters of aliphatic alcohols are unsuitable as plasticizers because their volatility is too high. Among the phthalyl glycolates, butyl phthalyl butyl glycolate is an interesting product in that it can be added to plastic compounds based on cellulose esters in quantities up to 100% on the film former. Compatibility with PVC and a viscosity similar to that of DOP allow this product to be used in the preparation of PVC plastisols. Its advantage over standard plasticizers such as DOP lies in the gasoline resistance of the phthalyl glycolate.

5.7.10 Polyesters (polymeric plasticizers)

The polyesters, known as polymeric plasticizers, come in very widely varying consistencies. There are low viscosity polyesters which have a viscosity of 100 mPa·s at 20 °C and extremely high viscosity polyesters with viscosity values up to 300000 mPa·s. Viscosity is determined not only by molecular weight but also by the esterifiers used in the polyester structure. Propane- and butanediols are among the preferred diols; adipic, azelaic and sebacic acid among the most commonly used aliphatic dicarboxylic acids. For reasons of economy, polymeric plasticizers based on phthalic acid are also made.

In the preparation of PVC plastisols with polymeric plasticizers either low viscosity types are used or high viscosity types in combination with monomeric plasticizers. For PVC compounds destined for extrusion or calendering, high viscosity polyesters can be used on their own. The choice of a suitable product is usually assisted by manufacturers' data sheets which supply information on heat aging loss, extraction by petroleum, vegetable or animal oils and fats, gasoline, alcohol, water, soap and synthetic detergent solutions, and on migration behavior.

5.7.11 Chlorinated hydrocarbons (chloroparaffins)

The normal liquid chlorinated paraffins used as plasticizers for PVC have viscosities ranging from 100 to 40000 m Pa·s at 20 °C. Depending on the base paraffin used and the chlorine content, low to high viscosity products may be obtained. Products with chlorine contents ranging from 30 to 70% are on the market. With increasing chlorine content, compatibility with PVC is increased but the plasticizing effect is reduced.

The low viscosity products (chlorine content 30 to 40%) are secondary plasticizers for PVC. They have a stabilizing effect on viscosity in plastisols. A maximum of 25% chlorinated paraffin, of the total plasticizer content of the PVC plastisol, can be used without risk of exudation.

At a chlorine content of 50%, chlorinated paraffins come very close to being primary plasticizers and are then sufficiently compatible with PVC to be used alone, i.e. without primary plasticizers, in semi-rigid compounds under certain circumstances. The plasticizing action of products with extremely good compatibility is low. As a rough guide, it may be said that approximately twice the amount of chlorinated paraffin is required to replace DOP as far as possible.

Low viscosity chlorinated paraffins with a low chlorine content are considerably more volatile than high viscosity products with a high chlorine content. Increase in bitumen and gasoline resistance and decrease in migration tendency follow in the same direction. In the manufacture of floorcoverings, additions of a suitable chlorinated paraffin reduce soiling of the surface. As chlorine-containing substances, these plasticizers also have a flame retarding effect.

5.7.12 Hydrocarbons

Hydrocarbons are popular as extenders, particularly in the manufacture of PVC plastisols which must maintain as stable a viscosity as possible for relatively long periods of time (dip and rotational molding). The petroleum industry offers imprecisely defined products with an aliphatic-aromatic structure for this purpose. They are added to plastisols in small amounts only as viscosity regulators. Dibenzyl toluene serves the same purpose. Extenders of this type remain in the plasticized material. Their disadvantages are low stability to light and a risk of exudation if excessive quantities are used.

5.7.13 Benzoic esters

Diglycol benzoates are liquid plasticizers rather like BBP as regards their plasticizing action in PVC. In the manufacture of PVC floorcoverings, the benzoates offer the advantage of a highly soil resistant surface. In this respect, they are superior to BBP which itself performs well here.

The relatively high cost of the benzoates precludes their general use. For special purposes, neopentyl glycol dibenzoate – a solid product with good PVC compatibility – is promising. The 2-ethylhexyl-p-oxybenzoate is suitable for plasticizing polyamides.

5.7.14 Trimellitic esters

Trimellitic esters are distinguished by high thermal stability and low volatility. In these properties, they come close to some polymeric plasticizers based on phthalic acid. The plasticizing action of the trimellitates in PVC is low. These esters are not suitable for applications requiring good low-temperature resistance. Among the most important commercial products are tri-2-ethylhexyl trimellitate, tri-iso-decyl trimellitate and tri-iso-tridecyl trimellitate. Because of the high price of the esters, their use remains restricted to PVC compounds exposed to high thermal stress, mainly cable compounds.

5.7.15 Sulfonic esters and sulfonamides

Greater industrial importance has been attained by alkylsulfonic esters of phenol and cresol (ASE). These are excellent PVC plasticizers whose properties match those of DOP in virtually every respect. An advantage is that they are difficult to saponify. Like DOP they can be used for certain articles covered by food regulations in accordance with the recommendations of the BGA (German Federal Office of Health).

Toluenesulfonamide and toluenesulfonethylamide are special powder plasticizers for plastic compounds based on cellulose esters. Benzenesulfonbutylamide is compatible with polyamide 11 and 12 and with copolyamides.

5.7.16 Alcohols, ethers, ketones

Plasticizers for polyvinyl alcohol should be organic substances with good hydrophilic properties. Glycol, diglycol and triglycol, polyethylene glycols, glycerin, propane-, butane- and hexanediols as well as 1,3,5-hexanetriol come into consideration. Neopentyl glycol and trimethylol propane are available as solid products.

Ethers are hardly represented at all as plasticizers; only diphenyl ether and di-(p-tetramethyl butyl)phenyl ether are suitable for plasticizing plastic compunds based on ethylcellulose.

Of the ketones, camphor is the sole representative, maintaining its position in celluloid manufacture.

5.7.17 Abietic esters

It should be mentioned that abietic esters are not used in the plastics sector. They are tackifiers and plasticizers for nitrocellulose and rubber.

5.8 Abbreviations for plasticizers and polymers according to DIN 7723, ASTM D 1600 and ISO 1043–1978

Abbreviation	Plasticizer
ASE	alkylsulfonic ester of phenol and cresol
BBP	butyl benzyl phthalate
DBP	dibutyl phthalate
DCHP	dicyclohexyl phthalate
DCP	dicapryl phthalate
DEP	diethyl phthalate
DHXP	dihexyl phthalate
DIBP	di-i-butyl phthalate
DIDA	di-i-decyl adipate
DIDP	di-i-decyl phthalate
DINP	di-i-nonyl phthalate
DIOA	di-i-octyl adipate
DIOP	di-i-octyl phthalate
DITP	di-i-tridecyl phthalate
DMP	dimethyl phthalate
DOA	di-2-ethylhexyl adipate
DOIP	di-2-ethylhexyl-isophthalate
DOP	di-2-ethylhexyl phthalate
DOS	di-2-ethylhexyl sebacate
DOTP	di-2-ethylhexyl terephthalate
DOZ	di-2-ethylhexyl azelate
ELO	epoxidized linseed oil
ESO	epoxidized soybean oil
TCP	tricresyl phosphate
TIOTM	tri-i-octyl trimellitate
TOP	tri-2-ethylhexyl phosphate
TOTM	tri-2-ethylhexyl trimellitate
TPP	triphenyl phosphate
CA	cellulose acetate
CAB	cellulose acetate butyrate
CAP	cellulose acetate propionate
CN	cellulose nitrate
CP	cellulose propionate
PVC	poly(vinyl chloride)
PVCA	poly(vinyl chloride acetate)

5.9 List of trade names and manufacturers of plasticizers

The following list contains 103 trade names of plasticizers and gives information on the chemical structure of the products. It makes no claim for completeness. The symbols *A–R* used to denote plasticizer classes have the following meanings:

A: phthalic esters,
B: phosphoric esters
C: esters of adipic, azelaic, glutavic or sebacic acid,
D: fatty acid esters (oleates, stearates, ricinoleates) or pentaerithrol
E: fatty acid esters, epoxidized (epoxy plasticizers),
F: citric esters,
G: esters of acetic, propionic or butyric acid,
H: esters of ethylbutyric or ethylhexanoic acid,
I: glycol esters,
J: polyesters (polymeric plasticizers),
K: halogenated hydrocarbons,
L: hydrocarbons, aliphatic and aromatic (extenders),
M: benzoic esters,
N: trimellitic esters,
O: sulfonic esters, sulfonamides, anilides,
P: alcohols, ethers, ketones,
Q: polymers, polycondensates,
R: abietic derivates

Trade name	Compound class																		Manufacturer/supplier
	A	B	C	D	E	F	G	H	I	J	K	L	M	N	O	P	Q	R	
Abalyn																		×	Himont U.S.A. Inc., Wilmington, DE 19894, USA
Abitol																		×	
Abracol									×			×				×			Bush Boake Allen Ltd. Walthamstow, London, GB
Acetin							×												Bayer AG, Leverkusen, D; Mobay Corp., Pittsburgh, PA 15205, USA
Adimoll			×																
Adipol			×																FMC Corp., Philadelphia, PA 19103, USA; C-I-L Inc., Willowdale, ON M2N 6H2, Canada
Admex					×														Nuodex Inc. Piscataway, NJ 08854, USA
Alaiflex											×								Pechiney-Saint-Gobain, Neuilly sur Seine, F
Amoco-Plast. AMS														×			×		Amoco Chemicals Corp., Chicago, IL 60601, USA
Baymod																×			Bayer AG, Leverkusen, D; Mobay Corp., Pittsburgh, PA 15205, USA
Bendol			×																Inolex Chemical Co., Philadelphia, PA 19148, USA
Benzoflex													×						Velsicol Chem. Corp., Chicago, IL 60611, USA
Bisoflex	×		×	×										×					BP Chemicals, London, W1X 6AY, GB
Carbowax																×			Union Carbide Corp., Specialty Polymers & Composites, Danbury, CT 06817, USA
Cereclor											×								ICI Petrochemicals and Plastics Division Imperial Chemical Industrie P.L.C. Wilton, Middlesbrough, Cleveland TS6 8JE, GB
Cetamoll															×				BASF, Ludwigshafen, D
Chlorowax											×								Diamond Shamrock Chemicals Co., Irving, TX 75061, USA
Chlorparaffin Hoechst											×								Hoechst AG, Frankfurt/M. 80, D

Trade name	Compound class																		Manufacturer/supplier
	A	B	C	D	E	F	G	H	I	J	K	L	M	N	O	P	Q	R	
Citroflex						×													Morflex Chemical Co. Inc., Greensboro, NC 27403, USA
Citrofol						×													J.A. Benkiser, Ludwigshafen, D
Clorafin											×								Himont U.S.A. Inc., Wilmington, DE 19894, USA
Conoco												×							Conoco Chemicals Co., Houston, TX 77001, USA
Crestapol										×									Scott Bader & Co., Ltd., Wollaston, GB
Cryoflex			×																Arco Specialty Chemicals, West Chester, PA 19073, USA
Dellatol BBS															×				Bayer AG, Leverkusen, D; Mobay Corp., Pittsburgh, PA 15205, USA
Deplastol									×										Henkel, Düsseldorf, D
Diolpate										×									B + T Polymers, Stockport, Cheshire, SK 125 BR, GB
Diplast	×		×																Destillerie Italiane, Milano, I
Disflamoll		×																	Bayer AG, Leverkusen, D; Mobay Corp., Pittsburgh, PA 15205, USA
Dow		×															×		Dow Chemical Co., Midland, MI 48640, USA
Drapex					×					×									Argus Div., Witco Corp., Brooklyn, NY 11231, USA
Edenol	×		×	×	×					×									Henkel, Düsseldorf, D
Elastosan			×																Rhône-Progil, Neuilly, F
Elvaloy																	×		DuPont de Nemours, Düsseldorf, D; DuPont de Nemours, Genf, CH
Emery				×															Unichema Chemie B.V., Gouda, NL;
Estol								×											Emery Chemicals, Cincinnati, OH 45202, USA

Trade name	Compound class																		Manufacturer/supplier
	A	B	C	D	E	F	G	H	I	J	K	L	M	N	O	P	Q	R	
Estabex					×														Akzo Chemie GmbH, Düren, D
Estaflex						×													
Estynox					×											×			CasChem Inc., Bayonne, NJ 07002, USA
Eviplast	×																		Chemol Impex, Budapest, H
Flexol	×	×	×		×			×	×	×			×		×				Union Carbide Corp., Specialty Polymers & Composites Div., Danbury, CT 06817, USA
Flexricin				×															CasChem Inc., Bayonne, NJ 07002, USA
Fliximel	×		×	×		×		×		×									Melle, Bezons, F
Garbeflex	×	×	×	×						×				×					Atochem, Paris La Défense, F
Genomoll	×	×																	Hoechst AG, Frankfurt/M. 80, D
Hallco			×						×	×									C.P. Hall Co., Chicago, IL 60638, USA
Hatcol	×		×		×														Hatco Chemical Corp., King George Post Rde, Fords., NJ 08863, USA
HB 40												×							Monsanto Deutschland, Düsseldorf, D; Monsanto Co., St. Louis, MO 63167, USA
Hercoflex	×			×						×									Himont U.S.A. Inc., Wilmington, DE 19894, USA
Hercolyn																		×	
Hexaplas	×		×											×					ICI Deutschland, Frankfurt/M., D; ICI Petrochemicals and Plastics Div., Imperial Chemical Industries, Cleveland, GB; C-I-L Inc., Willowdale, ON M2N 6H2, Canada
Hycar																	×		B.F. Goodrich Chem. Prod., Cleveland, USA
Isaplast															×				Th. Böhme KG, Geretsried, D
Induflex	×		×																Cepsa, Compañia Española de Petroleos S.A., Madrid, Spain
Jayflex	×		×																Deutsche Exxon Chemical GmbH, 5000 Köln 1, D; Exxon Chem. Americas, Houston, TX 77001, USA

Trade name	Compound class																		Manufacturer/supplier
	A	B	C	D	E	F	G	H	I	J	K	L	M	N	O	P	Q	R	
Karion																×			E. Merck, Darmstadt, D
Kenflex												×							Kenrich Petrochemicals Inc., Bayonne, NJ 07002, USA
Kenplast							×												
Kesscoflex			×	×			×												Stephan Chem. Co., Maywood, NJ 07607, USA
Ketjenflex															×		×		Koninklijke Zwavelzuurfabrieken v/h Ketjen N.V., Amsterdam, NL
Kodaflex	×		×				×			×				×					Eastman Internat. AG, Zug, CH; Eastman Chem. Products Inc., Kingsport, TN 37662, USA
K-Flex					×														Kalama Chemical Inc. 1296 N.W. Kalama WH 98625, USA
KP	×	×		×					×										FMC Corp., Philadelphia, PA 19103, USA
Kronitex		×																	
Lankroflex					×														Lankro Chemicals Ltd., Eccles-Manchester, M300 BH, GB
Lindol		×																	Stauffer Chem. Co., Westport, CT 06881, USA
Linevol-Ester	×																		Shell Chemie, Frankfurt/M., D
Linplast	×		×											×					Condea Chemie GmbH, Hamburg, D
Lipinol				×								×							Hüls AG, Marl, D
Lutonal																	×		BASF AG, Ludwigshafen, D; BASF Corp., Chemicals Division, Persippany, NJ 07054, USA
Mesamoll															×				Bayer AG, Leverkusen, D; Mobay Corp., Pittsburgh, PA 15205, USA
Mirflex	×		×																Rio Rodano, Capitan Haya 1, Madrid 20, Spain
Mollan	×		×																Chemie Linz AG, Linz, A

Trade name	Compound class																		Manufacturer/supplier
	A	B	C	D	E	F	G	H	I	J	K	L	M	N	O	P	Q	R	
Monoplex	×		×		×														C.P. Hall Co., Chicago, IL 60638, USA
Morflex	×		×	×						×									Morflex Chemical Co. Inc., Greensboro, NC 27403, USA
Natroflex			×																Harwick Chemical Corp., Akron, OH 44305, USA
Nevillac												×							Neville Chem. Co., Pittsburgh, PA 15225, USA
Nuoplaz													×						Nuodex Inc., Piscataway, NJ 08854, USA
Palatinol	×																		BASF AG, Ludwigshafen, D; BASF Canada Inc., Montreal, H4L4V8, Canada; BASF Corp., Chemicals Division, Persippany, NJ 07054, USA
Paraplex				×	×					×									C.P. Hall Co., Chicago, IL 60638, USA
Paricin				×															CasChem Inc., Bayonne, NJ 07002, USA
Phosphlex		×																	Dover Chemical Corp., Sub ICC Industries Inc., Dover, OH 44622, USA
Plastigen										×									BASF AG, Ludwigshafen, D
Plastimer	×																		Plastimer SA, Clichy, F
Plastomoll			×				×			×									BASF AG, Ludwigshafen, D; BASF Corp., Chemicals Div., Persippany, NJ 07054, USA
Plaxter										×									COIM S.p.A. 20019 Setimo Milano, I
Pluronic																×			BASF AG, Ludwigshafen, D; BASF Corp., Chemicals Div., Persippany, NJ 07054, USA
Pluracol																×			
Polycizer	×		×													×			Harwick Chem. Corp., Akron, OH 44305, USA
Priplast			×				×			×									Unichema Chemie B.V., Gouda, NL; Emery Chemicals, Cincinnati, OH 45202, USA

Trade name	Compound class																		Manufacturer/supplier
	A	B	C	D	E	F	G	H	I	J	K	L	M	N	O	P	Q	R	
PX	×	×	×	×															USS Chemicals, Div. U.S. Steel Corp., Pittsburgh, PA 15230, USA
2-Pyrol																×			GAF Corp., Wayne NJ 07470, USA
Quadrol															×				BASF Corp. Chemicals Division, Persippany, NJ 07054, USA
REA				×															Wacker Chemie, München, D
Reofos		×																	Ciba-Geigy AG, Basel, CH; Ciba-Geigy Corp., Hawthorne, NY 10532, USA; Ciba-Geigy Industrial Chemicals, Manchester, GB
Reomol	×		×																
Reoplex										×									
Reproxal	×													×					Deutsche Texaco AG, Hamburg, D
Ricatyl				×															Reichhold-Chemie AG, CH 5212 Hausen Brugg, CH
Rucoflex	×			×					×							×			Rucopolymer Corp., Hicksville, NY 11802, USA
Saib								×											Eastman Chemical Product Inc., Kingsport, TN 37669, USA
Santicizer	×	×	×						×		×				×				Monsanto Deutschland GmbH, Düsseldorf, D; Monsanto Co., St. Louis, MO 63167, USA
Scandiol	×		×																Scandiflex A/S, 1864 Copenhagen, DK
Sicol	×		×																Enichem S.p.A., 20097 San Donato Milano, I
Staflex	×		×	×						×									Reichhold Chemicals Inc., White Plains, NY 10603, USA
Synolate	×																		Synres N.V., Hock van Holland, NL

Trade name	Compound class																		Manufacturer/supplier
	A	B	C	D	E	F	G	H	I	J	K	L	M	N	O	P	Q	R	
Triacetin							×												Bayer AG, Leverkusen, D; Mobay Corp., Pittsburgh, PA 15205, USA
Tributyrin							×												
Ultramoll										×						×			
Unichlor											×								Neville Chemicals Co., Pittsburgh, PA 15225, USA
Uniflex			×		×					×									Union Camp Corp., Chemical Div., Savanah, GA 31402, USA
Unimoll	×																		Bayer AG, Leverkusen, D; Mobay Corp., Pittsburgh, PA 15205, USA
Uraplast										×									DSM Resines B.V., 3151XN Hock, NL; Scado-Archer-Daniels, Rühle, D
Uresin																	×		Hoechst AG, Frankfurt/M., D
Vestinol	×		×							×									Hüls AG, 4370 Marl, D
Vulcanol												×			×				Bayer AG, Leverkusen, D
Wacker	×		×																Wacker Chemie GmbH, 8000 München 22, D
Warflex			×																Acro Specialty Chemicals, West Chester, PA 19073, USA
Weplast										×									COIM S.p.A., 20019 Setimo Milano, I
Witamol	×		×							×									Dynamit Nobel AG, Troisdorf, D
Witco	×		×	×															Witco Organic Div., Chicago, IL 60638, USA
Wolflex	×									×									Victor Wolf Ltd., Bedlington, Northcumberland NE22, 7DH, GB
–	×																		Neste OY, S 44401 Stenungsund, S
–	×		×							×									Reichhold Chemie AG, CH 5212, Hausen Brugg, CH
–	×		×																S.I.S.A.S Societe Serie Acetica Sintetica, Milano, I
–	×		×												×				UCB Div. Spécialtités, Chemiques, Bruxelles, B

5.10 References

[1] ASTM D 833, 1966, Book of ASTM Standards, Part 27, p. 381.
[2] *O. Fuchs, H.H. Frey:* Kunststoffe 49 (1959) pp. 213/216.
[3] *H. Gnamm, W. Sommer:* Die Lösungsmittel und Weichmachungsmittel, 7th ed., Wissenschaftl. Verlagsgesellschaft mbH, Stuttgart, 1958, pp. 516/517.
[4] *R. Franck, H. Mühlschlegel:* Kunststoffe im Lebensmittelsektor, Carl Heymanns Verlag KG, Köln, Berlin, Bonn, 1978.
[5] DIN 58361, part 1 to 5; DIN draft 58361, part 6, September 1980.
[6] *H. Thinius:* Chemie, Physik und Technologie der Weichmacher, VEB Deutscher Verlag für Grundstoffindustrie, Leipzig, 1963, pp. 34/38.
[7] *P. Schmidt:* Kunststoffe 41 (1951) p. 23 and 42 (1952) p. 142.
[8] *A. Hartmann, F. Glander:* Koll. Z. u. Z. Polym. 137 (1957) pp. 79/86.
[9] Technical information, Bayer AG, Leverkusen/FRG.
[10] *F. Stühlen, L. Meier:* Kunststoffe 61 (1971) p. 867/874.
[11] Weichmacher in der Kunststoffindustrie, Chemische Werke Hüls AG, Marl, 2nd ed., 1970, pp. 22/23.
[12] Hostalit, technical information, Hoechst AG, December 1974, p. 45.

6 Lubricants and Related Auxiliaries for Thermoplastic Materials

Dr. Walter Brotz, Hoechst AG, Werk Gersthofen,
Federal Republic of Germany

6.1 Introduction

6.1.1 Lubricants and their applications in the processing of plastics

Thermoplastics are macromolecules of generally high molecular weight whose melts are commonly highly viscous. Virtually all thermoplastics are processed in the molten state under shear, since the molten polymer must pass through narrow slots, channels, pipes, and dies in the shortest possible time. To attain high production levels with minimum machinery wear and minimum energy consumption, it is essential to decrease melt viscosity to the lowest possible level. This aim can be achieved in two ways: firstly, by keeping the temperature as high as the material will tolerate and secondly, by adding auxiliaries – so-called lubricants – which decrease internal and external friction.

The allowable increase of temperature at a specific shear rate*) is limited. At higher temperatures, the long chain molecules of the polymer may break. The viscosity of the melt drops and the depolymerized material loses its strength. High thermal stress, despite the presence of stabilizers, may lead to the degradation of molecules with the formation of undesirable reaction products and strong discoloration or even carbonization. A well-known example is the thermal degradation of PVC due to the elimination of hydrogen chloride. Here, the only way to solve the problem is by the use of efficient lubricants.

By definition: Lubricants are additives which facilitate the processing of plastics by improving flow properties and reducing the adherence

*) The shear velocity gradients of different processes are:

Compression molding	10^0 to 10^1 s^{-1},
calendering	10^1 to 10^2 s^{-1},
extrusion	10^2 to 10^3 s^{-1},
injection molding	10^3 to 10^4 s^{-1}.

of the melt to machine parts. Their specificity increases in parallel with the difficulty and the sophistication of the processing technique involved.

Besides their main task – to improve flow properties – lubricants also perform other functions: they can act as melt promoters; antiblock, antitack and antistatic agents; as well as color and impact improvers.

Lubricants are basically organic compounds, some of natural origin, such as petroleum, animal or vegetable fats and waxes, and some are synthetic products, such as low molecular weight polyethylene. They are offered mostly as free-flowing, non-agglomerating powders with melting points up to ca. 150 °C, but also as liquids. Paste-like products are not used due to difficulties in handling

6.1.2 The technical and economic significance of lubricants

If one consults the classical handbooks on plastics on the subject of lubricants, it is soon apparent how neglected this part of the additives range was until the 60's. The first information about lubricants appeared in the technical brochures of lubricants' producers and continued sporadically in the professional periodicals, primarily when discussing difficult PVC processing. The relationship between lubricants and plastics may be compared with a motor car and its electrical system. As with the latter, the lubricant appears to be a subordinate appendix but without the electrical system the car won't run and so without lubricants, the commercial production of PVC articles would not be possible. Other thermoplastics, such as polystyrene or polyolefins, which initially could be processed without lubricants, now more and more require the use of these additives to overcome problems caused by high production rates. Although the use of lubricants in specialty plastics is seemingly less spectacular, it is none the less rather important.

It is not only the improvement in flow properties, and consequently production rates afforded by lubricants which makes them irreplaceable, but also the need to obtain articles of high quality, especially with respect to surface appearance. This is important both for esthetic *and* technical reasons, e.g. with respect to gloss and smoothness of household articles and packaging films, as well as to prevent inside surface roughness of bottles and pipes. The enormous development and use in the last decade of plastics processing machinery, such as high performance calenders, blow and injection molding machines and extruders, has been made possible only by the parallel development of specific lubricants.

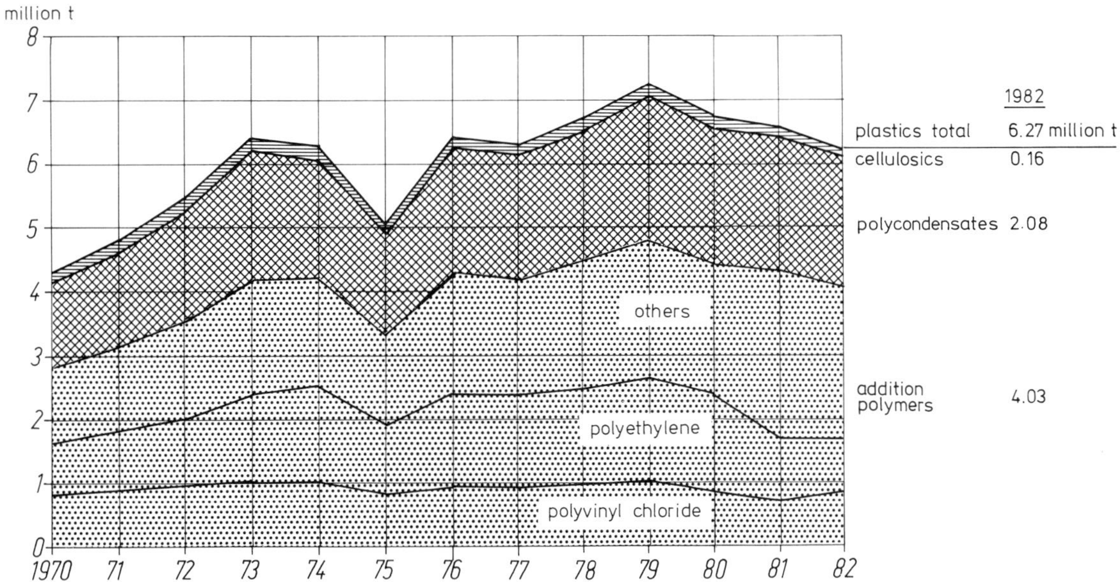

Fig. 1. Plastics production in West Germany 1970 to 1982.

Due to their relatively low concentrations in plastics materials, lubricants have already been called "the technical pinch of salt." Their concentration in plastics lays within the range 0.01 to 4.0%, but commercial use levels are between 0.5 to 1.0%, depending on the type of plastic, the processing equipment and the final article.

The consumption of lubricants is directly proportional to the production of plastics. The production volumes for thermoplastics in West Germany may be an indicator, therefore, of the production volumes of lubricants in the Western World (figure 1). The consumption of lubricants for specific polymers is shown in table 1. Table 2 is a breakdown of consumption by country.

Table 1. Consumption of lubricants for thermoplastics in Western Europe

Thermoplastic	1970	1980
Total consumption	*25,000 tons*	*70,000 tons*
PVC	58%	65%
PS and copolymers	21%	18%
Polyolefins	15%	11%
Others	6%	6%

Table 2. Consumption of lubricants in Western Europe

Country	1970	1980
Total consumption	*25,000 tons*	*70,000 tons*
West Germany	33%	29%
France	16%	19%
Italy	18%	17%
Benelux	10%	8%
Great Britain	16%	11%
Scandinavia	4%	9%
Spain	3%	7%

One kilogram of lubricant, in powder form, costs between approximately $ 0.5 U.S. for paraffins and $ 4.5 U.S. for the high performance montan waxes. In general, processors make every effort to minimize the cost of additives like lubricants, stabilizers, fillers, etc. in a formulation. However, with lubricants, it is not always more economical to replace high-priced products by cheaper ones since a small improvement in production rate, energy consumption or machinery wear more than outweighs the extra cost of the more expensive, but more efficient additives.

At an average price of approximately $ 2 U.S./kg, lubricants represent a sales volume of approximately 120 million U.S. dollars in Western Europe in 1980.

6.2 Lubricants: materials and effects

When the processing of PVC began, about 50 years ago, the technologists tried to stabilize this material with metal compounds like white lead. Soon, they turned their attention to relatively low melting metal soaps which blend in and process more easily and were therefore essentially more efficient. These products not only stabilized PVC but they also improved processing; the longer the hydrocarbon chains of the metal soaps, the more effective they were as lubricants. Based on this experience, other materials were found to be effective lubricants. Among these were glycerol esters, fatty acids, waxes, and paraffins. The lubricant became an indispensible ingredient in PVC formulations.

As with a pharmaceutical that helps in a desired way, yet may cause at the same time undesirable side effects, so these additives also caused frequent and disturbing side effects. The lubricant thus became the most difficult part to select in the art of thermoplastics formulation. There developed a kind of black art which was not superseded until scientific methods could be devised with the help of sophisticated testing equipment. Then industry became able to supply products with constant properties and a precisely defined area of efficiency.

The performance criteria for lubricants are clear. They must facilitate the processing of a plastic in such a way that the production runs economically without interruption, leading to final articles which are technically and esthetically acceptable. Awareness of harmful side effects which appear a long time after manufacture are just as important as the possible harmful side effects which may appear during processing. Such examples of the former are: the heat distortion temperature of pipes or the life expectancy of gutters and siding.

To avoid undesirable side effects, a well considered choice of materials is necessary; a decision which can seldom be made without compromise. Summarizing we can say, lubricants should obey the following rules:

- Not be volatile and not fume at the processing temperature;
- not exude from the plastic during processing or cause plate-out by acting as a carrier for pigments, especially titanium dioxide and dyestuffs; in this case, deposits build up on calender rolls

or extruder dies which destroy the smooth surfaces and cause interruptions in productions;
- not lead to discoloration and degradation by reducing the thermostability of the compound;
- not affect the mechanical properties of the polymer;
- not affect light and outdoor stability (improvements can often be achieved);
- not exude during storage thereby spoiling the surface for subsequent printing, welding, or coating.

Lubricants must, moreover, be physiologically approved and must not contravene food additive laws of the different countries. Most lubricants meet the regulations of the American Food and Drug Administration (FDA) and of the German "Bundesgesundheitsamt" (BGA).

6.2.1 Mechanism of lubrication

All lubrication results in a decrease in friction. During blending and processing the lubricants coat the surfaces of the polymer particles. Thus the powder is lubricated and flows easier in the cold parts of the processing machine. With increasing temperature, the polymer begins to soften and the lubricant melts and penetrates the polymer. At which point in the barrel of the extruder this occurs is determinant for the flow properties. Thus, the melting point as well as the melt viscosity of the lubricant have an influence on its effect. It is obvious that a low viscosity melt will behave in a different way to a highly viscous one. There are even special agents, so-called "processing aids" (mainly polymethacrylates or polymethylstyrenes) which, in their effect, are very similar to lubricants. Due to their highly viscous melt, they adhere strongly to the polymer powder and thereby greatly accelerate plasticization. At this point, we only want to mention them as they are discussed separately in another chapter of this book (cf. chapter 7).

The rate of penetration of the lubricant into the polymer is governed by its solubility, which depends on the structure of the lubricant molecule as well as its polarity in relation to that of the polymer. The full spectrum of solubility relationships is possible from total solubility to almost complete insolubility. In the case of the former, we would refer to the lubricant as a solvent which, because of its strong solvation, actually hinders flow and moreover embrittles the material (e.g. small quantities of plasticizer in PVC). On the other hand, less strong association between lubricant and the long chain molecules creates a kind of ball bearing effect.

They are able, therefore, to arrange themselves in the direction of flow and thereby slide from each other. This is the typical function of an *internal* lubricant.

By variation of bond strength between polymer and plasticizer fine-tuning is possible: the weaker the bond strength, the more movable and incompatible the lubricant molecules. Under high pressure they are squeezed out of the compound. They cover the surface and act as a lubricant between the thermoplastic and the walls of the machine. In the extreme, and undesirable case, they build up a release film which totally prevents the indispensible adherence to the wall. This is a disadvantage during processing for it disturbs the plasticization and smooth flow as a result of "over-lubrication". The art of effective processing consists of finding the right system of lubricants for a specific polymer and processing machine. This right system provides the correct ratio of internal and external lubrication at the right place and in the right concentration. The complexity of these processes makes the application of lubricants difficult and has had the consequence – in order to anticipate a new trend – of encouraging the producers, more and more, to relieve the processors of troublesome testing by offering special lubricant systems for specific applications. They consist of well adjusted blends of suitable components.

The labelling of lubricants as either *internal* or *external* is a common but somewhat crude classification, for there is practically no product which acts completely internally or externally. We should say more correctly: predominantly internal or predominantly external. Lubricants which act internally at low concentrations, but have a sharply defined limit to their compatibility, behave like an external lubricant at higher concentrations. A typical example is the behavior of stearic acid in PVC.

Lubricant molecules (paraffins excepted) always comprise polar groups and hydrocarbon chains of varying length. The ratio between the two determines the lubricating action. Figure 2 shows the relationship between molecular structure and lubricating effect in PVC. The placing of a compound on a scale between internal and external lubrication is, of course, different for all types of polymers. A relatively polar lubricant, such as glycerol monostearate, is fairly soluble in PVC and therefore acts internally, whereas it is a strongly external lubricant for the non-polar polyolefins. On the other hand, polyethylene waxes and paraffins act as strong external lubricants in PVC, with which they have only limited compatibility, and as almost ideal internal lubricants for polyolefins.

The polymer/lubricant system becomes even more complicated if the formula contains other lubricants and additional products such as heat

Chemical class	Chemical structure*)	Polar group	Origin of the non-polar group	Effect
fatty alcohol	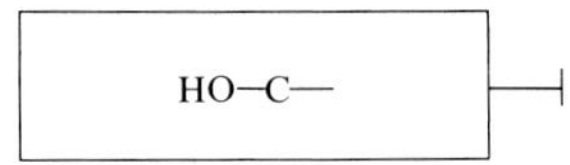	alcohol	fatty alcohol ($C_{14}-C_{18}$)	internally effective lubricants, predominantly polar, clearly soluble in PVC
dicarboxylic acid ester	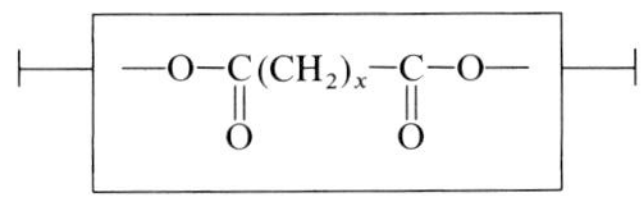	ester	fatty alcohol ($C_{14}-C_{18}$)	
fatty acid ester	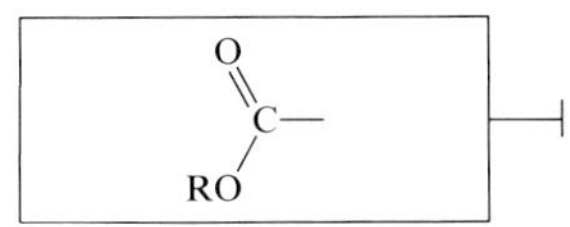	ester	fatty acid ($C_{14}-C_{18}$)	
fatty acid	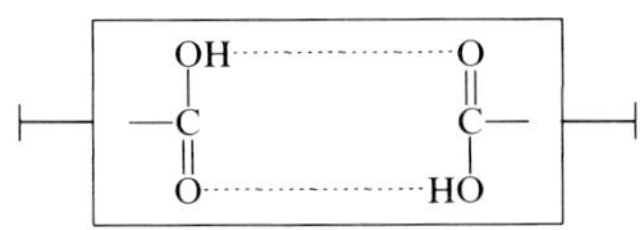	acid	fatty acid ($C_{14}-C_{18}$)	
fatty acid soaps	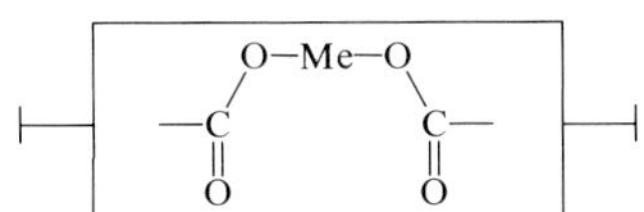	soap	fatty acid ($C_{16}-C_{18}$)	
fatty acid amide	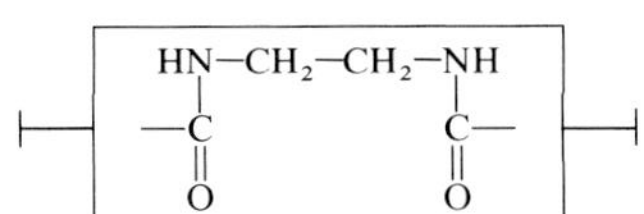	amide	fatty acid ($C_{16}-C_{18}$)	

soap	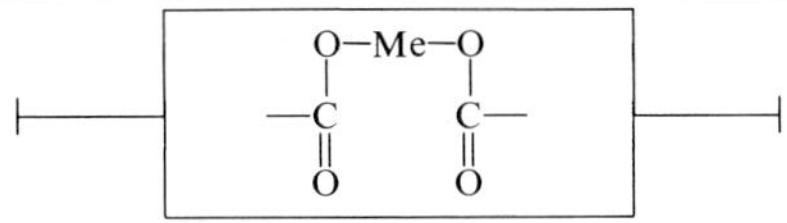	soap	synth. wax acid ca. C_{30}	balance between internal and external lubrication
ester, partially saponified	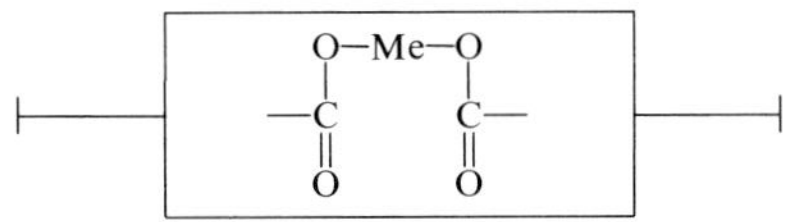	soap + ester	montanic acid ($C_{26}-C_{32}$)	
ester	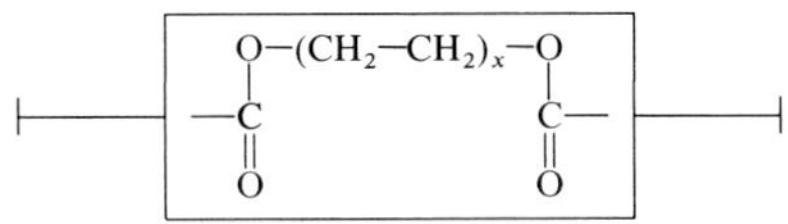	ester	montanic acid ($C_{26}-C_{32}$)	
wax acid	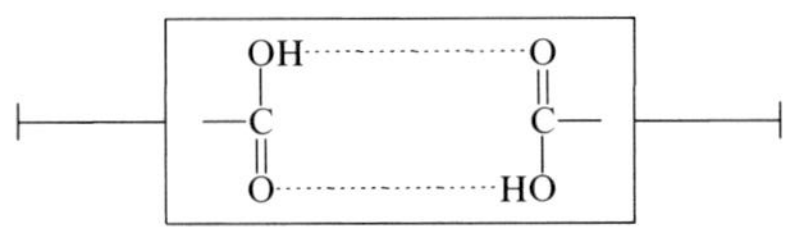	acid	montanic acid ($C_{26}-C_{32}$)	
paraffin, liquid	short, branched chain 	—	paraffin	externally acting lubricants, non-polar, little, compatibility with PVC, nevertheless not always causing turbidity
paraffin, solid	short, straight chain	—	paraffin	
high density polyethylene waxes (mw ~ 3000–9000)	long, straight chain	—	polyethylene	
low density polyethylene waxes (mw ~ 2000)	long chain, somewhat branched 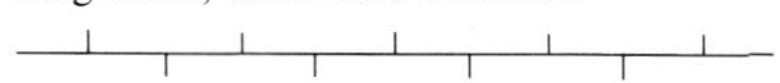	—	polyethylene	

*) The box signifies the polar group, ——— signifies the hydrocarbon chain.

Fig. 2. Schematic view of the relationship between molecular structure and effect of various lubricants in PVC.

and light stabilizers. Thereby strong synergistic effects may appear which can be used to improve flow and save energy during processing.

Such effects show up if a metal soap and a hydrocarbon are used in combination. Without changing the pressure and temperature of a capillary rheometer or the lubricant level in the compound, the output can be altered infinitely just by changing the ratio of lubricants. This corresponds well with the effect in an extruder or injection molding machine. Figure 3 shows such an interaction between a low molecular weight polyethylene and a montan wax soap. Evidently it is important, not only to balance the lubricants themselves but also to consider their relationship to the additives in the compound. What is considered useful for injection molding may yet be of disadvantage for extrusion. The effect might arise, that one of the components of the lubricant wets the surface of the molten polymer and, as in the case of PVC, its discrete particles, whereas the other component is squeezed out and becomes a mobile and highly efficient external lubricant.

Effective lubrication means lower energy consumption and therefore less internal heating during processing. Consequently less thermal and me-

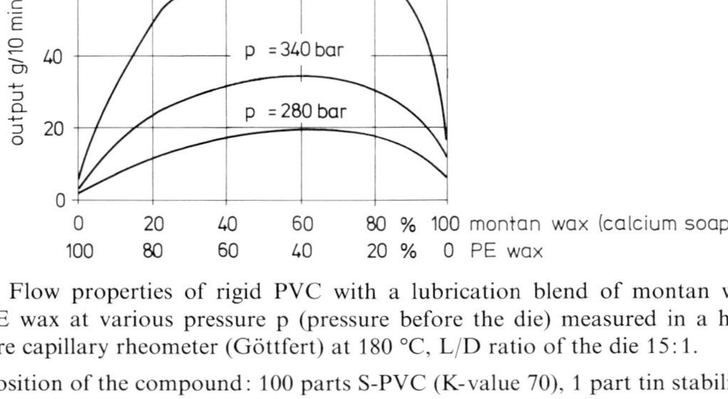

Fig. 3. Flow properties of rigid PVC with a lubrication blend of montan wax and PE wax at various pressure p (pressure before the die) measured in a high pressure capillary rheometer (Göttfert) at 180 °C, L/D ratio of the die 15:1.

Composition of the compound: 100 parts S-PVC (K-value 70), 1 part tin stabilizer (Hostastab SnS 61) and 2 parts lubricants (montan wax and PE wax).

chanical stress is applied, giving higher stability to the melt. So the lubricant acts concurrently as a stabilizer. An example of this is: S-PVC, without stabilizer or lubricant, plastified and kneaded on a twin-roll mill at 180 °C, decomposes after approx. 2 minutes; the material containing 3% of a montan wax runs, under the same condition, for approx. 20 minutes. The end point is marked by a sudden darkening of the film. This milling time is already sufficient for many processes. In practice a greater degree of stability is necessary especially when minimal discoloration is specified.

It has already been mentioned that lubricants are mostly marketed as non-agglomerating, free-flowing powders, occasionally as liquids and never, because of customer resistance on handling grounds, as pastes. We have also shown how the melting point of a solid lubricant influences the time and place at which it becomes effective. This means, of course, that real effectiveness begins only with the liquid phase, powder lubrication is of less importance.

We also pointed out the importance of the viscosity of the molten lubricant on the plasticization of the compound. In other words, the spectrum of viscosity (at 150 °C) of molten lubricants extends from approx. 5 to 30,000 mPa·s. Viscosities above this range are normally associated with the previously mentioned "processing aids".

The viscosity behavior of polymer melts lays in the region between a viscous liquid and a rubber. The internal lubricant, which undertakes lubrication in the region of the long chain molecules, thereby influences to a high degree the viscoelastic behavior of the melt and, therefore frequently a desirable side effect, the pronounced reduction of die-swell.

6.2.2 Classes of material and composition

The assumption for the lubrication effect of a lubricant molecule is always a hydrocarbon chain of varied length, more or less branched. This variation already opens up a lot of possibilities for adaptation to the problem at hand. The possibilities are still further increased by the introduction of polar groups into such chains, thereby varying even more the interaction with the polymer relative to solubility and compatibility. This enables us, therefore, to influence, in a desirable way, the lubrication and release properties together with the described side effects.

The polar groups are mostly formed with oxygen, often by oxygen and metals or oxygen and nitrogen and sometimes by halogens. The whole lubricant range can therefore be classified under the following headings:

Hydrocarbons,
alcohols,
carboxylic acids,
carboxylic acid esters,
ketones,
metal salts of carboxylic acids,
amides,
halogenated hydrocarbons.

6.2.2.1 Raw materials and origin

The majority of lubricants are of *natural* origin but few can be used in the form in which they are found or harvested. Because of variations in place of origin, climate, method of exploration, and content of impurities, the compositional deviations may be great. The material must be adapted, to an accepted standard, by refining.

Firstly, we mention the *natural paraffins* obtained from crude oil deposits, from the slow burning of brown coal and from bituminous deposits. After refining and distillation, they are classified according to their melting point. They are relatively pure and well-defined industrial products.

The most important raw material sources for lubricants are *vegetable and animal fats*. They furnish the *natural fatty acids* which form the basis of most lubricants syntheses; of special significance is the mixture of C_{16} to C_{18} acids (viz. palmitic, stearic, and oleic), of which stearic is the most important. The shorter chain acids, such as lauric (C_{12}) are occasionally used but their higher volatility seriously affects their ability to meet the requirements of a lubricant. The higher homologues, such as arachic (C_{20}) and behenic (C_{22}), are highly desirable but are not available in sufficient quantities to be used commercially as lubricants.

Oxystearic acid and its triglyceride are especially favorable lubricant components. Both are easily obtained by hydrogenation of castor oil, further hydrogenation of the fatty acids leads to the corresponding alcohols. The most useful of which, either for use alone or for syntheses, is technical stearyl alcohol and its companion product, palmityl alcohol.

Technically, however, the most important, albeit more expensive, lubricants are waxes with a hydrocarbon chain of approx. 30 carbon atoms. The most important representative is *montan wax,* obtained by solvent extraction from brown coal (lignite). It is a material of practically unlimited availability. It is, even in its original form, perfectly suitable as an additive for PVC, but being black it contributes undesirable color to the polymer. Almost colorless, technical montanic acid (C_{28} to C_{32}) is obtained by chromic acid oxidation of deresinified crude wax. With this product, and modified waxes based on it, we gain additives with constant quality which contribute to trouble-free polymer processing.

The chemical processes leading to new semi-synthetic lubricants are based on the use of the previously mentioned fatty or wax acids. They are reacted with the indicated coreactants to synthesize the products shown: with alcohols to form esters, with metal oxides or hydroxides to form soaps, with amines to form amides.

Totally synthetic products are also used regularly in formulations. Such products are the Fischer-Tropsch paraffins made from water gas, and the low molecular weight polyethylene and polypropylene waxes. Synthetic polar lubricants are made from waxlike copolymers based on ethylene and either vinyl acetate, acrylic acid, or crotonic acid. More important in this class, however, are the oxidized polyethylene waxes.

6.2.3 Lubricant and polymer

From the classes of materials shown in table 3, we are able to select for every polymer, products which, according to their polarity and solubility, will be suitable as internal or external lubricants or a combination of both.

Although the number of individual products actually used is very large, it may be reduced to a reasonable amount when processing high volume plastics.

6.2.3.1 Lubricants for PVC

PVC cannot be processed without lubricants and therefore has the largest requirements of all thermoplastics for these additives. Figure 2 has already shown the chemical structure of suitable compounds and their effect in PVC. PVC formulating is complex and difficult, therefore, additional specialty products of the chemical industry, in addition to those cited in table 3, are used.

According to the specific processing technique – compression molding, calendering, extruding, extrusion blow molding, injection molding – the formulation must take account of the shear velocity range and the right balance between flow, antitack and plasticization time. This is achieved by finding the correct relationship between internal and external lubrication.

Table 3. Properties of common lubricants

Chemical class	Substance	Technical data (dp: drop point, sp: solidification point, mp: melting point)	Characteristics
Hydrocarbons	paraffin oil	highly branched, liquid at 20 °C	Non-polar, colorless and unreactive materials, characterized by molecular weight and degree of branching.
	natural paraffins	dp: 52 to 75 °C	
	synthetic paraffins	dp: 100 to 110 °C	
	polyethylene waxes of low and high density	dp: ca. 100 to 140 °C density: 0.92 to 0.97 g/cm^3	
	polypropylene waxes	dp: ca. 160 °C viscosity (at 170 °C) of the polymer waxes: 100 to 35,000 mPa·s	
Alcohols	cetyl alcohol	sp: 46 to 49 °C	Polar products which help stabilization in addition to the lubricating action.
	stearyl alcohol	sp: 55 to 57 °C	
	tallow fatty alcohol (mixture of cetyl and stearyl alcohol)	dp: 43 to 57 °C	
Ketones	distearyl ketone	dp: ca. 88 °C	The ketone is surface-active in polar plastics due to the polar carbonyl group in the middle of the molecule.

Table 3 (continued)

Chemical class	Substance	Technical data (dp: drop point, sp: solidification point, mp: melting point)	Characteristics
Carboxylic acids	lauric acid (pure)	dp: 43 °C	Two molecules can form a hydrogen bond via the highly polar carboxylic groups, and in this way double the hydrocarbon chain. In polar plastics this leads to a better internal and external lubrication. Molecular weight is the most important feature. High molecular weight acids (e.g. technical montanic acid) show in non-polar plastics an excellent wetting of metal surfaces and therefore give high surface finish. Due to relatively high volatility, lower molecular weight acids are infrequently used and only then in compounds requiring low processing temperatures. Because of their high polarity, oxy acids are more effective as internal lubricants in polar plastics than their respective carboxylic acids. Oxidized polyethylene waxes contain alcohols, esters and ketones in addition to carboxylic acids, and are therefore suitable for the fine-tuning of compounds.
	myristic acid (pure)	dp: 54 °C	
	palmitic acid	dp: 62 °C	
	stearic acid (tech.), also called stearine (consisting of ca.: 55 to 67% stearic acid, 40 to 30% palmitic acid, and 3% myristic acid)	dp: 52 to 60 °C viscosity (at 60 °C): ca. 10 mPa·s	
	12-hydroxy stearic acid	dp: 72 to 78 °C viscosity (at 100 °C): ca. 14 to 16 mPa·s	
	arachic acid	sp: ca. 75 °C	
	behenic acid	sp: ca. 79 °C	
	montanic acid (tech.)	dp: ca. 84 °C viscosity (at 100 °C): ca. 20 mPa·s	
	oxidized polyethylene waxes	dp: 100 to 115 °C viscosity (at 120 °C): 30 to 600 mPa·s	

Table 3 (continued)

Chemical class	Substance	Technical data (dp: drop point, sp: solidification point, mp: melting point)	Characteristics
Metal salts of carboxylic acids	calcium stearate zinc stearate lead stearate (neutral) calcium montanate calcium salts of oxidized polyethylene waxes or synthetic wax acids	mp: 130 to 160 °C mp: ca. 120 °C mp: ca. 100 °C mp: ca. 123 °C dp: 90 to 110 °C	Metal salts act as lubricants, as well as release agents and stabilizers. Because of their basicity, they may react during processing with acidic ingredients. Stabilizers based on metal salts are, in most cases, also efficient lubricants. Metal salts account for approx. 23% of the lubricant market, of which 15% is calcium stearate alone.
Carboxylic acid amides	monoamides: oleamide erucamide stearamide diamides: ethylene-bis-stearamide	 mp: ca. 70 °C mp: ca. 95 °C mp: ca. 140 °C	Carboxylic acid amides, particularly the high melting diamides with their highly polar group in the middle of the molecule, exhibit high slip and antiblocking properties. As lubricants, they are suitable for almost all thermoplastics. With a market share of 25%, they are the most important group of lubricants.
Carboxylic acid esters	ethyl stearate n-butyl stearate	liquid at 20 °C liquid at 20 °C	In this group of esters, the possible permutations with respect to polarity, solubility,

Table 3 (continued)

Chemical class	Substance	Technical data (dp: drop point, sp: solidification point, mp: melting point)	Characteristics
Carboxylic acid esters	iso-butyl stearate	liquid at 20 °C	compatibility, melting point, and melt viscosity are extremely high. There exists, therefore, the possibility to adjust the balance of internal and external lubrication precisely to the requirements of the polymer and the processing equipment. The esters mentioned here, are products already used in practice. The number of speciality products offered by manufacturers is, however, far wider.
	iso-octyl stearate	liquid at 20 °C	
	iso-tridecyl stearate	liquid at 20 °C	
	cetyl palmitate	mp: ca. 49 °C	
	cetyl stearate	mp: 56 °C	
	stearyl stearate	mp: ca. 53 °C	
	ethylene glycol monosterate	mp: ca. 61 °C	
	glycerol monooleate	liquid at 20 °C	
	glycerol monoricinoleate	liquid at 20 °C	
	glycerol monostearate	mp: ca. 69 °C	
	glycerol mono-12-hydroxy stearate	mp: ca. 70 °C	
	glycerol tristearate	mp: 50 to 69 °C	
	glycerol tri-12-hydroxy stearate	mp: ca. 70 °C	

Table 3 (continued)

Chemical class	Substance	Technical data (dp: drop point, sp: solidification point, mp: melting point)	Characteristics
Carboxylic acid esters	glycerol tribehenate	mp: ca. 78 °C	
	glycerol trimontanate	mp: ca. 70 °C	
	pentaerythritol tetrastearate	mp: ca. 63 °C	
	pentaerythritol tetrabehenate	mp: ca. 73 °C	
	mixed esters (e.g. made from adipic acid, pentaerythritol, and stearic acid)	dp: ca. 50 to 70 °C viscosity (at 100 °C): 100 to 200 mPa·s	
	montanic acid esters	mp: ca. 80 to 82 °C viscosity (at 100 °C): 5 to 35 mPa·s	
	partially saponified montanic acid esters	dp: ca. 100 °C viscosity (at 100 °C): ca. 300 mPa·s	

The first step in determining the correct relationship is to place the lubricant on the scale between internal and external lubrication with respect to the polymer to be processed. The calculation of this point with the help of dielectric constant and molecular weight has not proven reliable, although the dielectric constant may still be helpful in the construction of a crude classification. A useful method for PVC is the determination of turbidity caused by the addition of the lubricant. The greater the turbidity, the more external is the lubrication. Exceptions to this rule are shown by unbranched, high density PE waxes with molecular weights above 6000 as well as highly branched liquid or low melting paraffins. Although these materials are strongly external in their effect, they do not interfere with the transparency of the PVC.

For injection molding and extrusion, it is advantageous to use as basic lubricant a pronounced internally – *and* externally – acting product, e.g. montan wax, in combination with a mainly internally acting lubricant (such as low molecular weight fatty acid esters); for compression molding or calendering highly external, anti-adherent lubricants (such as PE or amide waxes) are used in addition to the basic internal/external lubricant.

In 1977, the consumption of PVC in Western Europe was 3.26 million metric tons. Of this quantity, 1.34 million metric tons (41%) was used in the production of plasticized PVC floor coverings, cable, hoses, film, etc.

While the processing of rigid PVC normally requires the use of 1% lubricant, and in special cases up to 4%, plasticized PVC requires only 0.5% or less. Secondary plasticizers, such as chloroparaffins or liquid glycerol esters, fulfill this requirement to a large extent. By reason of the "carrier effect" of the plasticizer, precautions should be taken to insure that low molecular weight compounds do not exude, thereby causing the buildup of an objectionable surface layer. In such circumstances, the relatively high molecular weight montan waxes and the calcium salts of synthetic wax acids are particularly suitable. Amide waxes are also very useful should a high degree of slip be desired, but they might interfere with printability and weldability.

Typical formulations for rigid PVC processing

Pipes (lead stabilized)

100.0 parts*)	S-PVC (K-value 67),
0,5 parts	tribasic lead sulfate,
0.8 parts	dibasic lead stearate,
0.3 parts	neutral lead stearate,
0.3 parts	calcium stearate,
0.2 parts	stearic acid,
0.2 parts	hydrocarbon wax.

*) parts always means: parts by weight

Pipes (calcium stabilized)

100.0 parts	S-PVC (K-value 67),
1.0 parts	calcium stearate,
0.2 parts	stearic acid,
0.8 parts	hydrocarbon wax.

Pipes (tin stabilized)

100.0 parts	S-PVC (K-value 67),
0.6 parts	di-butyltin thioglycolate,
0.7 parts	paraffin (dp: 70 °C),
0.5 parts	part. saponified montan wax.

Glass-clear pressed sheets

100.0 parts	S-PVC (K-value 60),
2.0 parts	tin stabilizer,
0.5–1.0 parts	mixture of fatty acid ester/part. saponified montan ester/high density PE wax (mw ca. 9000), 50:35:15.

Shades (lead stabilized)

100.0 parts	S- or M-PVC (K-value 67),
2.5 parts	dibasic lead phosphite,
1.5 parts	dibasic lead stearate,
0.5 parts	calcium stearate,
0.4 parts	stearyl alcohol,
0.4 parts	Ca-salts of synth. wax acids,
5.0 parts	chalk (coated),
2.0 parts	titanium dioxide.

Transparent containers (tin stabilized)

100.0 parts	S- or M-PVC (K-value 57),
1.2 parts	di-octyl tin stabilizer,
0.2 parts	part. saponified montan ester,
0.1 parts	high density PE wax (mw ca. 9000),
0.2 parts	glycerol monostearate.

Transparent impact modified containers (tin stabilized)

100.0 parts	S- or M-PVC (K-value 57),
14.0 parts	impact modifier (e.g. MBS),

1.2 parts di-octyl tin stabilizer,
0.3 parts part. saponified montan ester,
0.2 parts montan ester,
0.2 parts glycerol monostearate.

Technical injection-molded articles (lead stabilized)

100.0 parts M-PVC (K-value 55 and 67),
3.5 parts tribasic lead sulfate,
0.4 parts calcium stearate,
0.4 parts montanic acid ester,
0.4 parts part. saponified montan ester,
1.0 to 3.0 parts fatty acid ester.

Calendered films
(tin stabilized, high temperature process)

100.0 parts M-PVC (K-value 57 and 62),
S-PVC (K-value 58 to 62),
1.2 parts di-octyl tin stabilizer,
0.5 parts montanic acid ester,
0.1 parts amide wax (dp: ca. 140 °C),
0.3 parts glycerol monostearate.

Calendered films, impact modified
(tin stabilized, high temperature process)

90.0 parts M-PVC (K-value 57 and 62),
S-PVC (K-value 57 to 60),
10.0 parts impact modifier (e.g. MBS),
1.2 parts di-octyl tin stabilizer,
0.8 parts montanic acid ester,
0.1 parts amide wax (dp: ca. 140 °C),
0.3 parts glycerol monostearate.

Calendered films
(metal-free stabilizers, low temperature process)

100.0 parts E-PVC (K-value 70),
1.5 parts amino crotonic acid ester,
0.5 parts di-phenyl thiourea,
3.5 to 4.0 parts part. saponified montanic acid ester,
2.0 parts titanium dioxide.

Blown film (tin stabilized)

100.0 parts	S- or M-PVC (K-value 55 to 58),
2.0 parts	di-octyl tin stabilizer,
0.2 parts	part. saponified montan ester,
0.1 parts	high density PE wax (mw ca. 9000),
0.3 parts	fatty acid ester,
2.0 parts	processing aid (e.g. polymethyl methacrylate).

Typical formulations for plasticized PVC processing

Profiles (lead stabilized)

100.0 parts	S-PVC (K-value 70),
65.0 parts	plasticizer,
20.0 parts	filler,
1.0 parts	tribasic lead sulfate,
1.0 parts	dibasic lead stearate,
0.6 parts	Ca-salts of synth. wax acids/PE wax (mw ca. 2000), 20:80.

Hoses (barium/cadmium stabilized)

100.0 parts	S-PVC (K-value 70),
65.0 parts	plasticizer,
1.0 parts	barium/cadmium stabilizer (solid),
3.0 parts	epoxy plasticizer,
0.5 parts	chelator (org. phosphite),
0.5 parts	montanic acid ester.

Cable compounds (lead stabilized)

100.0 parts	S-PVC (K-value 70),
40.0 parts	plasticizer,
3.0 parts	tribasic lead sulfate,
0.5 parts	Ca-salts of synth. wax acids/PE wax (mw ca. 2000), 20:80.

Calendered films (barium/cadmium stabilized)

100.0 parts	S-PVC (K-value 70),
40.0 parts	plasticizer,
1.5 parts	barium/cadmium stabilizer (solid),
0.3 parts	chelator (org. phosphite),
0.2 parts	PE wax (mw ca. 3000).

Injection-molded articles
(e.g. shoes, barium/cadmium stabilized)

100.0 parts S-PVC (K-value 70),
70 to 100.0 parts plasticizer,
2.0 parts barium/cadmium stabilizer (liquid),
0.5 parts montanic acid ester.

PVC pastes for coatings (barium/cadmium/zinc stabilized)

100.0 parts paste-PVC (K-value 75),
68 to 80.0 parts plasticizer,
3.0 parts barium/cadmium/zinc stabilizer (liquid),
0.2 parts montanic acid ester.

6.2.3.2 Lubricants for styrene-containing polymers (PS, IPS, ABS, SAN)

For styrene-containing polymers, the use of lubricants is confined to a small range of products, namely:

predominantly internal lubricants	butyl stearate paraffin oil liquid polybutene (mw ca. 500)
predominantly external lubricants	amide wax (dp: ca. 140 °C) zinc stearate
especially for ABS	PE wax (low molecular weight) part. saponified montan ester wax

It is normally sufficient, for the lubrication of polystyrene and its copolymers, to use 0.1 to 0.5% of an external lubricant. The finely powdered lubricant is applied to the polymer granules by tumble-mixing. Up to 1.5% of an internal lubricant (e.g. of butyl stearate) may be incorporated.

Contrary to PVC, where the processor or granulate manufacturer chooses and incorporates the lubricant, the producers of styrene polymers offer special grades already containing lubricants.

6.2.3.3 Lubricants for polyolefins (LDPE, LLDPE, HDPE, PP)

Polyolefins have inherently good processing characteristics, therefore, the addition of lubricating substances is not absolutely essential. However, the desirability of high production rates and the tendency to use higher molecular weight polymers (paper-like HDPE films) has led to the need for such additives. In addition to essential improvements in flow and surface quality which can be achieved with lubricants, undesirable die-swell can also be greatly reduced by suitable lubricants.

Internal lubricants are preferably polyethylene and crystalline polypropylene waxes (molecular weights ca. 2000), the former for PE and the latter for PP. They are also used frequently as carrier materials in pigment concentrates and for this reason may be already present in the polymer. (PE waxes are not fully compatible with polypropylene and therefore act, to a certain extent, as external lubricants.) With large injection moldings the waxes not only improve flow during processing, but also greatly enhance surface gloss and environmental stress crack resistance. Both polyethylene and polypropylene tolerate up to 2% lubricants without showing any change in properties. With regrind, up to 5% of polyethylene wax may be added in order to adjust the melt index to the desired level.

Amide wax (ethylene-bis-stearamide; dp: ca. 140 °C) acts *internally* and *externally*. It increases extruder output and reduces die-swell, especially in combination with calcium stearate (approximately 1:1). The addition level is about 0.25 to 1.0%.

Glycerol monostearate and partially saponified montan esters act as *highly external* lubricants in the production of slit film.

Calcium and zinc stearates may function not only as flow improvers, but also as stabilizers. On the other hand, light stabilizers and antioxidants may have a definite lubricating effect in polyolefins.

6.2.3.4 Lubricants for other thermoplastics

Vinyl chloride/vinyl acetate copolymers

for pressing of records:

montan esters,
amide wax (dp: ca. 140 °C, addition level of each ca. 0.5%).

Polyamides

for flow improvement, especially for the production of monofilaments:

di-octyl sebacate,
montan esters (addition level ca. 0.1%).

Polycarbonate

stearyl stearate,
calcium stearate,
montan esters (addition level: 0.1 to 1.2%).

Polyacetals

amide wax (dp: ca. 140 °C, addition level ca. 0.5%).

Thermoplastic polyurethanes

montan ester (part. saponified, addition level ca. 0.5%).

6.3 Lubricant testing

The choice of the correct lubricants is one of the most difficult problems in the processing of plastics. Production machine trials are very costly. Consequently, as the first step, laboratory equipment had to be developed which could simulate and measure the rheological processes in such a way that the results could be transposed to the conditions of a production-scale machine. In spite of the highly sophisticated testing equipment now used, it seldom happens that a laboratory-developed compound runs immediately trouble-free on a production scale. Often, it must be further modified either by changing the lubricant components or concentrations, or by altering the machine parameters (especially the temperature profile).

It is, of course, easier to replace an existing formulation with a new one containing other components than to develop a formulation for totally unknown processing conditions. The properties of proven compounds can be measured very exactly and these test values can be rather easily recreated with other raw materials.

6.3.1 Testing physical and chemical properties of lubricants

The following methods are suitable for testing the physical and chemical properties of lubricants:

Property	Test method
Drop point dp	Deutsche Einheitsmethoden*) M-III 3 (75), DIN 51801, ASTM D 566
Solidification point sp	Deutsche Einheitsmethoden*) M-III 4a, DIN 51556, ASTM D 938–47T
Heat of fusion Melting range Melting point Decomposition	Differential thermal analysis (DTA) and differential scanning calorimetry (DSC)
Viscosity	Deutsche Einheitsmethoden*) M-III 8, ASTM D 1824–66
Consistency (needle penetration)	Deutsche Einheitsmethoden*) M-III 9b, ASTM D 1321–61T
Color measurement	Deutsche Einheitsmethoden*) M-III 6; Gardner color number, photometry
Thermostability	Heating in a test-tube in an oven at 140 °C and measuring color after 1, 4, 8, 16, 24 hours
Volatility	Thermogravimetric analysis (TGA)
Particle size analysis	Sieve analysos DIN 53477; ASTM E 11, D 185
Flow properties	DIN 53492, DIN 53916
Caking tendency	Chem. Ing. Techn. (1968) pp. 191/192
Acid number	Deutsche Einheitsmethoden*) M-IV 2 (57), DIN 53402, ASTM D 1386
Saponification number	Deutsche Einheitsmethoden*) M-IV 2 (57), DIN 53401, ASTM D 1387

*) published by the Deutsche Ges. für Fettwissenschaft, Wissenschaftliche Verlagsges. mbH, Birkenwaldstr. 44, D-7000 Stuttgart

6.3.2 Testing of the lubricating effect in plastics

6.3.2.1 Laboratory twin-roll mill
(plasticization time and thermostability)

The twin-roll machine consists of two thermostatically controlled rolls which can be adjusted relative to nip, rotational speed, and friction (figure 4). The twin-roll mill was one of the first machines used for testing lubricants. The validity of its test results are, however, of a rather limited value. The reproducibility of the results is highly dependent on the operator since the compound in sheet form must be turned manually.

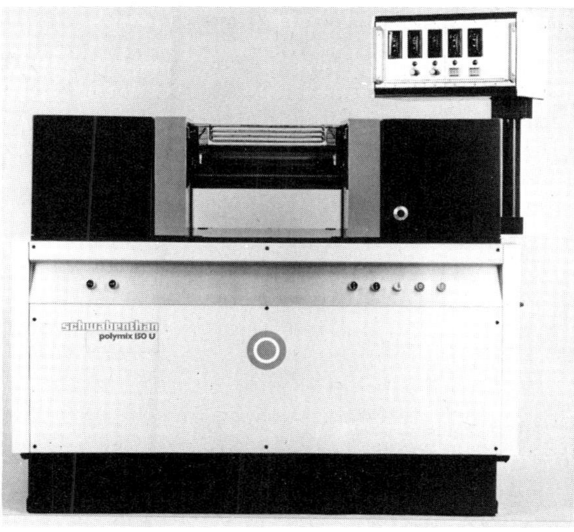

Fig. 4. Laboratory twin-roll mill (photo: Schwabenthan).

The laboratory twin-roll mill is indispensible for the production of a well mixed and plasticized test sample from which test specimens (e.g. compression molded sheets) for mechanical testing or evaluating transparency, light stability, etc. are made. In addition, a fairly accurate measurement of plasticization and dynamic thermostability can be obtained with this equipment. For this purpose, approximately 200 g of the powder blend are poured into the nip of the rolls, which are maintained at an exact test temperature (deviation ± 1 °C). The time is noted at which a homogeneous, completely plasticized sheet is formed. The nip (ca. 0.5 mm) as well as the friction are kept constant. While the machine is running, the sheet is continually cut with a brass knife and turned until it adheres so strongly that it sticks or until it discolors, degenerates, or crumbles. Every ten minutes a small sample is cut from the sheet and attached to a control card. We judge the stability and life-time of the compound under approximately similar processing conditions by assessing the degree of discoloration at specific times. Plasticization time and the time-lapse before sticking also give information as to the internal and external effect of the lubricant. This test is mainly used with PVC.

6.3.2.2 High pressure capillary rheometer (flow properties)

To test its flow properties in the high pressure capillary rheometer the plastified mass is forced under constant pressure through circular precision dies which are defined as to length and diameter (viz. with defined L/D ratios). The output during a specific time-lapse is measured. It is also possible to operate at constant piston speed and measure instead the variations in pressure before the die. This equipment allows exact

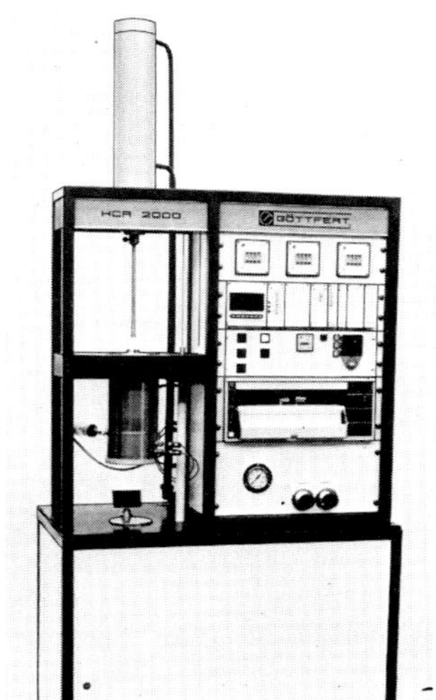

Fig. 5. High pressure capillary rheometer (photo: Göttfert Feinwerktechnik).

rheological properties to be measured. A typical example of a commercial machine is shown in figure 5.

Although the principle of the capillary rheometer appears to be very simple, to reproduce the results is very difficult. Essentially, it depends on the condition of the metal surfaces, which must be cleaned extremely carefully after every test. The reasons for widely differing results, with the same composition of test sample, are often variable particle size, inhomogeneous plasticization – or more important – variable moisture content (for water is a very active lubricant). Liberated water vapor tends to build up a cushioning effect between the wall and the thermoplastic material thereby giving the wrong impression of a strongly efficient external lubrication.

This method is most suitable, therefore, either for scientific purposes or production control of the same material in order to check deviations, less so for developing new lubricant formulations.

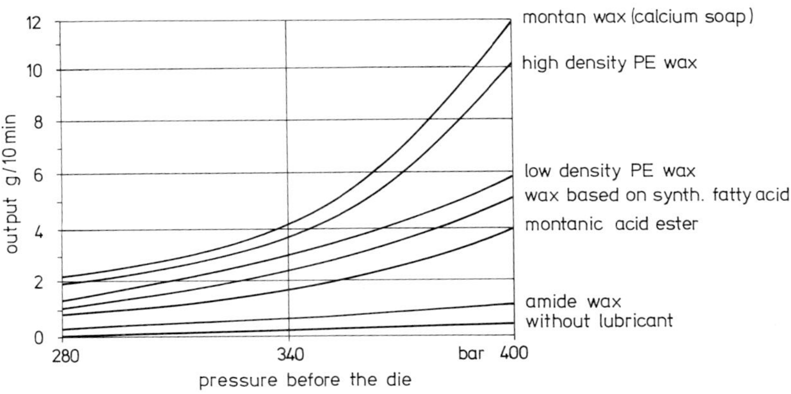

Fig. 6. Flow properties of rigid PVC, with the addition of different lubricants, as measured by the high pressure capillary rheometer (Göttfert); temperature 180 °C, L/D ratio 15:1.

Composition of the compound: 100 parts S-PVC (K-value 70), 1 part stabilizer (Hostastab SnS 61), 2 parts lubricant.

Figure 6 shows characteristic flow curves of PVC compounds lubricated by some typical lubricants (montan, PE, amide and synthetic waxes).

6.3.2.3 **Extrusiometer** (complete lubricating effect)

The extrusiometer is a laboratory extruder which normally has a screw of diameter 20 to 30 mm and length of 20 D. Additionally, it also has equipment to control temperature and pressure of the melt along the barrel – particularly before the die – as well as back pressure of the screw, torque and rpm (figure 7).

The extrusiometer becomes, therefore, the most useful instrument for measuring the effect of lubricants. In addition to its ability to measure exact rheological data, which is obtained by the high pressure capillary rheometer, it also gives information on the speed of plasticization, the relationship between internal and external lubrication (given by torque/output values), the surface quality (onset of melt fracture), the thermostability (discoloration of the extrudate) and finally the viscoelastic behavior (die-swell on leaving the die). Figure 8 shows the characteristic diagram of such a trial. The trends in the main data, such as torque, bulk pressure and output, should the thermoplastic compound be either under- or over-lubricated, are indicated in figure 9.

As with the high pressure capillary rheometer, the reproducibility of tests depends strongly on the uniformity of the barrel, screw and die surfaces. If we desire to test a number of different compounds consecutively, each compound must be run until all traces of the preceeding one have been eliminated and the data stabilizes – a process which may last for hours. It is also indispensible to clean the screw and die daily, preferably overnight, with suitable solvents. In order to standardize the equipment, it is necessary to have a standard compound, which gives invariable data even after long running times.

The extrusiometer is particularly suited to the solving of extrusion problems but can also be used sucessfully to optimize formulations for other processes e.g. calendering or injection molding.

6.3.2.4 **Laboratory kneader**
(plasticization time, torque, thermostability)

The kneader consists of a thermostabilized chamber which is closed by a loadable piston and in which the kneader-blades rotate. The torque of these blades is measurable (figure 10). The chambers is filled with a defined quantity of the thermoplastic compound, closed with the piston, and pressure is applied. While the chamber temperature is kept constant, the temperature of the compound changes according to the progress of plasticization, reaching an end value, which increases further with the onset of decomposition. Figure 11 represents such a "plastogram"

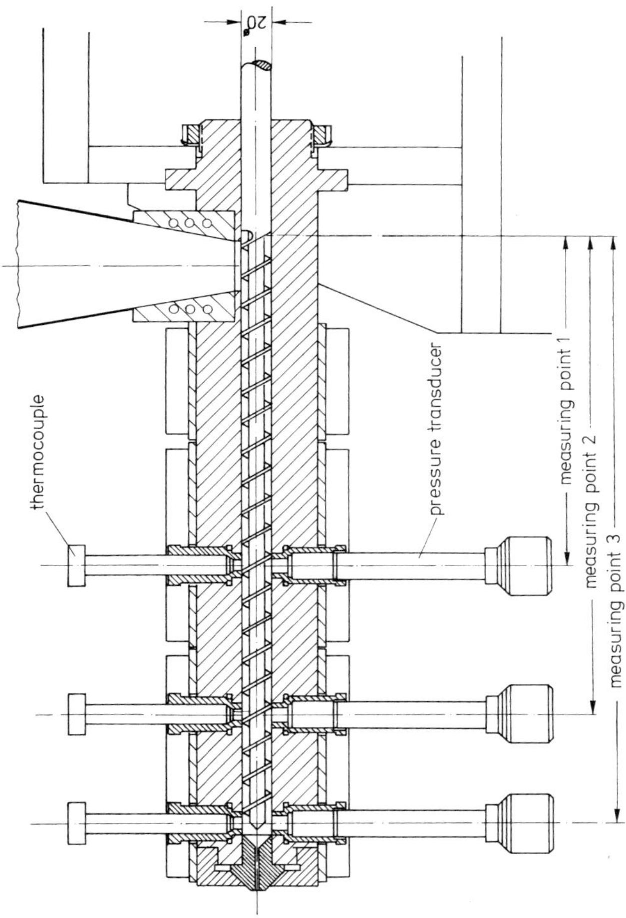

Fig. 7. Extrusiometer, schematic (illustration: Göttfert Feinwerktechnik).

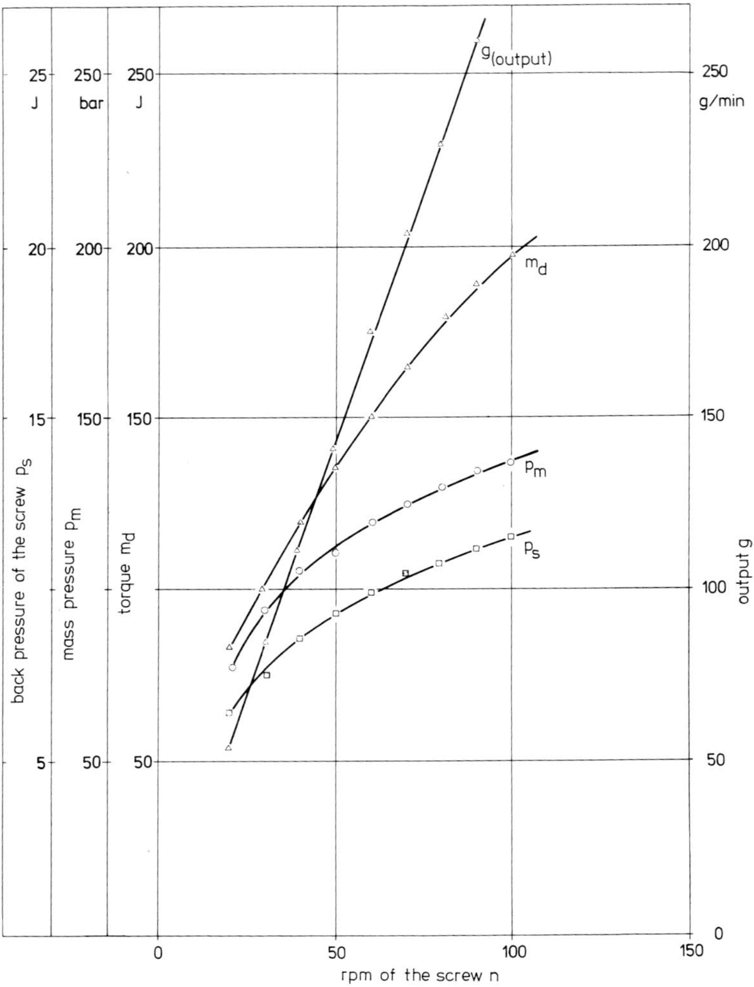

Fig. 8. Typical extrusiometer measurement curves for a trouble-free working PVC compound. Mass temperature before the die 180 °C, diameter of the screw 30 mm.

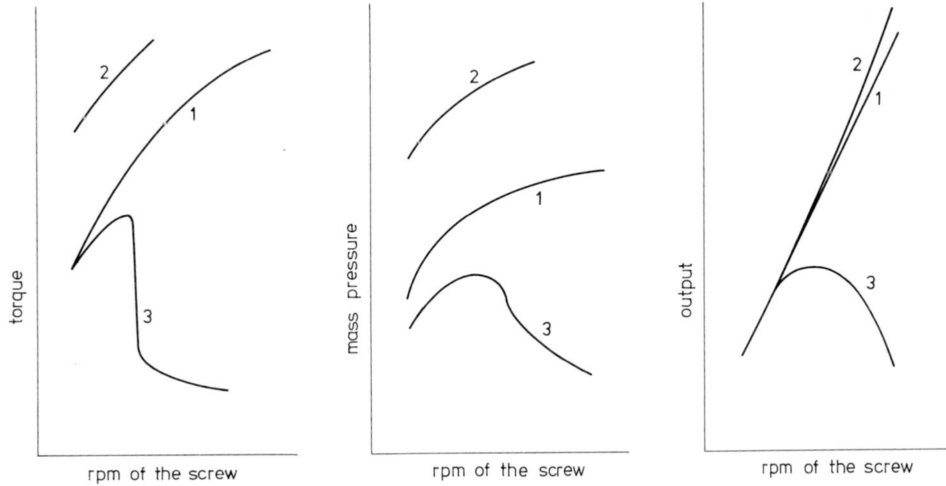

Fig. 9. Examples for the interpretation of extrusiometer curves.

1: normal, *2*: insufficient internal lubrication, indicates also lack of lubricant, *3*: over-lubricated.

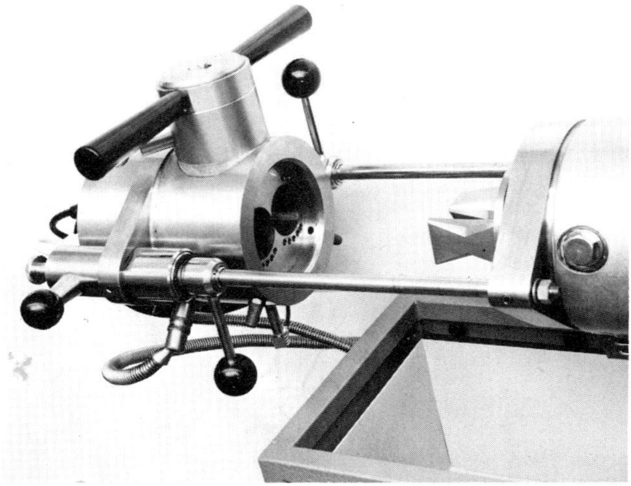

Fig. 10. Kneader open (photo: Göttfert Feinwerktechnik).

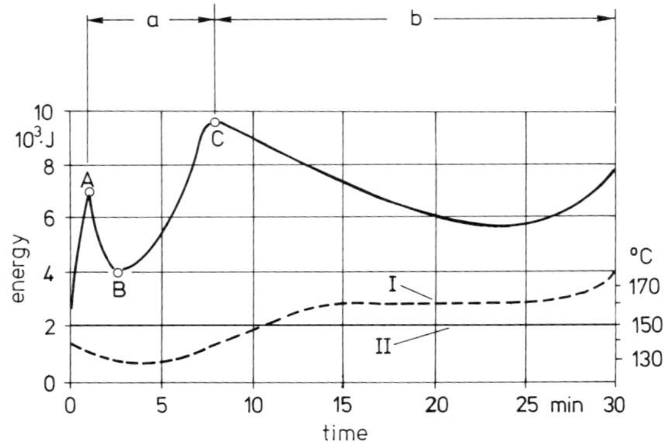

Fig. 11. Plastogram: development of torque as a function of time.

a: gelling time, *b:* heat stability;
A: beginning of gelation, *B:* minimum torque during gelation, *C:* end of gelation;
I: temperature of the gelated PVC mass, *II:* kneader temperature.

with its characteristic maxima and minima. The onset and termination of the gelling period, which may be measured only approximately with a laboratory twin-roll mill, can be determined fairly accurately, therefore, with the kneader. A consideration of the torque in relation to the temperature curve in section (b) gives a rough guidance on flow properties. These question may, however, be answered much more accurately by an extrusiometer. The decomposition of a compound is clearly shown by the rise in torque and temperature curves in section (b).

6.3.2.5 **The spiral test** (flow properties)

The spiral test is a method of determining the flow properties of lubricant-containing thermoplastic materials destined for injection molding. It cannot be considered a laboratory test as an injection molding machine is necessary. Under production conditions and at relatively high cost, the test compound is converted into a spiral, the length of which is measurable. Pressure, temperature and time are the variables.

With every effort being made to minimize the cost of materials testing in laboratories, the spiral test has become outmoded. Satisfactory predictions about the flow properties of a suitable injection molding compound can also be made with the help of the high pressure capillary rheometer.

6.3.2.6 Transparency

The transparency of plastics materials – for special applications such as bottles, packaging films, containers and sheets – is of great importance to appearance and function. It can also be used, however, as a standard for measuring internal and external efficiency of a lubricant in any given material. This is especially important for originally transparent PVC. Transparency is measured photometrically using red light ($\lambda_{max} = 741$ nm) – a wavelength that eliminates discolorations, which do not cause turbidity. Suitable test specimens are made as follows: polymer, stabilizer and lubricant are mixed and homogenized for 10 minutes on a twin-roll mill. The sheet is cut into pieces that fit into a square hole (40×40 mm) cut in a 2 mm thick metal frame. This metal frame is then sandwiched between chromium-plated sheet metal and placed in a heated press. Any surplus material flows away. In this way, we obtain a specimen of precise thickness and smooth surface. The test specimen normally contains 1% of lubricant and a non-opacifying stabilizer (e.g. an organo-tin compound for PVC). It is advisable to measure a series of test specimens with different lubricant contents (e.g. 0.5 to 5.0%) as often opacity does not run in parallel with concentration but appears suddenly after the solubility limit is exceeded. Transparency is expressed as percentage of the light transmitted without sample specimen. As an example: 1% of a high-melting synthetic paraffin gives, according to this method, test plates similar in appearance to porcelain and with a transparency of approx. 5%, whereas glycerol monostearate gives plates with approx. 90% transmission, which appear glass-clear to the human eye.

6.3.2.7 Static thermostability

The test is carried out in an oven normally at 180 °C although other temperatures may be used should special requirements be needed. Firstly, we again have the 10 minutes plasticization and homogenization period on the laboratory twin-roll mill. From the 0.5 mm thick sheet so produced, 6 × 2 cm test specimens are cut. These are wrapped in aluminum foil and tempered in an oven for various periods of time (30 min, 1 hr, 2 hrs etc. until degradation). By attaching these test specimens to an inspection card, a visual assessment of the thermostability can be made. The aging of the specimens in the oven should, if possible, be carried out consecutively. If such a series of wrapped specimens is tested simultaneously and single items taken out one after the other, the consequent cooling of the oven will lead to erroneous results. Packaging in aluminum foil is necessary in order to protect the specimens from decomposition products, which catalyze the process of degradation.

6.3.2.8 Exudation during processing (plate-out test)

The slow buildup of deposits on rolls, dies, calibrating equipment and molding parts has a disturbing effect on production because frequent shut-downs are necessary in order to clean the machinery.

Such deposits are composed of low molecular weight constituents of the polymer together with part of the stabilizers, lubricants, plasticizers and pigments. The initiators are mostly the lubricants or stabilizers but could also be other components of the formulation.

The following test – preferably used in PVC – allows predictions to be made about the behavior of a compound: To the batch to be tested (200 g), we add 2 g of a red pigment (Naphthalin Red Dark RT 539-D/Du Pont). The specimen is then mixed on a twin-roll (at 180 °C/20 minutes/20 rpm) without friction following which the sheet is taken off and any residues removed. During the milling time of the compound, the material which has exuded forms a red-dyed layer on the rolls. In order to remove this layer, a standard white-pigmented sheet is put on the machine and rolled at 12 rpm for a period of 10 minutes at 160 °. This cleaning sheet has the following composition:

100.0 parts	S-PVC (K-value 70)
33.0 parts	di-octyl phthalate
1.0 parts	tin stabilizer
4.0 parts	titanium dioxide
0.5 parts	wax lubricant.

The degree of red coloration of the original white sheet is a measure of the degree of exudation (viz. plate-out).

6.3.2.9 Exudation after processing (blooming test)

The exudation of lubricants during storage or use from plastic articles, may last for weeks, months or even years. The originally clean surfaces acquire an offensive layer, which in extreme cases looks like a layer of frost. This phenomenon is, however, confined almost exclusively to plasticized plastics – especially to plasticized PVC.

For this test, the mixture is homogenized as usual on a twin-roll mill then formed by hot-pressing into 1 mm thick sheets. From this sheet 40 × 20 mm plates are cut, which are then placed between two plates of mirror-glass. This sandwich is tempered in an oven for 24 hours at 90 °C, after which the PVC is removed and the glass plates allowed to cool. If a compound has no exudation tendency, the glass surface remains absolutely clean. If there are layers present, we may conclude that exudation will occur when storing under normal conditions. The degree of exudation is related to the thickness of the layer on the glass. Exudation interferes strongly with printability and weldability of a plastics compound. It is quite possible, for example, that a film, which is printable and weldable immediately after production, fails after some weeks in storage. If the exudation test is positive, greatest caution should be observed and the test repeated after a certain period of storage.

6.3.2.10 Testing of mechanical properties, light stability, gas and water vapor permeability

The test corresponds to those used for testing the polymer itself. It is only necessary to determine the values with and without lubricant, in order to study their influence.

6.4 Related auxiliaries for thermoplastics

6.4.1 Mold release agents

Whenever, during the processing of plastics, the highly viscous mass is injected under pressure into molds and subsequently cooled, there is a possibility of sticking or adhesion to the surface. This makes release of the formed article difficult or almost impossible. This is particularly relevant for injection molding, blow molding and compression molding. In these cases additives must be used which form a thin and highly effective release film and which cause the molded article to fall out unaided. Mold release agents can be classified as either *internal* or *external* depending on how they are applied.

Internal types are mixed into the polymer and migrate to the surface in a similar way to a strongly external lubricant. As an example, stearic acid, which is an internal release agent, is preferentially used in HDPE (at approx. 0.1% addition). Silicones are now frequently recommended as internal release agents. At first, a concentrate of liquid silicone oil and suitable polymer is produced. This concentrate is then added in portions to the base polymer. Silicones are, however, mainly used as the active ingredient in *external* release agents, which are sprayed, in the form of a solution or aqueous dispersion, intermittently into the mold. It the plastics surface is to be further treated, e.g. lacquered or printed, then mostly silicone-free, wax-containing preparations have to be used, otherwise the wearisome task of complete removal of the silicone film would be necessary to avoid printing defects.

6.4.2 Slip, antislip and antiblocking agents

Whereas all of the previously mentioned additives have aided in the processing of thermoplastic materials, these help in the conversion to final articles particularly with respect to films.

Films are often so smooth and slippery that when stored in bulk, e.g. in the form of bags, they adhere strongly to one another or to the surface of machinery. In such cases, they can no longer be separated or opened and thereby cause serious disruptions during filling and transport during packaging. This is a serious problem with LDPE and PP films. To prevent such effects, *slip agents* in quantities of 0.05% are added to the polymer prior to production of the film. Very efficient are oleamide for LDPE and erucamide for PP. Additive producers offer ready-to-use concentrates in the form of granules. In technical jargon, these slip agents are often called "lubricants" because they improve the surface slip of thermoplastics. They should not, however, be confused with the lubricants which act as processing aids.

If, on the other hand, films are so smooth that bags made from them slide apart when stored or packed on pallets, an *antislip additive* must be incorporated or sprayed on the surface. Incorporated antislip agents roughen the surface whereas sprayed-on products produce a limited tackiness. Packaging bags can, in this way, remain tightly bonded when stacked, but easily separated when desired. For this purpose, special PVA dispersions or solutions of amorphous PP in white spirit (with small additions of wax to regulate adhesion) are suitable.

Antiblocking agents are used to prevent the blocking of films, coated foils or sheets when subjected to pressure and increased temperature. They are very similar, therefore, to the slip agents which also aid release. Frequently, both agents are combined. Blocking occurs, preferentially in rolls and piles under the effect of pressure. At first, efforts were made to prevent this by dusting the surface with e.g. corn starch or pyrogenic silicic acid. This process had to be abandoned because of unacceptable dust levels. The antiblocking agent is now melt-incorporated into the thermoplastic. Particularly effective is chalk or, even better, siliceous chalk at a concentration of approx. 0.1%. The particles (ca. 20 μ) give a spacing effect between the films. In plasticized PVC, amide wax (mp: ca. 140 °C) in concentrations of approx. 1% also acts as a good antiblocking and slip additive providing that the finished articles have not to be printed or sealed. Amide wax is especially suited for plastisols used to improve sealing of bottle and container closures.

6.5 Technological trend and outlook

The choice of a suitable lubricant demands considerable testing expenditure and the services of an adequately and expensively equipped laboratory. For this reason, this task passes more and more from the user to the additive producer. His equipment and experience enable him to arrive more easily at an optimum solution.

Even if the price of a well balanced and reliably working lubricant system would be somewhat higher than a less efficiently working but cheaper system, the profit to the processor is nevertheless higher. Even small improvements in quality and output yield much greater savings than the relatively minor difference in cost between a ready-made lubricant system and a home-made composition. It is not, however, necessarily the case that ready-made systems will always be more expensive. The producer can often utilize synergistic effects (figure 2) and thereby equalize costs.

The trends in lubricant technology may be stated as follows:

- In large scale production, a transfer of formulation development from processor to producer.
- Switching away from individual products to complete systems viz. from *synergistic* and *partial systems,* which fulfill one or two functions (e.g. lubrication and stabilization) to *complete systems* formulated for specific applications (e.g. pipes and containers), which contain all necessary additives such as stabilizers, processing aids, pigments, impact modifiers, slip/antiblock/antistatic agents etc., in addition to lubricants. They are ready-mixed and packaged in quantities suitable for the dosing of certain quantities of polymer.
- From the curve showing plastics production in West Germany (figure 1), it can be concluded that lubricant consumption in the last few years has either stagnated or dropped. On the other hand, we can also say, that for technical reasons already discussed, polymers, which processed perfectly satisfactorily until now without lubricants, now require such additives more and more. This concerns, in particular, the field of polyolefin processing, where we can expect, perhaps, an increasing consumption in the future.

6.6 Producers of lubricants and related additives

The catalogue of producers, types and trade marks of lubricants and related additives is not entitled to be complete. Producers of paraffins are not included.

Producer	Types and trade marks for lubricants and related additives
Abril Industrial Waxes Ltd., London, GB	Abril Waxes
Akzo Chemie GmbH, Düren, D	lubricants and slip agents; Armoslip, Armid
Allied Corp., A-C Polyethylene Dept., Morristown, NJ, USA	polyethylene waxes; A-C Polyethylenes
American Hoechst Wax Dept., Somerville, NJ, USA	Hoechst Waxes; amide waxes, polyethylene waxes; Hostalub
BASF AG, Ludwigshafen, D	BASF Waxes
Chemische Werke Hüls AG, Marl, D	Vestowax
Chemische Werke München Otto Bärlocher, München, D	Bärolub
Ciba-Geigy Marienberg GmbH, Bensheim, D	Irgawax
Commer, Milano, I	Vinlub
Degussa, Frankfurt, D	antiblocking agents, pyrogene silicic acid; Aerosil
Eastman Chemical Products Ind., Kingsport, IN, USA	polyethylene waxes; Epolene
Gabriel-Chemie GmbH, Wien, A	lubricants, antiblocking/antislip agents; Maxithene, Slip, Eramide
Glyco Inc., Greenwich, CT, USA	amide waxes; Acra-Adva Waxes, Glycolube
Peter Greven, Fettchemie, Bad Münstereifel, D	fatty chemicals; Liga, Ligalub, Ligablend
Grindsted Products, Grindstedvaerket A/S Brabrand, DK	fatty chemicals; Grindtex

Producer	Types and trade marks for lubricants and related additives
Henkel u. Cie. GmbH, Düsseldorf, D	slip/antiblocking agents; Loxiol
Henkel Inc., Hoboken, NJ, USA	fatty acid esters; Loxiol
Hoechst AG, Gersthofen, D	partial/complete systems; Hoechst Waxes, Hostalub
Kahl & Co., Wachsraffinerie, Trittau, D	lubricants
Lankro Chemicals Ltd. Plastics Division, Manchester, GB	Lancroplast
Mallinckrodt Inc., St. Louis, MO, USA	metal stearates
Dr. L.C. Marquart GmbH, Chemische Fabrik, Bonn-Beuel, D	metal stearates
Meister AG, Möhlin, CH	mold release agents, release agents, lubricants
Lucas Meyer, Hamburg, D	mold release agents; Trenbest
Oleofina S.A., Bruxelles, B	fatty chemicals; Radiasurf, Radia
Erwin Reinke, Chem. Techn. Produkte, Walldorf, D	release agents
Spies, Hecker GmbH, Köln, D	release agents; Vetrophen
Unilever-Emery N.V., Gouda, NL	fatty chemicals; Uniwax, Unislip
Union Carbide Corp., Danbury, CT, USA	silicones, slip agents, mold release agents, antiblocking agents
Wacker-Chemie GmbH, München, D	release agents, silicones, HDK (highly dispersed silicic acid)

6.7 References

The literature concerning problems of the effect of lubricants in polymers and its measurement is shown chronologically.

[1] *W. Hechelhammer, W. Backofen:* Kunststoffe 47 (1957) pp. 69/73.
[2] *U. Jakobson:* Br. Plast. 34 (1961) pp. 328/333.
[3] *J.J. Gouza, G.G. Freygang:* SPE J. 17 (1961) pp. 1211/1215.
[4] *C.A. Russel:* SPE Tech. Pap. Baltimore-Washington Sect. 10 (1964) pp. 268/285.
[5] *V. Oakes, B. Hughes:* Plastics (1966) pp. 1132/1134.
[6] *G. Illmann:* SPE J. 23 (1967) pp. 71/76.
[7] *W. Hufnagel:* Plastverarbeiter 9 (1967) pp. 579/590.
[8] *A.R. Berens, V.L. Folt:* Trans. Soc. Rheol. (1967) pp. 95/111.
[9] *H. Lang:* VDI-Ber. 130 (1968) pp. 13/118.
[10] *R.D. Dworkin:* SPE Tech. Pap. 14 (1968) pp. 276/278.
[11] *A.R. Berens:* Polym. Eng. Sci. 8 (1968) pp. 5/10.
[12] *J.E. Bosselmann, R.L. Jalbert, W.E. Mendham:* SPE J. 24 (1968) pp. 65/70.
[13] *G.J. van Veersen:* Kunststoffe 59 (1969) pp. 180/182.
[14] *B.I. Marshall:* Br. Plast. 42 (1969) pp. 70/78.
[15] *O. Fredriksen:* J. Appl. Polym. Sci. 13 (1969) pp. 69/80.
[16] *R.C. Gross, W.M. Coaker, J.D. Hincher:* Vinyl Tech. Newsletter IV (1969) pp. 425/429.
[17] *C.L. Sieglaff, C.C. Vinson:* Polym. Eng. Sci. 9 (1969) pp. 73/80.
[18] *H.G. Sternagel:* Kunststofftechnik 9 (1970) pp. 428/432.
[19] *M. Gale:* Plast. Polym. 38 (1970) pp. 183/191.
[20] *K.E. Andrews, C. Bottens, B.J. Wain:* Br. Plast. (1970) pp. 88/90, pp. 97/101.
[21] *R.C. Gross, W.M. Coaker:* SPE Tech. Pap. 16 (1970) pp. 1/10.
[22] *P.L. Shah:* SPE J. 27 (1971) pp. 49/52.
[23] *K. Worschech, K. Wolf:* Kunststoffe 61 (1971) pp. 645/647.
[24] *P.L. Shah:* SPE Tech. Pap. 17 (1971) pp. 321/325.
[25] *A. Riethmayer:* Seifen Öle Fette Wachse 98 (1972) pp. 193/196, pp. 227/229, pp. 322/325, pp. 399/404.
[26] *L.F. King, F. Noel:* Polym. Eng. Sci. 12 (1972) pp. 112/119.
[27] *D.M. Detweiler, M.T. Puvis:* SPE Tech. Pap. 19 (1973) pp. 647/650.
[28] *J. Nagy, S. Ferenczi-Gresz, R. Farkas, T.T. Nagy, B. Pukanszky:* Plaste Kautsch. 21 (1974) pp. 919/921, 20 (1973) pp. 916/918.
[29] *J.E. Hartitz:* Polym. Eng. Sci. 14 (1974) pp. 392/398.
[30] *P.L. Shah:* Polym. Eng. Sci. 14 (1974) pp. 773/777.
[31] *D.H. Foelsch:* Plast. Eng. 32 (1976) pp. 44/47.
[32] *P.L. Shah:* SPE Tech. Pap. 22 (1976) pp. 327/329.
[33] *B. Pukanszky sen., B. Pukanszky jun.:* Plaste Kautsch. 23 (1976) pp. 503/505.
[34] *P. Bolli, P. Zeller:* Kunstst. Plast. 7 (1976) pp. 6/16.
[35] *G. Pfahler, T. Riedel:* Kunststoffe 66 (1976) pp. 694/698.
[36] *B. Pukanszky:* Plast. Mod. Elastomères 28 (1976) pp. 57/63, pp. 70/74.

[37] *J. Zechinati, B. Harvey, D.R. Despain:* SPE Ann. Tech. Pap. 24 (1978) pp. 740/746.
[38] *A. Rothenpieler:* Gummi-Asbest Kunstst. 31 (1978) pp. 81/84.
[39] *B. Pukanszky jun., B. Pukanszky sen., S. Jozsa:* Angew. Makromol. Chem. 79 (1979) pp. 107/124.
[40] *M.S. Logan, C.I. Chung:* Polym. Eng. Sci. 19 (1979) pp. 1110/1116.

7 High-Polymeric Processing Aids for PVC

Dipl.-Ing. Klaus-Dieter Böhme, Röhm GmbH, Darmstadt,
Federal Republic of Germany

7.1 Fundamentals of the PVC compound

7.1.1 General remarks

PVC is a very special type of thermoplastic. This special character lies above all in the particularly pronounced thermal instability of the pure polymer, in its rheological behavior and the readiness with which it accepts high quantities of plasticizers and fillers. Therefore, a PVC compound ready for processing is always a mixture of various components, in which the pure polymer is not necessarily the chief constituent in quantitative terms. The expression "PVC compound" as used in this paper shall, unless otherwise stated, be the mixture of a single PVC polymer with its various additives. Compounds containing several PVC polymers are the exception.

Due to this special position PVC has always been regarded as a "custom-tailored" plastics material, and as such it anticipated a development which has meanwhile reached the entire field of synthetic resins and resin mixtures.

7.1.2 The historical development of high-polymeric processing aids for PVC

In terms of stabilization and thermoplastic processing, plasticized PVC was highly developed at a fairly early stage, and, for the last 30 years at least, it has been regarded as a synthetic resin characterized by ease of processing. The development of safely manageable compounds of rigid PVC, on the other hand, took several decades. In Europe, the general shortage of raw materials, due to economic difficulties and the war, accounted for the fact that the Germans in particular tackled the processing problems of rigid PVC mainly from the equipment angle. As a result of these intense efforts, technically perfected processing machinery has been available for the last 20 years or so.

In contrast, the U.S. activities in this field concentrated more on the formulation of the PVC compound. From this development work resulted the high-polymeric processing aids. The first product to be used on a large technical scale, in 1957, was the PMMA-based Acryloid K 120 (Rohm and Haas Co.). Soon afterwards, the importance of high-polymeric PVC processing aids was also widely acknowledged in Europe, especially when it was found that these products not only improved the rheology but imparted other important properties to PVC which could not be achieved by technological means.

7.2 Description

7.2.1 Definition

High-polymeric processing aids for PVC are thermoplastic resins added to the PVC compound in small quantities in order to improve its processing characteristics on the way from the raw material to the semifinished and finished article.

The presently known high-polymeric processing aids for PVC normally possess one or several of the following properties:

- Acceleration of the melting process.
- Improvement of the rheological properties in the thermoplastic state.
- Improvement of the mechanical properties in the thermoelastic state.

From these improvements result a considerable number of advantageous secondary qualities.

Other high-polymeric additives which serve above all to increase the impact strength and notched-bar impact strength (cf. chapter 8: High-polymeric additives as impact modifiers for thermoplastics) or the heat distortion resistance may provide PVC with one or several of the aforementioned basic properties as secondary effects, but usually in a less efficient manner.

High-polymeric PVC processing aids are sometimes called "flow improvers". This term only considers one particular aspect of these modifiers and conveys the, usually wrong, idea that they reduce the viscosity of the PVC melt. Moreover, it does not set them unmistakeably apart from other additives which influence the rheology of PVC, e.g. lubricants or plasticizers, whose molecular weights are much lower than those of high-polymeric PVC processing aids. The latter, however, are substances with molecular weights ($\bar{M}_W$) between $1.2 \cdot 10^5$ and $2.5 \cdot 10^6$ g/mol. These molecular weights exceed those of most thermoplastic resins by far and thus also those of the PVC grades into which they are incorporated. PVC processing aids are distinctly high-polymeric resins with all the typical properties of the latter.

7.2.2 Mode of action

7.2.2.1 Acceleration of the melting process

On the vast majority of processing machines for thermoplastic resins the latter are melted through the conversion of mechanical energy, e.g. the energy which drives the extruder screw, into heat. This is achieved by friction between the resin particles (pellets, regrind, powder) and the contact areas of the melting unit or other resin particles as well as by shearing of these particles in the solid, thermoelastic and partly also in the thermoplastic state. The hot contact areas of the melting unit transmit only a small amount of the melting energy to the resin. Such a melting process presupposes a comparatively high coefficient of friction

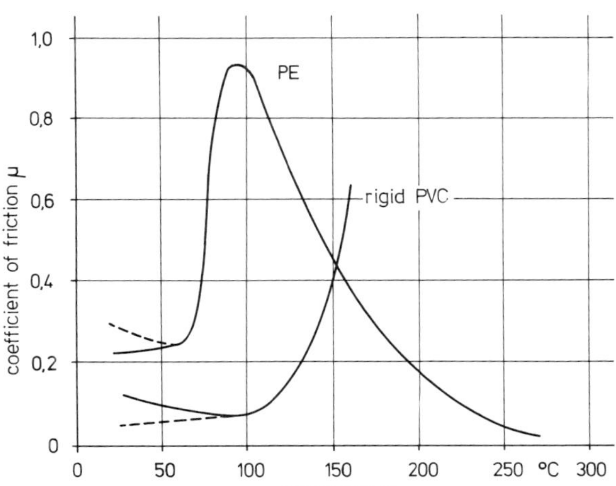

Fig. 1. Coefficient of friction plastics/steel (PE and rigid PVC).

between resin and contact surface. This requirement is ideally met by so-called *wall-adhering* plastics (e.g. polyethylene, polystyrene, polycarbonate, polymethyl methacrylate).

Laboratory investigations have shown [1 to 3] that pure rigid PVC, but often also rigid PVC to which only stabilizers and lubricants have been added, behaves over a wide range of the usual processing temperatures as a *wall-slipping* resin (figure 1).

Block flow [4 to 8] (also called plug flow) or particle flow [9 to 14] occurs up to a temperature of about 190 °C, which contrasts with the shear flow (or intermolecular flow) of wall-adhering resins. Shear flow first predominates from a temperature of about 210 °C onward. The consequence is a slow, inhomogeneous melting process with occasional thermal overloads. This in turn results in unsatisfactory mechanical properties (inhomogeneous solid state), non-uniform distribution and thus impairment of the activity pattern of stabilizers and other additives, as well as in a reduced light stability due to irregular thermal stressing of the melt (non-uniform residual stability).

This behavior is observed with all emulsion-type, suspension-type and bulk-polymer-type PVC. As far as the stabilizers are concerned, it is most pronounced with lead systems, less so with barium-cadmium and calcium-zinc systems, and least pronounced – although still clearly noticeable – with organotin stabilizers [3, 5, 7, 8]. Lubricants and costabilizers also exert a considerable influence.

High-polymeric processing aids for PVC as strongly wall-adhering substances impart this property to rigid PVC in the solid, thermoelastic and thermoplastic state. The improvement of the coefficient of friction or the imparted wall adhesion depend on the quantity of substances added, their own melting behavior and solubility in rigid PVC. The melting behavior and solubility in rigid PVC are influenced by the composition of the high-polymeric processing aids (usually copolymers of methyl methacrylate (MMA) and other acrylates and methacrylates), their molecular weight, grain size and granular shape [3].

For the formulation it must be borne in mind that the high-polymeric processing aid, due to its solubility in PVC, may change the solubility of the other additives. Particularly important in this connection is its influence on the efficiency of the lubricants. Thus it may happen, for example, that the changed solubility and resultant shifting of the saturation limit induce an internal lubricant to act like an external lubricant or that the compound is overlubricated. In other words, a high-polymeric processing aid may increase the wall slippage of a given formulation or delay its wall adhesion, indeed it may turn an already wall-adhering

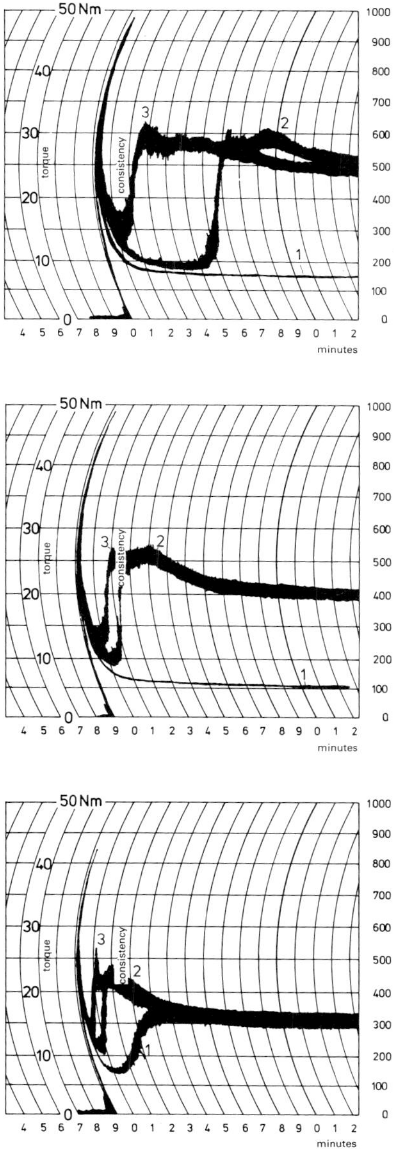

Test A
Temp. of the kneader chamber: 160 °C
speed: 40 min^{-1}
Basic formulation: 100 parts Solvic 229
2 parts Irgastab 15 MS
1 part Edenox OSSG
1 part OP wax
Compound 1:
Basic formulation without processing aids
Compound 2:
95% basic formulation
5% PMMA-co-EA $\eta_{sp}/c \approx 300\ cm^3/g$
Compound 3:
95% basic formulation
5% PMMA-co-EA $\eta_{sp}/c \approx 200\ cm^3/g$
Compound temperatures at end of test:
Compound 1: 142 °C (unmelted)
Compound 3: 166 °C
Compound 3: 167 °C

Test B
Temp. of the kneader chamber: 180 °C
speed: 40 min^{-1}
Basic formulation: 100 parts Solvic 229
2 parts Irgastab 15 MS
1 part Edenox OSSG
1 part OP wax
Compound 1:
Basic formulation without processing aids
Compound 2:
95% basic formulation
5% PMMA-co-EA $\eta_{sp}/c \approx 300\ cm^3/g$
Compound 3:
95% basic formulation
5% PMMA-co-EA $\eta_{sp}/c \approx 200\ cm^3/g$
Compound temperatures at end of test:
Compound 1: 155 °C
Compound 2: 176 °C
Compound 3: 176 °C

Test C
Temp. of the kneader chamber: 200 °C
speed: 40 min^{-1}
Basic formulation: 100 parts Solvic 229
2 parts Irgastab 15 MS
1 part Edenox OSSG
1 part OP wax
Compound 1:
Basic formulation without processing aids
Compound 2:
95% basic formulation
5% PMMA-co-EA $\eta_{sp}/c \approx 300\ cm^3/g$
Compound 3:
95% basic formulation
5% PMMA-co-EA $\eta_{sp}/c \approx 200\ cm^3/g$
Compound temperatures at end of test:
Compound 1: 186 °C
Compound 2: 187 °C
Compound 3: 187 °C

Fig. 2a–c. Reduction of the melting period by addition of high-polymeric processing aids.

compound into a wall-slipping one. In these cases it is necessary to adjust the lubricant content or to choose a different type of lubricant [3, 17].

The conversion of a PVC melt by these processing aids into a wall-adhering melt with shear flow invariably results in a longer residence time of the layers nearest to the wall of the processing unit than those further away. This in combination with the higher shear in the outer layers (more pronounced head dissipation) results in increased thermal stressing of the latter. Normally, this unfavorable influence on the thermostability is amply compensated by the better distribution of the thermostabilizers. Empirically only, this impairment of the thermal stability can be kept within tolerable limits by the carefully measured addition of optimally suited high-polymeric PVC processing aids (moderate increase in the coefficient of friction and late onset or prevention of pure shear flow with wall adhesion), if necessary in combination with lubricants that will only hesitantly display their full activity as external lubricants. The distribution of the stabilizers improves accordingly. On many occasions the improved melting behavior results in a higher throughput of the processing machine.

Figure 2a, b and c illustrate for different wall temperatures how fast a PVC compound is converted into a homogeneous melt dependent on the molecular weight of the high-polymeric modifier. In figure 2a and b the PVC compound without PMMA (compound 1) remains in the wall-slipping state without being homogeneously melted.

7.2.2.2 Improvement of the rheological properties in the thermoplastic state

7.2.2.2.1 Surface gloss

By addition of high-polymeric processing aids based on PMMA a wall-slipping melt of rigid PVC (block or particle flow) can be converted into a wall-adhering melt (shear flow). This change in the rheological behavior not only matters for the melting process but has a major influence on the surface quality of semifinished and finished PVC products.

A melt of rigid PVC which emerges from the die of the processing machine in a wall-slipping state gives the strand a dull to silky matt surface. Contrary thereto, a melt of rigid PVC made wall-adhering by a temperature of over 210 °C emerges as a brilliant strand. In many cases, however, such high temperatures cannot be tolerated for reasons of thermostability. By addition of high-polymeric processing aids it becomes possible for PVC melts to become wall-adhering and thus emerge with a brilliant surface at a temperature of only 160 °C. In extrusion processes and during blow molding of hollow bodies this may be an important aspect.

Sometimes the melt is in a rheological intermediate state. The strand emerging from the die then shows both dull and brilliant lines in the machine direction, the brilliant area appearing rather expanded and the dull area rather shrunken (for explanation cf. 7.2.2.2.3). On occasion it may be advantageous to impart complete wall slippage by adding a somewhat higher amount of external lubricants. On the other hand, however, a totally brilliant surface can be achieved by adding minor quantities of high-polymeric PVC processing aids or by increasing their quantity according to need.

7.2.2.2.2 Avoidance of melt fracture

Contrary to many other thermoplastics, the melt of rigid PVC shows very little strength at moderate shear rates, so that melt fracture may occur at the most economical throughput rates. It shows in a rugged surface of the extrudate (at comparatively low melt temperatures) or a very pronounced optical disturbance of the surface (at increased melt temperatures). This latter phenomenon often occurs in the form of striation and must not be confused with the dullness or lines caused by wall slippage. The addition of a high-polymeric processing aid improves the mechanical properties of the melt, i.e. prevents melt fracture, allows higher flow rates and thus ensures a higher throughput.

7.2.2.2.3 Promotion of die swell

When the melt of wall-adhering plastics flows in channels it is exposed to shear. The shear forces draw the macromolecules apart (orient them) in the direction of flow. At the transition to channels of larger cross-section the stretched macromolecules relax to a greater or lesser extent. This relaxation is particularly pronounced when the melt emerges from a channel (e.g. the die) and is not exposed to any major shear in the open space. The result is the familiar die swell. Since the non-wall-adhering melt of rigid PVC is exposed to moderate shear, the die swell is correspondingly reduced. This may be an advantage in profile extrusion where the cross-section of the profile does not differ noticeably from the cross-section of the die. If unsuitable former systems are used, however, a dull to silky matt profile surface must be accepted in this case. Occasionally high-polymeric additives permit working with dies initially used for wall-adhering resins.

The die swell may be of particular importance in extrusion blow molding, viz. to avoid an intolerable lengthening of the parison under its own weight. By the calculated use of high-polymeric additives, rather than by an often unfavorable variation of the melt temperature, it is also possible to control the die swell of parisons based on different formulations. This eliminates the need for tools (die ring and mandrel) of different dimensions.

In injection molding, and precisely during mold filling, it is important to prevent the formation of the so-called free jet. A wall-adhering melt is to be preferred in this case, especially because it impedes the separation of partially cooled PVC from the cavity wall and thus avoids the formation of dull areas (frosting). Melts showing a pronounced die swell then favor the desired swell flow, because they show a more pronounced tendency to cling to the cavity wall when emerging from the gate.

Figure 3 shows the dependence of the die swell on the molecular weight of the high-polymeric additive.

7.2.2.2.4 Reduction of the melt viscosity

Most high-polymeric processing aids for PVC increase the viscosity of the PVC melt, even though only slightly. This can be tolerated in view of the improvement of other properties. In a few special cases, however,

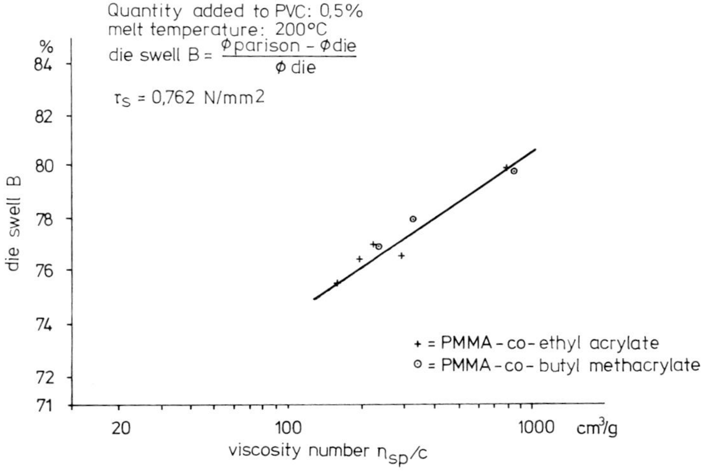

Fig. 3. Increasing of the die swell by PMMA additives.

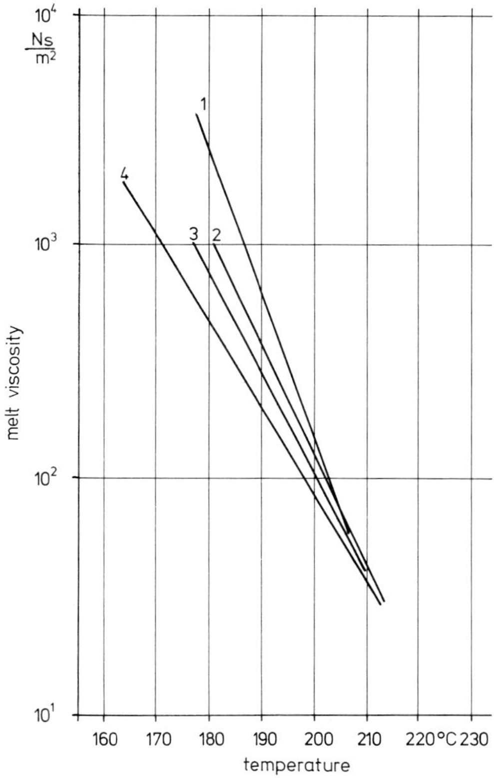

Fig. 4. Reduction of the melt viscosity of rigid PVC by the addition of PMMA of $\bar{M}_w$ $3 \cdot 10^5$ g/mol.
1: 100 parts PVC (without PMMA), *2*: 100 parts PVC + 5 parts MMA-butyl methacrylate (BMA) copolymer, *3*: 100 parts PVC + 10 parts MMA-BMA copolymer, *4*: 100 parts PVC + 20 parts MMA-BMA copolymer.

it is desirable to lower the viscosity of the PVC melt. To this effect one uses copolymers with molecular weights ($\bar{M}_w$), between $1.2 \cdot 10^5$ and $3.8 \cdot 10^5$ g/mol.

Figure 4 shows the melt viscosity of compounds of suspension-type PVC (K-value 65) and a PMMA copolymer with an $\bar{M}_w$ value of about $3 \cdot 10^5$ g/mol. Comparatively large quantities of the latter are required to achieve a noticeable viscosity reduction, and this only up to a temperature of about 210 °C. In other words, such additions produce an effect only at low melt temperatures and are thus mainly used to enable processing temperatures between 160 and 180 °C.

7.2.2.3 Improvement of the mechanical properties in the thermoelastic state

Additions of PMMA not only improve the mechanical properties of rigid PVC in the thermoplastic state but also in the thermoelastic state, and here in particular, the tensile strength and elongation at break (cf. 7.3.2.6).

7.2.2.4 Influence on the properties of rigid PVC in the solid state

The favorable melting behavior achieved with processing aids ensures a more homogeneous end product and thus often results in a marked improvement of the following properties: tensile and flexural strength, elongation at break, impact and notched-bar impact strength, creep strength, crazing resistance. In the case of compounds with high-impact additives, on the other hand, care must be taken not to produce on overhomogeneous melt as this may result in a reduction of the impact strength.

The improvement in the aforementioned properties achieved by the processing aid is more or less the same as achievable by homogeneous melting of a corresponding PVC compound without processing aid, which would, however, suffer severe thermal damage in this case. Under these same conditions, additions of up to 5% slightly improve the tensile strength, flexural strength and indentation hardness and may moderately reduce the impact and notched-bar impact strength. Dependent on the type of PVC and its K-value, the heat distortion temperature increases maximally by 2 °C. The chemical resistance remains unchanged up to additions of about 10%.

A PVC dry blend to which a high-polymeric processing aid based on PMMA has been added normally flows much more freely. Since PMMA-based processing aids are easily soluble in PVC, they do not affect the clarity of clear PVC blends. Slight turbidity, if at all, may first be observed above a molecular weight of about $1.5 \cdot 10^6$ g/mol, and especially if the additive contains comparatively little PMMA. If high-impact PVC additives are used whose refractive index differs from that of the unmodified compound, the clarity can be largely restored by adding a suitable PMMA copolymer.

7.2.3 Physiological harmlessness

High-polymeric processing aids for PVC based on PMMA are synthetic resins and as such governed by Recommendation XXII of the Federal German Health Office [15]. Most commercially available products of this group comply with this recommendation, both in terms of composition and production process. According to Recommendation II of the same authority [16], high-polymeric PMMA processing aids are permitted for PVC if they conform to Recommendation XXII. As a rule, PVC articles in which they are contained do not affect the odor or taste of foodstuffs that come into contact with them (as provided by Article 5, Sec. 1, Subsection 1 of the Federal German law on foodstuffs and goods of consumption, dated August 15, 1974).

Table 1. Characterization of high-polymeric processing aids, suspension and emulsion polymers.

Composition		Chief constituent: polymethyl methacrylate Comonomers: ethyl acrylate, butyl acrylate, butyl methacrylate, styrene	
		Suspension polymers	Emulsion polymers
Molecular weight ($\bar{M}_w$)	(g/mol)	$1.2 \cdot 10^5$–$3.8 \cdot 10^5$	$5 \cdot 10^5$–$2.5 \cdot 10^6$
Viscosity number η_{sp}/c (in chloroform)	(cm^3/g)	50–130	190–600
Heterogeneity U $U = (\bar{M}_w - \bar{M}_n)/\bar{M}_n$		0.85–0.95	1–33
Vicat softening temperature (method B in oil)	(°C)	45–110	45–110
Density	(g/cm^3)	1.18–1.19	1.18–1.19
Refractive index n_D^{20}		1.49	1.49
Particle size	(mm)	0.15–0.40	0.005–0.100
Powder density	(g/cm^3)	0.65–0.70	0.30–0.40
Moisture	(%)	$\leqq 1$	$\leqq 1$
Volatiles (residual monomers)	(%)	$\leqq 0.5$	$\leqq 0.5$

Most PMMA additives for PVC furthermore comply with FDA Regulation 21 CFR 178.3790 (Polymer modifiers in semirigid and rigid vinyl chloride plastics), which means that their physiological harmlessness is also acknowledged in the United States.

7.2.4 Characterization

High-polymeric PVC processing aids on the basis of PMMA are available as suspension or emulsion polymers. Their characteristic data are presented in table 1.

7.2.5 Product groups

According to their molecular weights, polymeric processing aids for PVC can be divided into three groups. A fourth group comprises products whose properties vary widely from those of the first three groups.

7.2.5.1 Products from the lower molecular weight range

These are exclusively suspension polymers with a molecular weight $\bar{M}_w$ between $1.2 \cdot 10^5$ and $3.8 \cdot 10^5$ g/mol (η_{sp}/c 50 to 130 cm^3/g) and characterized by the following properties:

- Excellent melting action,
- reduction of the melt viscosity of PVC,
- moderate die swell,
- moderate improvement of thermoelastic properties.

They are mostly used for better and faster melting.

7.2.5.2 Products from the medium molecular weight range

They are all emulsion polymers with $\bar{M}_w$ between $5 \cdot 10^5$ and $1.3 \cdot 10^6$ g/mol (η_{sp}/c 190 to 330 cm^3/g). Their property spectrum is characterized as follows:

- Good melting action,
- slight increase of the melt viscosity of PVC,
- pronounced die swell,
- good improvement of thermoforming properties.

These products show an optimum property spectrum and are widely used.

7.2.5.3 Products from the upper molecular weight range

This group includes only emulsion polymers with $\bar{M}_w$ between $1.3 \cdot 10^6$ and $2.5 \cdot 10^6$ g/mol (η_{sp}/c 330 to 600 cm^3/g). These are the outstanding properties:

- Good melting action,
- moderate increase of the melt viscosity of PVC,
- pronounced die swell,
- perfect improvement of thermoforming properties.

These processing aids from the upper molecular weight range may slightly affect the clarity of the end product. Major fields of application are blow molding and the production of rigid PVC foams.

7.2.5.4 Product group polymethyl methacrylate, styrene copolymers

They are PMMA copolymers with a high percentage of copolymerized styrene. Such products are at the lower end of the medium molecular weight range. They are not intended to improve the melting behavior but are used where an increase in the coefficient of friction and pronounced wall adhesion are undesirable (calender, special extrusion processes). Their properties are the following:

- No or retarding influence on the melting process,
- no influence on the viscosity of the PVC melt,
- moderate die swell,
- moderate adhesion to metal,
- favorable influence on thermoforming behavior.

7.3 Application

7.3.1 General remarks

The use of high-polymeric processing aids is particularly recommended for PVC in the medium and upper molecular weight ranges (from a K-value of 60 onward). Processing should always start from the dry blend. Additions vary from 0.5 to 5% and are normally limited by cost.

7.3.2 Processing techniques

7.3.2.1 Extrusion

For extrusion, additives of low to medium molecular weight are used, quantities vary from 1 to 5%.

Pipes are normally produced on twin-screw extruders, which operate with forced conveyance and allow good heat transmission by large contact areas and frequent changes of contact area. Therefore, high-polymeric processing aids are rarely needed for melting and, indeed, may not be particularly effective in the presence of the frequently used lead stabilizers with their pronounced external lubricating action. On the other hand, however, PMMA modifiers are required for high gloss and the avoidance of melt fracture.

For sheet and profile extrusion, single-screw extruders are as popular as twin- or multiple-screw extruders. PMMA additives are advantageously used on the former. Occasionally one will deliberately do without them, e.g. when extruding profiles of intricate geometry (die swell), or use styrene-containing PMMA copolymers instead (cf. 7.2.5.4). The changes compared with the unmodified material can be described as follows:

- Easier processing of the dry blend,
- better flow in the hopper,
- faster and gentler melting,
- thermally and mechanically more homogeneous melt,
- better utilization of stabilizers and lubricants,
- lower processing temperatures,
- higher output due to higher screw speeds (acceleration of the melting process, prevention of melt flow fracture),
- extrudate with a smooth, glossy, and even surface,
- end product of increased mechanical strength (due to the more homogeneous melt).

7.3.2.2 Injection molding

Processing occurs exclusively on single-screw injection machines and normally starts from the dry blend. Problems are the melting of the dry blend, the free jet formation, the separation of the melt from the cavity wall and gate smudge during mold filling. Therefore, additives from the medium and high molecular weight ranges are preferred for this process. Thin-walled moldings benefit from a reduction of the melt viscosity, which is achieved by additives of low molecular weight. Where weld line strength is particularly important, e.g. in the case of fittings, copolymers with a high styrene content will prove advantageous. The usual quantities for injection molding lie between 1 and 3%. Compared with the unmodified state these additives effect the following:

- Better flow of the dry blend in the hopper,
- easier processing of the dry blend,
- possibility of using PVC with a higher K-value,
- faster and gentler melting,
- wider processing range,
- reduction of the melt viscosity (with some products),
- better melt flow to wall thickness ratio (with some products),
- more homogeneous melt,
- higher injection speed,
- avoidance of free jet,
- smooth and glossy surface (no dull areas),
- increased mechanical strength.

7.3.2.3 Calendering

In the compounding equipment from which the melt is fed into a calender, PMMA additives promote the melting process and ensure a more homogeneous melt. On the calender, the improved mechanical properties of the melt enable higher roll speeds, fewer optical disturbances and better gloss. Despite a somewhat stronger adhesion to the rolls, these same properties ensure that the film or sheet separates easily and uniformly from the last roll.

For calendering, one will normally use PMMA additives of medium molecular weight. Where perfect clarity is not required, the even more effective polymethyl methacrylate/styrene copolymers can be used. Additions vary from 1 to 3%; 5% and more are used if the thermoforming behavior shall be markedly improved.

The addition of high-polymeric processing aids may be advantageous for the following reasons:

- Higher calendering speeds,
- more homogeneous melt,
- surface gloss,
- wider processing range,
- easy separation from the rolls,
- improved thermoforming behavior.

7.3.2.4 Rolling

Some processors still use roll mills for preparation of the melts. In this case the good adhesion to the rolls is an advantage. The improved mechanical properties of the melt still guarantee the particularly easy separation of the sheet from the rolls.

7.3.2.5 Blow molding

A particularly homogeneous melt is required for blow molding hollow articles. The parison must be very brilliant. Therefore, PMMA additives are indispensable for this process. When used in impact-modified PVC compounds, they avoid deposits of the impact-modifying component at the land of die ring and mandrel, which would otherwise result in striation. The die swell can be controlled via the type and quantity of the processing aids, which may be of medium or high molecular weight. The latter are preferred for optimization of the die swell. Additions vary from 0.5 to 3%. The advantages of PMMA additives in blow molding are these:

- Better flow in the hopper,
- easier processing of dry blend,
- faster and gentler melting,
- more homogeneous melt,
- high surface gloss,
- more favorable blow ratio,
- less optical disturbance,
- less pronounced thickness variations,
- optimum die swell,
- better processing of impact-modified PVC compounds,
- less pronounced variations in height of fall,
- better printability.

7.3.2.6 Thermoforming

The improvement of the thermoforming behavior first becomes noticeable at additions of 1 to 3%. In special cases, however, the quantity is raised to 5% and more. For instance, the addition of 5% PMMA processing aids to a calender formulation may improve the properties in the thermoelastic range as follows: the tensile strength by 30%, the elongation at break by 50%, the deep drawing ratio by 60%.

Films containing PMMA additives show the following advantages in thermoforming:

- Higher gloss,
- fewer optical disturbances,
- increased drawing depth,
- more uniform wall thickness,
- wider processing range,
- sharper contours,
- less tearing,
- better printability.

7.3.2.7 Production of PVC foams (rigid and semirigid)

During the last few years the production of PVC foam has become ever more important (e.g. Celuka process). High-polymeric additives on a PMMA base in quantities of 5 to 12% offer the following advantages:

- More finely dispersed bubbles,
- more uniform cell structure,
- reduced density of the foam (up to about 30%),
- smoother and glossier surface,
- better mechanical properties.

Some processing aids are combined with impact modifiers and thus also increase the impact strength of the foam.

7.3.3 Compounding

High-speed compound mixers are preferred for processing. Usually the high-polymeric processing aid is added together with the other powdery additives. For more effective removal of possible volatile constituents of the dry PVC blend (residual monomers, moisture) the mixer must be provided with a vent.

7.4 Technological trend and outlook

The developments of the last few years have been going in the direction of an ever increasing variety of copolymers, with methyl methacrylate making way for other methacrylates, acrylates and comonomers of a different nature. Moreover, there is a marked preference for products with a distinctly higher molecular weight.

Contrary thereto, the development of PVC grades progressed rather slowly. Indications are that any major improvements in the processing technology and properties of PVC will come from the additives and in particular from high-polymeric processing aids. These need not necessarily be based on PMMA but surely PMMA-based additives will continue to play an important part in the foreseeable future.

7.5 List of trade names and manufacturers or suppliers of PVC processing aids

Completeness is not claimed for the following list of trade names and manufacturers.

The capital letters have the following meaning: (A) processing aids based on PMMA from the lower molecular weight range, (B) aids from the medium molecular weight range, (C) aids from the upper molecular weight range and (D) polymethyl methacrylate/styrene copolymers.

Trade name	Manufacturer/supplier
Bärorapid (A, B, C)	Chemische Werke München Otto Bärlocher GmbH D-8000 München 50
Degalan (A)	Degussa AG D-6000 Frankfurt/M. 1 Degussa Corp. Teterboro, NJ 07608/USA
Diakon (B)	Imperial Chemical Industries Ltd. Plastics Division GB-Welwyn Garden City Herts AL7 1HD ICI Americas, Inc. Wilmington, DE 19897/USA
Irgamod (B, C, D)	Ciba-Geigy Marienberg GmbH D-6140 Bensheim 1

Trade name	Manufacturer/supplier
Kane Ace (C)	Kanegafuchi Chemical Industry Co. Ltd. Osaka / Japan
Metablen (B, C)	Mitsubishi Rayon Co. Plastics Division Chuo-ku, Tokyo, Jap
Modarez (B)	Protex S.A. F-92305 Levallois-Perret
Paraloid (B, C, D)	Rohm and Haas Company Philadelphia, Pa. 19105 / USA
Plexigum (A, B)	Röhm GmbH D-6100 Darmstadt 1
Supercryl (B)	Tenneco Chemicals Inc. Intermediates Div. Piscataway, N.J. 08854 / USA
Vestiform (B, C, D)	Chemische Werke Hüls AG D-4370 Marl
Vinnapol (B)	Vinyl Products Ltd. GB-Carshalton, Surrey

7.6 References

[1] *G. Menges, P. Klenk:* Kunststoffe 57 (1967) pp. 598/603.
[2] *K.-P. Klenk:* Plastverarbeiter 21 (1970) pp. 537/544, 723/730, 819/821.
[3] *K.-D. Böhme:* Angew. Makromol. Chem. 47 (1975) pp. 243/256.
[4] *C.L. Sieglaff:* SPE-Trans. 4 (1964) pp. 129/138.
[5] *W.C. Johnson:* SPE-Techn. Papers 13 (1967) pp. 514/523.
[6] *J.L.S. Wales:* J. Poly. Sci. Symp. 50 (1975) pp. 469/485.
[7] *G. Mennig:* Rheol. Acta 15 (1976) pp. 199/205.
[8] *G. Mennig:* J. Makromol., Sci.-Phys. 14,2 (1977) pp. 231/240.
[9] *M. Mooney, W.E. Wolstenholme:* J. Appl. Physics 25 (1954) pp. 1098/1101.
[10] *A.R. Berens, V.L. Folt:* Trans. Soc. Rheol. 11 (1967) pp. 95/111.

[11] *A.R. Berens, V.L. Folt:* Polymer Engng. Sci. 8 (1968) pp. 5/10.
[12] *E.A. Collins, C.A. Krier:* Trans. Soc. Rheol. 11,2 (1967) pp. 225/242.
[13] *E.A. Collins, A.P. Metzger:* Polymer Engng. Sci. 10,2 (1970) pp. 57/65.
[14] *H. Münstedt:* Angew. Makromol. Chem. 47 (1975) pp. 229/242.
[15] 123. Mitt., Bundesgesundheitsblatt 20 (1977) p. 28.
[16] 126. Mitt., Bundesgesundheitsblatt 20 (1977) p. 230.
[17] *G. Mennig:* Kunststoffe 71 (1981) pp. 359/362.

8 High-Polymeric Additives as Impact Modifiers for Thermoplastics

Dr. Gerhard Menzel, Chemische Werke Hüls AG, Marl,
Federal Republic of Germany

8.1 Introduction

Impact strength constitutes a weak point in the spectrum of properties of some thermoplastics. For example, the impact strength and notched impact strength of polyvinyl chloride (PVC), polystyrene (PS) and polypropylene (PP), in particular at low temperatures, are too low for many applications. However, the notched impact strength of plastics components manufactured from the above thermoplastics can be substantially increased by adding high-impact modifiers to these thermoplastics (figure 1). Impact strength is essentially governed by two factors:

1. the amount of modifier and
2. the processing conditions used in the production of the moldings.

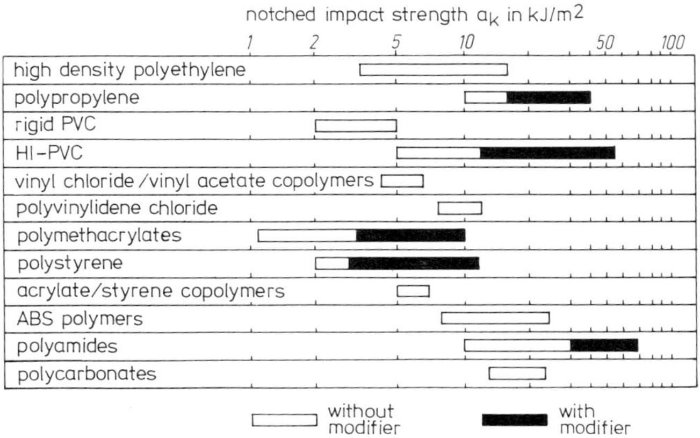

Fig.1. Notched impact strength a_k of various polymers at 23 °C (U notch).

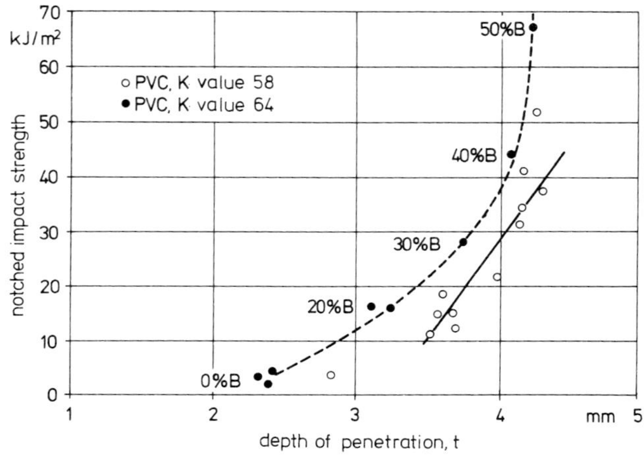

Fig. 2. Correlation between notched impact strength and depth of penetration of PVC containing 15 parts by weight of MBS per 100 parts by weight of PVC. %B: butadiene content of the MBS.

Impact strength and notched impact strength according to DIN, ISO and ASTM regulations correspond to work per cross-section, and are stated in kJ/m^2. One of the main methods of characterizing the toughness of PVC is to measure the notched impact strength a_k at 0 °C in accordance with DIN 53453 (ASTM D 256–73, ISO R 179–1981), using an impact pendulum. This parameter gives an indication of the susceptibility of a polymeric material to break when notches, cracks or other areas of highly localized stress are already present; hence, it also indicates the compatibility and the degree of dispersion of the modifier. In practice, the notched impact strength is measured at room temperature and is preferably stated for temperatures of 20 °C and 23 °C.

Another proposed method of assessing the impact strength is the dropweight test (DIN 53443, Sheets 1 and 2, ASTM D 3029–72). At low impact strength, the dropweight test is more informative, while at very high impact strength the notched test (giving the notched impact strength) is appropriate. In the middle range, the notched impact strength correlates well with the buckling depth determined by the dropweight test (figure 2) [1]. For this comparison, measurements were carried out on PVC modified with methacrylate/butadiene/styrene (MBS) polymers containing various amounts of butadiene; the impact strength of this polymer is known to increase as the butadiene content increases.

A further important method of characterizing impact-resistant plastics is the determination of their dynamic-elastic properties. The dynamic modulus of shear and the mechanical loss factor at constant frequency (e.g. 100 Hz) are determined as a function of temperature, and provide an insight into the internal structure of plastics (DIN 53445, ISO 537–1980).

The heat distortion temperature (according to DIN 53461, ISO 75–1974, ASTM D 648–61) can also be influenced by the addition of high polymers, and the Vicat softening temperature is therefore also determined in order to characterize impact-resistant plastics. The Vicat number is an integer obtained from the degree of softening and the flow behavior of a material.

Essentially, there are three possible ways of modifying brittle thermoplastics. A well-established method is the addition of plasticizers, another method being copolymerization. Graft polymerization and blending of two polymers which are substantially incompatible with one another and have different transition temperatures also give impact-resistant plastics; this method has become very important industrially over the past few years.

The methods used for modifying the impact strength differ greatly from one thermoplastic to another, and the individual plastics will therefore be discussed separately.

8.2 Impact modifiers – structures and theoretical considerations

The term "modifier", which is applied to high-polymeric additives used for improving certain properties, is rather vague.

The production of impact-resistant polymer blends or modified thermoplastics is based on the simple concept of combining various high-polymeric materials having different, complementary properties in such a way that novel plastics with a broad spectrum of well-balanced properties are obtained.

As a rule, high-polymeric materials in solution, as a melt or in the solid state are compatible with one another only to a very limited extent if at all, and therefore cannot generally be mixed homogeneously to give a molecular dispersion. Hence, multicomponent systems are usually not true polymer mixtures but in fact heterogeneous mixtures consisting of separate phases. As a result, the properties of the multi-component systems under consideration are determined not only by the chemical composition and the molecular structure of the individual components but to a large extent by the morphology of these multi-component systems.

An essential morphological characteristic, as well as a basic prerequisite with regard to the specific properties of impact-resistant polymers, is the presence of a multi-phase structure as a result of substantial incompatibility of the hard thermoplastic phase with the soft component, i.e. the elastomeric phase.

When finished articles made of impact-resistant polymers are subjected to shock or impact, the associated mechanical energy is first absorbed by the coherent matrix, i.e. the hard phase. If brittle fracture is to be avoided, the energy must be transferred immediately to the elastomeric phase embedded in the matrix. If the energy absorbed cannot be transferred to the elastomeric phase, it remains in the continuous hard phase, with the result that stress may be localized at unfavorable points, eventually causing fracture.

Structural investigations of impact-resistant polymeric systems having different structures have revealed two types of systems which differ fundamentally in their structure and hence in their fracture mechanism:

- Systems with a honeycomb, network-type dispersed elastomeric phase (e.g. PVC/EVA, PVC/CPE [1]).
- Systems with spherical elastomer particles in a rigid polymer matrix (e.g. PVC/PAE, PVC/ABS, PVC/MBS).

The phase boundaries must exhibit good adhesion since they are subject to stresses of various origin.

The matrix and the elastomeric phase have different thermal expansion coefficients, with the result that shrinkage stresses, so-called internal stresses, occur in the rubber particles when cooling the polymers after processing. Above the glass transition temperature of the matrix, the coefficients of expansion of the two phases are virtually the same, while below this temperature the coefficient of expansion of the matrix decreases and that of the elastomer remains essentially unchanged. As a result, cooling causes greater shrinkage of the rubber than of the matrix, so that tensile stress occurs between the matrix and the elastomer particles.

---

[1]) Abbreviations cf. tables 1 and 3

In order to ensure a large stress differential, the matrix should have low elasticity, i.e. a high modulus of elasticity, in agreement with the requirement that there should be a large difference between the modulus of the elastomeric phase and that of the matrix.

8.3 High-impact PVC (HI-PVC)

Polyvinyl chloride (PVC) is a predominantly amorphous thermoplastic having a syndiotactic molecular structure in the crystalline phase, which consists of spherical particles (globules) about 0.2 to 1 μm in diameter. The degree of crystallinity can be as high as 10%.

PVC is very versatile in terms of processing and possible applications, and cannot be matched by any other plastic. An important extension of the range of uses of PVC can be achieved by modifying it with elas-

Table 1. Impact modifier for PVC

Product name, abbreviation	Elastomer component/VC graft base polymer	Manufacturer
For outdoor use		
EVA	ethylene-vinyl acetate copolymer as grafting base	Bayer, Mobay
EVA/VC	ethylene-vinyl acetate/vinyl chloride graft copolymer	Hüls, Wacker
PAE/VC	polyacrylate/vinyl chloride graft polymer	BASF, Hüls, Solvay, Wacker
ACE	acrylate/methyl methacrylate graft polymer for PVC polyblends	M & T Chemicals, Rohm & Haas
CPE	chlorinated polyethylene for PVC polyblends	Dow, Hoechst
For indoor use only		
MBS	methyl methacrylate-butadiene-styrene graft polymer	Kaneka, Rohm &Haas
ABS	acrylonitrile-butadiene-styrene terpolymer	Bayer, Borg-Warner

tomers. This makes it possible to convert PVC, which in itself is brittle, into an impact-resistant material, without sacrificing its rigidity, hardness, heat distortion temperature and, where relevant, its transparency. In addition, the modifiers influence the rheological behavior, making the material easier to process.

8.3.1 Impact modifiers for PVC

The elastomers listed in table 1 are used as impact modifiers [21].

Impact-resistant PVC grades can be produced in a variety of ways:

- *Powder compounding (dry blend):* PVC and the elastomeric component are mixed in powder form to obtain a polyblend, which is processed directly.
- *Melt compounding:* The PVC/elastomer mixture and other additives are melted and worked in a Banbury or roll mill to give a polyblend, and this is then granulated.
- *Graft polymerization:* The elastomeric component is dissolved in monomeric vinyl chloride and converted to a PVC/elastomer graft polymer. Although the degree of grafting is low, the adhesion between the PVC and the elastomeric component is certainly improved. The good pourability of the resulting HI-PVC powder is advantageous for processing.

Figure 3 shows the effect of various concentrations of different impact modifiers on the notched impact strength (a_k) and ball indentation hardness of rigid PVC. The optimum concentration is marked 0 in the figure; at this concentration, the characteristic properties of PVC are substantially retained while the toughness is enhanced.

8.3.2 The morphology of impact-resistant PVC

Rigid PVC/EVA polyblends and EVA/VC graft polymers are interpenetrating networks which are also referred to as "honeycomb" structures. The opaque appearance of such two-phase systems indicates the presence of phase separation and that the dispersed phase has reached dimensions exceeding the wavelength of visible light.

Using transmission electron microscopy (TEM) and ultra-thin sections, the "network" of the two-phase system can be made visible by selective contrasting of the EVA phase with osmium tetroxide. The PVC is present

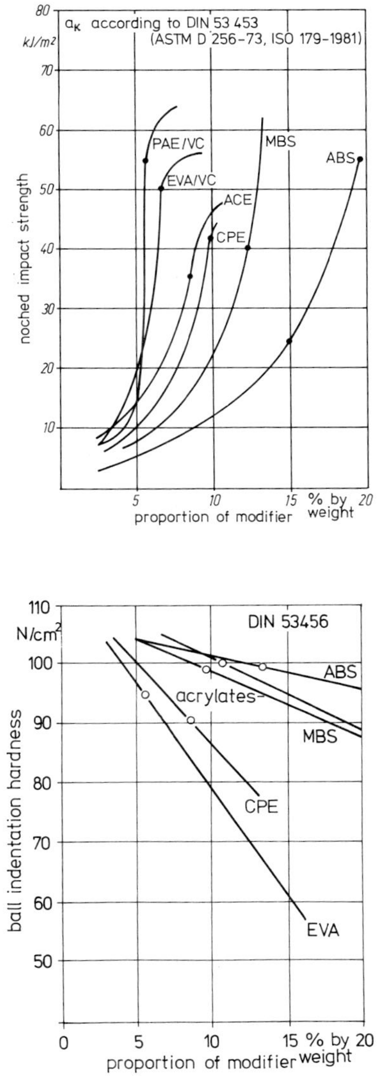

Fig. 3. Comparison of various impact modifiers for rigid PVC
○ = optimum concentration.

in larger amounts and forms the coherent phase, while the EVA copolymer forms the disperse phase [3].

During processing of the melt, impact-resistant two-phase systems undergo various morphological phase changes:

- Fusion of secondary PVC particles (50 to 200 μm) by the soft phase.
- Formation of an elastomer network with honeycomb structure while the particle structure of the PVC is retained. This structure gives the highest achievable notched impact strength (figure 4).
- Production of a finely divided dispersion of the elastomer after fusion of the PVC particles to form a homogeneous PVC matrix. This gives a lower notched impact strength (figure 5).

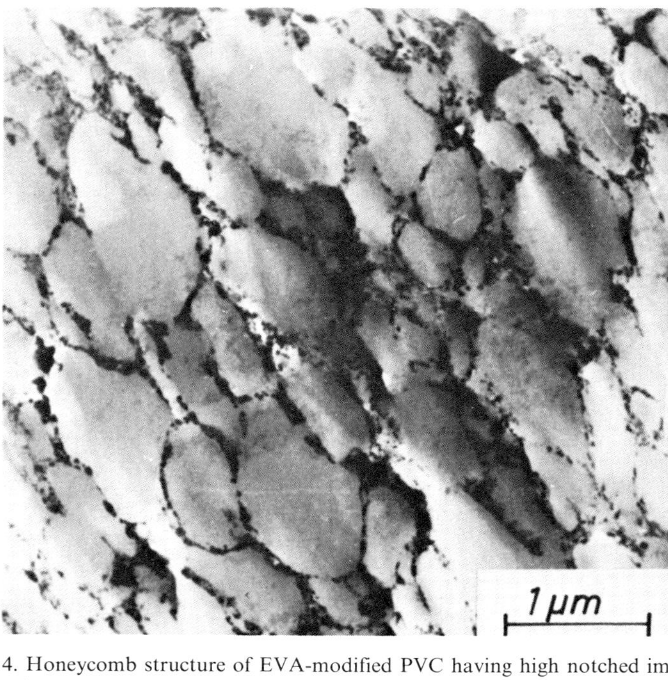

Fig. 4. Honeycomb structure of EVA-modified PVC having high notched impact strength.

The effect of phase formation on the notched impact strength of EVA-modified PVC is clearly evident from figures 4 and 5, and is valid generally. Under constant processing conditions, the formation of the optimum morphology depends on the following factors:

- EVA concentration,
- melt viscosity of the EVA phase,
- compatibility of the EVA component with PVC.

The dependence of the notched impact strength on the elastomer content (EVA) and on thermal stress (milling time) is shown in figures 6 to 8. Accordingly, in order to achieve maximum notched impact strength, the processing conditions must be optimized so as to retain the physical properties of PVC.

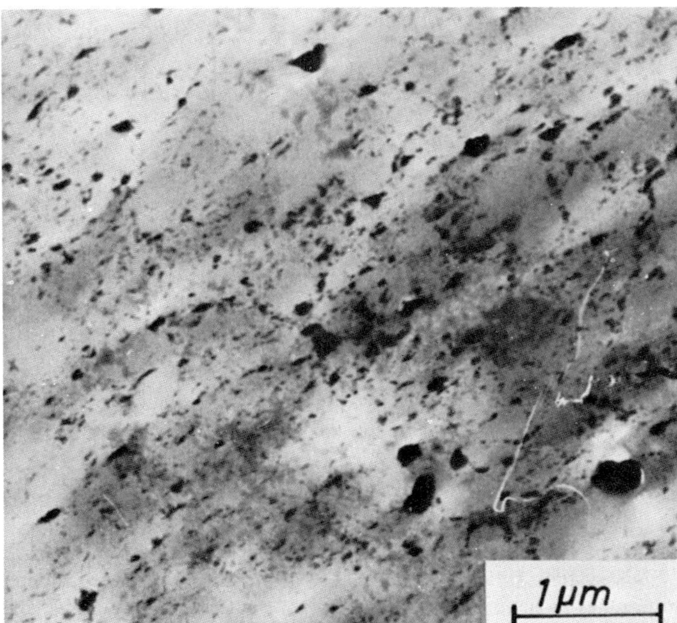

Fig. 5. Structure of EVA-modified PVC after excessive homogenization; low notched impact strength.

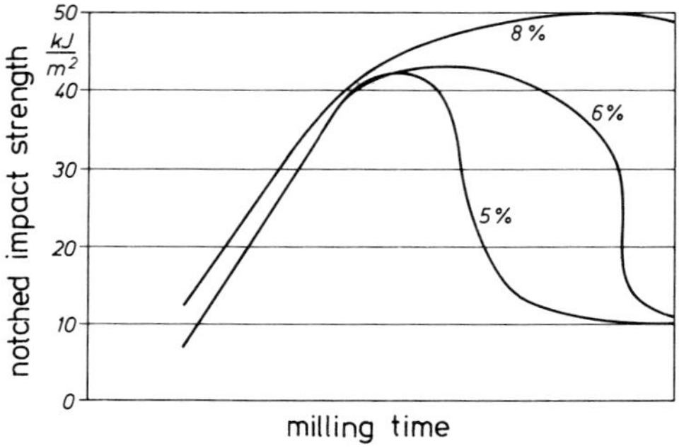

Fig. 6. Dependence of the notched impact strength of EVA-modified PVC on the EVA content (45% of VA) and on the milling time during the manufacture of test sheets.

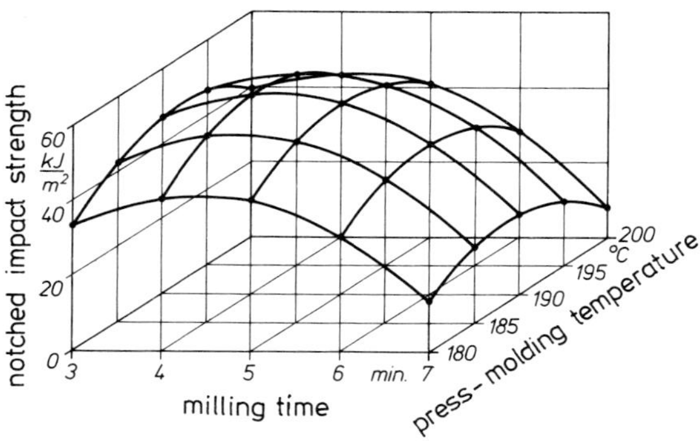

Fig. 7. Notched impact strength of EVA-modified PVC (6% of EVA containing 45% of VA) as a function of milling time and press-molding temperature; milling temperature 190 °C, molding pressure 170 bar.

The structure of the PVC/CPE combination is similar to that of the PVC/EVA system. In this case, too, a two-phase network with the layers at right angles to the direction of flow is obtained during thermoforming. When the material is exposed to an excessive thermal load (frictional heat), this structure breaks down. This structural transformation as a result of an excessive increase in temperature and/or excessive shear stress

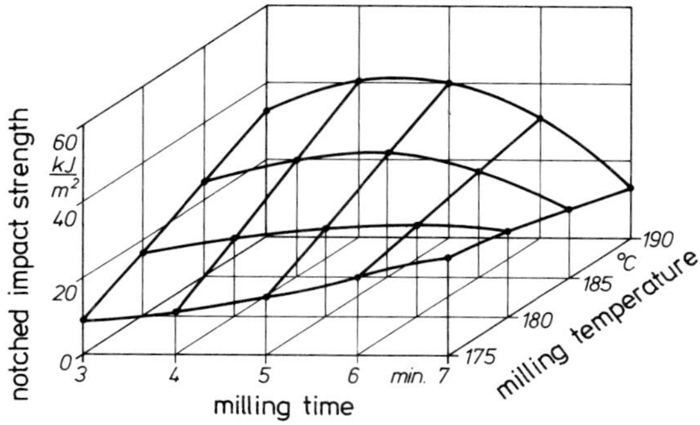

Fig. 8. Notched impact strength of EVA-modified PVC (6% of EVA containing 45% of VA) as a function of milling time and milling temperature; press-molding temperature 180 °C; molding pressure 170 bar.

in the melt causes the impact modifier to become finely dispersed, resulting in a distinct decrease in impact resistance.

During processing of PVC/EVA and PVC/CPE molding materials, the mass temperature and shear force must therefore be kept within relatively narrow limits, since the particle size and shape of EVA and CPE are not predetermined but have to be established during processing [3, 4]. In contrast, when ABS or MBS is used as a modifier, their particle size and shear strength can be controlled during the manufacturing process. After proper processing, the spherical ABS or MBS rubber particles are very homogeneously dispersed in the PVC matrix.

Regarding particle size and shear strength, polyacrylate (PAE) impact modifiers are similar to ABS and MBS (figures 9 and 10).

During processing of PAE-modified PVC, an initial partial agglomeration of the PAE occurs. This is followed after a short time by a dispersion process in which the PAE is divided into spherical particles of approximately equal size, the maximum particle size being 200 µm and the minimum about 50 µm. The resulting morphology is referred to as an "island structure". Such structures are very stable to dispersion, and contrast electron micrographs have shown that the same island structure appears regardless of the thermal load and shear forces to which the PAE-modified S-PVC is subjected [5].

The reason for the stability of these structures is the cross-linking of the PAE molecular chains. Thus, towards shear forces, PAE particles

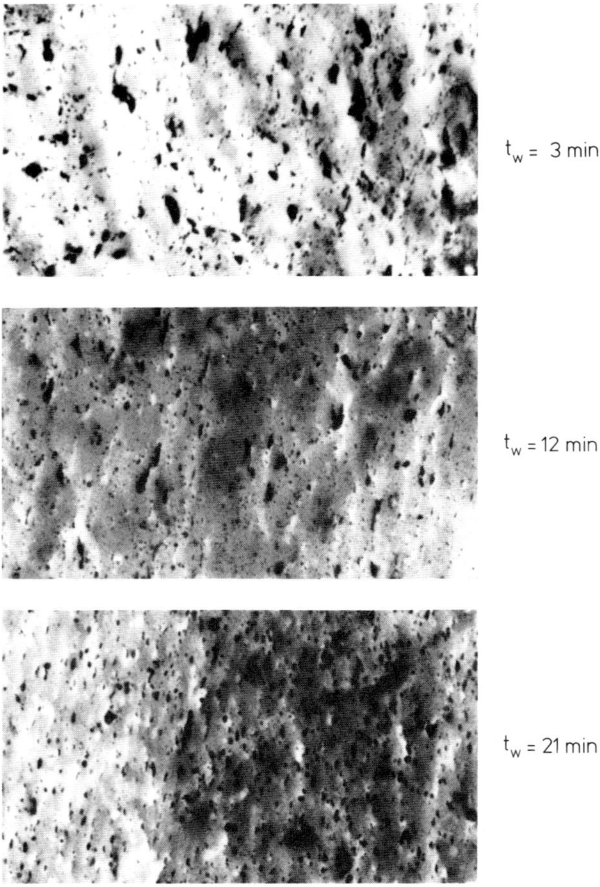

Fig. 9. Contrast electron micrograph of PAE-modified PVC after various milling times t_w at 180 °C. PVC matrix: pale, PAE particles: dark.

E. Röhrl: Kunststoffe 70 (1980) p. 43.

react like elastic spheres which are able to agglomerate but cannot be readily broken up into particles smaller than a certain size.

Modification of PVC with a cross-linked polyacrylate allows the material to be processed in a large variety of ways without having to sacrifice

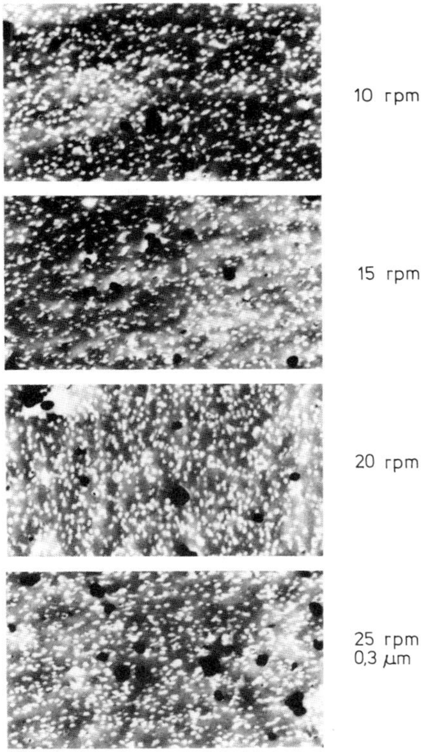

Fig. 10. Contrast electron micrograph of PAE-modified PVC profiles extruded at different screw speeds; Ba/Cd stabilized; 4 parts by weight of TiO_2; 2 parts by weight of chalk. PVC matrix: grey; PAE particles: white; TiO_2: black.

E. Röhrl: Kunststoffe 70 (1980) p. 43.

its high impact strength; the spectrum of properties of the impact-resistant PVC molding is also advantageously affected in other respects, as discussed below.

8.3.3 Polyacrylates as modifiers for PVC

Because of their almost ideal properties as impact modifiers for PVC, polyacrylates (generally abbreviated to ACE or PAE) have become very important.

Polyacrylate-modified high-impact rigid PVC for outdoor use is manufactured by VC polymerization in the presence of a polyacrylate dispersion. Butyl acrylate and 2-ethylhexyl acrylate are preferably used for the production of this modifier because these compounds are readily available and give polymers having a low glass transition temperature. The modification of rigid PVC with a polyacrylate can be carried out in a large variety of ways; this permits the selective formulation of a two-phase system (PVC/modifier) with the following properties and characteristics:

Structure:

- Spherical elastomer particles which are firmly bonded to the PVC matrix at the phase boundary and have optimum compatibility with PVC,
- the PAE particles are highly stable to dispersion as a result of cross-linking,
- homogeneous dispersion of the elastomeric phase in the PVC matrix.

Processing:

- Structural stability of the elastomer particles to high shearing forces and temperature effects (frictional heat), resulting in high reliability in processing (figure 11) [5].

Mechanical properties:

- High notched impact strength regardless of
  - the processing temperature,
  - output of the processing machine,
  - screw speed.
- Modulus of elasticity: close to that of rigid PVC.
- Ball indentation hardness: in the upper range of rigid PVC.

Physical properties:

- Heat distortion temperature (HTD) can reach about that of rigid PVC,
- Vicat softening temperature in the upper range of rigid PVC,
- low mold shrinkage and post-shrinkage of the moldings, resulting in low total shrinkage and advantageous shrinkage characteristics during processing of semi-finished products.

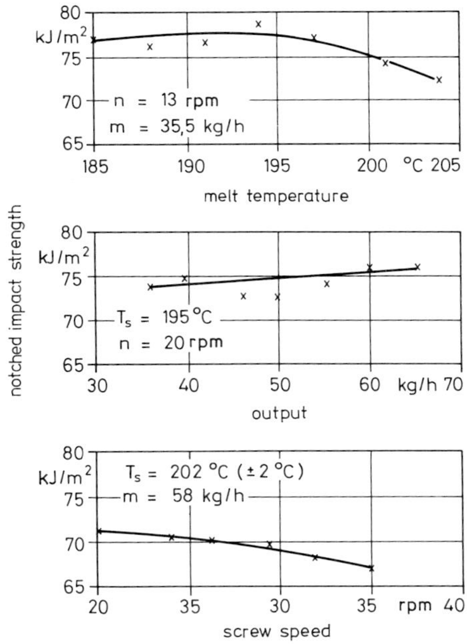

Fig. 11. Notched impact strength of extruded profiles of PAE-modified PVC (double V notch, r = 0.25 mm, 23 °C).

top: Effect of the melt temperature T_s at a screw speed n = 13 rpm and an output $\dot{m}$ = 35.5 kg/h

center: Effect of the output at a melt temperature of 195 °C and a screw speed of 20 rpm

bottom: Effect of the screw speed at a melt temperature of 202 +/−2 °C and an output of 50 kg/h.

E. Röhrl: Kunststoffe 70 (1980) p. 44.

Physicochemical properties:

- Low water absorption, as in the case of rigid PVC,
- chemical stability corresponds to that of rigid PVC,
- the polyacrylate modifier is very stable to light and to weather,
- high aging resistance (stability to thermal oxidation),

- retention of physicochemical properties during long-term outdoor exposure,
- resistance to attack by, for example, salt water and microorganisms.

The above properties can be achieved only by using certain cross-linked polyacrylates which modify the PVC by forming a graft polymer. Other advantages of such polyacrylate/VC graft polymers include the possibility of using a large variety of PVC stabilizers and the high pigmentability, so that the formulations can be prepared without restrictions.

8.3.3.1 Stability of PAE/VC-modified PVC to light and heat

In addition to depending on the amount of the elastomeric phase and its manner and degree of dispersion, the resistance of high-impact PVC, particularly to weather, essentially depends on the bond between the hard phase and the elastomeric phase. Optimum bonding at the phase boundary is achieved only by graft polymerization. Of course, the resistance of the individual components to light and weather, and their capability of being stabilized, are likewise of decisive importance.

Polyacrylates are among the most weather-resistant polymers, so that the stability of PVC to light and weather can be very favorably influenced by graft polymerization with polyacrylates if certain conditions are satisfied. Here is where certain cross-linked polyacrylates and a special polymerization process are used. Even polyacrylate/VC graft polymers can be difficult to stabilize and consequently exhibit poor resistance to light and weather if the properties of the graft polymer have not been optimized.

The outstanding light stability and weather resistance of an impact-resistant rigid PVC are shown in figures 12a, b and c. To obtain these measurements, a commercial PVC graft polymer was mixed with 6% of PAE in the prescribed manner, the mixture was processed to moldings and these were exposed in a Weather-Ometer (Xenon arc and water spray).

8.3.4 Acrylic rubbers as modifiers for PVC

Acrylic polymers of rubberlike consistency are difficult to incorporate into a PVC compound and, in this form, have not become important in practice because they produce only minor improvements in the impact strength. However, if they are converted to graft polymers with methyl

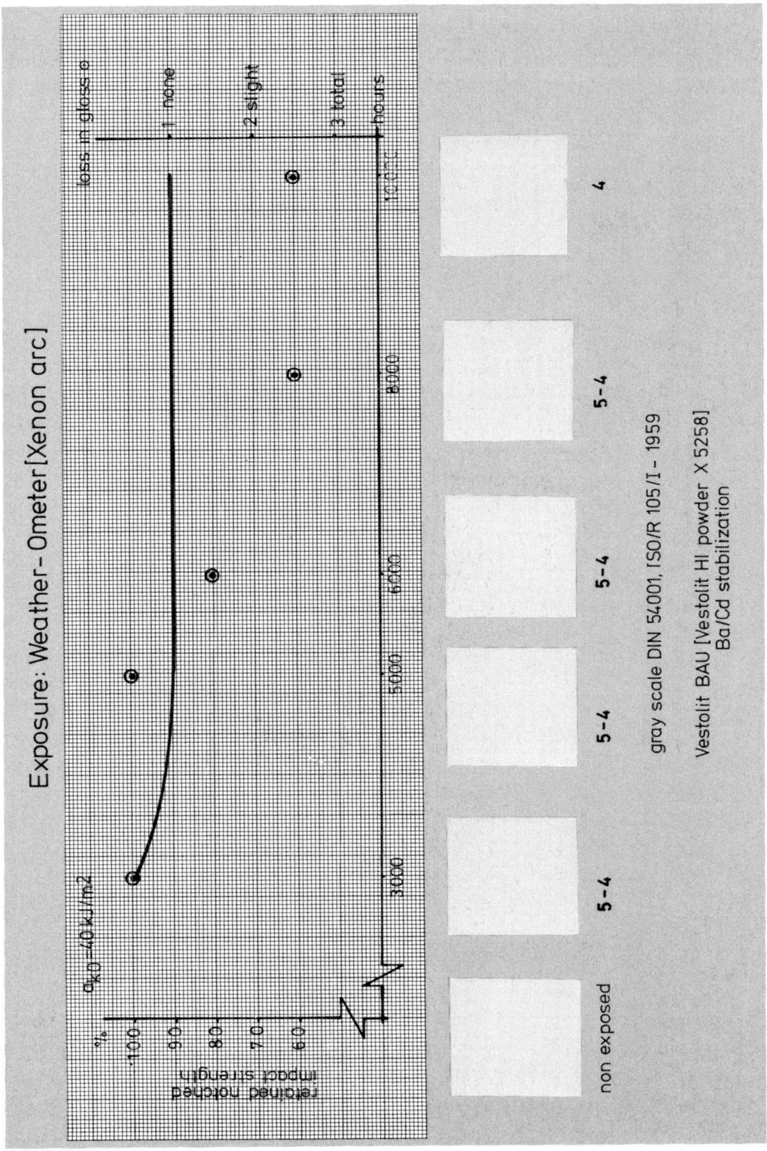

Fig. 12. Light stability and weather resistance of a PVC graft polymer containing 6% of PAE

a) Ba/Cd stabilization. Pigment: 3% of TiO_2 (rutile)

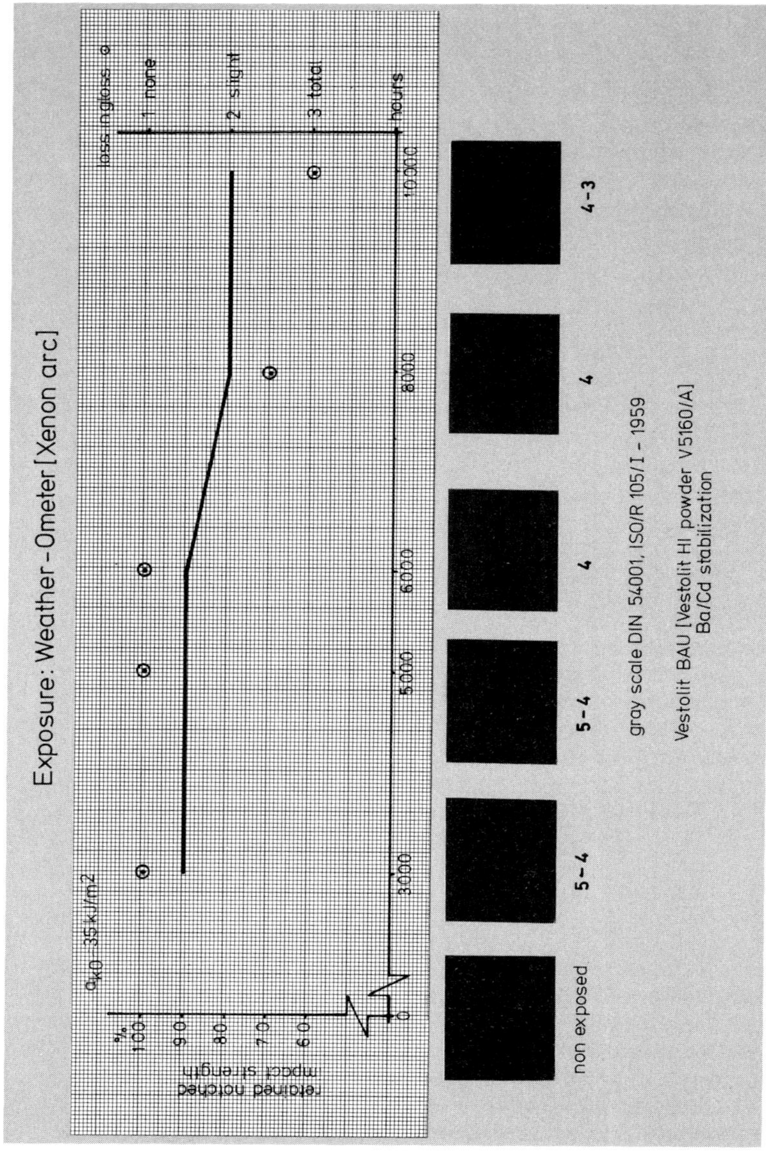

b) Ba/Cd stabilization. Pigment: 0.5% of Cromophtal Brown

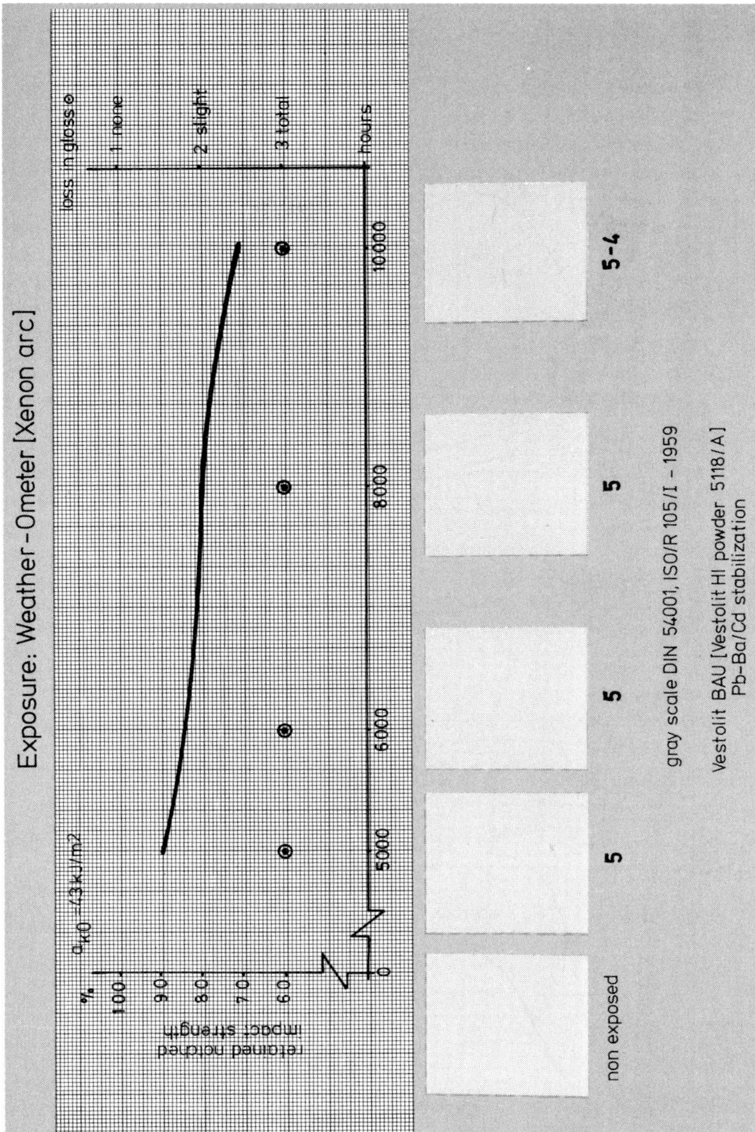

c) Pb-Ba/Cd stabilization. Pigment: 3% of TiO_2 (rutile).

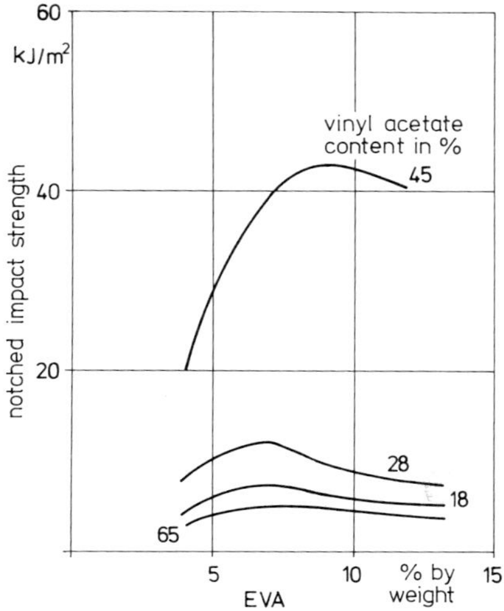

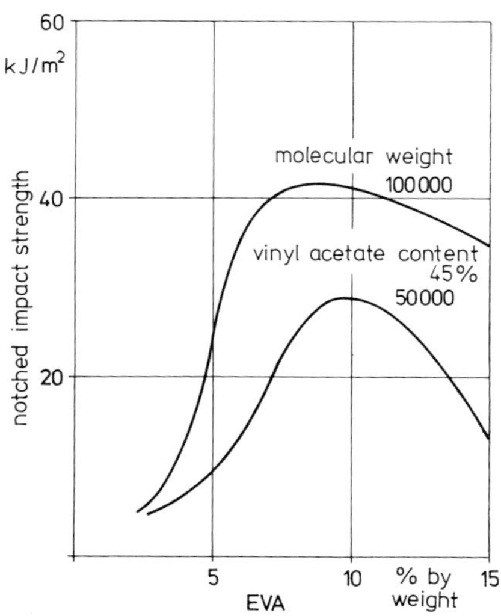

Fig. 13. Effect of the vinyl acetate content and the molecular weight of EVA on the notched impact strength of PVC (according to Charpy; tests were conducted by Mobay Chemical Corp.).

methacrylate (MMA), the resulting products are free-flowing and can be readily incorporated into PVC.

When such polymers are added in an amount of only 10%, the tensile strength, the flexural strength and the moduli are reduced by about 15%, and the resulting polyblends have good impact strength. Articles manufactured from such polyblends are opaque and are suitable for outdoor use; any conventional processing method can be used.

8.3.5 Ethylene/vinyl acetate copolymers as modifiers for PVC

As impact modifiers, ethylene/vinyl acetate copolymers are preferably used as EVA/VC graft polymers. To produce these, EVA is dissolved in monomeric vinyl chloride and the solution is polymerized by suspension polymerization. The resulting primary particle structure corresponds to that of conventional S-PVC. PVC/EVA graft polymers used in practice have EVA contents of 6, 10, 12 and 50%; the preferred graft polymer contains 6% of EVA and is free-flowing and hence very suitable for processing.

The impact strength achievable using ethylene/vinyl acetate copolymers depends on the vinyl acetate content and the molecular weight. From figure 13, it can be seen that EVA copolymers containing about 45% by weight of vinyl acetate give the highest notched impact strength. The harder copolymers, which contain 10 to 30% by weight of vinyl acetate, are not sufficiently compatible with PVC, while the softer copolymers, containing 50 to 70% by weight of vinyl acetate, exhibit increasing compatibility with PVC owing to the polar acetate groups, but behave like primary plasticizers; as a result, the hardness and the heat distortion temperature decrease considerably.

8.3.5.1 Stability of EVA-modified PVC to light and heat

Ethylene/vinyl acetate copolymers as such are thermally stable at the conventionally used processing temperatures of 180 to 220 °C and do not require the addition of heat stabilizers or antioxidants.

EVA-modified PVC is highly stable to light and weather, as shown by the outdoor weathering test carried out in the desert climate of Arizona (figure 14). Increasing the proportion of EVA results in a marked improvement in the stability to light and weather [3, 4].

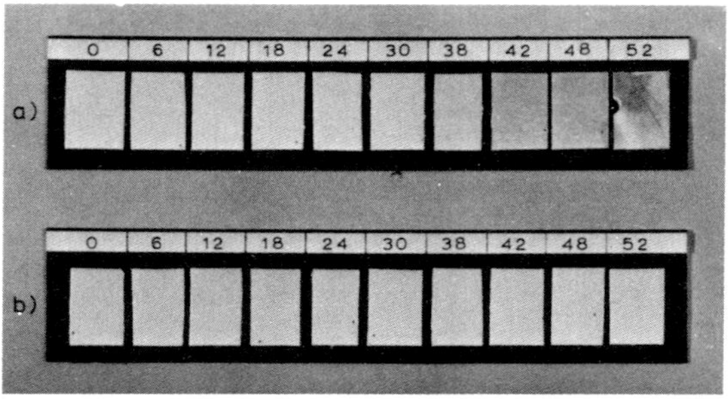

Fig. 14. Outdoor weathering of EVA-modified impact-resistant PVC in the desert climate of Arizona (USA), time of exposure in months. The surface temperature of the samples measured at 12.00 noon was 65 °C. Stabilization: 3% of a Ba/Cd fatty acid salt; pigmentation: 3% of titanium dioxide; EVA content: (a) 6% and (b) 50%.

8.3.6 Chlorinated polyethylene as a modifier

The morphology described in the case of PVC/EVA, and its dependence on the processing conditions, also apply in principle to chlorinated polyethylene (CPE). This is produced by chlorinating high density polyethylene, the product being obtained as a dry, free-flowing powder. The chlorine content is from 25 to 48%, and optimum products for use as impact modifiers for PVC contain 36% by weight of chlorine and have a sufficiently high molecular weight and low crystallinity.

The notched impact strength of PVC/CPE blends depends to a very great extent on the processing conditions, as shown in figures 15, 16 and 17. In order to achieve high-quality products, it is therefore necessary to maintain the processing parameters exactly [7, 8].

CPE is a chlorine-containing polymer and must therefore be protected from heat and light. Although the irregular structure (statistical distribution of chlorine) does not give rise to a "zipper effect" during dehydrochlorination as in the case of PVC, CPE does exhibit color changes (pale yellow–brown–dark) during processing and at the same time loses some of its advantageous mechanical properties. Hence, CPE also requires stabilizers which prevent the elimination of hydrogen chloride and the formation of double bonds. The additives which can be used in this case are virtually the same as those used for PVC (heat stabilizers,

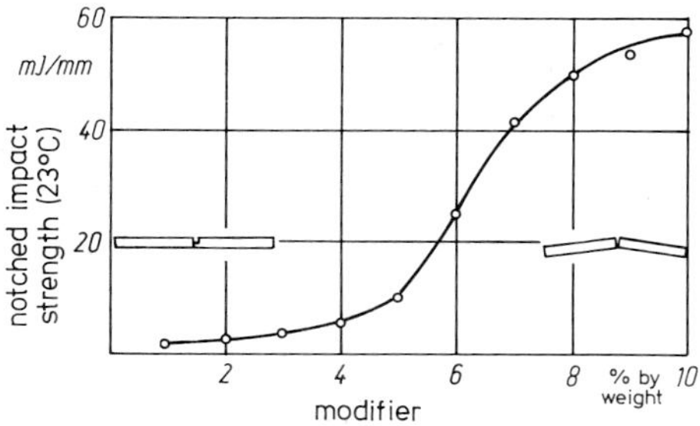

Fig. 15. Dependence of the notched impact strength of PVC/CPE blends on the modifier concentration.

D. Fleischer: Maschinenmarkt 86 (1980) p. 453.

lubricants, u.v. absorbers, antioxidants and pigments). The presence of titanium dioxide, u.v. absorbers and antioxidants in the CPE-modified PVC blend results in a product with weather resistance satisfactory for many uses. For example, CPE-modified PVC profiles containing a white pigment have shown scarcely any color changes after outdoor weathering

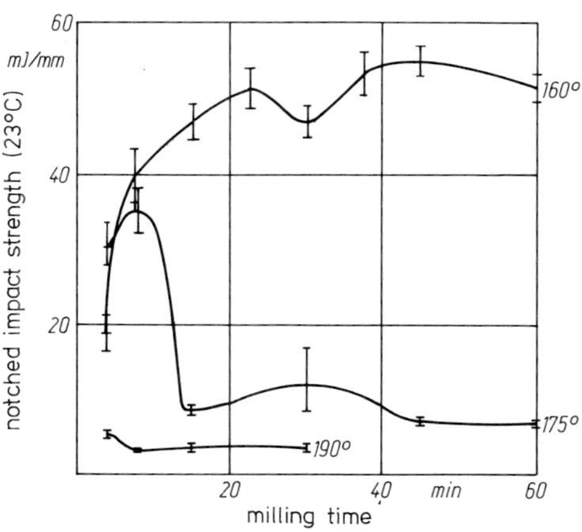

Fig. 16. Dependence of the notched impact strength of PVC/CPE blends (mixing ratio 9:1) on processing temperature and processing time.

D. Fleischer, F. Kloos, J. Brandrup: Angew. Makromol. Chem. 62 (1977) p. 76.

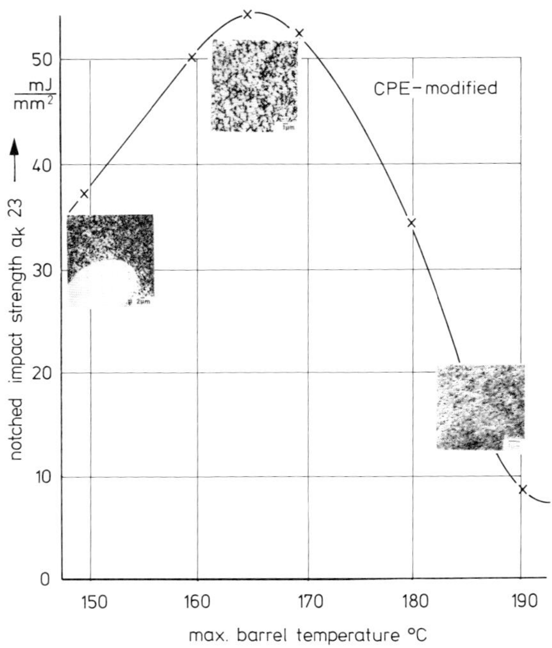

Fig. 17. Notched impact strength (a_k 23) of PVC/CPE blends as a function of the temperature of a double-screw extruder during extrusion of profiles.

D. Fleischer: Maschinenmarkt 86 (1980) p. 456.

for three years under the severe climatic conditions in Casablanca (Morocco). CPE-modified and EVA-modified PVC's may be regarded as being identical in properties [6].

8.3.7 Modifiers for transparent PVC moldings

Transparent impact-resistant PVC products, such as hollow articles (bottles) and packaging films, are manufactured mainly from methyl methacrylate/butadiene/styrene graft polymers (MBS).

These modifiers are pulverulent, and their physical properties are adapted to PVC so as to give highly transparent PVC mixtures. Good impact strength can be achieved by adding as little as 4 to 12% by weight of the modifier, depending on the type of the latter.

Owing to the butadiene content, and to the double bonds which are therefore present, the material is not suitable for outdoor use.

8.3.8 Modifiers for opaque or translucent PVC products

Suitable modifiers for opaque or translucent PVC products are acrylonitrile/butadiene/styrene terpolymers (ABS). The range of ABS polymers suitable for use as PVC modifiers is very broad. Depending on the elastomer content, a very tough PVC can be obtained using concentrations as low as those used in the case of MBS. Certain ABS polymers, when used in larger amounts, not only improve the toughness of PVC but substantially increase the heat distortion temperature of the finished article.

ABS-modified products are not sufficiently weather resistant and are therefore suitable only for indoor use.

8.3.9 Examples for the use of impact-resistant PVC

Machine and apparatus construction:

Containers for the chemical industry, linings and sheets.

Building industry:

Rain gutters, window profiles, door frames, corrugated sheets, cladding elements, antiglare fences on motorways, garden fence profiles, lining profiles for buildings, furniture profiles, skirting boards, etc.

Electrical industry:

Insulating pipes, cable conduits.

Packaging industry:

Bottles for oil and beverages, disposable cups, blister and skin packaging.

8.4 Polystyrene and styrene copolymers

8.4.1 Impact-resistant polystyrene (IPS) and related thermoplastics

There are two ways of improving the impact resistance of polystyrene:

- Mixing rubber latex with polystyrene latex, with subsequent precipitation, isolation and granulation, *or* thermoplastic incorporation of the rubber component at temperatures normally used for processing polystyrene.
- Graft polymerization of styrene, or of styrene and acrylonitrile, in the presence of a synthetic rubber which is either dissolved in the monomer or present as a latex in the polymerization in aqueous phase.

In practice, only the second method is important today [9, 10, 13].

The most important synthetic rubbers for modified styrene polymers are polybutadiene (high-cis or medium-cis) and EPDM, the latter giving more weather-resistant copolymers.

Table 2. Impact strength and notched impact strength of PS, IPS, SAN and ASA

Properties	Unit	Test method (DIN, ASTM)	Standard polystyrene (PS) mainly used for		
			Injection molding	Injection molding	Inj. mold. extrusion
Density	g/cm³	DIN 53479, ASTM D 792	1.05	1.05	1.05
Melt index MFI 200/5	g/10 min	DIN 53735, ASTM D 1238	>8	4 to 8	<4
Impact strength 20 °C	kJ/m²	DIN 53453, ASTM D 256	10 to 16	15 to 20	20 to 24
−40 °C	kJ/m²		–	–	–
Notched impact Strength 20 °C	kJ/m²	DIN 53453, ASTM D 256	approx. 2	approx. 2	approx. 2
−40 °C	kJ/m²		–	–	–

8.4.2 Transparent and translucent impact-resistant polystyrene

The two phases of impact-resistant polystyrene possess different refractive indices, so that the opacity varies depending on the proportion of the elastomeric phase (4 to 10%). Higher transparency is achieved either by bringing the refractive indices of the two phases close together or by using single-phase systems, e.g. copolymers of styrene with butadiene. In their basic properties, such products correspond to impact-resistant polystyrene (table 2).

8.4.3 ABS plastics

ABS plastics are two-phase thermoplastic molding materials based on polybutadiene-modified acrylonitrile/styrene copolymers [11, 14].

Finished articles made from these polymers possess excellent impact strength. Most varieties are opaque and have a more or less intense inherent yellow color so that this plastic is generally used only in colored formulations.

Table 2 cont.

Impact-resistant polystyrene (IPS)			Styrene-acrylonitrile copolymer (SAN)		ASA
Semi-impact resistant	Impact resistant	High-impact resistant	Not re-inforced	Reinforced by 35% by weight of glass fibers	
1.04	1.04	1.04	1.08	1.36	1.07
1.5 to 20	3 to 20	3 to 12	–	–	–
40 to 60 35 to 50	60 to 80 50 to 70	no break 70 no break	20 to 25 –	10 to 18 –	no break 20 to 50
5 to 6 3 to 5	5 to 8 4 to 6	8 to 14 6 to 12	2 to 3 –	4 to 5 –	7 to 14 –

Because of the polar nature of the C≡N groups of the acrylonitrile in ABS polymers and the relatively low viscosity of the hard SAN phase, ABS molding materials can be combined with certain other thermoplastics. Commercially, ABS/PVC, ABS/polycarbonate and ABS/polyurethane combinations are available.

Table 3. Elastomer-modified SAN copolymers

Product name	Elastomer component/grafting base
ABS	polybutadiene
ABS	butadiene-acrylonitrile polymers
ABS	styrene-butadiene block copolymers
ABS	butadiene-acrylate polymers
ASA	acrylate polymers
AES	terpolymer from ethylene, propylene and a diene (EPDM)
ACS	chlorinated polyethylene

In addition to ABS, a number of other elastomer-modified SAN copolymers are produced (table 3), and the ABS series also presents certain variations of the elastomeric component. For example, polybutadiene rubber can also be replaced by butadiene/acrylonitrile polymers or styrene/butadiene block copolymers. In every case, the rubber is grafted to the SAN in order to achieve the required bonding to the rigid SAN component.

In contrast to impact-resistant polystyrene, ABS may contain substantially larger amounts (15 to 25%) of the elastomeric component, depending on the process used. As a result, the toughness can be increased well above that of impact-resistant polystyrene. This combination of great toughness and rigidity together with excellent surface properties and good chemical resistance puts ABS in the class of engineering plastics. The specific properties of ABS result from the presence of the polar C≡N group which gives rise to strong intermolecular attractive forces. This accounts for the high stress crack resistance and the relatively low directional dependence of the strength. The smaller rubber particles possible in the case of ABS, and the consequently higher surface gloss, are also of advantage.

8.4.4 Weather-resistant elastomer-modified SAN copolymers

The polybutadiene rubber component of ABS is unstable to heat and sunlight. An increasing tendency to fracture and yellowing are observed after prolonged exposure to sunlight. This substantially restricts the use of ABS outdoors.

In order to obtain elastomer-modified SAN copolymers which are more weather resistant, the polybutadiene rubber is replaced by elastomers which are more light stable; this leads to a remarkable improvement in the outdoor weathering behavior. Elastomers of this type include the ASA, AES and ACS polymers (see table 3), of which the ASA and AES plastics have become commercially important.

The properties of ASA polymers are very similar to those of the ABS products. Outstanding characteristics include improved weather resistance and improved thermal stability during processing and when exposed to prolonged thermal stress in use.

After exposure to light, impact strength and color of ASA resins are retained about ten times as long as with ABS resins. This applies both to exposure to artificial light, e.g. fluorescent lamps, and to exposure to sunlight.

ASA polymers are used wherever the product is required to have properties similar to those of ABS resins but are to be exposed to light or prolonged thermal stress when in use.

Products containing polychloroprene or ethylene/vinyl acetate copolymers are known, but these are not employed industrially.

8.4.5 AES polymers

In AES polymers, the elastomeric component is an ethylene/propylene/diene copolymer (EPDM) [9]. The diene component permits grafting to SAN, by means of which the required compatibility with the SAN rigid phase is achieved.

The weather resistance of AES polymers is better than that of impact-resistant polystyrene or ABS by a factor of from 10 to 20.

8.4.6 ACS polymers

ACS polymers contain chlorinated polyethylene as the elastomeric component. A disadvantage of these products – as in the case of PVC –

is their high thermal instability during processing; this restricts the range of uses of these products.

8.4.7 Examples for the use of impact-resistant polystyrene and ABS

Impact-resistant polystyrene and ABS are used in particular for the production of tough industrial components having good surface gloss.

Precision engineering and electrical engineering:

Housings for radio and television sets, tape recorders, photographic and film equipment and spools.

Household equipment:

Housings for electrical household appliances, containers in refrigerators, refrigerator door linings, disposable dishes, toilet articles, drawers.

Other fields of use:

Toys, all types of packaging, furniture components.

ABS can also be put to a number of other interesting industrial uses where usually the properties of impact-resistant polystyrene are inadequate.

Motor vehicle construction:

Bodywork, dashboards, chromium-plated trim strips, first-aid boxes, steering column housings, radiator blinds.

Furniture industry:

Chairs, seat shelves, cabinet elements.

Household appliances:

Housings, operating knobs.

Other uses:

Mechanical toys, suitcases, shoe heels, hulls of boats, surfboards, material for sanitary installations, protective helmets.

8.5 Elastomer-modified polypropylene

For certain applications at low temperatures, the impact strength of polypropylene is insufficient. By modification with elastomers, however, the low-temperature impact resistance of this material can be improved substantially [15].

Elastomer-modified PP grades have been in use for a number of years. Polypropylene modified with ethylene/propylene rubbers in particular have increased greatly in importance, whereas polyisobutylene and butyl or butadiene rubbers are scarcely used as modifiers today. EPM and EPDM are used as elastomeric components, and the corresponding PP blends have proved very useful in the automotive industry for protecting components against shock.

The rigidity and toughness of elastomer-modified polypropylenes depend on the amount of elastomer added. High notched impact strength is always associated with low rigidity, while hardness and heat distortion temperature of polypropylene are markedly reduced by the addition of elastomers (see table 4). The toughness of these blends can also be influenced by the addition of carbon black, certain pigments and inorganic fillers and reinforcing agents. Carbon-black-containing PP/elastomer mixtures are very weather resistant.

The technique employed in blending the elastomer and polypropylene components also has a considerable influence on the properties. As with all thermoplastic/elastomer blends, it is important here too that the structure and viscosity of the matrix should be matched with those of the dispersed phase.

Important criteria in selecting a rubber suitable for blending with PP are properties such as mutual compatibility, tackiness, thermal stability and processability. Rubbers in the form of crumbs or granules are easier to incorporate and, when premixed with PP, can be extruded directly or processed to moldings by the injection molding method. The rubber added should not have a substantial effect on the advantageous properties of the polypropylene.

For use in the automotive industry, properties such as resistance to chemicals, stress crack resistance, resilience and good weather resistance are important.

The heat aging resistance, i.e. service life until embrittlement of the material occurs at a certain temperature, depends on proper stabilization, as does the weather resistance.

Table 4. Properties of EPDM-modified polypropylene

Properties	Method of testing	Test temperature (°C)	Unit	Polypropylene (melt index 190/5 = 5.4 g/10 min)				
				without modifier	with EPDM as modifier			
					10 (% by weight)	20 (% by weight)	30 (% by weight)	40 (% by weight)
Izod notched impact strength	ISO R 180 Method A ASTM D 256	−40 23	J/m	4 38	60 200	55 560	55 710	64 1150
Modulus in flexure	DIN 53457 ASTM D 790	23	N/mm^2	1950	1800	1510	1340	630
Vicat softening temperature VST/A/50	DIN 53460 ASTM D 1525	–	°C	154	152	148	126	114

PP/elastomer blends can also be provided with an antistatic or light-resistant finish.

For outdoor use, black-pigmented varieties are recommended; for colored and white formulations, sterically hindered amines should be used to improve the stability to light.

The materials can be readily coated using suitable primers; weather-resistant surface coatings improve the service life, and allow a colored surface.

8.5.1 Examples for the use of EPDM-modified polypropylene

Motor vehicle construction:

Radiator blinds, bumper covering, side panelling, steering wheel housings, dashboards, shelves, consoles, handles and wheel-well linings.

Shoe industry:

Heels.

Sanitary sector:

Flexible filter elements, flexible tubes and hoses.

Electrical industry:

Cable drums, cable insulations (for 1 kV cables).

8.6 High-impact polyamides

Recently, polyblends have also been obtained by blending polyamides with elastomers. For example, the incorporation of EPDM results in very-high-impact polyamides, and the use of acrylic rubbers as impact modifiers has also been proposed.

With regard to environmental stress cracking (ESCR), chemical resistance, behavior under prolonged stress, abrasion resistance, lubricity, etc., modified polyamides exhibit the good properties of the standard materials. The toughness properties, i.e. the notched impact strength as well as the low-temperature impact strength, are greatly improved.

Table 5. Properties of impact-resistant PA6 and PA66 (dry)

Properties	Unit	Test method	PA6 standard type	PA66 standard type	PA6/PE	PA66/ elastomer	PA6/ elastomer
Density	kg/m^3	DIN 53479	1.14	1.14	1.10	1.09	1.087
Yield point	MPa	DIN 53455 ASTM D 638	85	90	65	50	60
Tensile strength	MP a	DIN 53455 ASTM D 638	60	90	–	45	45
Elongation at break	%	DIN 53455 ASTM D 638	50	40	100	70	100
Modulus in tension	MP a	DIN 53457 ASTM D 638	3200	3400	2650	2100	2100
Flexural stress	MP a	DIN 53452 ASTM D 790	120	130	90	75	85
Notched impact strength (+23 °C)	kJ/m^2	DIN 53453 ASTM D 256	3.2	4	9	50	50 to 60
Izod notched impact strength (+23 °C, 1/8″)	J/m	ASTM D 256 ISO R 186	50	–	–	910	1000
Vicat softening temperature VST/B	°C	DIN 53460 ISO R 306 ASTM D 1525	>200	>200	191	–	185 to 190
Ball indentation hardness	N/m^2	DIN 53456	130	150	110	90	90
UL 94 flammability test			V 2	V 2	–	HB	HB

A remarkable feature is the fact that the notched impact strength is independent of the thickness of the finished article (see table 5).

Modification has an adverse effect on the heat distortion temperature (about 8 °C lower than for the base resin), the dimensional stability and the flammability. However, these properties are only slightly impaired, and substantial compensation can be achieved by the use of suitable additives (heat stabilizers, flame-retarding agents).

8.6.1 Examples for the use of high-impact polyamide

Equipment construction:

Chain wheels, cutting blades, technical articles subject to severe mechanical stress.

8.7 Conclusions

High polymeric additives for improving impact properties of inherently brittle polymers, have become very important in the development of thermoplastics over the past ten years. A number of new fields of application have been opened up, and conventional materials have been replaced, especially for outdoor use. Excellent results have been achieved in the case of impact-resistant PVC window profiles and automotive components made of EPDM-modified polypropylene. In the building and the automotive industry, further developments are in progress, especially with respect to moldings with a long service life and weight reduction by replacement of conventional materials.

8.8 References

[1] *V. Bugdahl, G. Morlock:* Kunststoffe 66 (1976) pp. 367/368.

[2] *V. Sahajpal:* Impact Modifiers for PVC, Development in PVC Technology. Applied Science Publishers, London, 1973.

[3] *G. Menzel:* Angew. Makromol. Chem. 47 (1975) pp. 181/199.

[4] *G. Menzel, G. Hundertmark, A. Polte:* Kunststoffe 67 (1977) pp. 339/344.

[5] *E. Röhrl:* Kunststoffe 70 (1080) pp. 43/45.

[6] *H.H. Frey:* Kunststoffe 67 (1977) pp. 706/711.
[7] *D. Fleischer:* Maschinenmarkt 86 (1980) p. 25.
[8] *D. Fleischer, F. Kloos, J. Brandrup:* Angew. Makromol. Chem. 62 (1977) p. 76.
[9] *H. Willersinn:* Makromol. Chem. 101 (1967) pp. 296/319.
[10] *H. Jenne:* Kunststoffe 66 (1976) pp. 581/590.
[11] *N. Platzer:* Chem. Tech. 10 (1077) pp. 634/641.
[12] *H. Oertel:* Plastverarbeiter 8 (1979) pp. 433/438.
[13] *U. Reichert:* Kunststoffe 70 (1980) pp. 613/617.
[14] *K. Schneider:* Kunststoffe 70 (1980) pp. 617/618.
[15] *G. Heufer:* Kunststoffe 68 (1978) pp. 145/148.

9 Fillers and Reinforcements

Dr. A.W. Bosshard and Dr. H.P. Schlumpf, Plüss-Staufer AG,
Oftringen/Switzerland

9.1 Introduction

Fillers are defined as additives in solid form that differ from the plastics matrix in respect to their composition and structure. The materials in question are generally inorganic by nature and less frequently organic. Inert fillers or extender fillers increase the bulk and lower the price [1], active fillers produce specific improvements in certain mechanical or physical properties and are thus also known as reinforcing fillers. The term reinforcement is not specifically defined, since reinforcement is understood to mean, for example, both an increase in tensile strength and an increase in flexural modulus.

The extensive range of fillers and reinforcing agents used nowadays indicates the major significance that these materials have attained. Although their original purpose was to lower the cost of the molding compounds, prime importance now attaches to selective modification of the properties of a specific plastic.

9.1.1 Theory of the action of fillers and reinforcements

The action of active fillers can be attributed to three causes. Some fillers form chemical bonds with the material to be reinforced, carbon black, for example, produces cross-linking in elastomers by means of radical reactions.

Other fillers act mainly through the volume that they take up. In the presence of the filler particles, the chain molecules of the material to be reinforced cannot assume all the conformational positions that are basically possible. Moreover, there are a number of basic reasons for assuming that in a certain zone around the filler particles the polymer material differs in structure and properties from the polymer matrix on its own. The polymer segments attached to the filler surface by secondary or primary valence bonds in turn cause a certain immobilization of adjacent segments and a possible orientation of the polymer matrix [2].

The reason for this immobilization lies in the high degree of dimensional stability of the filler particles. The increases in the glass transition temperature [3] observed in filled polymers – resulting from the limitations on mobility in the boundary zone of the polymer material – can be regarded as confirmation of the theoretical concept presented above.

The zone directly at the filler surface, whose structure would appear to be ordered, thus causes a stiffening of the material as a whole. The lower deformability and higher strength are also due to this composite nature.

A homogeneous distribution of fillers in the matrix is basically important, in order that as many polymer chains as possible can be bound to the free filler surface. The free surface energy and the polarity of the bond between filler and matrix are also important factors in this regard.

The third mode of action of active fillers results from the fact that when the polymer molecules are subject to stress with absorption of energy, they can slide off the filler surface [2]. The impact energy can thus be more uniformly distributed and impact strength increased as in the unplasticized PVC/chalk system. In some cases, microcraze formation that destroys the impact energy at the filler/polymer interface is also facilitated [31].

The theory of the reinforcement of thermoplastics has not yet reached a very advanced stage. On the one hand, there is the simple theory propounded by *Nielsen* [4], which states that the mechanical stress to which filled plastics are subject is mainly absorbed by the matrix. In his opinion, the stress-strain behavior of the compound is unchanged. The other extreme is represented by the so-called principle of combined actions in which the elongation of plastic and filler is the same when the system is subject to load. Tensile stress at break subsequently increases as the filler content rises. According to this theory, impact strength, too, can increase if the separation energy between filler and matrix is taken into account.

With a thermoplastic reinforced by short fibers with low orientation, it can no longer be assumed, as is the case with the *Voigt* model [5], that fibers and matrix are stretched to the same extent and that the stiffness levels of the components in accordance with their percentage volume act in parallel with one another, so that matrix and filler are extended to the same degree. The other model (*Reuss* model) [6] with its assumption that the different stiffness levels act in series (homogeneous stress) is too highly simplified. All these theories thus have to be qualified when applied to practical conditions.

For a more detailed insight into the field of fillers and reinforcements, the reader is referred to the literature [25 to 28].

9.1.2 Properties of filled and reinforced plastics

The main difference between inert and reinforcing fillers lies in the fact that modulus of elasticity and stiffness are increased to a greater or less extent by all fillers – also by the spherical types such as chalk or glass spheres – whereas tensile strength can only be appreciably improved by a fiber reinforcement. Heat deflection temperature, i.e. stiffness at elevated temperatures, cannot be increased by spherical additives to the same extent as by fiber reinforcement. On the other hand, fillers in flake form, such as talc or mica, likewise produce a marked improvement in these values.

The use of *extender fillers* can result in the following changes in the properties of thermoplastics:

- Increase in density,
- increase in modulus of elasticity, as well as in compressive and flexural strength (stiffening),
- lower shrinkage,
- increase in hardness and improvement in surface quality,
- increase in heat deflection temperature and lowering of the temperature dependence of the mechanical properties, there being generally no improvement in mechanical and physical properties compared with unfilled polymers,
- cost reduction.

Reinforcing fillers, on the other hand, produce the following improvements in thermoplastics:

- Increase in tensile strength and tensile stress at break, as well as in compressive and shear strength,
- increase in modulus of elasticity and stiffness of the composite material,
- increase in heat deflection temperature and lowering of the temperature dependence of the mechanical values,
- lower shrinkage,
- improvement in creep behavior and bend-creep modulus, reduction in the viscoelastic yield under load; there is also a partial improvement in impact strength.

Disadvantages of the reinforcing fillers are mainly attributable to the generally anisotropic (directional) effect of the reinforcement.

Two discrete phases are always present in reinforced plastics. The discontinuous filler phase should exhibit higher tensile strength and a higher modulus of elasticity than the polymer matrix, whereas the continuous plastics phase should possess higher elongation at break than the filler. For this reason, fibers are particularly suitable as reinforcing agents. When fiber reinforced material is subject to tensile load, local tensile stresses are transferred to the plastic/fiber interface by shear forces and distributed over the fiber surface. For this purpose, the fiber must adhere well to the polymer and possess a specific length, since otherwise it slips out of the matrix material. The higher the modulus of elasticity of the matrix, the smaller can be the minimum length of fiber. Adhesion can be considerably increased by coupling mechanisms between filler and plastic.

9.1.3 Application criteria for fillers in thermoplastics

A large number of different aspects have to be taken into account when employing fillers, e.g.:

- Optimum particle distribution curve,
- possible catalytic activities at the filler surface,
- dispersibility and linkage with the plastics matrix,
- abrasive action of the filler in processing machines,
- properties of the compound,
- problems of industrial safety due to dust etc.,
- cost.

An important criterion of a filler and/or reinforcement is cost. Reinforcing a polymer is only justified when a distinct improvement of mechanical properties or cost reduction compared to the non-reinforced base polymer is found, or when a specific combination of properties is not achievable by other means.

Requirements on the properties of fillers and reinforcements needed for a specific application and their influences on the composite properties are summarized in table 1.

Table 1. Influence of filler/reinforcement properties on composite properties.

Requirements to be met by the composite	Filler/reinforcement properties required
Manufacture of the composite	
Storage of the filler prior to compounding	low moisture absorption, high bulk density
Optimum compounding	mean particle size, good wettability through the polymer matrix, no static charge, good dispersion behavior, no shortening of the reinforcing fibers
Low viscosity during compounding	particles as round as possible, with small specific surface and low absorptivity, low surface energy
High compounding speed	low specific heat, high thermal conductivity
Processing of the composite	
Low shrinkage Low internal stresses, no cracks, quicker demolding	low specific heat, high thermal conductivity, low thermal expansion, uniform filler distribution in the matrix, good adhesion between reinforcement and plastic
No abrasion problems in the processing machines	low degree of hardness, small, round particles, good thermal stability of any surface treatment agent that may be used
Properties of the composite	
Mechanical properties:	
High tensile strength and elongation	high strength in comparison to the matrix, high length/diameter ratio and good fiber/matrix adhesion in the case of reinforcing fibers, good distribution in the matrix
High flexural strength	same requirements as for high tensile strength, it is important to obtain a smooth surface in the finished article
High compressive strength	low compressibility, small, round filler particles for partly crystalline polymers
High modulus of elasticity (high stiffness)	high length/diameter ratio in fibrous or lamellar reinforcements, high modulus of elasticity in comparison to the matrix, high orientation, good adhesion to the matrix

Table 1 (continued)

Requirements to be met by the composite	Filler/reinforcement properties required
Good impact strength	long fibers, adhesion to the polymer matrix should not be perfect
Good long-term behavior (fatigue)	permanent particle/matrix bond, no corrosive impurities in the filler, small, round particles producing a smooth surface in the finished article
High degree of hardness	high degree of hardness, particle size distribution suited to a high packing density, nucleating properties in partly crystalline polymers
Low abrasion and good slip characteristics	low degree of hardness, good slip characteristics (formation of a slip film)
High density	high density and good packing properties
Thermal properties:	
High heat deflection temperature	good thermal stability, other properties as under tensile strength
High resistance to thermal shock	high thermal conductivity, good filler/matrix contact
Low thermal expansion	small coefficient of thermal expansion, anisotropic, hard particles, good adhesion to matrix
Good flame and glow resistance	favorable chemical composition, good thermal conductivity, high thermal stability
Electrical properties:	
Low dissipation factor, favorable dielectric constant, good insulating properties	low moisture and plasticizer absorption, good adhesion to matrix, suitable electrical properties, insulating properties, nucleating action in partly crystalline polymers
Chemical properties:	
High chemical resistance	suitable chemical composition
Good weathering properties	high light resistance, good chemical resistance, water resistance, permanent bond with the matrix
Low water absorption	low moisture absorption, permanent bond with the matrix

Table 1 (continued)

Requirements to be met by the composite	Filler/reinforcement properties required
Optical properties:	
Good transparency	suitable refractive index
Good shade	suitable chemical composition
Good surface quality	small, round or lamellar filler particles
Cost:	
Maximum possible degree of filling	shape, size and distribution of the particles, low filler cost per unit volume
Low processing costs	see manufacture of the compound

9.2 Economic importance of fillers and reinforcements

The realization that reinforced and filled plastics are becoming an increasingly important factor in technical applications is being recognized to an ever growing extent. This also benefits the large-tonnage plastics.

It is unfortunate that outside the glass fiber sector virtually no figures have been published up to now on a worldwide basis on the economic importance of fillers. The following estimate has been made of filler consumption in the USA [32]:

Table 2. Estimated filler consumption in plastics (USA).

Filler/reinforcement	1980 (1 000 t)	1985 (1 000 t)	1990 (1 000 t)	Mean annual growth rate 1980/1990 (%/a)
Carbonates	1 100	1 780	2 900	10.2
Silicates				
Talc	110	180	300	10.0
Asbestos	235	300	365	4.5
Kaolin	76	100	130	5.5
Mica	0.5	0.5	0.6	3.0
Miscellaneous	8	13	21	10.0
Silicon dioxide	40	75	140	13.6
Various minerals	11	13	15	2.5
$Al(OH)_3$	150	210	295	7.0
Carbon black	28	32	36	2.7
Organic fillers (nut shells, wood flour, corn cobs etc.)	105	140	187	6.0
Glass fibers	455	670	980	8.0
Glass spheres, hollow glass spheres etc.	7	15	31	15.7
Carbon fibers, aramide fibers, whiskers etc.	1.2	2.5	5.5	16.0
Total	2 326.7	3 531.0	5 406.1	8.8

According to these figures, consumption of fillers and reinforcing agents is expected to more than double in the period from 1980 to 1990.

Sales turnover of fillers and reinforcing agents in the USA in 1980 is estimated as follows:

Mineral fillers	US $ 125 million
Organic fillers	US $ 170 million
Reinforcing fibers	US $ 500 million
Total	US $ 795 million

These data make it possible by extrapolation to arrive at an approximate figure for world consumption, if it is assumed that the percentage share of the USA in world plastics production amounts to around 25% and that Japan and the Federal Republic of Germany each account for a further 12.5% [30].

Table 3. European market for reinforced thermoplastics (according to Owens-Corning Fiberglas Europe S.A.).

Plastic	1970 (t)	1975 (t)	1980 (t)
Polyamides	15100	47200	82500
Polyolefins	1710	9000	31500
Polystyrenes	428	4500	13500
Polyacrylates	3000	10500	18000
Polyacetals etc.	1070	3750	4500
Total	21308	74950	150000

Table 4. European market for reinforced thermoplastics by field of application (according to Owens-Corning Fiberglas Europe S.A.).

Field of application	1970 (t)	1975 (t)	1980 (t)
Automobile industry	6000	26300	68500
Office machinery and household appliances	8750	25500	40500
Electrical industry	4500	14300	24000
Others	2058	8850	17000
Total	21308	74950	150000

Consumption of reinforced thermoplastics in Europe amounted in 1975 to 75000 t and by 1980 had already reached 150000 t (see tables 3 and 4), the main share being accounted for by glass-fiber-reinforced thermoplastics.

In contrast to most other filled or reinforced plastics, relevant figures are available for glass-fiber-reinforced products. These show that in Western Europe, 76000 t glass-fiber-reinforced thermoplastics were used in 1980 and in the USA 111400 t in 1979. European market expectations by country are summarized in table 5.

Table 5. Glass-fiber-reinforced thermoplastics market in Western Europe.

	1981 (t)	1985 (t)	1990 (t)
West Germany	35000	52000	80000
France	18500	27000	45000
Italy	11000	16000	25000
Benelux	1500	3000	10000
UK	11000	17000	25000
Western Europe (non EC)	10000	15000	25000
Total	87000	130000	210000

The Federal Republic of Germany accounted for almost 50% in 1981. The above production volume for glass-fiber-reinforced thermoplastics corresponds to 20 to 25000 t pure glass. Although the glass-fiber-reinforced thermoplastics account for only about 10% of the total market for glass-fiber-reinforced plastics, they show appreciably higher growth rates than the thermosetting plastics reinforced with glass fibers.

High growth rates are attributed to the lamellar fillers with a good stiffening action such as talc and mica, as well as to the exotic reinforcements (e.g. whiskers). It can nevertheless be assumed that calcium carbonate fillers will maintain their share of over 50% of the market for fillers in polymers. In view of the continuing rises in the cost of raw materials, both fillers and reinforcements will play an increasingly important rôle.

9.3 Description of the individual fillers and reinforcements

9.3.1 Calcium carbonates

Naturally occurring calcium carbonates are sedimentary rocks of marine origin. A rough distinction is made between:

Chalk: Loosely coherent, consisting of fine calcite crystals and fragments, originating from shells and skeletons of nannofossils. Large chalk deposits can be found in Europe, in particular on the Channel coast and in the Paris basin.

Limestone: More highly coherent, e.g. facies of Urgonian limestones in southern France, consisting of macrofossils of sea shells, as well as limestone from the Jura formation, which are generally older.

Marble: Formed by metamorphosis of a sedimentary rock, i.e. recrystallization in the earth's interior at high pressures and temperatures of 600 °C. The resultant rock is relatively hard with a dense, coarse-grained structure.

9.3.1.1 Natural calcium carbonates

Naturally occurring calcium carbonates in the form of chalk, limestone or marble for use in thermoplastics should show the following values:

$CaCO_3$	98.5 to 99.5%
$MgCO_3$	up to 0.5%
Fe_2O_3	up to 0.2%
Al-silicate (colloidal)	up to 1.0%
Loss on ignition	43.3 to 43.8%
Moisture content (DIN 53198, ISO/DIS 787/2–1979)	up to 0.2%
Density	up to 2.7 g/cm^3
Mohs hardness	3
Degree of whiteness (Elrepho green filter, MgO = 100%)	85 to 95%
pH (DIN 53200, ISO/DIS 787/9–1979)	9.0 to 9.5
Oil absorption (DIN 53199)	9 to 21 g/100 g powder
DOP number (ASTM D 281-31)	9 to 33 g/100 g powder
Specific surface	1 to 15 m^2/g

Quality of carbonate fillers

There is an unfortunate tendency to judge carbonate fillers solely from the cost aspect. Location of the quarry, chemical purity, i.e. quantity and type of impurities, treatment method and subsequent surface finishing are, however, of utmost importance, particularly as regards use in the plastics industry.

Leading firms are nowadays able to supply custom-made fillers for individual fields of application and to eliminate undesirable side effects on processing machinery (wear) and on the mechanical properties of the finished articles.

An additional factor is that a lipophilic character can be imparted to calcium carbonate by a suitable surface treatment (coating). This appreciably facilitates its use in thermoplastics.

Use of calcium carbonates in thermoplastics

In terms of weight, ground calcium carbonate is the most important filler used in plastics. Ultrafine ground chalks are frequently employed, particularly in PVC. The inert filler calcium carbonate plays an important rôle in the search for a balance between a low price and the retention of physicomechanical values. High-quality carbonate fillers are distinguished by the following properties:

- High chemical purity, no heavy metal ions that can catalyze the aging process in polymers,
- no tendency to form agglomerates,
- calcitic structure with relatively low specific surface, without any great absorptive effect on plasticizers and other additives,
- high degree of whiteness, hence the possibility of partial substitution of expensive white pigments,
- coloration in pastel shades is possible, also at high filler contents,
- non-abrasive, gentle on machine parts (Mohs hardness ~ 3),
- good dispersibility (in particular coated grades) with only slight influence on the mechanical and electrical properties of the final products,
- increase in stiffness and modulus of elasticity,
- reduced shrinkage, higher color fastness,
- improvement in the surface of the article,
- lower plate-out (deposition of mixture ingredients),
- increased impact strength, in particular through the use of coated grades,

- improvement in stability and aging resistance, particularly when using coated grades,
- non-toxic, odorless, tasteless, heat-resistant up to 600 °C,
- low price, cheaper than precipitated calcium carbonate, hence beneficial increase in weight and volume.

9.3.1.2 Synthetic, precipitated calcium carbonates

These carbonate fillers also known as CCP (calcium carbonicum praecipitatum) possess the following properties:

Mean particle diameter	approx. 0.004 to 0.07 μm
Degree of whiteness (Elrepho green filter, MgO = 100%)	95 to 96%
Oil absorption	35 to 40 g/100 g powder
Specific surface	32 to 40 m^2/g

The special properties of precipitated, surface-treated calcium carbonates result from their small particle size, their high degree of purity, their aragonite structure. As a consequence of the surface treatment in question, good dispersibility in the polymer matrix is thereby obtained.

Like the ground carbonates, precipitated calcium carbonates are frequently used in thermoplastics such as PVC and make possible the following improvements in their properties:

- Increased impact strength (especially in unplasticized PVC),
- high surface gloss in the finished products,
- good elongation at break, tear propagation resistance and tensile strength,
- higher modulus of elasticity,
- reduced plate-out,
- good weathering resistance.

Compared with the natural, ground calcium carbonates, the synthetic, precipitated calcium carbonate fillers, do, however, possess the following disadvantages:

- More expensive than ground chalk,
- due to the larger surface, the shearing forces during processing are appreciably higher; high filler additions are thus not possible,
- greater absorptive effect on plasticizers, stabilizers, lubricants etc.

9.3.1.3 Dolomite

Dolomite is a double carbonate formed from calcium and magnesium by metamorphosis. The $CaMg(CO_3)_2$ content amounts to 98 to 99.5%. Ground products with the following composition can be produced from high-grade dolomite with a pronounced crystalline structure:

$CaCO_3$	54.0%
$MgCO_3$	45.6%
Al_2O_3	0.03%
Fe_2O_3	0.2%
Density	2.85 g/cm^3
Mohs hardness	3
Degree of whiteness (Elrepho green filter, MgO = 100%)	85 to 95%
pH	10
Specific surface	1 to 10 m^2/g
Oil absorption	13 to 20 g/100 g powder

Use of dolomite in thermoplastics

Ground dolomite is used for similar applications to those of ground calcium carbonate. Dolomite is noted for good wettability and dispersibility, as well as for its moderate oil and plasticizer absorption and high degree of whiteness. However, since the magnesium salts possess higher solubility in water than the calcium salts, weathering resistance is likely to be inferior to that of the calcium carbonates. Dolomite is much more abrasive than calcium carbonate.

9.3.2 Silicates

9.3.2.1 Fiber glass

The engineering trend towards lightweight constructions calls for the use of materials with high strength, combined at the same time with low weight and corrosion resistance. Although plastics as such possess many advantages, their strength properties and especially their stiffness are often inadequate. In order, in such cases, to be able to obtain a solution that is acceptable both from the technical and economic aspects, it is necessary to switch to composites and to incorporate materials such as reinforcing fibers in the plastic. Glass fibers have thus attained major importance as reinforcements for thermoplastics.

Textile glass

Glass grades for technical applications consist of oxidic network formers and silicon, boron or aluminum oxide as skeleton substances. From the melts of these naturally occurring minerals, textile glass fibers mainly 10 to 20 μm in diameter are produced by means of the mechanical drawing method for continuous-filament glass fiber and the rod or blast drawing processes for staple fiber.

The fibrous products include:

Strands: Approximately aligned continuous strands of 100 to 400 single filaments without twist. They are gathered together into rovings of 30 to 60 strands.

Filament yarns: Spun from continuous filaments with twist.

Staple fibers: They consist of randomly oriented, comparatively short monofilaments (textile fibers) with a fleecy handle.

Staple yarns: Yarns produced with low-twist staple fibers.

The two-dimensional structures are subdivided into glass fiber fleeces, mats, coverings, fabrics, felts and sheets.

Glass grades

A glass: Alkali-containing, with and without additions of boron, moderately soluble in alkalis, used in articles not greatly subject to stress and exposed neither to the action of weathering nor moisture. Use in apparatus engineering etc.

D glass: Dielectric grade, limited to electrical insulating materials.

E glass: Most commonly used grade, aluminum-boron-silicate glass. The alkali oxide content must not exceed 0.8%, highly soluble in acids, moderately soluble in alkalis, used in the electrical industry etc.

M glass: High modulus glass for products subject to extremely high stress.

R glass: $SiO_2/Al_2O_3/MgO$ system with high strength and good thermal stability.

S glass: High strength for aircraft and rocket construction.

Glass fibers differ according to density, refractive index and chemical resistance. With 6.1, E glass possesses the lowest dielectric constant and with tan δ $9 \cdot 10^{-4}$ the lowest dissipation factor. The most important properties of the different types of glass fiber are compiled in table 6.

Table 6. Properties of different types of glass fibers.

Property	E glass	A glass	A glass with boron addition	R glass	S glass	M glass
Density (g/cm^3)	2.54	2.48	2.5	2.5	2.49	2.89
Strength of monofilament (N/mm^2)	3500	2450	2900	4750	4900	3500
Modulus of elasticity (N/mm^2)	73000	45000	60000	83000	87000	124000

Glass fibers are rot-resistant, reasonably resistant to aggressive chemicals, heat-resistant up to 650 °C, possess high tensile strength and good elasticity. The greatest drawback to the use of glass fibers as reinforcement lies in their very high brittleness.

Sizing of glass fibers

During the production of glass fibers, the glass surface must be protected against physical and chemical stress, as well as against the influence of moisture during storage. This is done by applying a size (in the case of continuous filament) or a lubricant (in the case of staple fibers).

9.3.2.1.1 Coupling agents

The glass surface itself generally possesses no or only little affinity with the plastic. In order that full advantage can be taken of the reinforcing action, it is necessary to ensure a good bond between matrix and glass by other means. This is achieved by applying coupling agents to the glass surface. The coupling agents generally used consist of an inorganic component reacting with the glass surface and an organic component reacting or at least having good compatibility with the polymer matrix [7].

The coupling agents between the silicates or silica, containing the silanol (Si—OH) groups, and the polymers to be reinforced can be roughly subdivided as follows:

- Chrome complexes,
- silanes,
- titanates,
- other coupling agents.

Chrome complexes

Chrome complexes, which are supplied under the trade name Volan, and historically speaking were the first among the coupling agents [34], have today been largely replaced by the silanes. The main field of application for chrome complexes are glass-fiber-reinforced, unsaturated polyester products.

Silanes

Particularly in recent years, fiber glass reinforcement of thermoplastics has made considerable progress as a result of the silane coating of glass fibers [7, 19 to 22]. The silanes normally used are tetrafunctional molecules with the general formula $YR'—Si(OR)_3$. They usually contain three methoxy or ethoxy groups at the central silicon atom, which are converted by hydrolysis into Si—OH groups. These Si—OH groups can bond with the silicate or silica surfaces by chemisorption. The fourth group Y bound to the Si atom can form a primary valence bond to heat-curable plastics – and increases thereby the adhesion between plastics matrix and reinforcement; in the case of thermoplastics, this generally takes place by purely physical forces (adhesion). The silanes most commonly used nowadays are listed in table 7.

No chemisorption takes place between silanes and alkaline filler surfaces (alkaline-earth carbonates, metallic oxides and hydroxides). Moreover,

Table 7. Selected silanes and their field of application.

Silane	Plastic
Vinyl-triethoxysilane Vinyl-tris (2-methoxy) silane γ-methacryloxypropyltrimethoxy silane	Unsaturated polyesters, polyolefins, PVC, PS, ABS, SAN, PMMA
γ-aminopropyltriethoxy silane	Epoxy, phenolic and melamine resins, PA, PUR, PP, PE, PMMA, PC
γ-glycidoxypropyltrimethoxy silane	Epoxy, phenolic and melamine resins, unsaturated polyesters, PUR, PVC, PC, PS, ABS, SAN, PA, PP, PE, polysulfides
γ-mercaptopropyltrimethoxy silane	Phenolic and epoxy resins, PVC, PS, EPDM, polysulfones, polysulfides

silanes are destroyed by the alkaline surfaces. Their application is thus restricted to filler surfaces containing OH groups.

Titanates

Like the silanes, titanates are tetrafunctional compounds. They were largely developed by Kenrich Petrochemicals Inc., Bayonne, N.J., USA [23, 24]. Their mode of action is similar to that of the silanes. Chemisorption is likewise only possible on filler surfaces containing OH groups. The titanates have not yet been able to compete commercially with the silanes.

Other coupling agents

Many of the coupling agents mentioned in the patent literature are based on esters. Their mode of action lies solely in better dispersion or wetting of the fillers or reinforcements with the polymer matrix and a resultant improvement in the mechanical properties. Most of the polymeric coupling agents such as maleic anhydride-propylene copolymers can be considered from the same aspect.

Fiber glass in thermoplastics

Fiber-glass-reinforced thermoplastics are manufactured by two different processes. In the first process, glass fiber strands are impregnated with the melt of a thermoplastic and after solidifying, the strands are cut into rodlike granules. The fibers are, therefore, of the same length as the rods and oriented in the latter's longitudinal direction. In the second process, the thermoplastic in powder form is blended with the ready-cut short fiber glass, extruded and granulated. In Europe, the major proportion of glass-reinforced plastics are sold as granules in ready-to-process form. Only in a few instances are glass fiber and plastic brought together in the feed hopper of the machine and mixed there in accordance with American practice.

The material mainly used for the reinforcement of thermoplastics and thermosetting plastics is the alkali-free E glass, which in comparison with the alkaline A glass is noted for its good electrical properties also in damp surroundings. The lower brittleness of E glass additionally contributes to the fact that the original length of the fibers is retained during manufacture of the granules.

Short fibers are more readily dispersible. The sprues can be more easily removed during injection molding. This is of importance in the case of single-gating and multi-gating. Moreover, short-fibered granules are cheaper than long-fibered material. The advantages of the short fibers are opposed by a number of disadvantages. Impact strength, for example, does not match the values of the long-fibered product, heat deflection temperature is 5 to 6 °C lower. Flexural strength and flexural modulus are also lower than with the products reinforced with long fibers.

The reinforcing action of glass fibers is so great that an addition of 30% generally results in a doubling of tensile strength, modulus of elasticity is trebled, coefficient of linear thermal expansion and shrinkage are reduced to a third. At the same time, however, a limited decrease in notched impact strength and a somewhat more pronounced directional dependence of shrinkage in injection moldings occur.

Reinforcing action of fiber glass

The reinforcing action of fiber glass is decisively influenced by the following properties:

- The length/diameter ratio of the reinforcing fibers should be large,
- the fiber should, where possible, be oriented in the direction of the force profile,
- optimum adhesion between fiber and matrix is desirable (size, coupling agent),
- intimate wetting of the fiber surface can be achieved by intensive blending operations, however, this may have an unfavorable influence on fiber length, the fibers become shorter and assume the character of a filler.

Strength and stiffness increase to a maximum level as the glass content rises and then fall again. This maximum is reached when the plastics matrix no longer suffices to transfer the forces between the fibers and to fill cavities. This influence is much greater with tensile and flexural stress than with compressive stress. In thermoplastics, short-fibered reinforcement additionally results in an appreciable lowering of elongation at break. Apart from the glass content, glass distribution has a decisive influence on the properties. This applies in particular to long-fibered reinforcement. Basically speaking it can be stated that the longer the fibers, the higher are the mechanical values.

Fiber-glass-reinforced thermoplastics combine the good properties of the plastics with those of the inorganic glass and in their property values approach the level of metals. Glass fibers metallized with aluminum are also supplied for the purpose of increasing electrical conductivity.

Property improvements through fiber glass reinforcements (20 to 40% fiber glass):

- Increase in tensile, compressive and flexural strength, as well as in the modulus of elasticity,
- appreciable increase in stiffness and flexural modulus; this property is of major importance, especially for technical applications,
- increase in tensile stress at break in the case of brittle base polymers and in notched impact strength at low temperatures, which usually drops sharply in non-reinforced thermoplastics,
- reduction in the coefficient of linear thermal expansion and in shrinkage, dimensional stability is considerably improved,
- resistance to cold flow is increased and creep lowered (smaller bend-creep modulus),
- improvement in heat deflection temperature,
- less noise development, also in the case of vibration,
- flexibility in design,
- material savings are possible, since as a result of the improved mechanical properties, many parts can be reduced in size or made thinner.

Disadvantages of fiber glass reinforcement:

- A slight reduction in impact strength can be expected in thermoplastics which by nature are impact-resistant,
- surface finish of the finished part is usually somewhat rougher than that of non-reinforced molding compounds,
- higher abrasion in processing machinery.

9.3.2.2 Glass spheres

Glass spheres are a relatively new form of fillers, both solid and hollow spheres are used.

Properties

Solid glass spheres are obtainable in diameters of 0.004 to 5 mm. Preference is given to spheres from 4 to 44 μm, their density is 2.5 g/cm^3.

Hollow glass spheres float on water, since their density ranges from 0.3 to 0.6 g/cm^3, their diameter ranges from 10 to 250 μm.

Use of glass spheres in thermoplastics

Both in thermoplastic and in thermosetting compounds, solid glass spheres are employed as fillers. They all have the same shape, are transparent, compression-resistant, have a controlled granulometry and high thermal stability. Many advantages of the glass spheres are directly attributable to the spherical shape. They act in the same way as the balls in a roller bearing and increase flowability. They also contribute to an improvement in stress distribution. Since a sphere has the smallest surface at a given volume, the shear forces between polymers and filler are low. Oil absorption is a factor of no significance with glass spheres. The favorable flow properties permit high filler contents. The action of glass spheres is decisively influenced by their wettability, silanes being therefore often used as coupling agents.

Fiber-reinforced plastics are known to possess two anisotropic shrinkage values, shrinkage in the fiber direction being lower than that perpendicular to it. By contrast, molded parts filled with glass spheres exhibit isotropic behavior, i.e. shrinkage in material reinforced with glass spheres is almost the same in all directions. This means that the shrinkage level can be specified in advance.

Glass spheres generally lead to an increase in tensile and compressive strength, as well as in the flexural modulus. Because of the spherical shape, stress cracking in finished parts occurs to a less extent than is the case with ground inorganic fillers.

Molding compounds containing glass spheres can be extruded and injection molded. Although hollow spheres lower the apparent density, the mechanical properties are usually impaired. They are chiefly used in foams. Solid glass spheres are cheaper than hollow spheres per unit of weight, but are more expensive in relation to volume.

Conductive glass microspheres coated with silver are offered as a special development. The silver coating lowers the volume resistivity of the molding materials down to 0.03 $\Omega \cdot$cm. Inexpensive, electrically-conductive molding compounds can be manufactured by this means.

9.3.2.3 Asbestos

Asbestos is a generic name for a number of fibrous, hydrated Mg and Na silicate minerals. Asbestos is not, therefore, the designation for a specific material, but is the commercial term for a large number of fibrous minerals of different composition, strength, flexibility and length. The physical and chemical properties of the individual types differ considerably. The non-flammable asbestos fibers represent single crystals that have grown naturally under specific conditions of pressure and temperature. They can thus also be described as naturally occurring whiskers. The long-fibered types are used in the form of fibers and yarn, the short-fibered as fillers, extenders and viscosity adjusting aids.

Asbestos minerals

The six known varieties of asbestos comprise the fibrous serpentine asbestos – known as chrysotile – and five minerals in the amphibole group. The properties of the individual types are compiled in table 8.

With a share of over 90%, *chrysotile* is the most frequently used asbestos type (white asbestos). As a result of its positive charge, softness, flexibility and low refractive index, this type contrasts markedly with the others. The fiber diameter is less than 30 nm at a mean strength of 500 to 1 000 N/mm^2 (steel wire approx. 1 100 N/mm^2, E glass fiber approx. 1 500 N/mm^2). The fibers are up to 500 mm in length. Chrysotile is attacked by acids, but it is highly resistant to alkali. Hardness corresponds to that of glass (2.5 to 4.0 Mohs).

Crocidolite is the asbestos type in the amphibole range which has recently been used to the greatest extent. Its needle-shaped structure imparts high strength to this blue asbestos from South Africa. It is used for those applications in which the blue color is not a disturbing factor. Stiffness and strength correspond to that of chrysotile. Crocidolite is resistant to acids and alkalis and in this respect is clearly superior to E glass fibers.

Anthophyllite is a complex silicate of iron and magnesium. The high thermal stability and considerable resistance to acids have, for example, led to applications in polyolefins.

Tremolite, actinolite and *amosite* are scarcely used in thermoplastics.

Use in thermoplastics

As filler and reinforcement, asbestos possesses the following advantages and disadvantages:

When added as a reinforcing fiber,

advantages: higher modulus of elasticity, higher tensile strength, higher dimensional stability under heat, higher dimensional stability, lower creep tendency, higher chemical resistance, higher hardness.

disadvantages: higher melt viscosity of the molding compound, lower electrical resistance, higher abrasion, negative influence on the light resistance and thermal stability of certain thermoplastics.

Table 8. Comparison of the properties of various asbestos minerals [8].

Property	Chrysotile	Anthophyllite	Crocidolite	Actinolite	Tremolite	Amosite
Composition	$Mg_3SiO_2O_5(OH)_4$	$Mg_7Si_8O_{22}(OH)_2$	$Na_2Fe_5Si_8O_{22}(OH)_2$	$Ca_2Fe_5Si_8O_{22}(OH)_2$	$Mg_5Ca_2Si_8O_{22}(OH)_2$	$(Fe, Mg)_7Si_8O_{22}(OH)_2$
Crystalline state	fine fibers	prismatic crystals and fibers	long, brittle fibers	prismatic crystals and fibers	prismatic crystals and fibers	prismatic crystals and fibers
Color	white, gray	gray-white	blue	green	white, gray-white, green	ash-gray, brown
Handle	soft to rough	rough	soft to rough	rough	rough	rough
Flexibility	very good	low	good	low	low	good
Hardness (Mohs)	2.5 to 4	5.5 to 6	4	6	5.5	5.5 to 6
Fiber diameter (nm)	20 to 40	60 to 90	60 to 90	60 to 90	60 to 90	60 to 90
Tensile strength ($N/mm^2 \cdot 10^3$)	2.1 to 5.6	2.0 to 2.8	2.8 to 4.2	<0.007	0.007 to 0.056	0.7 to 1.4
Modulus of elasticity ($N/mm^2 \cdot 10^6$)	0.16	–	0.19	–	–	0.17
Apparent density (g/cm^3)	2.4 to 2.6	2.8 to 3.1	3.2 to 3.3	3.0 to 3.2	2.9 to 3.2	3.1 to 3.3
pH	9.5 to 10.3	9.4	9.1	9.5	9.5	9.1
Electric charge in aqueous slurry	+	–	–	–	–	–
Acid resistance	non-resist.	resist.	resist.	resist.	resist.	resist.

When added as a reinforcing filler,

advantages: lower cost, high flexural strength, low water absorption depending on the plastic, higher dimensional stability under heat, improved chemical resistance.

disadvantages: lower impact strength, inherent color of asbestos, difficult to blend with the plastic, high stabilizer requirement.

The advantage of asbestos fibers in thermoplastics in comparison with glass fibers lies in the fact that stiffness of the asbestos fiber is twice as high at a similar level of tensile strength. The asbestos fiber differs from other fibers by virtue of its unusually large surface. This enables a large part of the mechanical load applied to the composite material to be transferred from the polymer matrix to the fiber. Asbestos fibers are more compactly bundled and have stronger interfibrillary bonds than glass fibers. On the other hand, asbestos fibers exhibit a higher degree of felting than glass fibers. In comparison to glass, this results in an improvement in the isotropy of the reinforcing effect in the finished product.

The hazards associated with exposure to asbestos dust led to the development of non-dusting grades that represent a much lower health risk.

9.3.2.4 Talc

Talc is a natural, hydrated magnesium silicate with the formula $3MgO \cdot 4SiO_2 \cdot H_2O$. Talc occurs in four particle shapes: fibrous, lamellar, needle-shaped and modular (steatitic); however, only the lamellar form is used in commercial applications. The composition of talc is approximately as follows:

SiO_2	40 to 62%
Al_2O_3	0.2 to 11%
Fe_2O_3	0.1 to 1.5%
FeO	0.1 to 6%
CaO	0.3 to 1%
MgO	30 to 33%
H_2O	16 to 17%
pH	8.6 to 9.9
Oil absorption (depending on particle size)	28 to 51 g/100 g powder
Density	2.9 g/cm^3

The basic constituent of talc, the oxide and hydroxide of magnesium, exists as a sandwich between two silicon oxide layers. This imparts to

each lamella a certain degree of water repellancy and high chemical inertness. The good slip properties of the talc are largely determined by the planar form of the lamellae and the low hardness.

Use of talc in thermoplastics

The lamellar structure of talc has recently also resulted in specific applications in the polymer sector when stiffness problems are in the forefront. In this regard, the action of the filler is not only dependent on the degree of filling, but on the fineness of grind and the chemical nature of the talc grades concerned.

In contrast to other silicates, surface treatment of talc has not up to now produced any appreciable improvement in the mechanical properties of the compound materials in question.

Talc is used in thermoplastics to improve the following properties:

- Increase in stiffness, flexural modulus and modulus of torsion,
- increase in flexural strength,
- reduction in the creep tendency, also at elevated temperatures,
- improvement in the bend-creep modulus,
- improvement in the heat deflection temperature and in dimensional stability,
- improvement in shrinkage and warpage, combined at the same time with a lower coefficient of thermal expansion,
- improvement in thermal conductivity,
- increase in the surface hardness of molded parts.

However, talc possesses the following disadvantages when used in thermoplastics:

- Decrease in tensile strength and elongation at break,
- decrease in notched impact strength, increase in brittleness,
- lowering of the melt index of the compound,
- anisotropic reinforcing and shrinkage behavior,
- impairment of seam strength,
- change of shade in the composite material, depending on the natural color of the talc used,
- depending on type, unsuitable for contact with food (asbestos content, heavy metals),
- pure products are relatively expensive,
- negative influence on the thermal stability of certain thermoplastics.

9.3.2.5 Kaolin

Kaolin, the end product of the weathering of granite and feldspar, is found in very many parts of the earth, but rarely in the required degree of purity. Kaolin is also known under the name porcelain earth and china clay. It consists of primary and secondary kaolinites. In their lamellar structure, the primary kaolinites have a ratio of length to thickness of 10:1; the crystalline form of the individual lamellae is hexagonal.

Properties

Kaolins are hydrated aluminum silicates possessing a clearly determinable crystal lattice with a plate-like, hexagonal structure. The typical chemical composition of a hard kaolin with the molecular formula $Al_2O_3 \cdot 2SiO_2 \cdot 2H_2O$ is: Al_2O_3 (38.8%), SiO_2 (45.4%), MgO (0.2%), K_2O (0.97%), Na_2O (0.07%), FeO (variable).

Kaolins generally possess a high degree of whiteness, are nonconductors of electricity, highly resistant to chemicals and have good resistance even towards strong acids. Hard kaolin has a slightly acidic reaction with a pH of 5.0 to 5.6.

Hard and soft kaolins

Apart from its use in the paper industry, which accounts for 50% of kaolin consumption, this mineral is the second most important filler in the rubber industry after carbon black. The rubber chemist also distinguishes between "hard" and "soft" kaolins, depending on whether the finished product has a high or low modulus. From the morphological aspect, the main difference lies in the mean particle size: in hard kaolin, at least 75% of all particles are less than 2 μm, soft kaolins are coarser.

Use of kaolin in thermoplastics

Kaolin is very frequently used in thermosetting compounds, e.g. UP premix, BMC and SMC (unsaturated polyester premix, bulk molding compounds and sheet molding compounds) as well as in low-profile resins. In thermoplastics, kaolin is used, for example, as filler for wire insulating materials, since it improves the electrical properties.

In general, kaolin contributes to an improvement in chemical resistance and in the electrical properties, as well as to a reduction in water absorption. It lowers the crack tendency in the finished article and improves shock resistance and surface quality. The lamellar structure additionally results in an improvement in the surface hardness of the finished product.

Calcined kaolin

This form is appreciably harder than natural kaolin, but considerably improves the electrical properties in polymers (e.g. higher dielectric strength in cable insulating materials made from plasticized PVC or EPDM).

Kaolins are obtainable with various silane coatings that facilitate dispersion of the filler in the plastics.

9.3.2.6 Mica

The lamellar reinforcement mica is obtained from the minerals muscovite – $K_2Al_4(Al_2Si_6O_{20})(OH)_4$ – or phlogopite – $K_2(MgFe^{+2})_6(Al_2Si_6O_{20})(OH,F)_4$. As the molecular formulas show, the products in question are complex K/Al silicates. They have a density of 2.8 g/cm^3, a Mohs hardness of 2.5 to 4 and oil absorption values of 48 to 500 g/100 g powder. The types not suitable for the electrical industry are used in ground form as fillers and reinforcements. The decisive factor as regards reinforcing action is the ratio of diameter to thickness of the lamellae (aspect ratio).

Mica imparts the following properties to thermoplastics:

- High stiffness,
- high dimensional stability,
- good electrical properties,
- good dimensional stability under heat.

Silanization of the filler surface facilitates the incorporation of mica in plastics, the use of amino silanes, in particular, resulting in good mechanical properties of the compounds.

In the automobile industry, mica is used as an inexpensive substitute for glass fibers in polypropylene parts.

9.3.2.7 Feldspar and nepheline syenite

Feldspar (composed of white granite) and nepheline (composed of syenite rock) occur naturally as anhydrous alkali/aluminum silicates; their density is 2.6 g/cm^3, the Mohs hardness value of 6.0 to 6.5 is relatively high. The refractive index of 1.53 combined with low tinctorial strength results in good reflecting power. The mineral possesses high chemical resistance and exhibits a slightly alkaline reaction (pH 8 to 9). This filler is coarser than kaolin and talc. The low oil absorption (13 to 24 g/100 g powder) has led to applications in thermosetting plastics and PVC. These minerals are also supplied in silanized form. They have the same optical density as PVC and therefore cannot be seen in a PVC article.

9.3.2.8 Wollastonite

Wollastonite ($CaSiO_3$) is the only commercially available pure white mineral occurring in needle form. The length/diameter ratio can amount to 15:1. β-Wollastonite possesses high chemical resistance, a density of 2.9 g/cm^3, a Mohs hardness of 4.5 to 5.0 and a pH of 9 to 10. In thermoplastics it competes with talc and mica.

The needle-shaped structure results in high stiffness combined with good impact strength, as well as good electrical and thermal properties. Here, too, silanized grades are available.

9.3.2.9 Hollow silicate spheres

In addition to hollow glass spheres, other hollow spherical silicate fillers such as products consisting of flue dust (Cenospheres, Fillite) or of aluminum silicate (Armospheres) are nowadays also obtainable. Their low density, uniform size, low oil absorption and high chemical resistance have led to them being used in the PVC, plastisol and thermosetting sectors. During processing, however, these thin-walled fillers are sensitive to pressure and shear forces.

9.3.3 Silica

9.3.3.1 Natural silicas

Sand, quartz, quartzite, novaculite, tripoli and diatomaceous earth are naturally occurring silicas and differ relative to their particle size, degree of crystallinity and hardness. β-Quartz is, moreover, the hardest of the common minerals. Quartz has a density of 2.65 g/cm^3 and a Mohs hardness of 7.

9.3.3.2 Synthetic silica

In all the manufacturing processes, very fine, generally spherical primary particles are obtained, which as secondary particles form aggregates and agglomerates. Depending on the manufacturing process, the surface may be very large and attain values of 50 to 800 m^2/g.

Apart from the chemical composition (SiO_2), all synthetic silicas have another important property in common: they are not crystalline like natural quartz, but are shown by X-ray analysis to be amorphous. The silicosis risk is avoided thereby.

Silicas are likewise supplied with various silane coatings.

Use of silica in thermoplastics

In contrast to silicas of natural origin, which as fillers are mainly used in the form of quartz flour, synthetic silicas can already be employed in low quantities to achieve specific effects. In thermoplastics, silicas perform the following functions:

- Reduction in shrinkage and crack formation,
- reinforcement,
- prevention of film blocking,
- improvement in dimensional stability under heat,
- reduction in the coefficient of linear thermal expansion,
- improvement in electrical properties,
- increase in hardness,
- reduction of die swell during extrusion,
- thickening, rheological and thixotropic effects.

9.3.4 Barium sulfate/heavy spar

The most important mineral from which precipitated barium sulfate or naturally occurring heavy spar ($BaSO_4$) is obtained, is barite. Heavy spar is distinguished by a relatively high density of 4.3 to 4.6 g/cm^3 and a Mohs hardness of 2.5 to 3.5. The relatively expensive barium sulfate possesses high reflectivity. It is primarily used in foams with high apparent density and in floor coverings, as well as for protection against high-energy radiation.

9.3.5 Other mineral fillers

Among the minerals in the carbide family, silicon carbide (SiC) is used to increase the abrasive action of polymers. Among the sulfides, molybdenum sulfide (MoS_2) is employed to reduce sliding friction and zinc sulfide (ZnS) for plastics with a white coloration. Among the silicates, mention can be made of aluminum silicate (mullite) and zirconium silicate ($ZrSiO_4$) for improving slip properties, and among the titanates, barium titanate ($BaTiO_3$) for electrical applications.

9.3.6 Metals and metallic powders

Metallic powders consisting of aluminum, bronze, zinc, copper and nickel are always used in thermoplastics if products with very high thermal or electrical conductivity are required. Heavy metal powders additionally increase the resistance to neutron and gamma rays.

Metals in fiber form are also supplied as reinforcements.

Metallic oxides, too, are employed for selective changes in the properties of thermoplastics, e.g. aluminum oxide for improved electrical values, zinc oxide for weathering properties, UV stability, good thermal conductivity, resistance to mildew attack and because of its high density. Beryllium oxide is used to improve thermal conductivity, iron oxide for magnetic effects, lead oxide because of its high density, magnesium oxide to increase stiffness and for the thickening of unsaturated polyesters and finally zirconium oxide for stiffening purposes combined at the same time with high density.

9.3.6.1 Alumina trihydrate

Alumina trihydrate, also known as hydrate of alumina, is an intermediate manufactured in high quantities for the production of aluminum from bauxite and of aluminum salts, and for the manufacture of special ceramics. When heated to temperatures above 200 °C with considerable heat consumption, it has the characteristic of decomposing to form alumina (aluminum oxide) and water vapor. This reaction terminates at around 600 °C, up to 35% by wt. being liberated as water vapor.

Properties

Alumina trihydrate is a non-flammable white powder insoluble in water, with the chemical formula $Al(OH)_3$. In its properties, alumina trihydrate differs appreciably from the anhydrous aluminum oxide Al_2O_3. Alumina trihydrate is a non-abrasive filler with low density (2.4 g/cm^3) and a Mohs hardness between 2.5 and 3.5. The abrasive, anhydrous aluminum oxide, on the other hand, has a density of 4.0 g/cm^3 and a Mohs hardness of 9.

Use as a filler in thermoplastics

Alumina trihydrate is generally used on account of its flame-retardant properties. In view of the fact that decomposition already commences at 200 °C, difficulties are experienced with molding compounds that have to be processed above this temperature.

As with most mineral fillers, stiffness usually rises as the $Al(OH)_3$ component increases, while tensile and impact strength decrease. As a result of the lamellar structure, the good flow properties of alumina hydroxide enable excellent surfaces to be obtained.

Of importance to its application in the electrical industry is the improvement in electrical properties such as dielectric strength and tracking resistance. Special alumina trihydrates with low electrolyte content are now available for this field. Their aqueous dispersions have low electric conductivity.

9.3.7 Carbon

9.3.7.1 Carbon black

In the technical sphere, the term carbon black is understood to mean a special form of carbon obtained through partial combustion of liquid or gaseous hydrocarbons.

Carbon black has for centuries been used as a black pigment. Its use as a reinforcing filler did not commence until the start of this century. Carbon black is the most important reinforcement employed by the rubber industry.

Use of carbon black in thermoplastics

As regards their use in plastics, carbon blacks are classified according to their particle size and color intensity. Carbon blacks with small particle size are usually more difficult to disperse than coarser types.

Carbon black is also used to achieve volume conductivity and to improve heat deflection temperature. In the record industry, carbon black serves as a filler, in foams, use is also made of hollow carbon black spheres.

Carbon black is frequently utilized as a protection against UV radiation and as a black pigment in thermoplastics.

9.3.7.2 Graphite

Graphite powder is highly suitable for imparting self-lubricating properties to thermoplastics. Lubricating depends on the structure, purity and particle size of the graphite. Provided the type and quantity are correctly selected, graphite powder can be blended homogeneously with a large number of thermoplastics by the extrusion technique. The coefficients of friction and the abrasion factors of such composites lie between those of the basic material and formulations containing silicon or polytetrafluoroethylene.

Graphite-containing thermoplastics are mainly used in those applications where there is a requirement for self-lubricating properties when the material is exposed to the action of water. Under these conditions, graphite has proved to be a very good lubricant. High graphite contents (30% by wt.) also reduce processing shrinkage. With amorphous thermoplastics, it is possible in conjunction with mineral fillers to obtain ideal materials with close tolerances for parts exposed to abrasion that operate continuously in an aqueous environment (e.g. water meters).

The use of graphite fibers as reinforcement is discussed in conjunction with carbon fibers.

9.3.8 Fillers and reinforcements in flake form

In addition to mica (cf. 9.3.2.6) and silicon carbide flakes (cf. 9.3.5), glass flakes (decorative purposes, improved scratch and chemical resistance), aluminum diboride (AlB_2) flakes (high tensile strength and stiffness) and aluminum and steel flakes are now also supplied.

9.3.9 New developments in short inorganic fibers

Apart from asbestos (cf. 9.3.2.3), wollastonite (cf. 9.3.2.8) and whiskers (cf. 9.3.12.6), new products have also recently become available in the field of short fibers. These include the following developments: Processed Mineral Fibers (PMF), 75% of which consist of aluminum silicate with aluminum and magnesium oxides, from Jim Walter Resources Ind. Birmingham, Ala (USA) and the so-called Franklin-Fiber (calcium sulfate-hemihydrate) from Certain-Teed Corp. Valley Forge, PA (USA). Alcoa Pittsburg, Pa (USA) has introduced aluminum hydroxycarbonate under the name Dawsonite. These products have up to now only been available in trial quantities.

9.3.10 Organic fillers and reinforcements

In addition to wood flour, obtained by pulverizing wood, fibrous products such as α-cellulose, cotton, sisal and jute are also used, mainly in the processing of thermosetting molding compounds. The use of starch as a filler is also suggested for plastics articles that are allegedly biodegradable. In addition, industrial trials have also been successfully carried out with cork flour, lignin and ground nut shells or corn cobs. Silicon-containing products such as rice grain husks and similar materials are likewise being tested for industrial usage.

Wood flour is already being employed in thermoplastics on an industrial scale. Its use results in the following advantages:

- Cheap filler,
- improved shrinkage behavior,
- increase in modulus of elasticity and stiffness, falls less rapidly at elevated temperatures,
- tensile strength and tensile stress at break are scarcely affected.

The disadvantages of wood flour as a filler in thermoplastics are:

- Low adhesion between plastic and filler,
- lower impact strength,
- deterioration in thermal stability and weathering resistance,
- decrease in processability,
- stability problems in extruders and injection molding machines at temperatures above 180 °C.

9.3.11 Synthetic reinforcing fibers

Polyester fibers

Polyester fibers based on polyethylene terephthalate are often used together with glass fibers and produce a considerable improvement in impact strength, stiffness and abrasion properties.

Polyvinyl alcohol fibers

Treatment with formaldehyde renders PVA fibers insoluble in water. Although these fibers possess appreciably lower strength than glass fibers, they are distinguished by the following advantages:

- Good adhesion between fibers and many plastics,
- high impact strength,
- low density of 1.25 g/cm^3,
- coefficient of thermal linear expansion the same as most plastics,
- no fiber breakage during injection molding,
- no machine wear,
- good flexibility.

9.3.12 High-tenacity fibers and fibers with high thermal stability

9.3.12.1 Basalt fibers

Basalt fibers possess all the positive characteristics of glass fibers combined with even higher thermal stability, better adhesion to polar polymers such as epoxy resins and better alkali resistance.

9.3.12.2 Carbon and graphite fibers

A certain similarity with the single-crystal fibers is exhibited by the recent developments in the field of synthetic, organic fibers, which are treated by pyrolysis to form carbon or graphite fibers. Depending on the method of pyrolysis, a distinction is made between partly carbonized fibers, so-called carbon fibers, which have undergone extensive transformation, and a fiber completely converted into graphite, also in its crystallographic structure. The cross section of the fibers is polygonal or indented.

There are two main types of carbon-based fibers. It has become customary to refer to carbon fibers with a modulus of elasticity of up to 75000 N/mm^2 as carbon fibers and to those with a higher modulus of elasticity as graphite fibers.

Carbon fibers

Fibers produced from rayon or bitumen have a lower graphite content, but possess a larger surface, higher volume resistance, higher water absorption and lower thermal conductivity than fibers with a higher graphite content. Carbon fibers are mainly employed for thermal insulation and as reinforcing fibers in composite materials (excellent ablation behavior). Their comparatively low electrical resistance has resulted in them being used also in those applications where a flexible, non-metallic conductor is required.

Graphite fibers

Graphite fibers with their high strength/weight ratio result in composite materials which at the same weight possess three to nine times the stiffness of titanium, steel or aluminum. The fibers are generally produced from polyacrylonitrile or rayon fibers. The graphite fibers obtainable on the market offer the possibility of manufacturing products that either possess high strength or high stiffness. With other constructions, on the other hand, it is the low coefficient of linear thermal expansion that is utilized. Table 9 gives a survey of the properties of the commonly used carbon and graphite fibers.

Carbon fibers are supplied in the form of rovings, consisting of several hundred to over 10000 single filaments, or as staple fibers. Carbon fiber fabrics or felts are also obtainable.

Use of carbon fibers in thermoplastics

For technical applications, composites of carbon fibers with thermosetting plastics and both with thermoplastics are important. In both cases,

Table 9. Characteristic values of carbon and graphite fibers [9].

Property	Unit	Carbon fiber	Graphite fiber
Tensile strength	$N/mm^2 \cdot 10^2$	2.1 to 5.6	18 to 32
Modulus of elasticity	$N/mm^2 \cdot 10^4$	4.2 to 7.5	21 to 56
Elongation at break	%	1 to 2	0.25 to 1.2
Elastic recovery	%	100	100
Density	g/cm^3	1.3 to 1.8	1.4 to 2.0
Fiber diameter	μm	8 to 9	6 to 9
Graphite content	%	80 to 99	92 to 99.9
Water absorption (20 °C, 56% rel. humidity)	%	2 to 14	0
Volume resistance	$\Omega\ cm \cdot 10^9$	3.5 to 6	0.7 to 1.2
Forms supplied		filament yarn, fabric, felt, staple fiber	filament yarn, cords, staple fiber

the use of carbon fibers permits the manufacture of extremely lightweight components which at the same time possess high strength and stiffness (aircraft and rocket construction).

The use of carbon fibers offers the following advantages:

- Low density,
- high flexural modulus,
- high stiffness, the specific modulus of elasticity (modulus of elasticity/density) amounts to $4\,^1/_2$ times that of steel,
- high thermal conductivity,
- non-fusible,
- high strength also at elevated temperatures,
- electrically conductive.

Apart from these advantages, the use of carbon fibers in plastics also has the following disadvantages:

- There is still no effective coupling agent available,
- sensitivity to oxidation at high temperatures,
- high price (except fibers made from bitumen).

9.3.12.3 Aromatic polyamide fibers

These high-performance fibers also known as aramid fibers include, for example, the products sold under the names "Kevlar" and "Nomex".

They possess high strength and a high modulus of elasticity, their strength being somewhat higher than that of glass fibers and the modulus of elasticity twice as high as that of glass. Elongation at break is low and density is only 60% of that of glass. The flame-resistant fibers have good chemical resistance together with excellent electrical properties. Aromatic polyamide fibers are suitable for all applications in which importance attaches to high strength, stiffness and low weight. The properties in general lie between those of glass fibers on the one hand and graphite fibers on the other.

9.3.12.4 Boron fibers

The methods normally used for fiber manufacture, such as wire drawing etc., are not suitable in the case of boron. Up to now, the only promising method of producing boron fibers takes place via the deposition of boron on heated tungsten wires.

With a tensile strength of 3000 to 3500 N/mm^2, elongation at break of 0.7 to 0.9% and a modulus of elasticity of 2.5 to $4.0 \cdot 10^5$ N/mm^2, boron fibers lie midway between glass fibers and whiskers with regard to their properties. In comparison with glass fibers of approximately the same tensile strength, however, they attain far higher weight-related values for modulus of elasticity. Up to now, the main field of application has been the aviation sector (helicopter rotors).

9.3.12.5 Ceramic fibers

Ceramic fibers are made from metallic oxides and possess combinations of properties not previously known. They have extremely high thermal stability (1350 to 1650 °C) combined with a high modulus of elasticity and chemical resistance. The product range comprises 100% aluminum oxide fibers (e.g. Fiber FP or Saffil), aluminum-boron-silicate and aluminum-boron-chrome fibers, as well as combinations with Al_2O_3 or special quartz fibers. They are generally supplied in the form of fabrics or mats.

9.3.12.6 Whiskers

Whiskers are small single crystals in filament form, which can be used to manufacture composite materials with excellent mechanical and physical properties. Up to 1970, the main activity lay in the field of γ-Al_2O_3 whiskers (sapphire), but today interest is directed more towards the cheaper silicon carbide (SiC) whiskers.

Whiskers combine the outstanding properties of glass and boron fibers, they possess the elongation of glass fibers (3 to 4%) and the high modulus of elasticity of boron fibers (4 to $7 \cdot 10^5$ MPa). Apart from the high cost, the major disadvantage of whiskers lies in the fact that a completely new fiber technology has to be developed in order that the products can emerge from the experimental stage. The mechanical properties of some whiskers are compiled in table 10.

Table 10. Characteristic values of some whisker materials [10].

Whisker material	Density (g/cm^3)	Modulus of elasticity ($N/mm^2 \cdot 10^5$)	Theoretical tensile stress at break ($N/mm^2 \cdot 10^4$)	Experimental tensile stress at break ($N/mm^2 \cdot 10^4$)
Al_2O_3	3.96	4.22	4.22	1.97
SiC	3.19	8.55	8.40	1.12
BC	2.52	4.57	4.57	0.65
C	1.66	10.0	10.0	2.11
Fe	7.85	2.04	2.04	1.34
Ni	9.95	2.18	2.18	0.39
Cu	8.92	1.27	1.27	0.30
Wires				
Steel	7.75	2.11	–	0.41
Beryllium	1.83	2.46	–	0.13
Glass filament	2.94	0.91	–	0.49

Most whiskers are produced by vapor phase reactions. In order to achieve their maximum reinforcing effect, the fibers have to possess optimum length, shape and the right cross section. Provided the fibers are adequately wetted and are oriented in the melt, it is frequently possible for the tensile strength of the basic plastics material to be increased by ten to twenty times. The price of these reinforcing materials must, however, first of all be reduced to an economically acceptable level in order that the present fields of application such as space, weapons and deep sea technology can be expanded.

In table 11 are compiled in general form the changes in the physicomechanical properties of thermoplastics that can be achieved by means of the commonly used fillers and reinforcements.

Table 11. Diagrammatic presentation of the influence of fillers and reinforcements on thermoplastics [11].

	Glass fiber	Asbestos	Wollastonite	Carbon fiber	Whiskers	Synthetic fibers	Cellulose	Mica	Talc	Graphite	Sand/ quartz powder	Silica	Kaolin	Glass spheres	Calcium carbonate	Metallic oxides	Carbon black
Tensile strength	+ +	+		+	+ −			+	o					+			
Compressive strength	+								+		+			+	+		
Modulus of elasticity	+ +	+ +	+ +	+ +	+			+ +	+		+	+		+	+	+	+
Impact strength	− +	−	−	−	−	+ +	+	− +	−		−	−	−	−	− +	−	+
Reduced thermal expansion	+	+			+			+	+		+	+	+			+	
Reduced shrinkage	+	+	+	+				+	+	+	+	+	+	+	+	+	+
Better thermal conductivity		+	+	+					+	+	+	+			+		+
Higher heat deflection temperature	+ +	+	+	+ +				+	+				+		+	+	
Electrical conductivity				+						+							+
Electrical resistance			+					+ +	+			+	+ +			+	
Thermal stability			+					+	+		+	+	+			+	+
Chemical resistance		+	+					+	o	+			+	+			
Better abrasion behavior				+				+	+	+			+				
Extrusion rate	− +	+						+					+		+		
Machine abrasion	−	o			o	o	o		o	o	−			o	o		o
Price reduction	+	+	+				+	+	+	+	+ +	+	+	+	+ +		
	Fibrous fillers and reinforcements							Lamellar types			Spherical fillers						

+ + considerable influence, + less pronounced influence, o no influence, − negative influence

9.4 Fillers and reinforcements in thermoplastics

9.4.1 Polyethylene

9.4.1.1 Low density polyethylene (LDPE)

Low density polyethylene is mainly used in applications calling for very good toughness or high deformability of the material, also at low temperatures. These properties are, however, influenced to a greater or less extent by fillers.

Like high density polyethylene (HDPE), LDPE is also generally used in unfilled form. The requirements that have to be met by a filler for LDPE are the same as those for HDPE.

Fillers in low density polyethylene

Calcium carbonate

The following properties of LDPE are modified by the addition of calcium carbonate:

Low specific heat and high thermal conductivity of the filled LDPE result in higher throughputs during extrusion and injection molding. Hardness, stiffness, heat deflection temperature, slip resistance and resistance to stress cracking increase; weldability, printability and antiblocking properties are improved. Melt index, shrinkage and elongation, as well as water vapor and oxygen permeability decrease. Calcium carbonate has no adverse influence on the photooxidation of LDPE films and enables reductions to be made in the titanium dioxide content [33].

Talc/Mica

The addition of talc enables most of the mechanical properties to be retained over a fairly wide temperature range, the electrical values are favorably influenced.

Glass fibers

A 20% addition of glass fiber makes LDPE stiffer and reduces shrinkage. Adhesion to the matrix can be improved by alkaline treatment of the glass fibers.

Asbestos

Asbestos contents of 20 to 40% by wt. increase in particular the flexural strength and heat deflection temperature.

Other fillers used in LDPE are:

Carbon black: reinforcing action when the material is cross-linked by radiation, effective light stabilizer, increase in electrical conductivity.

Silica: increase in heat deflection temperature, antiblocking.

Wollastonite: improved impact strength, better electrical and thermal properties.

Aluminum trihydrate: flame retardancy.

The most important properties of filled LDPE grades are compiled in table 12.

9.4.1.2 High density polyethylene (HDPE)

Because of the high degree of crystallization of HDPE, the strength of the material at low temperatures is more strongly influenced by fillers than is the case with LDPE. As with all non-polar polymers, the critical factor is the adhesion between filler and matrix. Appreciably better adhesion and markedly higher mechanical values (modulus of elasticity, impact strength) can be achieved simply by using a vented extruder.

Fillers and reinforcements for HDPE

Glass fibers

Glass fibers are the reinforcements most frequently used, but unfortunately give rise to the above-mentioned adhesion problems. These have since been solved by means of silanes. Modulus of elasticity and shrinkage are favorably influenced, in presence of 30% glass fiber heat deflection temperature rises to 108 °C, compared with 50 °C in non-reinforced materials.

Calcium carbonates

The use of mineral fillers was until recently confined to paperlike films. However, blown articles are now also produced with calcium carbonate as filler. Here the advantages lie in higher stiffness and higher production outputs, resulting from the lower specific heat and improved thermal conductivity (more rapid cooling). Moreover, all the advantages obtainable with calcium carbonate in LDPE, e.g. better printability, are also achieved in HDPE [33].

Compared with most other fillers, the use of surface-coated calcium carbonate results, in particular, in higher impact strength. Melt index, elon-

Table 12. Influence of fillers on the properties of polyolefins [12].

Material	Filler content (% by wt.)	Tensile strength (N/mm^2)	Elongation (%)	Modulus of elasticity (N/mm^2)	Melt index 190/5 (g/10 min)	Heat deflection temperature (ISO/R75A, 1.85 N/mm^2) (°C)	Ball indentation hardness (N/mm^2)	Density (g/cm^3)
Low density polyethylene	–	10	500	210	1.7	35	16	0.92
Chalk	40	16	220	900	0.2	–	–	1.26
Glass fiber	30	24	65	1200	1.6	–	33	1.11
Asbestos	30	20	20	670	2.3	–	22	1.13
Talc	30	16	40	600	1.1	–	30	1.14
Mica	30	13	46	440	1.2	–	27	1.16
Glass spheres < 50 μm	30	10	73	290	2.6	–	19	1.11
Quartz powder	30	12	77	400	2.5	–	23	1.14
High density polyethylene	–	27	> 550	1400	22	50	46	0.95
Glass fiber	30	60	2	7800	6	108	68	1.16
Chalk	30	20	9	1900	18	–	48	1.17
Wood flour	30	28	6	2100	–	–	–	–
Kaolin	40	26	11	2000	2.5	–	–	1.24
Asbestos	40	30	10	2100	0.7	77	–	–
Polypropylene	–	31	620	1600	1.0	62	75	0.90
Short glass fiber	30	49	5	7500	6.0	120	100	1.20
Asbestos	40	45	7	5000	4.0	96	84	1.28
Talc	40	30	11	3100	3.0	95	83	1.21
Mica	30	30	–	6900	–	–	–	–
Chalk	40	25	180	3400	2.5	73	63	1.23
Wood flour	40	20	8	2200	–	–	–	–
Wollastonite	40	23	7	7000	–	85	88	1.24

gation, tear propagation strength, shrinkage, stress cracking and cold flow are reduced by the filler.

Talc/Kaolin

Silicates impart to HDPE higher tensile strength and stiffness than chalk.

Other fillers used in HDPE:

Carbon black: increase in conductivity and light stability.

Wood flour: improvement in acoustic properties.

Heavy metal powders: protection against neutron and gamma rays.

Table 12 contains details of the properties of filled HDPE grades.

9.4.2 Polypropylene (PP)

The use of fillers and reinforcements in PP makes it possible to utilize the latter's high chemical and stress cracking resistance also at high temperatures and to obtain greater stiffness at room temperature. The increasing brittleness of PP at temperatures below 0 °C is intensified by the use of fillers. Under these conditions, copolymers with ethylene possess higher impact strength and at the same time exhibit better adhesion between matrix and filler or reinforcement. Reinforcement thus takes place mainly in copolymers. The chemical purity of the fillers is very important; heavy metals such as copper, manganese or cobalt should not be present or if so only at minimum levels, since in general they have an adverse influence on thermal stability. Depending on the intended application, reinforced or filled polypropylene must also be stabilized against the deleterious action of UV light.

Fillers and reinforcements for polypropylene

As in most other thermoplastics, reinforcements in PP produce the following improvements in properties: increase in stiffness, modulus of elasticity, flexural modulus, bend-creep modulus, hardness, dimensional stability under heat, tensile stress at break and tensile strength, compressive strength, dimensional stability, density, alternate bending strength, thermal conductivity and melt viscosity.

On the other hand, a reduction in elongation, impact strength, notched impact strength, creep tendency, shrinkage, thermal expansion, abrasion resistance and melt index can usually be observed.

Glass fibers

In the USA, glass-fiber-reinforced PP is the most important of all the reinforced thermoplastics. The main advantages of fiber glass reinforcement lie in the high tensile strength over a wide temperature range, the high stiffness and in improved creep behavior. Short or long glass fibers in ground form are used. The short fibers are distinguished by high toughness combined with good stiffness and the long fibers, particularly when coupling agents are used, by improved tensile strength and torsional stiffness at elevated temperatures, as well as by better creep strength and flexural stiffness. In general, the use of coupling agents results in an improvement in mechanical values as well as in an increase in fatigue strength, good long-term properties and a higher bend-creep modulus. The disadvantages of the fiber glass reinforcement lie in the orientation effect, reduced impact strength, matt surface and increased abrasion in the processing machines.

Glass spheres

In PP, the less frequently used glass spheres result in higher stiffness at elevated temperatures, better compressive strength and dimensional stability.

Asbestos

Asbestos exhibits higher stiffness in PP, but also a more pronounced creep tendency than talc. Stiffness may be higher than with fiber glass reinforcement. Of the asbestos varieties, only chrysotile and anthrophyllite are suitable for PP. The orientation problems are generally not as great as with fiber glass reinforcement. PP must, however, be additionally stabilized against the heavy metals contained in asbestos.

The following fillers are also used in PP:

Wood flour: good acoustic properties, better than with talc filling (40 to 60% wood flour).

Wollastonite: low moisture absorption, high impact strength, good electrical and thermal properties.

Zinc oxide: imparts protection against UV light and attack by microorganisms, increases thermal conductivity.

Reinforcing fibers: fibers such as polyamide or polyester fibers increase tensile stress at yield and impact strength of PP.

Metallic powders: high heat deflection temperature, improved conductivity.

Mica: at a lower cost level, 40% mica imparts the same stiffness as 30% glass fibers. The addition of chlorinated paraffins enables a marked improvement to be obtained in adhesion to the matrix.

Talc

Talc is among the reinforcements and fillers for PP with the highest growth rates. As a result of its lamellar structure, talc is the classic filler that imparts higher stiffness, as well as an increase in the heat deflection temperature and dimensional stability. Apart from cost reduction, the higher stiffness permits, as with calcium carbonates, a lowering of wall thicknesses with the resultant increase in the cycle speeds on processing machines. Creep behavior and bend-creep modulus are better than at the same level of asbestos filling or short glass fiber content.

Talc is a good nucleating agent. Torsional stiffness of PP with 40% by wt. talc is higher than with the addition of 30% long glass fibers.

Disadvantages of talc filling in PP are reduced impact strength at low temperatures, lower tensile stress at break, matt surface, impairment of welding properties and negative influence on thermooxidative stability.

Calcium carbonates

Calcium carbonates as fillers in PP are in many respects distinguished by the same advantages as talc. Although the same high stiffness is not obtained at the same degree of filling, impact strength is appreciably higher. However, calcium carbonates (also in coated form) possess the following advantages over talc: easier dispersibility, higher melt index, better u.v. stability, completely free of asbestos, better surface of finished parts, suitable for contact with food, lower abrasion in processing machines [33].

In injection molding, the use of calcium carbonates results in articles with higher stiffness and dimensional stability at shorter cycle times (higher thermal conductivity). In PP films, the filler improves the deep-drawing properties. In PP cords and tapes, use of the filler makes it possible to avoid undesirable splicing and at the same time to obtain a cleaning action in the extruder die.

PP filled with calcium carbonate can be employed as an inexpensive substitute for polystyrene and ABS. Calcium carbonate is also very often used as a filler in atactic PP (carpet backing).

Table 12 gives a survey of the most important properties of reinforced PP grades.

9.4.3 Polybutene

In polybutene-1, carbon black is the filler mainly used to increase stiffness and reduce chalking, and also as UV stabilizer. However, grades with 20% by wt. glass fiber and 40% by wt. talc are also supplied.

9.4.4 Polyisobutylene

Polyisobutylene tends to exhibit cold flow, which can be reduced by additions of chalk, talc and polyethylene. Special PIB films for protection of buildings and corrosion prevention have a high filler content, in order to improve the mechanical properties.

9.4.5 Ethylene-vinyl acetate copolymers

The following fillers are used in EVA for technical applications:

- Chalk (for toys),
- heavy spar (resistance to X rays and for sound damping),
- carbon black (for electrically conductive articles),
- barium ferrite (up to 90% by wt. for magnetizable sealing strips),
- alumina hydrate (as flame retardant),
- talc (stiffening).

9.4.6 Polyvinyl chloride (PVC)

PVC is a largely amorphous, thermoplastic material with a diversity of processing methods and fields of application. Together with polyethylene, PVC is among the thermoplastics most frequently used. By changing the polymer, the plasticizer and the stabilizers, lubricants and fillers, the properties of the final products can be varied over a wide range. PVC is the thermoplastic which has already for a long time been successfully processed with fillers.

Fillers in PVC

In contrast to most other thermoplastics, fillers in PVC are generally employed for cost reduction reasons (extenders), since with the exception of impact strength and electrical properties, no improvement in the mechanical properties is achieved. In the case of the cable industry, the aim is to obtain the necessary characteristic values and mechanical properties as cheaply as possible. The most important application for fillers lies in the plasticized PVC sector.

9.4.6.1 Plasticized PVC

Fillers and reinforcements for plasticized PVC

Natural calcium carbonates

In terms of weight and usage, natural calcium carbonates are by far the most important fillers for thermoplastics. This also applies to PVC. Chalk grades are used both in untreated and in coated form. High-quality, finely divided chalk grades nowadays have a mean particle diameter of 1 to 1.5 μm with an upper limit of 5 to 10 μm. Good products possess a reproducible granulometry and are completely free of coarse particles and abrasive silica compounds. They occur mainly in the French Champagne region and in southern France. The following improvements can be obtained in the properties of plasticized PVC:

- Good dispersibility, especially in the case of coated products,
- only slight influence on the mechanical values,
- use of surface-coated products with a small surface results in low absorption of plasticizers and additives,
- high chemical purity, no heavy metals with catalytic influence on the aging properties,
- increase in output and better surface of finished products (in particular through the use of coated chalk grades), lower lubricant requirement,
- partial replacement of expensive white pigments,
- chalk serves as hydrochloric acid acceptor,
- improvement in aging resistance, heat deflection temperature and water resistance (reduction in stabilizers),
- improvement in electrical and dielectric properties, since the use of chalk permits reductions in the plasticizer content,
- lower shrinkage, lower plate-out, stiffening,
- improvement in dry handle, prevention of sticking to calendar rolls,
- unlike other fillers, the abrasion caused by fine chalk grades in processing machines is negligible,
- good storage stability,
- reduction in the cost of the final product, inexpensive increase in volume.

Surface-treated, ground calcium carbonates have a favorable influence on processability and improve the electrical properties. The cable industry is, therefore, among the most important users of chalk. Surface treatment of the calcium carbonate often results in a synergistic effect with the stabilizers. The improvement in thermal stability obtained thereby, permits a reduction in the quantity of stabilizers used.

Kaolin

Along with ground chalk, kaolin is mainly used in the cable industry, since it increases the specific resistance of the cable compounds. Soft kaolin is additionally employed in floor coverings and household articles, improvements being obtained in chemical resistance and electrical properties. A major disadvantage of kaolin, however, lies in its high plasticizer absorption.

Other fillers used in PVC are:

Asbestos: together with chalk, the use of chrysotile in floor coverings produces increased strength, hardness, thermal stability and better abrasion behavior than asbestos types of the amphibole group. Additional stabilization is however, necessary.

Carbon black: filler in records, considerable modification of the insulating properties, protection against UV radiation.

Silicates: synthetic silicates serve as ion scavengers in the cable industry, they additionally reduce plate-out and blocking.

Alumina trihydrate. increase in flame resistance, used in particular in backcoating of textile floor coverings.

Talc: mainly used in vinyl asbestos floor coverings, as also wollastonite.

Dolomite: same applications as ground chalk.

Barium ferrite: magnetic seals.

9.4.6.2 Unplasticized PVC

Fillers and reinforcements for unplasticized PVC

Natural calcium carbonates

As with plasticized PVC, the fillers mainly used in unplasticized PVC are natural chalk grades, although only the types with very small particle size and in surface-treated form are employed. The latest technological

Table 13. Natural chalk grades, technically the most important fillers in PVC for various fields of application [13].

Composition of compound		Cable cores	Cable sheathing	Filling compounds
Plasticized PVC				
PVC	(pts. by wt.)	S type, K70 (100)	S type, K70 (100)	S type, K70 (100)
Plasticizer	(pts. by wt.)	DINP (40)	DINP (40 to 60)	DOP (100)
Stabilizer*)	(pts. by wt.)	Pb (4 to 6)	Pb (2.5 to 3.5)	Pb (3 to 4)
Lubricant	(pts. by wt.)	– (0.5 to 1.5)	– (0.5 to 1.6)	– (1.5)
Filler**)	(pts. by wt.)	EX-H1 SP (30 to 60)	EX-H1 SP (70 to 100)	BSH (500)
		Pipes, profiles, high filler content	Roller shutters	Pressure and sewer pipes
Unplasticized PVC				
PVC	(pts. by wt.)	S type, K65 (100)	E + S type, K70 (100)	S type, K65 (100)
Stabilizer*)	(pts. by wt.)	Pb (3)	Pb (3 to 4.5)	Pb (1.5 to 2.5)
Lubricant	(pts. by wt.)	– (1)	(0.5)	(0.5)
Modifier	(pts. by wt.)	CPE (2 to 10)	–	–
Blowing agent	(pts. by wt.)	– –	–	–
Filler**)	(pts. by wt.)	95T (50)	95T (10 to 30)	95T (5)
		Imitation leather		Flooring
		Base coat	Top coat	Base coat
Plastisols				
PVC	(pts. by wt.)	Paste, K65–70 (100)	Paste, K65–70 (100)	Paste, K65–70 (100)
Plasticizer	(pts. by wt.)	– (50 to 70)	– (40 to 60)	– –
Stabilizer*)	(pts. by wt.)	Ba Cd (3.5 to 5)	Ba Cd (3.5 to 7)	Ba Cd (3 to 4)
Blowing agent	(pts. by wt.)	– –	– –	– –
Viscosity depressant	(pts. by wt.)	– –	– –	– –
Foaming agent	(pts. by wt.)	– –	– –	– –
Filler**)	(pts. by wt.)	Durcal 15 (20)	Millicarb (10)	Durcal 40 (500)

*) incl. costabilizers such as epoxy plasticizers
**) Omya products

Films	Tubes, profiles	Shoe soles, injection molding	Calendering, floor coverings	Vinyl-asbestos covering
S type,	S type,	S type,	S type,	S type,
K70 (100)	K70 (100)	K70 (100)	K70 (100)	K70 (100)
DOP (50)	DOP (60 to 70)	DOP (80)	DOP (40 to 60)	DOP (45)
Ba Cd (5)	Ba Cd (2 to 3)	Ba Cd (3.5 to 5)	Ba Cd (3.5 to 5)	Ba Zn (6 to 8)
– (0.5)	– (0.2 to 0.5)	– (0.5)	– (0.5)	– (1)
BSH (40)	BSH (50 to 60)	Durcal 5 (15)	Durcal 40 (200)	Durcal 65 (350)
0.90T (40)	0.90T (50 to 60)			Abestos (100)
Hydrocarb (40)				
Durcal 2 (40)				

Windows, outer profiles	Calendered sheeting	Rigid PVC foam	Injection molding	
S type,	S/M type,	S type,	S/M type,	
K65 (100)	K65–70 (100)	K60 (100)	K55–58 (100)	
Ba Cd (4 to 6)	Sn (2 to 4.5)	Sn (2.5 to 4)	Pb (3)	
(1 to 1.5)	(0.8 to 1.7)	(1 to 1.6)	(1.8 to 2.5)	
(6 to 12)	(5 to 10)	(6 to 10)	(1 to 2)	
		(0.3 to 1.0)		
HC95T (10 to 20)	95T (10 to 30)	95T (5 to 10)	95T (5 to 10)	

	Carpet backing		Plastisols	Dipping pastes
Top coat	Chemically foamed	Mechanical foamed	High filler content	
Paste,	Paste,	Paste,	Paste,	Paste,
K65–70 (100)	K65–70 (100)	K65–70 (100)	K65–70 (100)	K65–70 (100)
– (50 to 100)	– (50 to 70)	– (80 to 90)	– (130)	– (50 to 80)
Ba Cd (3.5 to 7)	Ba Cd (1 to 2)	Ba Cd (1 to 1.5)	Ba Cd (2 to 4)	Ba Cd (2 to 4.5)
– –	– (1 to 2)	– –	– –	– –
– –	– (1.5)	– –	– (5)	– –
– –	– –	– (4)	– –	– –
Millicarb (50)	Hydrocarb (80)	Calibrite (120)	Calibrite (475)	Calibrite (30)
			Omya BSH (50)	

developments have shown that the addition of impact modifiers allows a considerable increase in the use of fillers. Chlorinated polyethylene and other products influencing impact strength permit a filler content of 100 parts by wt. chalk to 100 parts by wt. PVC.

The use of coated chalk results in increased stiffness, lower shrinkage and reduced plate-out; notched impact strength is increased and aging resistance and heat deflection temperature improved in comparison with unfilled PVC. Extrudability is improved and the price per unit volume lowered.

Precipitated calcium carbonate

These are chiefly used in unplasticized PVC. However, the small particle size of this class of filler leads to a high increase in torque in the processing machines, with the result that the possible filler volume is far lower than with ground chalk grades. Precipitated chalk improves the surface, shade, degree of whiteness, surface gloss and plate-out. The modulus of elasticity rises, but the other mechanical properties are not greatly influenced.

In addition to the fillers mentioned in the case of plasticized PVC, the following products are also used in unplasticized PVC:

Glass fibers: reinforcement of PVC injection moldings, cold formable, glass-fiber-reinforced sheet.

Talc: increased modulus of elasticity, better flexural and tensile strength, but reduction in elongation at break and impact strength at low temperatures.

Wood flour: overlay sheets for chipboard, sheeting, profiles.

9.4.6.3 PVC plastisols

Natural calcium carbonates

Ground limestone and marble powders with favorable particle distribution and minimum plasticizer absorption are of importance both in coating and dipping pastes and in PVC foams. They consolidate and compact the plastisols and improve adhesion and bond strength on the substrate. The simplest way of adjusting to the desired thixopropic properties is to add variable percentages of very finely ground, microcrystalline, coated chalk grades.

Silicates

The use of synthetic silicates results in matt surfaces, an improvement in the penetration of coating pastes and an increase in thixotropy.

In table 13 are compared various formulations of PVC articles filled with natural calcium carbonates and in table 14 properties of filled unplasticized PVC molding compounds.

Table 14. Properties of filled rigid PVC [14].

Fillers	Filler content (pts. by wt.)	Tensile strength (N/mm^2)	Elongation at break (%)	Modulus of elasticity (N/mm^2)	Vicat softening point (°C)	Notched impact strength (Charpy) 23 °C (mJ/mm^2)	Density (g/cm^3)
S-PVC, unplasticized	–	60	6 to 10	2700	85	3	1.36
Calcium carbonate, natural	30 100	46	8	3200	94 116	9	1.53 1.78
Calcium carbonate, precipitated	15	30 to 47	6	3100	87	14	1.45
Talc	20	34	6	3500	–	–	1.48
Quartz powder	20	38	–	3100	–	–	–
Glass fiber	40	25	3	8000	85	–	–
Asbestos	30	33	1.2	5500	85	–	1.65
Wollastonite	20	25	5.4	–	–	3	1.47

9.4.7 Fluorinated polymers

The very high chemical resistance and very good electrical properties of most fluorinated polymers such as PTFE and PCTFE, as well as the service temperature range of −250 °C to +250 °C, unfortunately

Table 15. Mechanical properties of filled and reinforced plastics [15].

Plastic/filler	Filler content (% by wt.)	Tensile strength (N/mm^2)	Elongation (%)
Polyacetal (POM)	–	63	45
Glass fiber	30	104	4
Talc	30	61	3
Ethylene/tetrafluorethylene copolymer (PTFE)	–	40	200
Glass fiber	25	83	9
Carbon fiber	30	100	
Vinylidene fluoride PTFE copolymer			
Carbon fiber	20	86	3.5
Polystyrene	–	55	4
Glass fiber	30	95	2
Asbestos (chrysotile)	20	71	–
Calcium carbonate	30	15	2
Talc	40	39	1.6
Styrene-acrylonitrile copolymer (SAN)	–	70	5
Glass fiber	35	110	3
Mica	30	83	3
ABS	–	43	15
Glass fiber	30	70	3
Asbestos	20	70	–
Mica	50	65	3
Polymethyl methacrylate	–	80	5
Glass fiber	35	120	–

Modulus of elasticity (N/mm²)	Notched impact strength (N/mm²)	Heat distortion temperature (ISO/R75A, 1.85 N/mm²) (°C)	Ball indentation hardness	Density (g/cm³)
2700	9	101	155	1.41
7200	7.6	161	180	1.60
8600	2.6	–	160	1.64
840	–	74	–	1.75
8400	–	210	–	1.86
10000	9	240	–	1.73
7700	2.6	295	–	1.77
3800	2	86	165	1.05
11000	4	93	240	1.32
8200	–	91	–	1.26
2000	18	–	–	1.30
5600	–	–	–	–
3600	2	90	–	1.08
10000	4.5	100	250	1.35
16300	–	110	–	1.53
2600	10	83	95	1.06
6600	6	94	130	1.33
4100	–	97	–	–
14600	–	103	–	1.52
3200	5	75	195	1.18
12000	–	–	–	–

have to be set against a high creep tendency and high coefficient of linear thermal expansion that allow virtually no mechanical stress.

Fillers and reinforcements in fluorinated thermoplastics

Suitable fillers enable improvements to be made in the following properties:

Creep strength (factor approx. 10), abrasion resistance, stiffness (factor 4 to 5), thermal conductivity (5 to 10 times), compressive strength (3 to 4 times), hardness, creep tendency and heat deflection at high temperatures. Elongation at break and notched impact strength, however, are adversely affected.

Glass fibers

15 to 25% by wt. short glass fibers are used, sometimes together with small quantities of graphite.

Carbon fibers

Up to 20% by wt. carbon fiber are employed as reinforcement in polyvinylidene fluoride and ethylene tetrafluoroethylene copolymers.

Metals, metallic oxides and metallic compounds

No metallic fillers are allowed to be used in fluorinated polymers that come into contact with food (soluble metallic fluorides). Molybdenum sulfide (sometimes in conjunction with glass powder), graphite, carbon black and up to 68% by wt. bronze are occasionally used for bearing materials and seals.

Depending on field of application, graphite fibers together with glass fibers, and also talc, can be used in copolymers of tetrafluoroethylene with ethylene or hexafluoropropylene.

Chalk

After the chalk has been leached out, PTFE-filled chalk can be used as porous electrical material. Some data of these materials are listed in table 15.

9.4.8 Polyacetal (POM)

Among the acetal polymers, only polyoxymethylene obtained by polymerization of formaldehyde has attained technical importance. The highly crystalline POM is a stiff material with a high degree of hardness and good dimensional stability and is primarily used for technical applications such as sliding elements, bearings, machine parts etc. The good resilience of POM has also acquired technical importance, but is diminished by the addition of fillers. However, the high inherent stiffness of POM can be further increased by the use of fillers.

Fillers and reinforcements for polyacetal

Glass fibers

Since the material reacts sensitively to thermal and oxidative stress, the high heat deflection temperature achieved by means of fillers can generally only be utilized for a short period. The maximum service temperature is thus just over 100 °C. Here, fiber glass reinforcement has achieved the greatest importance. Both long and short glass fibers are used. Their favorable influence on tensile strength and stiffness can be enhanced by coupling agents. Compared with unfilled POM, fiber glass reinforcement of 20 to 30% by wt. results in an increase in tensile strength (2 times), flexural modulus (2 to 3 times), tensile stress at break (2 times), hardness and stiffness, as well as specific gravity. Fiber glass reinforced POM can be used in place of zinc, bronze and aluminum castings. It is used for those applications where stiffness, hardness, favorable creep behavior, also at elevated temperatures, low water absorption, high dimensional stability and low coefficient of expansion are of importance.

These properties are appreciably influenced by the fiber glass component. As disadvantages of fiber glass reinforcement, mention must be made of the decrease in toughness and notched impact strength as well as the higher price per unit volume.

Glass spheres

POM can be filled with up to 80% by wt. glass spheres, without the processing conditions having to be appreciably changed in comparison with unfilled material.

For bearing materials made from POM, optional use can be made of molybdenum sulfide, fluoroplastics and chalk. This results in a considerable improvement in slip properties and abrasion behavior. PTFE is also used in fibrous form and produces even better friction coefficients than MoS_2 or graphite.

On a smaller scale, use is also made of carbon black to increase UV stability and of metallic powders such as aluminum or bronze for high dimensional stability under heat and electrical conductivity.

Some of the properties of reinforced POM are compiled in table 15.

9.4.9 Polystyrene (PS)

Unmodified, general purpose PS is an amorphous, hard, brittle plastic with good electrical properties and resistance to aging and high dimensional stability. Adverse features are high brittleness and susceptibility to environmental stress cracking. Unfortunately these properties cannot be improved by fillers or reinforcements and the use of fillers in PS has thus up to now remained at a low level.

Use of fillers and reinforcements in polystyrene

Glass fibers

Glass fibers are used for those applications where increased strength, hardness and dimensional stability can be fully utilized and reduced impact strength is not a significant factor. The use of 30% by wt. fiber glass results in a trebling of the flexural modulus and an improvement in the mechanical properties (except brittleness and impact strength). Apart from general purpose PS, glass fibers are also used for reinforcement of impact PS. In expanded PS and impact modified PS, the use of glass fibers likewise enables an improvement in the physical properties to be achieved.

Glass spheres

In PS, glass spheres produce higher stiffness, in particular at elevated temperatures, higher compression resistance, improved dimensional stability and increased dimensional stability under heat, but reduce impact strength and crack formation.

Other fillers occasionally used are:

Asbestos: good processing properties, good dimensional stability and flame resistance.

Calcium carbonates: extender filler, can also be used in expanded PS as nucleating agent.

Talc: improved impact strength, adjustment of flow properties.

Alumina trihydrate: increased flame resistance.

In table 15 are listed the most important values for reinforced polystyrenes.

9.4.9.1 Styrene copolymers

The disadvantages of general purpose polystyrene (high brittleness and susceptibility to environmental stress cracking) can be improved by graft polymerization with elastomers. These tougher modifications are better suited as matrix material for fillers and reinforcements.

Styrene-acrylonitrile copolymers (SAN)

There are glass-fiber-reinforced grades of SAN copolymers. A considerable improvement in strength and a lower fracture tendency can be observed in comparison with unreinforced material. Elongation at break decreases. Glass spheres are also being used as filler in SAN.

Butadiene-styrene-acrylonitrile copolymers (ABS)

Glass fiber reinforcement of ABS improves flexural strength by three times and produces a considerable increase in the modulus of elasticity: special fiber-glass-reinforced grades for blow molding and ABS structural foam are already on the market. Glass spheres are also used for improving stiffness at elevated temperatures, for higher compressive strength and better dimensional stability.

Up to 30% by wt. chalk are used in Japan for the production of components for robots, in order to improve the mechanical properties of ABS without appreciable loss of impact strength.

The properties of some reinforced SAN and ABS grades are listed in table 15.

9.4.10 Polyamides (PA)

The toughness and wear resistance of this class of plastics are utilized in particular in technical applications. In general, polyamides are hard, hornlike and opaque with good abrasion resistance and high dimensional stability. These properties can be improved by fillers or reinforcements. Water absorption, however, is influenced to only a slight extent by fillers.

PA 6.6 has the highest hardness, stiffness, abrasion resistance and heat deflection temperature of all polyamides, whereas PA 6 is characterized by its higher toughness and PA 6.10 by lower water absorption and thus by better dimensional stability. Of all the polyamides, PA 11 has the highest impact strength and only a low water absorption tendency, although the latter is somewhat higher than that of PA 12. In addition, there are also the tough, very hard polyamide casting resins based on PA 6, which are produced by anionic polymerization.

Among the various fillers and reinforcements, the ones that have gained most importance are those which by virtue of their fibrous or lamellar structure enhance the advantages of polyamides or improve abrasion and wear behavior. The adverse properties of unfilled polyamides in comparison with metals, e.g. reduced modulus of elasticity, lower tensile and creep strength, pronounced dependence of the mechanical properties on temperature, high thermal expansion and water absorption can be considerably improved, in particular by fiber glass reinforcement.

Fillers and reinforcements for polyamides

Glass fibers

Fiber-glass-reinforced polyamides with up to 50% by wt. glass fibers in the case of PA 6 and PA 6.6 and 30% by wt. with PA 6.10, PA 11 and PA 12 represent the most important reinforced grades.

Fiber glass reinforcement (30% by wt.) produces the following property changes in PA:

- Modulus of elasticity increased 3 times, tensile strength 2 times, tensile stress 4 times, flexural stress (stiffness) 3 times, ball indentation hardness 1.5 times, stiffness increases with fiber length,
- considerable improvement in creep strength, reduction in creep rate,
- lower thermal capacity (shorter cycle times, higher heat deflection temperature, Martens temperature increased by 100 °C),
- lower coefficient of thermal expansion, approaching that of aluminum and tin alloys,
- lower water absorption, modulus of elasticity less dependent on water content,
- better chemical resistance,
- better hydrolysis stability when exposed to hot water.

The disadvantages of fibers lie in a certain anisotropy of the properties, in distortion problems (shrinkage differences due to fiber orientation) and abrasion in the processing machines. These problems are usually less pronounced in PA reinforced with short fibers than in long fibered grades.

Glass spheres

Up to 50% by wt. glass spheres are used in PA 6.6 and PA 12. They improve tensile and flexural strength, elongation at break, impact strength and heat distortion temperature. Risk of abrasion is less than with glass fibers. Glass spheres permit uniform shrinkage in all directions, close tolerances, good abrasion behavior and a smooth surface.

The glass spheres, which are generally surface-coated, can also be combined with glass fibers, thus enabling an increase in notched impact strength to be obtained.

Carbon and graphite fibers, carbon black and graphite

Up to 20% by wt. carbon fibers are used in PA 6, up to 40% by wt. in PA 6.6, PA 11 and PA 12. They yield the following advantageous properties:

- Smaller increase in density than with glass fibers,
- modulus of elasticity twice as high as with the same quantity of glass fibers,
- flexural strength, tensile strength, stiffness and creep behavior are at room temperature comparable with the values of tin, aluminum or manganese,
- better slip properties and lower abrasion in processing machines than is the case with glass fibers,
- improvement in thermal conductivity and electrical conductivity.

The disadvantages of reinforcement with carbon fibers lie in a deterioration of impact strength, in addition to which the fibers are still very expensive.

Graphite and carbon black are generally used only in quantities of up to 5% by wt.

Silicon oxide

Tensile stress at break is greatly increased, extrudability improved and a smooth surface obtained.

Silicates (mica, calcined kaolin, asbestos, wollastonite)

By addition of up to 40% by wt., lamellar mineral fillers such as mica and talc, in particular, increase stiffness, mechanical strength, hardness, thermal and dimensional stability, as well as yielding improvements in surface quality and distortion behavior. Moisture absorption and electrical and chemical properties are favorably influenced. As in other partly crystalline polymers, talc is also used as nucleating agent.

Calcium carbonate

Polyamides filled with calcium carbonate lower the raw material price and are used in applications in the electrical, household appliance, washing machine and automobile sectors where good dimensional stability without shrinkage problems is desired. Molding compounds filled with this material are distinguished by very good surface quality, high heat deflection temperature, good stiffness and toughness, as well as resistance to many solvents, fuels, oil and fats.

Metallic powders

Aluminum, bronze, steel, lead, zinc, copper and nickel in powder form are used to obtain a high heat deflection temperature and electrical conductivity. The quantity of bronze powder used is up to 95% by wt.

Molybdenum sulfide

In glass-fiber-reinforced grades, MoS_2 is employed for improving the slip properties.

Barium ferrite

PA filled with up to 80% by wt. barium ferrite is used for small magnets.

PTFE

Fluorinated polymers are likewise used in fiber-glass-reinforced PA for improving the slip properties (bearing materials as with MoS_2).

Table 16 gives a survey of the use of reinforcements in various polyamide types and the property changes obtained thereby.

Table 16. Mechanical properties of reinforced polyamides [16].

Material	Filler content (% by wt.)	Tensile strength (N/mm^2)	Elongation (%)	Modulus of elasticity (N/mm^2)	Notched impact strength (Nmm/mm^2)	Heat deflection temperature (ISO/R75A, 1.85 N/mm^2) (°C)	Rockwell hardness	Density (g/cm^3)
PA 6	–	64	220	1200	25	80	R 111	1.13
Short glass fiber	30	148	3.5	5500	16	208	E 55	1.37
Glass spheres	30	65	20	3000	14	–	R 120	1.35
Asbestos	30	123	3.0	8000	3	193	R 85	1.30
Carbon fiber	20	100	–	8000	–	–	–	–
Silicon oxide	10	57	140	1000	–	–	–	1.19
Chalk	30	50	30	3000	6	60	–	1.35
PA 6.6	–	63	60 to 300	1500	16	66 to 85	R 113	1.14
Short glass fiber	30	153	3.0	7200	15	204 to 249	E 60 to 70	1.37
Glass spheres	30	81	5.0	3700	11	74	R 120	1.35
Asbestos	30	133	–	9200	–	–	–	–
Carbon fiber	20	197	4	16900	–	257	–	1.23
Mica	30	39	–	6900	–	–	–	–
PA 6.10	–	60	85 to 300	1900	11	–	R 111	1.19
Short glass fiber	30	128	3.0	7800	12	204	E 30 to 40	1.30
Talc	20	60	5.0	40000	–	–	–	1.25
PA 11	–	58	325	1200	27	58	R 107	1.04
Glass fiber	30	93	4.0	6200	11	173	R 115	1.26
Bronze powder	90	34	4.0	5500	–	100	R 106	4.00
PA 12	–	60	270	1200	20	–	R 108	1.02
Glass fiber	30	83	6.0	3500	9	147	R 113	1.23
Glass spheres	30	45	25	2500	8	120	R 113	1.23

9.4.11 Polyethylene and polybutylene terephthalate (PET, PBT)

The slip and wear behavior of these saturated polyesters and their dimensional stability and creep strength are of particular importance for technical applications. Creep strength surpasses that of polyamides or polyacetals.

PET is supplied in amorphous, partly crystalline (degree of crystallinity of 30 to 40%) and filled form. Chain mobility is greatly hindered by the aromatic chain segments. The resultant slow rate of crystallization leads to problems during injection molding. These disadvantages can, however, be compensated by the use of fibrous fillers. The main fields of application of reinforced polyalkylene terephthalates lie in apparatus and mechanical engineering.

Fillers and reinforcements in PET and PBT

Glass fibers

The following properties are influenced by fiber glass reinforcement:

- High increase in stiffness,
- good slip and wear properties for guide and bearing elements,
- higher mechanical strength, good impact strength, lower creep,
- improved heat deflection temperature, outstanding water resistance, excellent long-term behavior, as well as good electrical properties and chemical resistance (except concentrated acids and alkalis).

Flowability, toughness, shrinkage, thermal expansion and surface brilliance decrease with increases in glass fiber content.

Glass spheres

The use of 20% by wt. glass spheres reduces the distortion tendency and results in high surface hardness, isotropic shrinkage, good surface and lower machine wear than with fiber-glass-reinforced types.

Mineral fillers

Asbestos and mica are used as fillers for special applications in the electrical industry. Although wollastonite increases the mechanical strength, the melt index is considerably reduced as a result of its large surface. Combinations of glass fiber and mica or wollastonite produce high toughness with lower distortion.

Table 17 shows the most important properties of reinforced polyalkylene terephthalate grades.

Table 17. Mechanical properties of reinforced engineering plastics [17].

Material/filler	Filler content (% bt wt.)	Tensile strength (N/mm^2)	Elonga-tion (%)	Modulus of elas-ticity (N/mm^2)	Notched impact strength (Nmm/mm^2)	Heat deflection temperature (ISO/R75A, 1.85 N/mm^2) (°C)	Rockwell hardness	Density (g/cm^3)
Polycarbonates	–	69	98	2150	28	142.5	R 122	1.22
Glass fiber	30	115	3.6	7100	11	145	R 125	1.43
Wollastonite	50	44	–	8100	–	266	–	1.44
Polyethylene terephthalate	–	66	160	2800	5	72	L 107	1.35
Glass fiber	30	165	4.0	11000	7	221	M 100	1.58
Carbon fiber	30	140	2.5	14300	–	221	–	1.47
Polybutylene terephthalate	–	60	120	2700	3.4	67	–	1.30
Glass fiber	30	130	2.0	8500	10	205	R 120	1.52
Glass spheres	20	50	4.0	3500	–	70	–	1.45
Wollastonite	50	60	–	6500	–	130	–	–
Aromatic polyesters	–	70	9	2100	22	–	–	1.2
Glass fiber	30	108	3.5	6900	8	–	–	1.44
Modif. polyphenylene oxide (Noryl)	–	68	20	2500	–	130	–	–
Glass fiber	30	120	2.5	8500	–	126	L 108	1.30
Polyphenylene sulfide	–	76	3.0	3400	1.6			1.34
Glass fiber	40	137	–	12000	4.3	218	R 123	1.6
Carbon fiber	30	190	–	17000	2.3	260	–	1.45
Polysulfones	–	75	75	2500	–	175	70	1.24
Glass fiber	30	115	3.0	7600	–	177	90	1.43
Carbon fiber	30	130	2.0	13000	–	177	–	1.37

9.4.11.1 Aromatic polyesters (APE)

These thermoplastics, which represent an interesting addition to the amorphous thermoplastics with high heat deflection temperature such as polycarbonates and polysulfones that are already available, are also supplied in glass-fiber-reinforced form.

As usual, glass fibers increase the stiffness and hardness of APE, whilst impact strength and elongation at break are lowered. Creep behavior is improved.

Mechanical values of glass fiber reinforced formulations are contained in table 17.

9.4.12 Polycarbonate (PC)

Polycarbonates are amorphous, thermoplastic, aromatic polyesters with excellent hardness, transparency, tensile strength and toughness. PC is distinguished in particular by its high toughness at low temperatures. Its high glass transition temperature (T_g) of 150 °C is a major factor as regards technical and mechanical applications, since for practical purposes the material is not suitable for use above T_g. Maximum service temperatures are around 115 to 140 °C. Although PC does not crystallize, oriented states of order are possible above T_g. This orientation is improved by nucleating agents and can also occur at the matrix-reinforcement interface.

Reinforced PC is suitable for parts subject to high static load and which have to meet high requirements in respect of dimensional accuracy, dimensional stability and heat deflection temperature. Among the special reinforced grades, those reinforced with glass fibers have acquired the greatest importance in the electrical, radio, phonographic, television and photographic sectors, as well as for various types of housings.

Fillers and reinforcements for polycarbonates

Glass fibers

After polyamides and polyolefins, polycarbonates are the third most important application for fiber glass reinforcement among the thermoplastics, short fibers being mainly used. Their adhesion can be improved by silane treatment. Alkali-free glass exhibits better adhesion between fibers and matrix. Fiber-glass-reinforced PC is employed either for injection molding or for laminates. A considerable increase in stiffness is generally obtained by lengthening the fibers.

Compared with the unfilled plastic, polycarbonates reinforced with 10 to 40% by wt. fiber glass exhibit the following property improvements:

- Doubling of tensile strength with lower elongation, doubling of the modulus of elasticity by 30% by wt. fiber glass, increased hardness, compressive strength and stiffness,
- lowering of the coefficient of linear thermal expansion to a third, reduction in specific heat, increase in heat deflection temperature by 10 °C,
- PC reinforced with 30% by wt. fiber glass can be substituted for non-ferrous metals or thermosetting plastics,
- higher dimensional stability, creep strength and dimensional stability under heat,
- alternate bending strength is increased and attains the level of non-reinforced polyacetals,
- creep tendency and cold flow are reduced, shrinkage and susceptibility to environmental stress cracking are lower,
- improvement in the electrical values and fire resistance properties.

The reduction in notched impact strength and increase in density must be regarded as disadvantages. Heat deflection temperature is also not appreciably increased by high filler contents.

Fiber glass reinforced blends of PC and ABS, so-called polyblends, have also recently become available.

Graphite fibers: the types introduced in the USA with 10 to 50% by wt. carbon fibers have not yet gained acceptance in the market.

Molybdenum sulfide, PTFE, graphite: with an addition of 15 to 20% by wt., minimum abrasion and an appreciably lower wear behavior is achieved.

Mineral fillers, asbestos: for high stiffness and good flame resistance, up to 30% by wt. asbestos are usually incorporated in PC together with glass fibers.

Wollastonite: improved stiffness.

Metallic powders: aluminum powder is used to increase chemical resistance and electrical conductivity.

In table 17 are compiled the most important values of a number of reinforced polycarbonates.

9.4.13 Polyphenylene oxide (PPO)

PPO is distinguished by high heat deflection temperature, good mechanical properties, stiffness, tensile strength, dimensional stability and favorable electrical properties. PPO and modified PPO (Noryl) can be included among the engineering materials such as PA, PC, ABS, POM etc. Modified PPO (Noryl) is supplied in glass-fiber and carbon-fiber-reinforced forms (20 to 40% by wt. glass fibers) and mainly used for technical parts.

Fiber glass reinforcement enables an increase in tensile strength, modulus of elasticity, flexural and shear strength, dielectric properties and dimensional stability to be obtained. The mechanical properties of reinforced PPO are similar to those of polycarbonate.

Properties of fiber glass reinforced polyphenylene oxide are contained in table 17.

9.4.14 Polyphenylene sulfide (PPS)

Stability to prolonged exposure to heat up to temperatures of 260 °C, high chemical resistance, good mechanical and electrical properties, high dimensional stability, flame resistance and low shrinkage are the outstanding properties of the highly crystalline PPS. Heat deflection temperature of PPS is even better than that of PPO and polyether sulfones.

PPS is employed with 40% by wt. glass fibers for injection molding, mechanical strength, stiffness, dimensional stability under heat and impact strength being increased thereby. A similar influence is exhibited by reinforcement with asbestos fibers. Reinforcement with 30% by wt. carbon fibers additionally enables the surface resistance to be reduced considerably and electrical conductivity to be obtained. With mineral fillers or chalk grades, better processability, lower shrinkage, more accurate demolding and improved surface gloss are obtained. The addition of graphite yields an improvement in stiffness and creep tendency and PTFE a more favorable slip and wear behavior. Properties of reinforced PPS are compiled in table 17.

9.4.15 Polysulfones, polyether sulfones

Polysulfones and polyether sulfones (PES) are heat-resistant thermoplastics that can be used at temperatures above 100 °C. Non-reinforced polysulfones have similar properties to polycarbonates. Reinforcement with glass and carbon fibers brings a considerable improvement in the mechanical, thermal and chemical properties, but the heat deflection temperature is not appreciably increased. Fiber reinforcement imparts a reduction in shrinkage and thermal expansion and the incorporation of 10 to 50% by wt. carbon fibers enables good electrical and thermal conductivity to be obtained. Some of the values of reinforced polysulfones are listed in table 17.

9.4.16 Polymethyl methacrylate (PMMA)

Glass fibers are the material most frequently used for reinforcement of PMMA. On the same principle as with polyester resins, it is thus possible to produce moldings with high mechanical strength, which also exhibit good mechanical values at 100 °C.

Mineral fillers such as chalk, kaolin, heavy spar etc. reduce shrinkage and increase hardness, density, modulus of elasticity and dimensional stability under heat, whereas the mechanical properties in general are not improved. Chalk is used to produce opaque formulations for illuminated advertising panels. Some data on reinforced PMMA are contained in table 15.

9.4.17 Cellulose esters

The cellulose esters most frequently used, such as cellulose acetates, cellulose acetate butyrates and cellulose propionates are employed for injection molding also in glass-fiber-reinforced form. Properties such as resistance to stress corrosion, sound attenuation, dimensional stability under heat and flame retardancy are improved by the addition of glass fibers, whereas impact strength is reduced.

9.5 Summary

Interest in fillers and reinforcing agents in the thermoplastics processing industry has risen considerably since 1974. The main reasons have been the increases in raw material costs and the industry's search for materials with good physicomechanical properties and lower densities than those of metallic materials. However, there are still many problems to be solved from the technological aspect. On the one hand, work is being carried out on the direct incorporation of filler and reinforcement blends with bypassing of the preceding compounding stage, while on the other hand, research activities are focused on the development of inexpensive solutions to technological problems, through combining cheap, spherical mineral fillers with more expensive lamellar or fibrous reinforcements for which a promising future is predicted.

9.6 List of trade names and manufacturers of fillers and reinforcements

For space reasons, only the manufacturers are listed. A detailed list of commercial firms and trade names can be found in *Lückert*: Pigment + Füllstofftabellen, M. and O. Lückert, Laatz, 1977.

US manufacturers are compiled in [29].

The list of trade names and manufacturers given here does not claim to be complete. Details of local agencies should be obtained from the manufacturers.

Trade names	Manufacturers
Alumina trihydrate	
Martinal	Martinswerke GmbH, Bergheim/Erft, D
–	Gebr. Giulini, Ludwigshafen/Rh., D
–	Vereinigte Aluminiumwerke AG, Schwandorf, D
Asbestos	
–	Asbestos Corp. GmbH, Nordenham, D
–	Grace GmbH, Worms, D
Barium sulfate/heavy spar	
Albaryt	Deutsche Baryt Industrie Dr. R. Alberti, Bad Lauterberg, D
Blanc fixe	Kali-Chemie AG, Hannover, D
Blanc fixe	Sachtleben Chemie GmbH, Duisburg, D
Dyobarit	Worlée-Chemie GmbH, Hamburg, D
Glanzweiss	Gebr. Giulini, Ludwigshafen/Rh., D
Schwerspat B	W. Priem & Co., Bielefeld, D
Schwerspatmehl	Sachtleben Chemie GmbH, Duisburg, D

Trade names	Manufacturers
Basalt fibers	
–	Basalan-Isolierwolle GmbH u. Co., Hiltrup, D
Boron filaments	
–	Avco System Division, Lowell (Mass.), USA
–	Composite Technolgy Inc., Broad Brook (Conn.), USA
Calcium carbonates, natural	
Calibrite	Omya GmbH, Köln, D
Calcit	VHZ, Langelsheim, D
Calcit	Ziegler u. Co., Wunsiedel, D
Carborex Durcal Hydrocarb Micromya	Omya GmbH, Köln, D
Microsöhl	Vereinigte Kreidewerke, Dammann KG, Söhlde, D
Millicarb Omya Omyalite	Omya GmbH, Köln, D
Ulmer Weiß	Ulmer Füllstoff Vertrieb GmbH, Ulm, D
–	Nordbayr. Farben- u. Mineralwerke, Hof, D
Amolit, Calcicoll	Alpha Calcit Füllstoff GmbH, Köln, D
Calcium carbonates, precipitated	
Socal	Deutsche Solvay-Werke GmbH, Solingen-Ohligs, D
–	Kreidewerke Schäfer, Diez/Lahn, D
Winnofil	Imperial Chemical House, Millbank, London SW 1, GB
Carbon/carbon black	
–	Cabot GmbH, Hanau, D
–	Columbian Carbon Deutschland, Hamburg, D
–	Degussa, Frankfurt/M., D

Trade names	Manufacturers
Carbon/graphite fibers	
Thornel	Union Carbide Europa S.A., Genf, CH
–	Courtaulds, Courtaulds, GB
–	Sigri Elektrographit GmbH, Meitingen, D
Dolomite	
Dolomit	Ziegler & Co., Wunsiedel, D
Microdol	Norwegian Talc, Bergen, N
Mikro-Dolomit	Blancs Mineraux de Paris, Chatou, F
Nofacal	Nordbayr. Farben- u. Mineralwerke, Hof, D
Polydol	Sopromine, Paris, F
Flake fillers and reinforcements	
Magnesium silicate	Frank u. Henne GmbH, Hof/Saale, D
Plastofil	
Silicon carbide	Carborundum Co., Niagara Falls (N.Y.), USA
Glass flakes	Owens/Corning Fiberglas, Wiesbaden, D
Aluminum diboride	Carborundum Co., Niagara Falls (N.Y.), USA
Glass fibers	
–	Bayer AG, Leverkusen, D
–	Gevetex Textilglas GmbH, Düsseldorf, D
–	Klöckner-Schott Glasfaser GmbH, Dortmund-Mengede, D
–	Owens/Corning Fibreglas, Wiesbaden, D
–	Saint Gobain Industries, Chambery, F
–	Vetrotex International, Genf, CH
Glass spheres, filled, hollow	
–	Potters-Ballotini GmbH, Kirchheimbolanden, D
–	3M, St. Paul (Mn), USA
Graphite powder	
Graphit KS	Lonza AG, Basel, CH
–	Graphitwerke Kropfmühl L. Struve & Co., Hamburg, D
–	M. Neven, Köln, D
Kaolin	
China Clay	English China Clays Sales Company Ltd., St. Austell, Cornwall, GB

Trade names	Manufacturers
China fill	Gesellschaft für Rohstoffveredlung mbH, Hirschau, D
Kaolin	Gebr. Dorfner OHG, Hirschau, D
Speswhite	English China Clays Sales, St. Austell, Cornwall, GB
Supreme	English China Clays Sales
Metals and metallic oxides	
Magnesium oxide	Lehmann & Voss & Co., Hamburg, D
Metallic oxides	C. Schlenk, Nürnberg, D
Metallic powders	G.M. Langer & Co., Ritterhude, D
Metallic powders	Liebau Chemie, Hamburg, D
Zinc oxides	VHZ, Langelsheim/Harz, D
Zinc oxides	Zinkweiß-Handelsgesellschaft, Oberhausen, D
Zinc dust	Lindur GmbH u. Co. KG, Köln, D
–	Stolberger Zincoli GmbH, Stolberg, D
Mica	
Concord Mica	Sciama SA, Paris, F
English Mica Glimmer	The English Mica Comp., Stamford (Cn.), USA
Micro-Mica	Norwegian Talc, Bergen, N
Mineral fillers, others	
Silicon carbide	Carborundum Co., Niagara Falls (N.Y.), USA
Molybdenum sulfide	DOW-Corning Corp., Brüssel, B
Zinc sulfide	Sachtleben Chemie GmbH, Duisburg, D
Organic fillers and reinforcements	
–	Schwarzwälder-Textilwerke Heinrich Kautzmann GmbH, Schenkenzell, D
Silica, natural	
Aktisil	R. Hoffmann & Söhne KG, Neuburg, D
Coloritquarz	Gesellschaft für Rohstoffveredlung mbH, Hirschau, D
Cristobalit	Quarzwerke GmbH, Frechen, D
Elmin	Quarzwerke GmbH, Frechen, D

Trade names	Manufacturers
Silica, natural	
Gloxil	Gesellschaft für Neuburger Kieselweiss, Neuburg, D
Grenette	Sanson S.A., Paris, F
Crystal quartz	Dörentruper Sand- und Thonwerke GmbH, Dörentrup, D
Crystal quartz sand	Dudinger Glassandwerk Bock u. Co., Dudingen, D
Quartz flour	Quarzwerke GmbH, Frechen, D
Quartz flour	Gesellschaft für Rohstoffveredlung mbH, Hirschau, D
Sigrono	Westdeutsche Quarzwerke, Dr. Müller GmbH, Dosten, D
Sikron	Quarzwerke GmbH, Frechen, D
Silbone	Quarzwerke GmbH, Frechen, D
WQO Quartz flour	Westdeutsche Quarzwerke
Silica, synthetic	
Aerosil	Degussa, Frankfurt/M., D
Cab-O-sil	Cabot GmbH, Hanau, D
Syloid	Grace GmbH, Worms, D
Silicate spheres	
Armospheres	Georg M. Langer u. Co., Ritterhude, D
Synthetic fibers	
–	Schwarzwälder Textil-Werke Heinrich Kautzmann GmbH, Schenkenzell, D
Talc	
Cascade, Mistron	Chemag, Frankfurt/M., D
Micro-Talc	Norwegian Talc, Bergen, N
Mikro-Talkum	J. Scheruhn, Hof, D
–	Naintsch Mineralwerke GmbH, Graz, A
OXO, OOS	S.A. des Talcs de Luzenac, Luzenac-sur-Ariège, F
Whiskers	
–	Electro Schmelzwerke, München u. Kempten, D
–	Lonza AG, Basel, CH
–	Silag, New York, USA

Trade names	Manufacturers
Wollastonite	
Casiflax	Quarzwerke GmbH, Frechen, D
Tremin	Quarzwerke GmbH, Frechen, D
Wollastonite 100, 500	Vereinigte Harzer Zinkoxyde, Langelsheim, D

9.7 References

[1] Techn. report no. 199, Omya GmbH, Köln/FRG.
[2] *H.G. Elias:* Makromoleküle, Hüthig and Wepf, Basel/CH, 1971, p. 360.
[3] *U. Zorll:* Gummi Asbest Kunststoffe 7 (1977) p. 436.
[4] *L.E. Nielsen, P.E. Chen:* J. Mater. Sci. 3 (1968) 2 p. 38.
[5] *W. Voigt:* Lehrbuch der Kristallphysik. B.G. Teubner, Leipzig, 1910, p. 38.
[6] *A. Reuss:* Z. Angew. Math. Mech. 9 (1929) p. 49.
[7] *R. Meyer Rosen:* J. Coat. Technol. 50 (1978) 644, p. 70.
[8] Gummi Asbest Kunstst. 8 (1975) p. 533.
[9] Plastverarbeiter 28 (1977) p. 57.
[10] Plastverarbeiter 28 (1977 p. 59.
[11] *J. Becht, H. Kausch, G. Schettler:* Report from Battelle Institut e. V., Frankfurt/M., report no. BF-V-3092-1.
[12] *J. Becht, H. Kausch, G. Schettler:* Report no. BF-V-30.921-2 from Battelle Institut e. V., Frankfurt/M., T23–T52.
[13] Techn. report from Omya GmbH and Plüss-Staufer AG.
[14] Techn. report from Omya GmbH, Plüss-Staufer AG, and report no. VF-V-30.921-2 from Battelle Institute, T95, 98 and 101.
[15] *J. Becht, H. Kausch, G. Schettler:* Report no. BF-V-30.921-2 from Battelle Institut e.V., Frankfurt/M., T53, 54, 107, 108, 110 and 111.
[16] *J. Becht, H. Kausch, G. Schettler:* Report no. BF-V-30.921-2 from Battelle Institut e.V., Frankfurt/M., T56, 57, 58, 62, 65, 66, 68, 71, 75, 77, 78, 80 and 81.
[17] *J. Becht, H. Kausch, G. Schettler:* Report no. BF-V-30.921-2 from Battelle Institut e.V., Frankfurt/M., T83, 84, 86, 87, 89, 90, 93, 94 and 104; Gummi Asbest Kunstst. 9 (1974) p. 752; Plastverarbeiter 28 (1977) p. 58.
[18] *J. Becht, H. Kausch, G. Schettler:* Report no. BF-V-30.921-1 from Battelle Institut e.V., Frankfurt/M., pp. 97/99.
[19] *J.G. Marsden* et al.: Br. Polym. J. 11 (1979) p. 199.
[20] *H. Ishida* et al.: Polym. Eng. Sci. 18 (1978) p. 128.
[21] *B. Aldes:* Chem. Tech. 7 (1977) p. 766.

[22] *P.J. Orenski* et al.: Ann. Tech. Conf. Reinforced Plastics/Composite Institute SPI (1973) Section 3-E, p. 1.
[23] *S.J. Monte* et al.: Ann. Tech. Conf. SPI (1980) p. 486.
[24] *S.J. Monte* et al.: Ann. Tech. Conf. Reinforced Plastics/Composite Institute SPI (1979) Section 16-E, p. 1.
[25] *R.D. Deanin, N.R. Schott:* Adv. Chemistry Ser. 134, 1974.
[26] *W.C. Wake:* Fillers for Plastics, The Plastic Institute, London, Butterworth & Co. Ltd., London, 1971.
[27] *W.V. Titow, B.J. Lanham:* Reinforced Thermoplastics, Applied Science Publishers Ltd., London, 1975.
[28] *Y.S. Lipatov:* Physical Chemistry of Filled Polymers, Rubber and Plastics Association of Great Britain (RAPRA), Shawbury, 1977.
[29] *H.S. Katz, J.V. Milewski:* Handbook of Fillers and Reinforcements for Plastics, Van Nostrand Reinhold Co., New York, 1978.
[30] *H. Domininghaus:* Gummi Asbest Kunstst. 5 (1979) 32, p. 294.
[31] *K. Dinges:* Kautsch. Gummi Kunstst. 32 (1979) pp. 748/756.
[32] Plastics Focus 12 (1980) p. 20.
[33] *H.P. Schlumpf, W. Bilogan:* Kunststoffe 71 (1981) pp. 820/823.

10 Colorants for Thermoplastics

Dr. E. Herrmann, Leinfelden-Echterdingen, and
Dr. Winfried Damm, BASF Farben + Fasern AG, Stuttgart,
Federal Republic of Germany

Part I: Pigments and Dyes

Dr. E. Herrmann

10.1 Basic physicochemical principles

10.1.1 Definitions and basic information

In discussing colorants for plastics it is important to make a distinction between dyes and pigments (cf. DIN 55944), the latter having by far the greater significance with regard to the plastics industry. Under the prevailing processing conditions, pigments are virtually insoluble in plastics, whereas dyes are soluble (cf. DIN 55949).

The term pigment is linked with a specific particle size range (~0.01 to ~1 μm), which differs from product to product and during production is capable of being influenced within certain limits. In the definition of pigment particles (see DIN 53206), a distinction is made between primary particles or single particles, aggregates and agglomerates (figure 1).

Because of their extremely small particle size, *primary particles* such as those usually obtained during production, exhibit a pronounced tendency to combine. This massing together in a single plane results in the primary particles forming *aggregates,* which thus possess a smaller surface than that corresponding to the sum of the surfaces of their primary particles. The massing together of primary particles and/or aggregates at their corners and edges results in the formation of *agglomerates,* the total surface of which deviates only slightly from the sum of the individual surfaces. If, without going into detail, one speaks of pigments particle size, one is usually referring to aggregates such as are present as a result

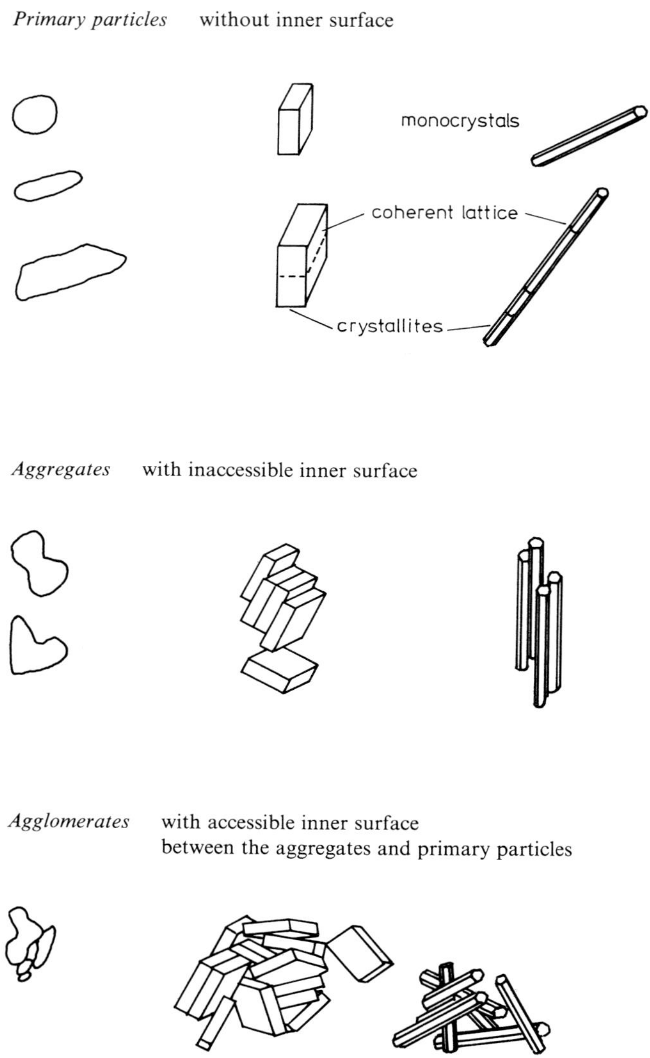

Fig. 1. Elements of a pigment.

of coloration. Their size and size distribution are responsible for the coloristic properties.

In pigments in powder form, aggregates are always massed together to form agglomerates. During the coloration process, the latter have to be broken down, wetted by the plastic and distributed homogeneously. These processes, which take place concurrently, are known as *dispersion*. By contrast, coloration with dyes involves a dissolving process, as a result of which the dye is present in molecular solution. Solubility must always be related to the respective plastic and the processing conditions.

Unlike inorganic pigments, individual organic pigments are occasionally found to be not wholly insoluble. This applies in particular to pigments with a simple structure and low molecular weight. Apart from the chemical structure of an organic pigment, other factors of significance are the nature of the plastic to be colored and the temperature at which the latter is processed. An example of such behavior is the coloration of polystyrene with C.I. Pigment Red 88 [9], which dissolves to an increasing extent as the processing temperature rises, i.e. it behaves like a dye (figure 2).

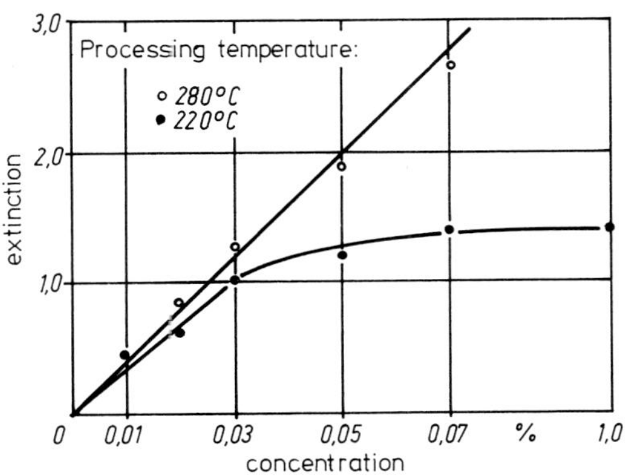

Fig. 2. Behavior of C.I. Pigment Red 88 in general purpose polystyrene [9].

For practical purposes, dyes can be adequately described by their chemical structure. Pigments of the same chemical constitution can, however, be manufactured in various crystalline forms, which may differ considerably from each other in their performance and coloristic properties. Typical examples of this are the important titanium dioxide, lead chromate, phthalocyanine blue and quinacridones. The pigment manufacturer also has the possibility of improving the service and performance properties by aftertreatment (coating) of the surface of the pigment particle. This applies in particular to increasing the ease of dispersion and improving the resistance to light, weathering and chemicals.

Since it is necessary during dispersion to overcome high cohesive forces between the pigment particles [10], which in plastics coloration may require considerable time and energy, various types of pigment preparations are available on the market [11, 12] and supplied in the following forms:

- *Liquid pigment concentrates:* the pigment is present in a completely dispersed form. It is thus only necessary to ensure that it is distributed homogeneously in the respective plastic. A typical example are plasticizer pastes for the coloration of PVC.
- *Solid pigment concentrates:* the pigment is fully dispersed in a solid carrier. Suitable carriers are certain resins, the plastic to be colored or another, possessing adequate compatibility with the intended medium. Here, too, a breakdown of the agglomerates is no longer necessary, but only a homogeneous distribution. This, however, requires more time than with liquid preparations. The particle size (as supplied) of the preparation can also have a certain significance in this connection.
- *Special pigment mixtures:* these are defined as combinations of readily dispersible, inorganic pigments and/or fillers with organic pigments of lower dispersibility. In this instance the organic agglomerates are already broken down and the aggregates formed are firmly attached to the inorganic particles. This type of special pigment mixture can be described as pre-dispersed. They are easier to process than organic pigments and in this respect are comparable with inorganic pigments. A detailed description is given in a subsequent section.

10.1.2 Optical principles

The color change resulting from the coloration of plastics is based on the wavelength-dependent absorption and/or scattering (reflectance) of light. Dyes can only absorb light and not scatter it, since the physical pre-requirement for scatter – a certain minimum particle size – does not exist in the case of dyes in molecular solution; these types of colored articles are, therefore, transparent. In so far as the transparency is attributable to the dye, complete absorption of the light results in black shades, selective absorption in colored shades.

The optical effect of pigments may in the same way be based on light absorption. If, however, the refractive index of the pigment differs appreciably from that of the plastic and, if a specific particle size is present, reflectance – the throwing back of light – takes place; the initially transparent plastic becomes white and opaque, or if selective light absorption takes place at the same time, colored and opaque. No reflectance occurs when the particle sizes are very small and none or only very little if they are very large. Optimum particle size for titanium dioxide in plastics is around 0.3 μm. In all colored pigments that selectively absorb and reflect, the shade is influenced by particle size. Examples of non-reflecting inorganic color pigments are transparent iron oxides, due to their small particle size, and ultramarine blue, due to its refractive index being largely identical with that of plastics.

For detailed information, reference must be made to the literature [1 to 5, 13 to 15].

10.1.3 General properties of the system colorant/plastic

Knowledge of pigment properties is of considerable value to the pigment user in anticipating behavior in plastics, but ultimately the only way of obtaining wholly meaningful information is to carry out a test in the plastic and under the prevailing conditions, i.e. determining a property of the system [16, 17].

10.1.3.1 Thermal stability

During the manufacture or processing of colored plastics, heat can be the cause of coloristic changes. The term "temperature stability", which is often used, is inaccurate since it is not only a question of the temperature level but also the duration of exposure and the accepted shade tolerance. It is more advisable, therefore, to use the terms "thermal stability" or "shade-temperature-time" stability. Only the performance test is meaningful. The time included in values such as "5 minutes/220 °C in unplasticized PVC", determined in the performance test, must contain a safety factor, since the material is often likely to be reprocessed. The thermal stability of a number of inorganic pigments is so high they always satisfy the requirements of thermoplastics. As examples of such products, mention can be made of titanium dioxide, nickel titanium yellow, chrome titanium yellow, iron oxide red, chrome oxide green, cadmium pigments, cobalt pigments, carbon black.

10.1.3.2 Lightfastness

Just like heat, light is a form of energy which can be a cause of degradation. Lightfastness is tested in accordance with DIN specifications (DIN 53389, ISO 4582–1980, ASTM E 187 as well as [6, 7]), the blue or wool scale serving as the standard of comparison. It constitutes a cumulative energy measuring device and is described in DIN 54003. It consists of 8 differently colored textile fabrics that differ in their resistance to the effects of light. "1" is the lowest value, the next higher digit corresponding in each case to twice the fastness level of the one below, i.e. double the length of time at the same level of light exposure, "4" thus being equivalent to 8 times the fastness level of "1". The highest value is "8", testing not normally being carried out beyond this level. In addition to these numerical ratings, use is also made of verbal descriptions. However it is inadvisable to employ them, since, for example, the lightfastness rating "5" which is often inadequate, is already termed as "good". Figure 3 shows the relationship between irradiated light energy (MLxh) and wool scale values.

As with thermal stability, the lightfastness of a colorant is a property of the system colorant/plastic. It can only be meaningfully tested in the

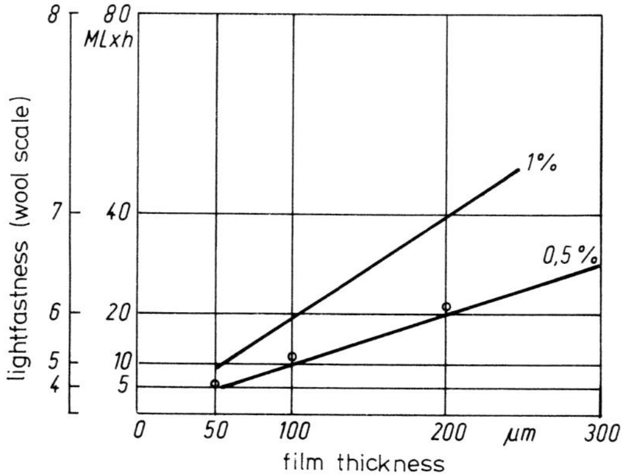

Fig. 3. Lightfastness of pigmented polyethylene films as a function of thickness [16].

Table 1. Weather resistance of unplasticized PVC, Ba/Cd stabilized. Change in NBS units [16].

Time, months	Unpig-mented	0.2% u.v. stabilizer	0.5% Phthalo-cyanine blue	0.5% Phthalo-cyanine green	2% Titanium dioxide (rutile)
18	6	0	0	0.5	0
24	12	2	1.5	3	12
30	–	4	3	6	–
36	–	9	9	12	–
48	–	12	12	–	–

system in question, since the considerable influence exerted by the various binders and plastics result in different fastness values being obtained. Similar to u.v. absorbers (see table 1), certain pigments can perform a protective function in plastics, others behave inertly, while individual products can accelerate the photooxidative degradation of plastics. The anatase form of titanium dioxide, for example, is known to do so (see section 10.2.1.1).

Determination of lightfastness is carried out in accordance with DIN 53888 (ISO 877–1976, ASTM E 187) in daylight behind glass. Accelerated testing methods employing artificial light are also customary. In these, the light spectrum is almost the same as that of daylight, but

Table 2. Lightfastness of pigment combinations in accordance with the blue wool scale [16].

	pure	1:1	1:20
C.I. Pigment Red 49/Mn	7 to 8		
with TiO_2(Rutile)		6	4
with C.I. Pigment Red 104 (Molybdate Red)		7	7
C.I. Pigment Red 52/Mn	7 to 8		
with TiO_2 (Rutile)		5 to 6	3 to 4
with C.I. Pigment Red 104 (Molybdate Red)		7	7
C.I. Pigment Red 88	8		
with TiO_2 (Rutile)		7 to 8	7
with C.I. Pigment Red 104 (Molybdate Red)		7 to 8	7

the light intensity is appreciably higher (DIN 53389; ISO 4892–1981; ASTM D 1494, D 2565, G 23). In transparent colorations, lightfastness is related to colorant content, assuming uniform layer thickness (figure 3). In combinations of mainly transparent and in particular organic colorants with white pigments (pastel shades), lightfastness decreases as the white pigment content rises [18]. It is especially important to avoid tinting with small quantities of colorants with poor lightfastness, since very marked coloristic changes can otherwise occur at a very early stage. The fastness levels of colored pigments in combination with a white pigment or an inorganic pigment differ appreciably (see examples in table 2).

10.1.3.3 Weather resistance

Exposure during the weathering test is influenced by a whole range of factors. Apart from visible light, the most important factors that can be mentioned are: high-energy ultraviolet radiation, heat, moisture, impurities in the atmosphere e.g. sulfur dioxide, hydrogen sulfide etc.

Weather resistance can only be determined with complete reliability in outdoor exposure tests, although even then the results strictly apply only to the climatic conditions prevailing locally. The test data should include the exact period of the test (not only the duration), the location (where possible stating climatic data) and information on the changes observed (see DIN 53386; ISO 4607–1980, 4582–1980; ASTM E 187, E 1435). A reference standard such as in the lightfastness test is not available. A test report might read as follows: "During exposure in Stuttgart on the periphery of the city in an industrial atmosphere, the specimen inclined at an angle of 45° towards the south exhibited no change in gloss over a period of 2 years (01.06.78 to 01.06.80), but a change in shade corresponding to step 4 in the grey scale".

The grey scale in accordance with DIN 54001 (ISO/R 105/I–1959) is an aid to describe the change in shade. Step 5 corresponds to no change, while in tests of this type, step 3 to 4 is generally still regarded as permissible.

Using a similar procedure and the same equipment as when determining lightfastness, accelerated tests are also carried out for weather resistance. The program (light/dark period, temperature changes, water spraying) can be adapted to the respective climate (see DIN 53387; ISO 4892–1981; ASTM D 1494, D 2565, G 23).

10.1.3.4 Migration

Migration is described as the mobility of colorants due to incomplete insolubility; no migration is observed with inorganic pigments; migration does not include plate-out, since the latter is not due to solubility.

A distinction is made between three types of migration:

- *Solvent bleeding,* i.e. the release of coloring constituents into an organic liquid. Dyes and some organic pigments exhibit this type of solvent bleeding, which apart from the plastic/colorant system, also depends on the respective solvent.
- *Contact bleeding* in a solid/solid system (e.g. stacked plastics films). The conditions basically correspond to those in solvent bleeding; here, too, it is not only a question of the plastic/colorant system, but also of the composition of the medium (plastic) coming in contact with it. This phenomenon is observed in particular with dyes and some organic pigments.
- *Blooming,* i.e. the migration of pigment particles during storage at room temperature from the plastic/pigment system to the surface; in extreme cases a bronze-like haze is observed. The reason is the formation of saturated solutions during the thermal processing of plastics and the subsequent crystallization of colored constituents from the solutions which at lower temperatures are then supersaturated. This effect can appear with a time-lag of days, weeks or months; blooming can occur spontaneously if the surface is polished or scratched, as well as through a fingerprint on the surface. If no supersaturated solutions are formed, as is generally true with plastics, no blooming takes place; the phenomenon is thus confined to a few organic pigments.

In pigmented plastics whose glass transition temperatures are above the storage temperature, no migration generally occurs, since in this instance the colorants are greatly restricted in their mobility (e.g. crystal polystyrene).

10.1.3.5 Plate-out

This refers to the deposition of pigment-containing plastics or plastics constituents on mold or die surfaces during the processing of plastics melts. With calendered films, plate-out on rolls causes surface defects and in extrusion dies can lead to variations in cross-section and finally even result in rejects when the deposited material sloughs off from the surface. This phenomenon is caused by inadequate compatibility of an ingredient (e.g. stabilizers or lubricants in PVC). Exudation of the ingredient already takes place at the processing stage, portions of the pigment being carried along at the same time. That this is not a genuine case of blooming, i.e. a migration process, is shown by the fact that inorganic pigments and fillers, which are known to be fast to blooming, likewise exhibit plate-out [19].

Depending on circumstances, plate-out phenomena can be eliminated by a variety of different methods:

- Improving the dispersing conditions,
- adding surface-active substances,
- modifying the formulation (stabilizers, lubricants),
- changing the pigment.

10.1.3.6 Influencing the rheological properties

It is well-known that the flow behavior of plastics melts can be altered by means of additives. Soluble dyes have virtually no influence on melt viscosity, when added in small quantities, but may have significant effects when used at higher concentrations. In the same way as fillers, pigments are able to increase viscosity. At equal parts by weight, the more finely dispersed pigments are generally more effective than fillers. However, because of the low concentrations in which they are used, pigments have been observed to influence melt viscosity only in the case of thin-walled articles which for coloristic reasons require a higher level of pigmentation.

10.2 Discussion of the colorants

10.2.1 White pigments

10.2.1.1 Titanium dioxide

Titanium dioxide is by far the most important of the white pigments (cf. table 3) and possesses all-round suitability. A characteristic feature of this pigment group is its ability to accelerate by photochemical activity the oxidative degradation of plastics surfaces and paints. The surface disintegration of the binder results in the phenomenon known as chalking. Of the two crystalline forms of titanium dioxide (chemical formula TiO_2) commonly used as pigment, only the rutile form is of significance to the coloration of thermoplastics. The oxidative action of the second

Table 3. Worldwide production capacities of synthetic inorganic pigments, [*L.K. Schmidt*: Chem. Ind. 34 (1982) pp. 452–456].

Pigment	Capacity [t/year]	Share [%]
Titanium dioxide	2700000	74.4
Iron oxides	530000	14.6
Carbon blacks	220000	6.0
Chrome yellows	90000	2.5
Molybdate reds	35000	1.0
Chrome oxide greens	35000	1.0
Mixed-phase pigments	10000	0.3
Cadmium pigments	9500	0.2
Total	3630000	100

crystalline form, anatase, on its embedding media is far greater than that of rutile. With rutile, this oxidation tendency can be reduced by selective treatment of the surface. The weather resistance of rutile grades of this kind is thus wholly adequate for practical purposes.

Anatase is recommended only if its lower degree of hardness, which in flexible articles also results in lower mechanical stress on the substrate, is of importance, e.g. car tires, delustering of polyamide fibers, abrasion of fiber processing equipment (thread guides etc.).

10.2.1.2 Other white pigments

Zinc white (ZnO) is a basic product that can neutralize acid degradation products, but is of no importance as a pigment in plastics.

Zinc sulfide (ZnS) is only seldom used in thermoplastics; the same applies to *lead white,* $2PbCO_3 \cdot Pb(OH)_2$, as a pigment.

10.2.2 Black colorants

10.2.2.1 Carbon blacks

Among black colorants, the carbon blacks enjoy by far the greatest importance (cf. table 3). They must be distinguished from the carbon blacks used as reinforcing fillers. The latter are much cheaper and produced in considerably larger quantities. Depending on the production process, they are characterized by different organic groups (carboxyl, lactone, phenol, lactol groups etc.), which influence important performance properties such as shade and wettability. A further criterion is particle size, the finer the particles, the smaller the already low degree of reflectance and the deeper the shade.

Carbon blacks possess high u.v. absorption, which among other things helps to maintain the elongation at break of polyethylene during natural weathering [25]. When exposed to severe thermal stress over a relatively short period, polyethylene yields volatile acidic oxidation products, a reaction that can be greatly reduced by adding increasing quantities of carbon black; in this respect, fairly fine pigment particles are more effective than coarser ones [26]. Commercial phenolic antioxidants are, however, considerably more effective at much lower concentrations. The high surface forces that result from the fineness of the particles (carbon blacks are more finely divided than most organic pigments) are the cause of difficulties during dispersion. In many cases this results in preference being given to employing pigment concentrates. Special grades of carbon black are recommended for plastics. The functional organic groups present at their surface renders them more readily dispersible.

10.2.2.2 Lamp blacks

Lamp blacks are also carbon pigments which are distinguished from the carbon blacks by a higher ash content and consequently a lower tinctorial strength. They enjoy less significance than the carbon blacks.

10.2.2.3 Oxidic black pigments

Iron oxide black, the oxide of bivalent and trivalent iron, Fe_3O_4, is, in spite of its limited thermal stability in thermoplastics, still suitable for use if exposure lasts for only a short period. Under certain conditions it accelerates the decomposition of PVC [21].

Spinel black is a mixed-phase pigment with the formula $Cu(Cr,Fe)_2O_4$, which, in spite of adequate thermal stability, has only minor importance for thermoplastics.

10.2.2.4 Organic black colorants

Helio Fast Black IR can be mentioned for its remarkably high reflectance in the infrared range.

Thermoplastic Black M consists of a blend of polycyclic dyes whose lightfastness is, however, only satisfactory in full shades. Its fields of application are plastics, the glass transition temperatures of which are above normal temperature, e.g. unmodified polystyrene, polymethyl methacrylate and unplasticized PVC.

10.2.3 Inorganic color pigments

They are generally characterized by their complete insolubility in plastics and thus do not exhibit migration on account of partial solubility. Their thermal stability may be high, but this is not necessarily always true. In the following discussion, main importance is attached to the chemical structure, individual data are contained in tables 4 to 7. Worldwide production capacities for inorganic color pigments are listed in table 3.

10.2.3.1 Oxidic color pigments

If the metals are present in their oxides at the highest valence state, the corresponding pigments have very high thermal stability, since no further oxidation is possible. If metals are present in pigments at a low stage of oxidation, e.g. bivalent iron in standard iron oxide black and brown, the thermal stability of such pigments is limited. Particle agglomerations that can lead to coloristic changes are in practice not an important factor in plastics coloration with this kind of pigments.

C.I. Pigment Red 101 (iron oxide red, Fe_2O_3) is supplied in an extensive range of colors; it is a very economically priced product, the use of which is restricted by its not very pure shade. The same product is available in a transparent form that results from its small particle size. Although it is very much purer in shade, this advantage is lost in blends with white pigments.

C.I. Pigment Yellow 53 is a titanium dioxide/nickel oxide/antimony oxide mixed-phase pigment [$(Ti,Ni,Sb)O_2$] and is distinguished by very high fastness properties, including excellent weather resistance, and general suitability. Its fields of application are limited by the inadequate purity of shade and low tinctorial strength.

C.I. Pigment Yellow 118 has a similar structure to that of C.I. Pigment Yellow 53, but contains chrome instead of nickel.

C.I. Pigment Yellow 119 (zinc iron pigment) is a spinel based on the oxides of iron, zinc and titanium and in its applications can be rated similarly to the titanium oxide mixed-phase pigments.

C.I. Pigment Green 17 (chrome oxide green, Cr_2O_3) is similar to iron oxide red, a standard pigment with all-round suitability, but with a not very pure shade.

C.I. Pigment Green 19 (cobalt green) is a spinel of the cobalt-aluminum-titanium-nickel-zinc oxide system. A number of shades are available, they are purer than that of chrome oxide green.

C.I. Pigment Blue 28 (cobalt blue) is a spinel of the cobalt oxide/aluminum oxide or chrome oxide system, in its performance properties it corresponds to cobalt green and is likewise produced in several shades.

C.I. Pigment Brown 6 [iron oxide brown, $(FeO)_x \cdot (Fe_2O_3)_y$] has limited thermal stability due to its bivalent iron content.

C.I. Pigment Brown 29 (chrome iron brown) is a mixed-phase pigment based on oxides of iron and chrome, which here are present in their highest oxidation states. This accounts for its high thermal stability. The product is distinguished by high tinctorial strength and opacity.

10.2.3.2 Pigments with hydroxyl content

C.I. Pigment Yellow 42 (iron oxide yellow, $Fe_2O_3 \cdot xH_2O$, FeO(OH), when $x = 1$) is an oxide-hydroxide of trivalent iron; the possible elimination of water is the reason why its thermal stability is not always adequate. The product is supplied in various shades and like C.I. Pigment Red 101 (iron oxide red) is also produced in a transparent form. Because of its needle-shaped structure it tends to be dichroic, i.e. the shade changes depending on the direction from which it is viewed. In application methods that influence the direction in which pigments are arranged, optical disturbances (different shades in different directions) become noticeable. This is observed, for example, in calendered films.

10.2.3.3 Sulfidic pigments

Since hydrogen sulfide is a weak acid, the acid resistance of sulfidic pigments is low. Some products can oxidize under the influence of atmospheric oxygen, thus forming sulfates and undergoing coloristic changes.

C.I. Pigment Red 108 [cadmium red, Cd(S, Se)] is a cadmium/sulfide selenide. Its color scale ranges from orange to dark red, purity of shade being extremely high [24]. Light and weather resistance satisfy the highest requirements, the darker types having the better weather resistance (highest oxidation stability) due to their higher selenium content.

C.I. Pigment Orange 20 (cadmium orange) corresponds chemically to C.I. Pigment Red 108, but because of its lower selenium content possesses lower weather resistance. This is, however, often adequate for many applications.

C.I. Pigment Yellow 37 [cadmium yellow, (Cd, Zn)S] is a cadmium or cadmium/zinc sulfide. Like cadmium red it is distinguished by all-round suitability in all plastics; however, it possesses only limited weather resistance, since oxidation can lead to the formation of sulfate. With cadmium red, this is prevented by the incorporation of selenium. Full shades are more weather-resistant than combinations with white or blue pigments. The range of shade extends from greenish to reddish shades.

C.I. Pigment Blue 29 (ultramarine blue, $Na_{6-8}Al_6Si_6O_{24}S_{2-4}$) is a sodium/aluminum silicate with polysulfide-like sulfur in the crystal lattice, which is produced in different shades, depending on composition and particle size. It is a transparent pigment with general suitability for use in plastics. Contradictory results have been obtained with regard to weather resistance. These must also be considered in relation to the aftertreatment customary with this pigment.

10.2.3.4 Chromates

Lead chromate exists in three different crystalline forms, of which the monoclinic and the rhombic are yellow and the form presumed to be tetragonal is red [20].

C.I. Pigment Yellow 34 (chrome yellow, $PbCrO_4$) is the monoclinic crystalline form of lead chromate, stable at normal temperature. This crystalline form has a tendency to form needles and causes dichroism, similar to iron oxide yellow. By using special manufacturing processes it has been possible to prevent the pronounced longitudinal growth of the crystals and, instead of needles, to obtain a cubical habit and thus avoid dichroism. Under the influence of light, lead chromates tend to change their color towards the dark, green direction. This can be prevented by different forms of surface treatment (stabilization). Thermal stability is also improved thereby (see table 4). Families of pigments with wide differences in their resistance to light, weather and heat are thus obtained in color ranges extending from greenish to reddish yellow. The top products possess high weather resistance, they also promote gloss retention at the plastics surface. At the same time, they are distinguished by being relatively inexpensive to use.

C.I. Pigment Red 104 [molybdate red, $Pb(Cr,Mo,S)O_4$] is the crystalline form of lead chromate presumed to be tetragonal. Technically it is produced by incorporating molybdenum and sulfur in the crystal lattice of lead chromate. Problems of poor lightfastness are similar to those in chrome yellow and are corrected in a similar manner. As regards performance properties, these pigments are very largely comparable to chrome yellow, also as regards the protection of film surfaces (gloss retention) in many plastics.

C.I. Pigment Green 48 (chrome fast green) represents a combination of chrome yellow and phthalocyanine blue or green. Its properties result from its components.

10.2.4 Organic color pigments

Inorganic and organic color pigments have approximately equal importance as regards the coloration of plastics. The former are generally employed when high processing temperatures are necessary and good opacity is called for. Organic pigments are mainly used when brilliance, high tinctorial strength and transparency are of decisive importance in a given application.

In contrast to inorganic pigments, a certain degree of solubility of organic pigments in plastics cannot always be precluded, as a result of which migration can occur (see section 10.1.3.4). The degree of solubility of an organic pigment is, among other things, dependent on the type of substituents of the base molecule and on the molecular weight of the colorant to the extent that solubility drops as molecular weight increases.

When large articles are manufactured from high density polyethylene, the use of individual organic pigments can result in distortion, which is ultimately attributable to the nucleating effect of such pigments.

10.2.4.1 Azo pigments

Characteristic of this largest class of compounds is the azo (—N=N—) group. A distinction is made between *monoazo* pigments, which contain only one azo group, *disazo* pigments, containing two azo groups, *disazo condensation* pigments, manufactured by a special process, and *salts of azo color acids (color lakes)*. The fastness properties of azo pigments differ greatly and range from medium to very high values. The benzimidazolone and disazo condensation pigments can be mentioned as examples of the latter. With the exception of color lakes, chemical resistance is very good, since readily saponifiable groups are absent. The color scale ranges from greenish yellow to dark bordeaux. Yellow azo pigments generally contain the acetoacetanilide group, the red pigments the naphthol group. Examples of structures are shown in figures 4 and 5.

Monoazo pigments: C.I. Pigment Red 171, 175, 176, 185; C.I. Pigment Yellow 97, 120, 151, 154.

These are yellow and red colorants (see tables 4 and 5).

Disazo pigments: C.I. Pigment Yellow 13, 81, 83.

Disazo pigments (if they are not condensation pigments) lie in the yellow range (cf. table 4). Improvement in solvent resistance, and hence in migration resistance, as well as tinctorial strength, has been achieved by increasing the molecular weight.

Disazo condensation pigments: C.I. Pigment Red 144, 166; C.I. Pigment Yellow 93, 94, 95, 128; C.I. Pigment Brown 23.

The color range is larger than with normal disazo pigments and includes red and brown pigments (see tables 4, 5 and 7). A special manufacturing process has made it possible to develop products with even higher molecular weights, which are consequently distinguished by their good migration resistance.

Salts of azo color acids (color lakes): C.I. Pigment Red 48/1, 48/2, 48/3, 48/4, 53/1, 57/1.

These comprise alkaline earth and heavy metal salts of soluble color acids, the salts of which are largely insoluble and are used as pigments. These are economically priced products, chiefly in the red range. The metals change the shade from orange to blue-red in the sequence barium, strontium, calcium, manganese. With manganese compounds there is a risk of the decomposition of polyolefins being accelerated, combined with the possibility of odor nuisance.

10.2.4.2 Non-azo pigments

This section covers a number of classes of compounds with widely differing structures, generally of the polycyclic type, which possess high fastness properties. Individual products in this group represent metallic complexes. The classes with greatest commercial importance are discussed below (structures see figure 6).

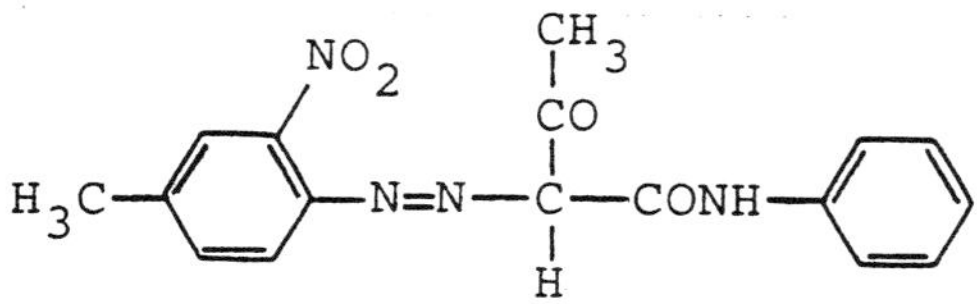

Monoazo pigment

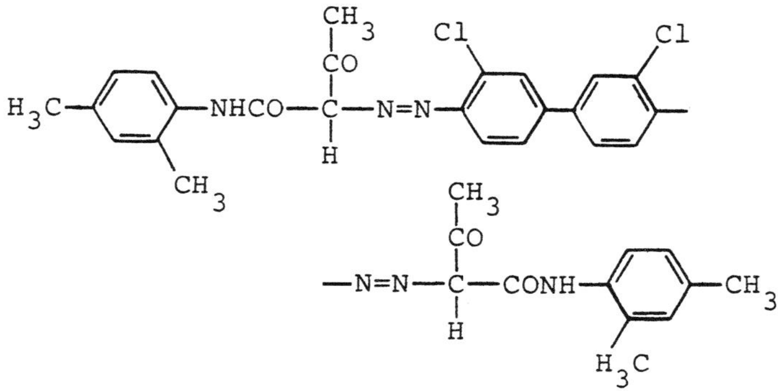

Disazo pigment

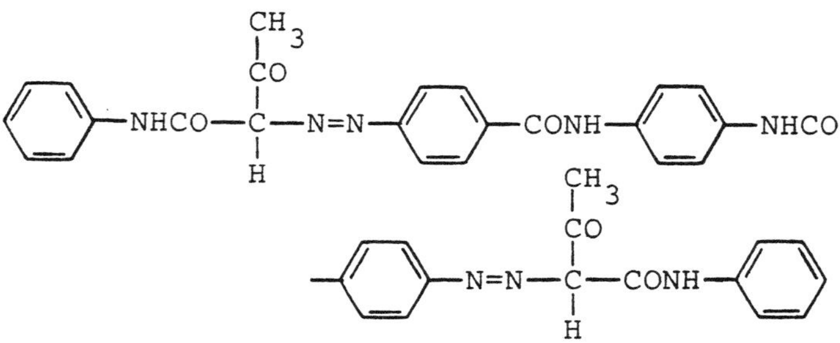

Example of a yellow azo condensation pigment

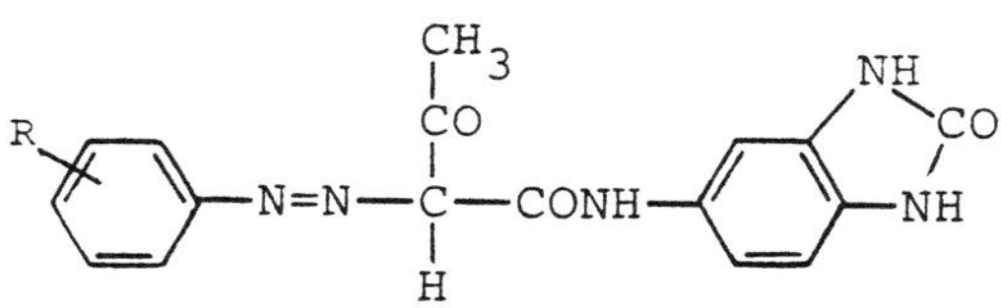

Example of a yellow azobenzimidazolone pigment

Fig. 4. Examples of the structures of yellow azo pigments.

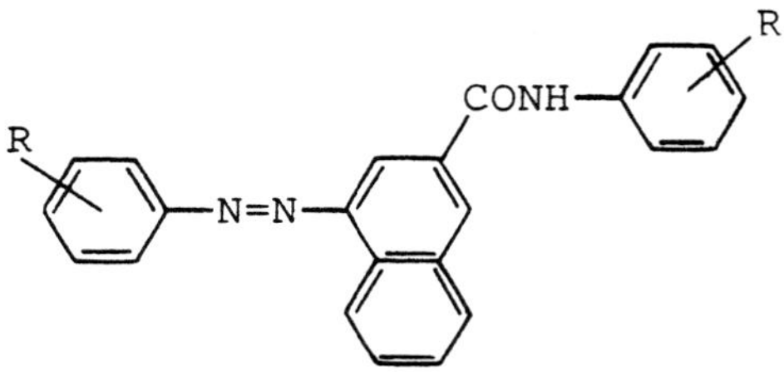

Monoazo naphthol AS pigment

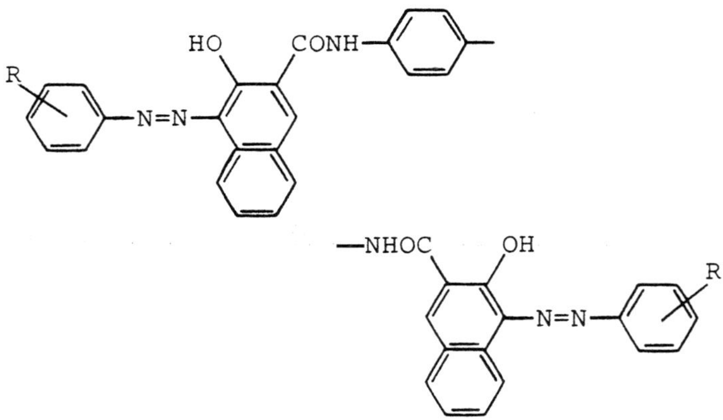

Structure of a red azo condensation pigment

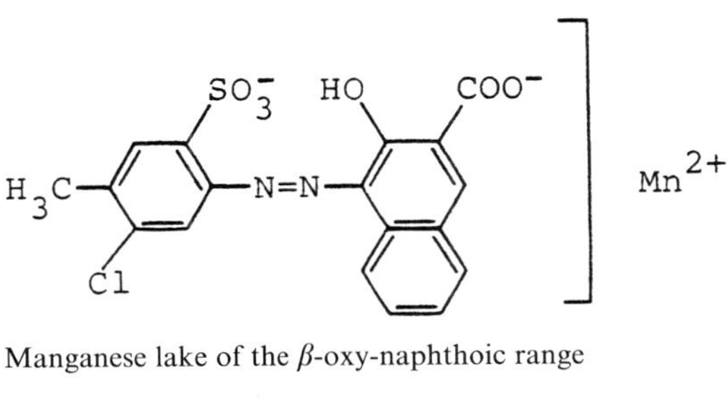

Manganese lake of the β-oxy-naphthoic range

Fig. 5. Examples of the structures of red azo pigments.

Phthalocyanine pigments: C.I. Pigment Blue 15:1, 15:3, 16; C.I. Pigment Green 7, 36.

Phthalocyanine blue is present in several crystalline forms, of which the stabilized ("partly chlorinated") α and the β forms are especially important, the β form being more greenish. These pigments for the most part contain copper bound in complexes, however, there are also some metal-free types available commercially. Phthalocyanine green is obtained by chlorination, and especially yellowish shades by substitution with chlorine and bromine or with bromine alone. Phthalocyanines are highly stable to light and heat (see tables 6 and 7), but may cause distortion and embrittlement when used in large parts made from high density polyethylene.

Quinacridones: C.I. Pigment Red 122, 209; C.I. Pigment Violet 19.

C.I. Pigment Violet 19 is a five-membered, linear quinacridone; it exhibits migration resistance which is higher than would be expected based on its comparatively low molecular weight of 312. *G. Kaufmann* [9] assumes that this is attributable to strong intermolecular hydrogen bonds. C.I. Pigment Red 122 and 209 represent substituted quinacridones (see tables 5 and 6).

Perylene pigments: C.I. Pigment Red 123, 149, 178.

They are derived from perylene tetracarboxylic acid (see table 5). Taking C.I. Pigment Red 149 as an example, *G. Kaufmann* [9] has pointed out the adverse effect on notched impact strength of polyolefins.

Naphthalene tetracarboxylic acid derivative: C.I Pigment Red 194.

In its structure it resembles the perylene pigments, if it is imagined that the five-membered cyclic system is replaced by a two-membered system.

Flavanthrone: C.I. Pigment Yellow 24.

This is an anthraquinone vat dye with excellent fastness properties, which has been formulated in pigment form. On exposure to light and in the presence of water, the reduced form develops (see figure 6). This differs coloristically and is less stable.

Tetrachloroisoindolinones: C.I. Pigment Yellow 109, 110.

Yellow pigments which in some cases exhibit lightfastness values of 7 to 8 even in extremely high reductions with more than 100 parts titanium dioxide (rutile) (see table 4).

Isoindolines: C.I. Pigment Yellow 139.

In their chemical structure they are close to the tetrachloroisoindolinones and have roughly comparable properties (see table 4).

Anthraquinone pigments: C.I. Pigment Red 177; C.I. Pigment Blue 60.

Because of their high solubility, anthraquinone and its simple substitution products are unsuitable as pigments. However, there are products in this range with higher molecular weights, which are distinguished by their good fastness properties. C.I. Pigment Yellow 123 is a phthalic acid derivative with two aminoanthraquinone moieties in the molecule; C.I. Pigment Red 177 is a dimerized aminoanthraquinone. C.I. Pigment Blue 60 (Indanthrone Blue), which likewise belongs to this group, is strictly speaking an indanthrone with a redder shade compared with phthalocyanine blue (see also under flavanthrone).

Anthanthrone pigment: C.I. Pigment Red 168.

This scarlet pigment is obtained by bromination of anthanthrone.

Dioxazine pigments: C.I. Pigment Violet 23.

Substitution products of dioxazine, the oldest of which is the very important C.I. Pigment Violet 23.

Nitrosonaphthol pigment: C.I. Pigment Green 8.

Iron complex of nitrosonaphthol, a relatively cheap product with modest fastness properties.

10.2.4.3 Daylight fluorescent pigments

The characteristic property of these products is their ability to convert ultraviolet light into color. This results in extremely high levels of light reflectance, which leads to very striking shades being obtained. The pigments are closely related to optical brighteners. With one exception (naphthaldazine), the products in question are fluorescent dyes embedded in plastics, as a result of which they become insoluble and yield colorants with a pigment character. They differ from conventional pigments by virtue of their swelling power and the fact that their solvent resistance is less than optimum; their inherently low lightfastness can be increased appreciably by adding large quantities of colorant. *H.-P. Beyerlin* [22] has reported in detail on this group.

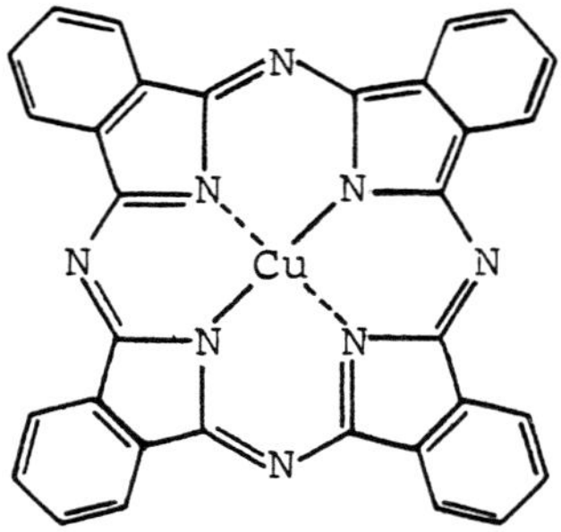

Cu-phthalocyanine (C.I. Pigment Blue 15, 16; C.I. Pigment Green 7, 36; C.I. Solvent Blue 70)

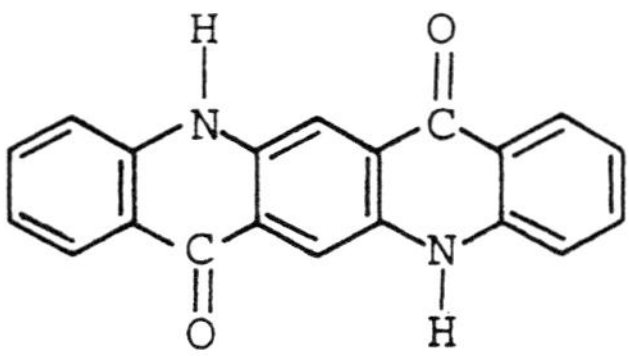

Trans-quinacridone (C.I. Pigment Red 122, 209; C.I. Pigment Violet 19)

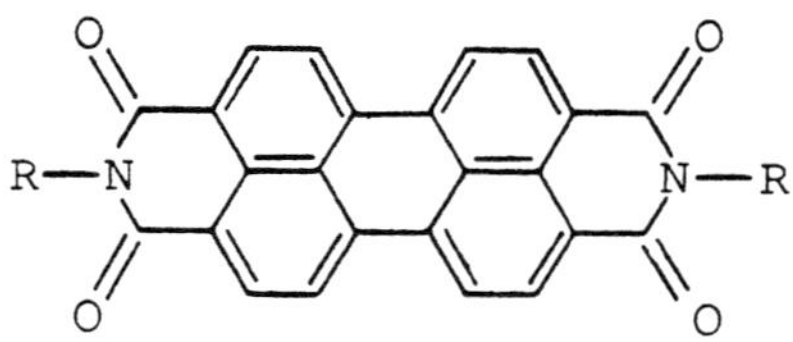

Perylene tetracarboxylic acid derivative (C.I. Pigment Red 123, 149, 178)

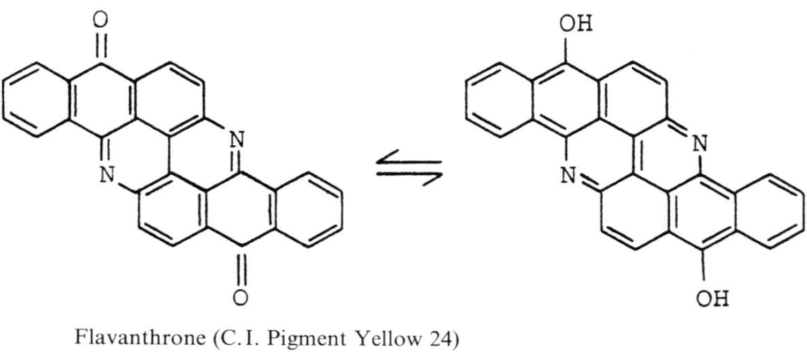

Flavanthrone (C.I. Pigment Yellow 24)

Fig. 6. Examples of the structure of various organic color pigments (non-azo pigments).

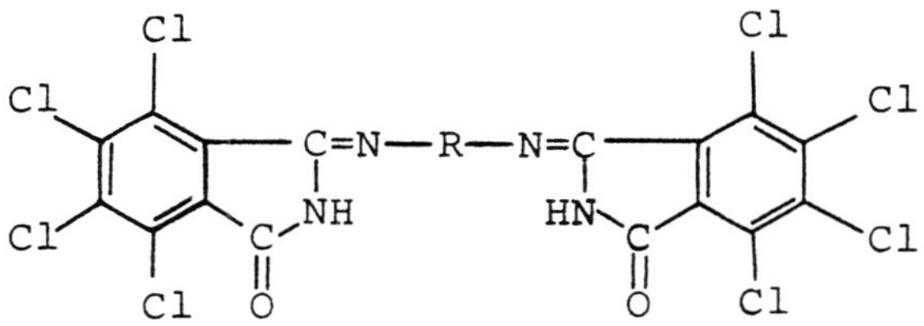

Tetrachloroisoindolinone pigment (C.I. Pigment Yellow 109, 110)

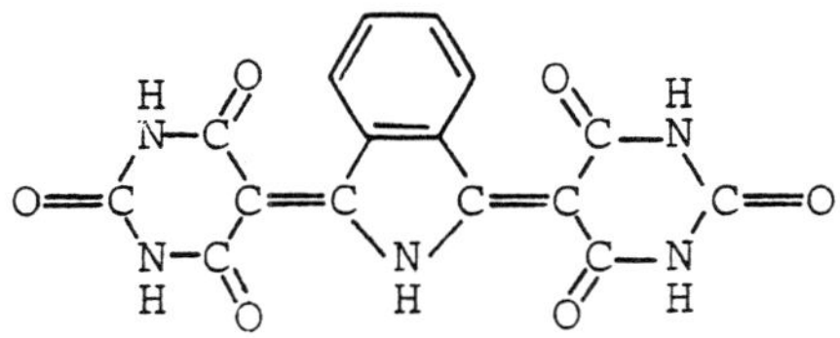

Isoindoline pigment (C.I. Pigment Yellow 139)

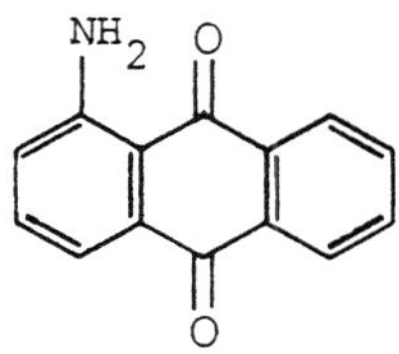

1-Aminoanthraquinone (C.I. Pigment Red 177; C.I. Pigment Blue 60)

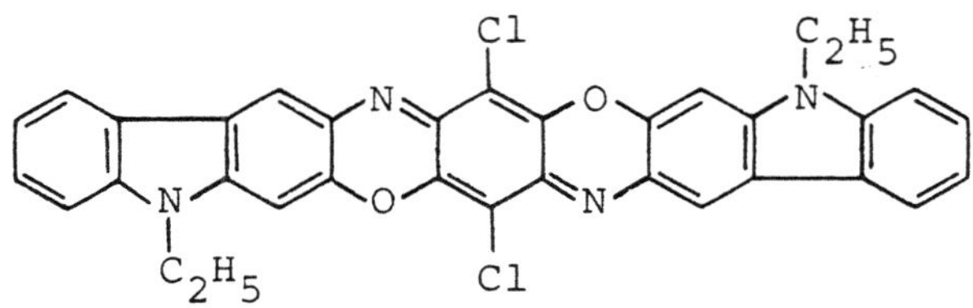

Dioxazine pigment (C.I. Pigment Violet 23)

Fig. 6 (continued)

10.2.4.4 Dyes

Among the various classes of dyes, those soluble in aliphatic and aromatic compounds are of special interest. These are generally azo or anthraquinone dyes with a simple structure. They are used for the coloration of unmodified polystyrene, polymethyl methacrylate and unplasticized PVC, i.e. plastics with high glass transition temperatures, thus limiting the migration of the dyes [23].

10.2.4.5 Remarks on the performance of colorants

In view of the limits placed on the scope of this study, it has not been possible to discuss the range of products in its entirety. In spite of the large number of colorants listed, it is inevitable, therefore, that many a practitioner in this field will note the absence of products of interest to him. However, the aim of providing a general survey has hopefully been attained. Selection of the colorants listed is not intended to represent an evaluation.

As a matter of principle it is advisable, also in cases of chemical identity, to use those colorants that are recommended for plastics, since they have been tested for this field of application.

In the *colorant tables* 4 to 7, the products are listed in the following order: inorganic pigments, organic pigments, dyes. In order to facilitate an understanding of the tables and the abbreviations given there, attention is drawn to the following:

Column 1: It contains in addition to the class of chemical compounds, also the C.I. (Color Index) designation [8] under which are combined products whose chemical structure is largely identical, but which can differ in their performance properties.

Column 2: The lightfastness values listed relate to full shade (pure) and to reductions 1:4 or 1:10 with a stabilized titanium dioxide (rutile). With the dyes – where lightfastness depends to an especially high degree on the plastic to be colored – detailed information has been given. Here the data on reductions do not refer to a specific reduction ratio, but to a (1/3) shade depth attained (standard depth of shade, SD).

All the tests were carried out in accordance with the relevant DIN specifications and the blue (wool) scale values have been stated.

Column 3: In order to characterize thermal stability, the temperature is stated at which no coloristic changes occurred during normal dwell times (approx. 5 minutes) in processing machines.

Column 4: In the data on migration, 5 is the best value, i.e. no migration at all was observed; values below 4 call for especially careful testing of the colorant in the plastic in question.

Column 5: The following abbreviations were used:

PS = polystyrene
PMMA = polymethyl methacrylate
PVC = polyvinyl chloride
H-PVC = unplasticized PVC
W-PVC = plasticized PVC
HDPE = high density polyethylene
LDPE = low density polyethylene
PO = polyolefins

+ = suitable/recommended
○ = limited suitability/recommended
– = not suitable/not recommended
() = signifies a qualification of the statement
*) = caution is needed in the case of HDPE articles sensitive to distortion

Column 6: The trade names listed are intended only as examples and not as a recommendation; the names of the manufacturers are given in abbreviated form in parentheses (). Key:

BASF = BASF AG, Ludwigshafen, and BASF Farben + Fasern AG, Stuttgart
Bayer = Bayer AG, Leverkusen
CG = Ciba-Geigy AG, Basel (Switzerland)
Heub = Dr. Hans Heubach, Langelsheim
Hoechst = Hoechst AG, Frankfurt/M.

A detailed list of manufacturers, commercial firms and trade names is contained in Lückert: Pigment- und Füllstofftabellen, *M.* and *O. Lückert,* Laatz, 1977 as well as in Modern Plastics Encyclopedia, McGraw-Hill, Inc. New York, 1983/84.

All data given in the colorant tables should be considered as general information and before use should be compared with original data from the manufacturer or confirmed by carrying out own tests.

Table 4. Yellow colorants (partly based on manufacturers' information, data have only limited comparability).

C.I. designation Chemical description	Fastness values			Remarks on performance	Trade names (manufacturer)
	Light	Heat °C	Migration		
Inorganic pigments					
Pigment Yellow 53 Nickel titanium yellow; $(Ti, Ni, Sb)O_2$	8 pure 8 1:4	300	5	general suitability	Light Yellow (Bayer); Sicotan Yellow K1011 (BASF)
Pigment Yellow 118 Chrome titanium yellow; $(Ti, Cr, Sb)O_2$	8 pure 8 1:4	300	5	general suitability	Light Yellow (Bayer); Sicotan Yellow K2011 (BASF)
Pigment Yellow 42 Iron oxide hydrate; FeO(OH)	8 pure 8 1:4	220 PVC	5	W-PVC + ; H-PVC ○; LDPE, HDPE, PS –	Iron oxide yellow (Bayer; BASF)
Pigment Yellow 37 Cadmium yellow; (Cd, Zn)S	8 pure 8 1:4	300	5	general suitability, weather resistance low	Cadmopur Yellow (Bayer); Sicotherm Yellow K1301 (BASF)
Pigment Yellow 119 Zinc/iron pigment	8 pure 8 1:4	300	5	general suitability	Sicopal Yellow K2395, Sicopal Brown K2595 (BASF)
Pigment Yellow 34 (Chrome Yellow) monoclinic, stabilized; $Pb(Cr, S)O_4$	6–7 pure 7 1:4	240 PO	5	PVC, LDPE + ; HDPE, PS –	Sicomin Yellow K1422 (BASF); Heucosin, Heucotron Yellow (Heub)
Pigment Yellow 34 (Chrome Yellow) monoclinic, highly stabilized; $Pb(Cr, S)O_4$	8 pure 8 1:4	220 PVC 230 PS 250 PO	5	PVC, LDPE + ; HDPE, PS ○	Sicomin Yellow K1630 (BASF); Heucosin, Heucotron Yellow (Heub)
Organic pigments					
Pigment Yellow 97 Monoazo pigment	8 pure 7 1:10	260 PS 260 PO	5	LDPE, PS + ; PVC –	Novoperm Yellow FGL (Hoechst)
Pigment Yellow 120 Monoazo benzimidazolone pigment	8 pure 7–8 1:10	260 PO	5	PVC, LDPE +	PV Fast Yellow H2G (Hoechst)
Pigment Yellow 151 Monoazo benzimidazolone pigment	8 pure 7–8 1:10	260 PO	5	PVC, LDPE + ; HDPE + *)	Hostaperm Yellow H4G (Hoechst)

Table 4 (continued)

C.I. designation Chemical description	Fastness values			Remarks on performance	Trade names (manufacturer)
	Light	Heat °C	Migration		
Pigment Yellow 154 Monoazo benzimidazolone pigment	8 pure 7–8 1:10	260 PO	5	PVC, LDPE, PS + ; HDPE ○ *)	Hostaperm Yellow H3G (Hoechst)
Pigment Yellow 13 Disazo, diaryl pigment	6 pure 5–6 1:10	220 PVC 240 PO	4	PVC, LDPE +	PV Yellow GR (Hoechst); Helio Fast Yellow GRN (Bayer); Graphtol Yellow GRBL (Sandoz); Sico Fast Yellow K1351 (BASF); Irgalith Yellow BAWP (CG)
Pigment Yellow 83 Disazo, diaryl pigment	8 pure 7 1:10	220 PVC 220 PS 250 PO	4–5	PVC, LDPE, PS + ; HDPE + *)	Novoperm Yellow HR (Hoechst); Sico Fast Yellow K1781 (BASF); Helio Fast Yellow HRN (Bayer); Graphtol Yellow RCL (Sandoz); Irgalith Yellow B3RS (CG)
Pigment Yellow 81 Disazo, diaryl pigment	7 pure 6 1:10	260 PS 260 PO	4–5	PS, PVC, LDPE + ; HDPE + *)	Permanent Yellow H10G, PV Yellow H10G (Hoechst)
Pigment Yellow 93 Disazo condensation pigment	8 pure 7 1:10	260 PO	5	PVC, PS, LDPE, HDPE +	Cromophtal Yellow 3G (CG)
Pigment Yellow 94 Disazo condensation pigment	8 pure 7 1:10	280 PO	5	PVC, PS, LDPE + ; HDPE + *)	Cromophtal Yellow 6G (CG)
Pigment Yellow 95 Disazo condensation pigment	6–7 pure 6–7 1:10	260 PO	5	PVC, PS, LDPE + ; HDPE (+)	Cromophtal Yellow GR (CG)
Pigment Yellow 128 Disazo condensation pigment	8 pure 7 1:10	260 PO	5	PVC, PS, LDPE + ; HDPE + *)	Cromophtal Yellow 8G (CG)
Pigment Yellow 24 Flavanthrone	8 pure 8 1:10	280 PO	5	PVC, PS, LDPE + ; HDPE + *)	Paliogen Yellow L1870 (BASF); Cromophtal Yellow A2R (CG)

Table 4 (continued)

C.I. designation Chemical description	Fastness values			Remarks on performance	Trade names (manufacturer)
	Light	Heat °C	Migration		
Pigment Yellow 109 Tetrachloroisoindolinone	8 pure 7–8 1:10	280 PO	5	PVC, PS, LDPE + ; HDPE + *)	Irgazin Yellow 2GLTE (CG)
Pigment Yellow 110 Tetrachloroisoindolinone	8 pure 7–8 1:10	280 PO	5	PVC, PS, LDPE + ; HDPE + *)	Chromophtal Yellow 2RLTS, Irgazin Yellow 3RLTN (CG)
Pigment Yellow 138 Quinophthalone	8 pure 8 1:10	220 PVC 280 PS 260 PO	4–5	PVC, PS, LDPE + ; HDPE + *)	Paliotol Yellow K0961 HD (BASF)
Pigment Yellow 139 Isoindoline	8 pure 7–8 1:10	220 PVC 250 PO	5	PVC, LDPE + ; PS ○; HDPE + *)	Paliotol Yellow K1841 D (BASF)
Dyes					
Solvent Yellow 93 Pyrazolone derivative	8 0.05% PS 7 0.05% PVC 8 1/3 SD PS	300 0.05% PS		PMMA, H-PVC, PS +	Thermoplast Yellow G (BASF)
Solvent Yellow 16 Fat-soluble azo dye	8 0.05% PS 3 0.05% PVC 6 1/3 SD PS	280 0.05% PS		PMMA, H-PVC, PS +	Sudan Yellow 3G (BASF)
Fluorescent polycyclic dye	7 0.05% PS 4 0.05% PVC 3 1/3 SD PS	300 0.05% PS		PMMA, H-PVC, PS +	Fluorol Green Gold K (BASF)

Table 5. Red colorants (partly based on manufacturers' information, data have only limited comparability).

C.I. designation Chemical description	Fastness values Light	 Heat °C	 Migration	Remarks on performance	Trade names (manufacturer)
Inorganic Pigments					
Pigment Red 101 Iron oxide; Fe_2O_3	8 pure 8 1:4	300	5	general suitability; H-PVC ○	Iron oxide Red (Bayer, BASF)
Pigment Red 101 Iron oxide, transparent; Fe_2O_3	8 pure 8 1:4	300	5	general suitability; H-PVC ○	Sicotrans Red K2915 (BASF)
Pigment Red 108 Cadmium/sulfide selenide; Cd(S, Se)	8 pure 8 1:4	300	5	general suitability	Cadmopur Red (Bayer); Sicotherm Red K3201 (BASF)
Pigment Red 104 Lead/chromate molybdate, highly stabilized Molybdate Red; $Pb(Cr,Mo,S)O_4$	7 pure 8 1:4	220 PVC 230 PS 250 PO	5	PVC, LDPE +; PS, HDPE ○	Sicomin Red K3030 (BASF) Heucotron Red (Heub)
Pigment Red 104 Lead/chromate molybdate, stabilized Molybdate Red; $Pb(Cr,Mo,S)O_4$	5 pure 6–7 1:4	240 PO	5	PVC, LDPE +; PS, HDPE –	Sicomin Red L3022 (BASF); Heucotron Red (Heub)
Organic Pigments					
Pigment Red 176 Monoazo naphthol AS pigment	7 pure 5–6 1:10	280 PO 280 PS	5	PVC, PS, LDPE +; HDPE (+)*)	PV Carmine HF3C (Hoechst)
Pigment Red 171 Monoazo benzimidazolone pigment	7 pure 7 1:10	240 PO 240 PS	4–5	PVC, PS, LDPE +; HDPE (+)*)	Novoperm Marron HFM, PV Fast Marron HFM (Hoechst)
Pigment Red 175 Monoazo benzimidazolone pigment	7–8 pure 7 1:10	260 PO 260 PS	5	PVC, PS, LDPE +; HDPE (+)*)	Permanent Red HFT, PV Fast Red HFT (Hoechst)
Pigment Red 185 Monoazo benzimidazolone pigment	7–8 pure 7 1:10	260 PO	5	PVC, PS, LDPE +; HDPE (+)*)	Novoperm Carmine HF4C, PV Carmine HF4C (Hoechst)

Table 5 (continued)

C.I. designation Chemical description	Fastness values			Remarks on performance	Trade names (manufacturer)
	Light	Heat °C	Migration		
Pigment Red 48/1 Ba salt of a monoazo color acid	3 pure 2 1:10	250 PO	4	PVC, PS, LDPE +; HDPE –	Lithol Scarlet K3700 (BASF)
Pigment Red 48/2 Ca salt of a monoazo color acid	3–4 pure 2–3 1:10	240 PO	4	PVC, PS, LDPE +; HDPE –	Lithol Scarlet K4460 (BASF); Irgalith Red 2BP (CG)
Pigment Red 48/3 Sr salt of a monoazo color acid	6 pure 4 1:10	260 PO	4–5	PVC, PS, LDPE +; HDPE + *)	Lithol Fast Scarlet K4160 (BASF); Graphtol Flame Red 3RL (Sandoz)
Pigment Red 48/4 Mn salt of a monoazo color acid	5–6 pure 5 1:10	240 PO	4	PVC, PS +; LDPE(○); HDPE –	Lithol Fast Scarlet K4260 (BASF); Novoperm Red H2BM (Hoechst)
Pigment Red 53/1 Ba salt of a monoazo color acid	3 pure 2–3 1:10	250 PO	4	PS, LDPE +; HDPE –; W-PVC, H-PVC ○	Lithol Bronze Red K3660 (BASF); Pigment Lake Red LC (Hoechst)
Pigment Red 57/1 Ca salt of a monoazo color acid	4–5 pure 3 1:10	250 PO	4	PVC, PS, LDPE +; HDPE –	Lithol Rubine K4631 (BASF); Irgalith Rubine 4BP (CG)
Pigment Red 144 Disazo condensation pigment	7–8 pure 7–8 1:10	280 PO 280 PS	5	PVC, PS, LDPE +; HDPE + *)	Cromophtal Red BRN (CG)
Pigment Red 166 Disazo condensation pigment	7–8 pure 7–8 1:10	280 PO 280 PS	5	PVC, PS, LDPE +; HDPE + *)	Cromophtal Scarlet RN (CG)
Pigment Red 122 substituted quinacridone	7–8 pure 7 1:10	240 PS 280 PO	5	PS, PVC, LDPE +; HDPE + *)	Hostaperm Pink E, PV Fast Pink E (Hoechst)
Pigment Red 209 substituted quinacridone	7–8 pure 7 1:10	240 PS 280 PO	5	PS, PVC, LDPE +; HDPE + *)	Hostaperm Pink ER, Fast Red EG (Hoechst)
Pigment Red 123 Perylene pigment	8 pure 8 1:10	220 PVC	4–5	PVC, LDPE, PS +; HDPE + *)	Paliogen Red K3871 (BASF)

Table 5 (continued)

C.I. designation Chemical description	Fastness values: Light	Heat °C	Migration	Remarks on performance	Trade names (manufacturer)
Pigment Red 178 Perylene pigment	8 pure 7 1:10	300 PS 300 PO 220 PVC	4–5	PVC, PS, LDPE + ; HDPE + *)	Paliogen Red K3911 HD (BASF)
Pigment Red 149 Perylene pigment	8 pure 7 1:10	280 PS 300 PO		PS, PVC, LDPE + ; HDPE + *)	Permanent Red BL, PV Fast Red B (Hoechst)
Pigment Red 168 Anthanthrone pigment	7 pure 6–7 1:10	220 PVC	4–5	PVC + ; LDPE ○ ; PS –	Hostaperm Scarlet GO (Hoechst); Helio Fast Scarlet EB (Bayer)
Pigment Red 177 Anthraquinone pigment	8 pure 7 1:10	280 PO	5	PS, PVC, LDPE, HDPE +	Cromophtal Red A3B (CG)
Pigment Red 194 Naphthalene tetracarboxylic acid derivative	8 pure 7–8 1:10	280 PO 280 PS	5	PS, H-PVC, LDPE + ; W-PVC –	Permanent Red TG (Hoechst)
Pigment Red 88 Tetrachlorothioindigo pigment	8 pure 7 1:10	260 PS 260 PO 220 PVC	4–5	PS, PVC, LDPE + ; HDPE ○ *)	Paliogen Red Violet K4980 (BASF); PV Red MR, Novoperm Red MR (Hoechst); Cromophtal Bordeaux RN (CG); Helio Fast Red Violet ERO6 (Bayer)
Dyes					
Solvent Red 1 Azo dye	5 0.05% PS 2–3 0.05% PVC 2–3 1/3 SD PS	260 0.05% PS		PMMA, H-PVC, PS +	Sudan Red G (BASF)
Solvent Red 138 Anthraquinone derivative	8 0.05% PS 7–8 0.05% PVC 8 1/3 SD PS	300 0.05% PS		PMMA, H-PVC, PS +	Thermoplast Red G (BASF)

Table 5 (continued)

C.I. designation Chemical description	Fastness values: Light	Fastness values: Heat °C	Fastness values: Migration	Remarks on performance	Trade names (manufacturer)
Fluorescent benzopyran dye	7 0.05% PS 5 0.05% PVC 5 1/3 SD PS	300 0.05% PS		PMMA, PS + ; H-PVC ○ (Shade may darken)	Fluorol Red GK (BASF)
Fluorescent benzopyran dye	6–7 0.05% PS 4 0.05% PVC 5 1/3 SD PS	300 0.05% PS		PMMA, PS + ; H-PVC ○ (Shade may darken)	Fluorol Red BK (BASF)
Azo dye	8 0.05% PS 7 0.05% PVC 7 1/3 SD PS	300 0.05% PS		PMMA, PS, H-PVC +	Thermoplast Red BS (BASF)
Azo dye	6–7 0.05% PS 5 0.05% PVC 3–4 1/3 SD PS	260 0.05% PS		PMMA, PS, H-PVC +	Sudan Red BB (BASF)

Table 6. Orange, violet and blue colorants (partly based on manufacturers' information, data have only limited comparability).

C.I. designation Chemical description	Fastness values Light	 Heat °C	 Migration	Remarks on performance	Trade names (manufacturer)
Orange pigments					
Pigment Orange 20 Cadmium/sulfide selenide, Cd(S,Se)	8 pure 8 1:10	300	5	general suitability; higher weather resistance with darker (redder) shades	Cadmopur Orange (Bayer); Sicotherm Orange K2801 (BASF)
Pigment Orange 194 Azo pigment	4–5 pure 3 1:10	240 HDPE	3	H-PVC +; PS, LDPE ○; HDPE ○ *); W-PVC –	Sico Fast Orange K2850 (BASF)
Pigment Orange 43 Perinone pigment	7–8 pure 6 1:10	260 PE 240 PS	4–5	PVC, PS, LDPE +; HDPE (+)*)	PV Fast Orange GRL (Hoechst)
Orange dyes					
Solvent Red 14 Azo dye	5–6 0.05% PS 2 0.05% PVC 2–3 1/3 SD PS	270 0.05% PS		PMMA, PS, H-PVC +	Sudan Orange R (BASF)
Fluorescent perylene dye	7 0.05% PS 4 0.05% PVC 5–6 1/3 SD PS	300 0.05% PS		PMMA, PS, H-PVC +	Fluorol Orange GK (BASF)
Violet pigments					
Pigment Violet 23 Dioxazine pigment	6 pure 5 1:10	250 PS 260 PO 220 PVC	3–4	PVC, LDPE +; PS ○; HDPE + *)	Paliogen Violet L5890 (BASF)
Pigment Violet 19 Quinacridone	7 pure 7 1:10	280 PE 260 PS	5	PVC, PS, LDPE +; HDPE + *)	Hostaperm Red E5B (Hoechst)
Pigment Violet 19 Quinacridone	7–8 pure 7–8 1:10	280 PE 260 PS	5	PVC, PS, LDPE +; HDPE + *)	Hostaperm Violet ER (Hoechst)

Table 6 (continued)

C.I. designation Chemical description	Fastness values			Remarks on performance	Trade names (manufacturer)
	Light	Heat °C	Migration		
Blue, inorganic pigments					
Pigment Blue 29 (Ultramarine) Na/Al silicate, sulfide-containing	8 pure 8 1:4	300 PS 300 PO	5	PVC, PS, LDPE, HDPE +	Ultramarine Blue (BASF)
Pigment Blue 28 Spinel; $CoAl_2O_4$	8 pure 8 1:4	300 PS 300 PO	5	PVC, PS, LDPE, HDPE +	Sicopal Blue K7310 (BASF); Light blue (Bayer); Cobalt blue (Degussa)
Blue, organic pigments					
Pigment Blue 15:1 Cu-phthalocyanine, stabilized α form	8 pure 8 1:10	280 PS 280 PO	5	PVC, PS, LDPE +; HDPE + *)	Heliogen Blue (BASF); Cromophtal Blue (CG)
Pigment Blue 15:3 Cu-phthalocyanine, β form	8 pure 8 1:10	280 PS 280 PO	5	PVC, PS, LDPE +; HDPE + *)	Heliogen Blue (BASF); Cromophtal Blue (CG)
Pigment Blue 16 Phthalocyanine, stabilized, metal-free	8 pure 8 1:10	240 PS 280 PO	5	PVC, PS, LDPE +; HDPE + *)	Heliogen Blue (BASF)
Pigment Blue 60 (Indanthren Blue) Anthraquinone pigment	8 pure 8 1:10	250 PS 270 PO	5	PVC, PS, LDPE +; HDPE + *)	Paliogen Blue L6470 (BASF); Cromophtal Blue A3R (CG); Sandorin Blue RL (Sandoz)
Blue dyes					
Solvent Blue 35 Fat-soluble anthraquinone dye	7 0.05% PS 2 0.05% PVC 4–5 1/3 SD PS	280 0.05% PS		PMMA, PS, H-PVC +	Sudan Blue 2 (BASF)
Solvent Blue 70 Phthalocyanine dye	6–7 0.05% PS 1–2 0.05% PVC 4–5 1/3 SD PS	260 0.05% PS		PMMA, PS, H-PVC +	Neozapon Blue FLE (BASF)

Table 7. Green and brown colorants (partly based on manufacturers' information, data have only limited comparability).

C.I. designation Chemical description	Fastness values			Remarks on performance	Trade names (manufacturer)
	Light	Heat °C	Migration		
Green, inorganic pigments					
Pigment Green 17 Chromoxide; Cr_2O_3	8 pure 8 1:4	300 PS 300 PO	5	general suitability	Chrome oxide green (Bayer); Sicopal Green 9996 (BASF)
Pigment Green 19 Spinel; $(Co, Ni, Zn)_2(Ti, Al)O_4$	8 pure 8 1:4	300 PS 300 PO	5	general suitability	Light green (Bayer); Sicopal Green K9710 (BASF); Cobalt green (Degussa)
Green organic pigments					
Pigment Green 7 Chlorinated phthalocyanine	8 pure 8 1:4	300 PS 300 PO	5	PS, PVC, LDPE +; HDPE + *)	Heliogen green K8681 (BASF); Cromophtal Green (CG)
Pigment Green 36 Halogenated phthalocyanine	8 pure 8 1:4	300 PS 300 PO	5	PS, PVC, LDPE +; HDPE + *)	Heliogen green K9360 (BASF); Cromophtal Green (CG)
Pigment Green 8 Fe-complex of nitrosonaphthol	3 pure 2–3 1:4	220 PO	4	PS, LDPE +; W-PVC ○; HDPE, H-PVC –	Paliotol Green K9781 (BASF)
Brown pigments					
Pigment Brown 6 Iron oxide brown; Fe_3O_4	8 pure 8 1:10	240 HDPE	5	PS, LDPE +; PVC, HDPE ○	Iron oxide brown (Bayer; BASF)
Pigment Brown 29 Chrome iron brown; $(Fe, Cr)_2O_3$	8 pure 8 1:10	300	5	general suitability	Sicopal Brown K2795 (BASF)
Pigment Yellow 119 Zinc iron pigment	8 pure 8 1:10	300	5	general suitability	Sicopal Brown K2595 (BASF)
Pigment Brown 23 Disazo condensation pigment	8 pure 8 1:10	260 PE	5	PVC, PS, LDPE +; HDPE + *)	Cromophtal Brown 5R (CG)

Part II: Coloring Technology

Dr. Winfried Damm

10.3 Introduction

On specification of certain delivery criteria, the polymer manufacturers supply mass-colored plastics in a form ready for processing, and indicate to the user the advantages that derive from the elimination of coloristic work and the reduction in the burden of responsibility that results from this. The colorant manufacturer points out that when carried out by the plastics processor, coloration is more efficient and leads to greater flexibility in production. Ultimately it is important to weigh both alternatives and determine where the advantage lays based on the conditions prevailing in the plant.

10.4 Selection criteria

From the viewpoint of the chemical industry, colorants for plastics are classified into the following groups:

- inorganic pigments,
- organic pigments,
- dyes,
- carbon pigments (carbon blacks),
- effect pigments.

The selection criteria applying to a colorant are closely related to processing behavior and to the performance properties required in the end product.

It is understandable, therefore, that the fastness requirements that have to be met by pigments for plastics are not comparable with those of paints or printing inks. If we disregard addition of colorants before or during polymerization, coloration is carried out during processing, i.e. either at the compounding stage (slow-running mixers/high-speed mixers) or during plasticization direct on the plastics processing machine (e.g. extruder, injection molding machine, PUR foaming unit etc.), the colorants being added by means of metering devices. The nature of the problems encountered in individual cases depends greatly on the type of plastic concerned (e.g. unplasticized or plasticized PVC, LDPE or HDPE, PP etc.). Selection of the correct colorant assumes that the prevailing technical requirements and the range of properties of the suitable colorants are known. A very important factor in these considerations is also a possible interaction between the colorants and the other components of the formulation. Compiling a formulation can become complicated if colorant combinations are necessary in order to achieve a certain final shade and if, in addition, specific optical properties (e.g. infrared reflectance) have to be adhered to in accordance with given specifications.

10.5 Dispersion

The tinctorial strength attainable is one of the decisive factors to be taken into account when determining the cost/performance relationship and is closely related to the dispersibility of a colorant. Dispersion of a pigment [27] is a physical process in which a number of separate steps take place at the same time. According to DIN 55943 (ISO 4518/1–1978), "dispersion" of a pigment in another medium is understood to mean the breaking down of the agglomerates in the powder into smaller particles (small agglomerates, aggregates and primary particles) and their simultaneous wetting by the medium. Dispersion also includes a randomly uniform distribution of the pigment particles formed. If the medium is of relatively low viscosity, the pigment particles already dispersed tend to mass together – also known as reagglomeration – due to the forces of attraction prevailing between them. This process results in loss of tinctorial strength. In order to prevent this, it is necessary to stabilize the state of dispersion attained. Generally speaking, dispersion of reagglomerates is scarcely possible under normal conditions. Dyes, on the other hand, are soluble in plastics and thus require no dispersing operation.

The dispersion process is thus subdivided into the following steps: breakdown, wetting, distribution and stabilization.

10.5.1 Breakdown

In the state in which they are delivered, powdered pigments are not present in the particle size desired. When a pigment is incorporated in a plastic, dispersion work thus has to be carried out, since it is only in certain particle size ranges that pigments fully develop their coloristic power and can be processed without difficulty (figure 7).

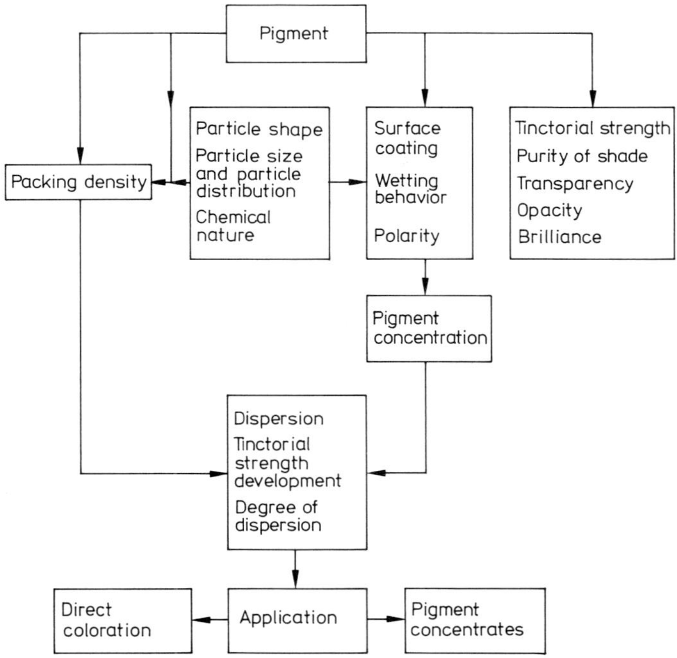

Fig. 7. Relationships between physical properties and pigment application.

According to *Rumpf* [28], in the breakdown of solid particles a distinction is made between the following processes:

a) breakdown between two solid surfaces,
b) by impingement,
c) by introduction of energy into a surrounding medium by mechanical means,
d) by natural oscillations and centrifugal forces.

The breakdown processes in a) and b) are carried out in equipment such as mills, crushers, high-speed mixers and edge runners. Process c), on the other hand, takes place in a liquid, pasty phase or in polymer melts. It is typical, for example, of the operation carried out on extruders, calenders, kneaders, triple-roll mills and attrition mills (shear stresses).

Process d) can take place through the action of ultrasonic or thermal energy. Breakdown processes in highly viscous media are more intensive when carried out under pressure than in pressureless modes of operation (e.g. triple-roll mills or kneaders). The results are dependent on the "strength" of the pigment particle and can be regarded as pigment-specific.

Inorganic pigments with their ionic lattice are generally more readily dispersible than organic pigments with their molecular lattice. Consequently, there are some organic pigments which cannot be used for direct coloration of plastics without special preparation.

10.5.2 Wetting

The interactions between the free pigment surface and the surrounding medium are an important factor in the wetting process. The energy relationships can be explained by means of the laws of thermodynamics, whereby surface tension and surface energy represent important auxiliary quantities [29, 30]. The kinetic aspect additionally describes the processes taking place during thorough wetting of the pigment powder and the agglomerates [31]. Wetting time is dependent on the pore geometry of the powder. It increases as the viscosity of the medium rises and decreases as wetting tension rises. Under practical conditions, wetting is on the one hand accelerated by raising the temperature, while on the other hand the shear stress forces (breakdown action) decrease as the temperature rises (lowering of viscosity).

Wetting volume can be regarded as a measure of the pore volume or packing density of a pigment. It is closely related to the relative tinctorial strength and thus indirectly to the quality of dispersion in the end product. This knowledge is utilized in the coloration of plastics by admixing low-molecular additives such as plasticizers or adhesion promoters with the medium (cf. figure 7).

The dispersion hardness DH of a pigment can be determined in accordance with DIN 53775, sheet 7, by means of a hot/cold rolling test carried out with plasticized PVC:

$$DH = \left(\frac{\frac{K}{S}\text{cold}}{\frac{K}{S}\text{hot}} - 1 \right) \cdot 100$$

K = coefficient of light absorption
S = scattering coefficient

The quotient K/S to be determined by colorimetry in accordance with DIN 53234 is a measure of tinctorial strength and directly proportional to the latter.

10.5.3 Distribution

Optimum distribution of the broken-down, wetted pigment particles is achieved by means of high plastics processing temperatures, low viscosity, turbulent flow and long dwell times in the processing unit. In practice, however, the latter is limited by the output required. An attempt has been made to increase breakdown action and dwell time by installing torpedoes, breaker plates, shear elements, kneading sections and by other design measures in injection molding machines and extruders.

10.5.4 Cohesive forces

Pigment particles exert attractive forces on each other (dipoles, hydrogen bonds, electrostatic forces, number of points of attachment), which during the dispersion process have to be overcome by the supply of energy from outside. A relationship is thus obtained between dispersibility and particle size distribution. Finely divided pigments have large surfaces, are thermodynamically unstable and exhibit a reagglomeration tendency, but because of the generally high viscosity of the plastics melts are prevented from rapidly interchanging their positions. In plastisols and solvent-containing coating compounds (e.g. PUR), however, the relationships are more critical (flocculation, floating and flooding). In this respect, the polarity of the liquid binder is also a decisive factor. The mutual attractive forces between the pigment particles can be reduced by means of specific coating agents applied to the surface of the pigment particles and thus help to avoid dispersion troubles.

10.5.5 Dispersion faults and their manifestation

Inadequate dispersion of a pigment can manifest itself in the following ways [32]:

- fluctuations in color intensity,
- deviations in shade,
- streaking,
- clogging of screen packs in the extruder,
- tearing of the bubble during blown film extrusion, due to pigment agglomerates which have not broken down,
- tearing of monofilaments and film tapes,
- printing problems due to inhomogeneities at the surface,
- drop in mechanical strength (notch effect),
- reduced abrasion resistance due to unsatisfactory coating of the pigments.

Finally, economic aspects also arise when the cost of pigment dispersion is compared with the color yield obtained.

10.6 Forms of supply

10.6.1 Single pigments

Single pigments are supplied with a greater or less degree of dusting and generally as non-free-flowing powders. They impose high requirements regarding dispersion. Inorganic pigments have an average particle size of approx. 1 μm, organic pigments approx. 0.1 μm. The latter thus have an appreciably larger surface, which in relation to that of inorganic pigments is 25–100 to 1. This numerical example explains the considerable amount of time and energy often needed for dispersion. Preference is thus frequently given to pre-dispersed pigment mixtures or fully dispersed pigment concentrates for the coloration of plastics. The use of single pigments requires that the processors have their own efficient color matching facilities.

10.6.2 Free-flowing pigment mixtures

Since the free-flowing properties of single pigments are generally inadequate, disturbances can arise on account of bridging in the feed hopper of processing machines. These show up as shade variations in the finished articles. Pigments for gravimetric or volumetric feed are, therefore, also supplied in the form of free-flowing powders. Although this type of preparation has a lower degree of dusting than the basic single pigments, the amount of dispersion required is virtually the same.

10.6.3 Pigment mixtures

When matching given shades, the plastics processor can experience difficulties due to extreme differences in the quantitative proportions of the individual components or when pigments with wide differences in dispersion behavior are used. In such cases, manufacturers of pigments and pigment preparations supply color-matched pigment mixtures in powder form, which are frequently produced by means of edge-runner mills. The agglomerates of the finely divided, hard pigments are broken down by pressure and friction in the powder phase. In the process they are taken up by larger, but readily dispersible particles of other pigments or fillers and ultimately behave like readily dispersible, inorganic pigments. This type of pigment preparation can also be favorably influenced by the addition of certain wetting agents.

Advantages of pigment mixtures are:

- reduction in coloristic work,
- the expensive dispersion process has already been carried out,
- efficient operating sequence.

Furthermore, inorganic pigments with high opacity can be selectively combined with organic pigments with high tinctorial strength.

10.6.4 Pigment pastes

Completely dustfree coloration is ensured by the use of pigment pastes. Pigment pastes are produced by means of three-roll mills, kneaders, dissolvers, grinders or flushing processes, using liquid carriers such as plasticizers, polyols, glycerol fatty acid esters etc. The paste viscosity can naturally be adjusted so that automatic metering at the processing machines is possible.

10.6.5 Pigment concentrates

These are special products in which the pigments are already optimally dispersed in a carrier material that is compatible with the plastic. Because of the high pigment content desired, extremely high energy consumption is necessary when manufacturing concentrates. Only highly efficient dispersing machines such as heated roller mills, kneaders, continuously operating twin-screw extruders with special screws or planetary screw extruders are used.

The pigment concentrates, which are usually matched to shade, are marketed as granules with irregular granulation or as powders in ground form.

For the processor, the following advantages accrue:

- optimum pigment dispersion,
- elimination of coloristic work,
- accuracy of shade,
- almost dustfree operation,
- reduction in cleaning costs.

A legitimate question that is often raised is whether pigment concentrates permit greater economy, since in relation to pigment content they are always more expensive than the corresponding powdered pigments. Although with concentrates the application costs are naturally higher, savings can be made in the time and energy required, in the procurement of expensive dispersing units as well as in personnel. In addition, the plastics processor is given greater assurance from the coloristic aspect and in many cases the coloration cost can be lowered as a result of the increase in tinctorial strength yield.

10.7 Coloration techniques

At the plastics manufacturer, plastics are colored in the molten state, the polymer to be colored being melted together with the colorants in kneaders or extruders; the homogeneously colored melt obtained by this means is discharged, cooled and granulated.

In order that optimum results are obtained when the plastics processor carries out his own coloration, preference is nowadays generally given to pigment concentrates the reasons being easier metering and better coloration quality in the finished products.

10.7.1 Coloration with single pigments

The colorants are dry blended with the plastics granules at room temperature in a mixer. The addition of an adhesion promoter [33] or a dispersing agent [34] may possibly be necessary. This premix can subsequently be processed directly on injection molding machines or extruders. Effect pigments such as bronzes cannot be incorporated by this means. In this case, pigment concentrates have to be used.

10.7.2 Coloration with pigment concentrates

This processing technique is the same as that described when using single pigments, but possesses the advantage that the pigments are already present in an optimally dispersed form. An optimum in tinctorial strength (resist-free) and accuracy of shade are thus obtained.

10.7.3 Machinery

A number of mixing units are available for the direct coloration of plastics and the production of premixes for pigment concentrates.

Mixers

A distinction is made between gravity mixers such as

- eccentric tumbler mixers,
- conical mixers,
- drum mixers (barrel), and

mixers with rotating blades such as

- ribbon mixers,
- ploughshare mixers,
- fluid mixers.

The use of mixers adapted to the requirements of the particular form in which the plastic is supplied (powder, granules) enables attaining maximum mixing quality within a short mixing time.

Equipment for manufacture of pigment concentrates

The manufacturer of pigment concentrates has available a large number of sometimes complex units. These include:

- mixing rolls,
- single-screw extruders,
- twin-screw extruders,
- multiscrew extruders,
- internal mixers,
- kneaders.

Pigment concentrates produced via plastics melts can be prepared for processing either by the cold granulating method employing pelletizers or rotary cutters or by die-face granulation [35]. The reader can obtain further details from the relevant literature. Table 8 summarizes the advantages and disadvantages of the various coloration methods.

Table 8. Methods of coloring PVC.

Preparation	Gravity mixer	Agitator mixer	Fluid mixer
Single pigment	problem-free	good	problem-free
Pigment mixtures	problem-free	good	problem-free
Paste, high-viscosity	not possible	not possible	good
Paste, low-viscosity	good	good	problem-free
Color concentrate, powder	difficult without adhesion promoter	difficult without adhesion promoter	difficult without adhesion promoter
Color concentrate, granules	problem-free	problem-free	too complicated
Usual mixing time	10 to 20 min	5 to 20 min	0.5 to 2 min
Shading	good	difficult	scarcely possible

10.8 Coloration of PVC

The stability requirements to be met by colored PVC sometimes differ considerably and are determined by the fields of application, e.g. articles with a short or long life time, indoor or outdoor application; possible methods for coloration of PVC are listed in table 9.

Table 9. Colorants for the various methods used to process PVC.

	I		II		III	IV		
	Single pigments		Pigment mixture matched to shade		Pigment pastes	Pigment preparations		
	normal	free-flowing	normal	free-flowing		granules	grains	powder 100 μm
Calendering of unplasticized PVC	+	not usual	+	not usual	○	–	–	++
Calendering of plasticized PVC	+	not usual	+	not usual	++	○	○	++
Extrusion and injection molding of unplasticized PVC powder mixture	+	++	+	++	○	–	++	++
Extrusion and injection molding of plasticized PVC powder mixture	+	++	+	++	+	++	++	++
Extrusion and injection molding of plasticized PVC granules	○	○	○	○	–	++ (cable)	○	+
Processing of PVC plastisol	○	–	○	–	++	–	–	–

++ = preferred + = suitable ○ = possible, but not usual – = not possible

10.8.1 Calendering

The further processing of edge trimmings from calendered films necessarily results in a multiple thermal stress at 180 to 220 °C. In addition to migration fastness, a high color-shade-temperature-time stability is therefore especially necessary. A further requirement that has to be fulfilled is ready dispersibility of the pigments, since pigments that are difficult to disperse inevitably result in deeper shades being obtained (release of tinctorial strength reserves) if several passes are necessary. Pigment selection depends on the field of application, organic pigments in the mono- and disazo range being the products mainly used in addition to chrome yellow, molybdate red and iron oxide pigments. Blue shades are matched with phthalocyanine pigments along with a few ultramarine products. Phthalocyanines are likewise used for green, combinations of yellow and blue pigments being also sometimes employed for coloristic reasons.

Pigments with very high fastness values are used in particular in PVC films for outdoor application. Nickel and chrome titanium yellow, chrome oxide green, cobalt green as well as cadmium red can be mentioned as important products among the inorganic pigments. Organic pigments of importance are mainly the condensed azo pigments, as well as monoazo compounds (yellow, red, brown), products in the isoindoline range (also halogenated), quinophthalones, perylenes and phthalocyanine blue and green.

10.8.2 Factors influencing the weather resistance of PVC [36]

The percentage of colored PVC used in outdoor applications is quite appreciable. Since, after a certain time, PVC like most thermoplastics initially undergoes a purely visual change on account to weathering, the fulfilment of very high fastness requirements necessitates that the following pre-conditions are met:

- correct selection of polymer and additives,
- optimum stabilization in type and quantity,
- optimum pigmentation in type and quantity,
- opacity of the pigments,
- proper preparation of the mixture,
- preparation of test specimens under conditions likely to be encountered in practice,
- low thermal stress during processing,
- good surface finish of the plastics products.

Further factors that are of influence as regards high weather resistance are:

- climate (temperature, humidity, precipitation),
- soiling due to atmospheric influences (aggressive media),
- geographical position and elevation (solar irradiation),
- assembly of the plastics products at exposure (heat buildup).

10.8.3 Weathering phenomena [37]

In the course of time, all the factors listed above cause weathering in plastics. This results from molecular degradation at the surface, which is, however, retarded when opaque organic and/or inorganic pigments are used. A change in shade manifests itself in a matte, whitish, opaque appearance (cracking), but has nothing to do with destruction of the pigment.

"Chalking" in plastics with high percentages of titanium dioxide as white pigment and/or fillers is a familiar phenomenon. This term describes a disintegration of the plastics surface that is so pronounced that embedded pigment particles are exposed and come to the surface.

10.8.4 Interaction between pigments and PVC stabilizers

Lead-containing PVC stabilizers interact chemically with sulfidic pigments (cadmium yellow, cadmium red) and in a humid atmosphere form the dark brown lead sulfide. Because of its chalking tendency, lead stabilization in the presence of colored pigments should be examined especially critically if at the same time the coloration is required to have good weather resistance. This change is rather rapid and marked when combinations of lead stabilizers and lead-containing pigments are used (chrome yellow, molybdate red). Sulfur-containing tin stabilizers, which by nature are poor stabilizers of PVC against the effects of weathering, likewise react chemically with lead-containing pigments to form dark lead sulfide.

10.8.5 Extrusion and injection molding

When these processing methods are employed, pigment selection must be carried out on a more differentiated basis, since the stabilizer exercises a considerable influence. For lightweight profiles used for interior applications, e.g. cable ducting, fillets, curtain rails etc., pigments with a low to medium fastness level are employed. The main shades are white, grey and beige. Heavy-duty profiles and panels such as are used for windows or facade cladding call for high-grade pigments. Apart from highly stabilized titanium dioxide (rutile), the products suitable for colored formulations are, in particular, nickel and chrome titanium yellow as well as high-grade organic pigments such as phthalocyanine blue and green, condensed azo pigments, halogenated isoindolinone pigments and flavanthrones. The chief requirement is for brown shades (imitation of wood). The shades yellow and brown are superior to red, blue and green as regards weather resistance. Furnace black is the black pigment recommended for outdoor application. The decisive factor is the level of pigmentation, which with inorganic pigments should be 3 to 5% and 0.5 to 1% with organic pigments. RAL shades[1]) produced with pigment mixtures are generally used for pipes. Here the level of pigmentation is very low at 0.02 to 0.3%. Weathering thus quickly gives rise to chalking.

For the coloration of PVC cable sheathing compounds, the only products suitable are selected, specially washed organic pigments with low electrolyte content, high tinctorial strength and moderate fastness properties, as well as lead chromate and lead molybdate pigments. Because of the very high extrusion rates, it is advisable to use the above-mentioned pigments in the form of pigment concentrates (masterbatches). These "cable sheathing concentrates" should possess high dispersibility, since pigment agglomerates have an adverse effect on the dielectric properties required.

When coloring plasticized PVC profiles and tubing, special attention should be paid to the migration resistance of the organic pigments.

10.8.6 Coating, casting, dipping

These techniques are employed for the production of imitation leather, covers and toys. Coloration is carried out almost exclusively with pigment pastes. Apart from lightfastness, the most important criteria to take into account are migration behavior and fastness to perspiration and saliva (toys).

[1]) RAL = Committee for terms of delivery and quality assurance in the German Standards Committee

10.8.7 Causes of faults during compounding

PVC is normally blended in a high-speed mixer at temperatures up to 120 °C. At peripheral speeds of 20 to 23 m/sec., very high friction arises. With the readily dispersible, *soft-grained* pigments such as chrome yellow or molybdate red, as well as cadmium pigments, pigment damage can occur which is manifested in a coloristic change and a fall in the fastness values (weather resistance, chemical resistance). This can be remedied by waiting until the temperature has reached 80 °C before adding the pigments in the high-speed mixer or later in the cooling mixer.

Hard-grained pigments such as oxidic pigments, on the other hand, soon cause metal abrasion in the mixing units. The carrying-over of heavy metals thus results not only in a coloristic change (greying, especially with white shades), but also in a reduction in weather resistance, since heavy metals, in particular iron, cause catalytic degradation of PVC by formation of metal chlorides. This can be remedied by adding these pigments at a relatively late stage in the mixing process. Agglomerates, especially in organic pigments, must be eliminated by means of a very high degree of dispersion, since tinctorial strength is otherwise "thrown away". Moreover, specky, inhomogeneous material would be obtained. In combinations with inorganic pigments, the use of compounded pigments (pigment mixtures) is to be recommended. Attention should be drawn to a possible deterioration in the mechanical properties due to pigment agglomerates.

10.9 Coloration of polyolefins

10.9.1 Coloration criteria

The major importance and wide range of applications of polyolefins are well-known. When selecting the colorants it is necessary to take into account the type of polyolefin and method of processing. For the processor carrying out his own coloration, a wide range of colorants with different fastness levels is available. Preference is given to inorganic and organic powdered pigments and pigment concentrates in granule, grain and powdered form. However, the use of pigment pastes with sufficiently compatible carriers is also feasible for small injection-molded parts and hollow moldings whereby processing is completely dustfree. In view of the diversity of processing techniques, it is necessary to ensure that the pigments have the required high level of heat stability. In injection molding this can lie in the vicinity of 280 °C; with the processing of LDPE, however, a heat stability of approx. 240 °C very often suffices. When using organic pigments, it is expedient to check the migration resistance, since an organic pigment can be sufficiently stable in HDPE or PP, but fail in LDPE. With regard to this property, dyes, too, are therefore unsuitable, although in PP satisfactory migration resistance was determined in a few cases. Manganese lakes (e.g. C.I. Pigment Red 48/4), exhibit highly critical behavior particularly with thin-walled products such as plastics film and if they are added in large quantities. This can lead to a catalytic degradation of polyethylene even when light, moisture and heat are excluded. As a consequence, polyethylene loses its typical properties and becomes brittle. In spite of this risk, coloration with manganese lakes continues to be of interest for articles with a short life time. This can be attributed to the highly favorable cost/performance relationship and the coloristic aspect (deep red shade). However, it is being replaced to an increasing extent by the more stable strontium lake. Manganese and barium lakes are unsuitable for the coloration of utility articles with a short life time.

Among organic pigments, the blue and green phthalocyanines are highly rated due to their high level of stability. However, especially in the injection molding of large articles such as bottle crates, this product class causes problems, since phthalocyanines accelerate crystallization (nucleation), in particular of HDPE, with the result that stresses arise in the injection-molded part in the longitudinal and transverse directions. This leads to distortion and a lack of dimensional stability. Even fairly small injection-molded parts with narrow dimensional tolerances (can lids) can be problematical. These phenomena apply not only to phthalocyanines. A number of other organic pigments such as perylenes or disazo pigments can also cause similar types of disturbances.

On the other hand, when homopolymers and copolymers of polypropylene are used, the pigments mentioned have no adverse effect on the dimensional stability of the finished products. Inorganic pigments cause absolutely no distortion in polyolefins.

The abrasive action of hard-grained pigments may be a disadvantage. Chrome oxide green, titanium dioxide, nickel and chrome titanium pigments can be mentioned as typical products in this group. During the mixing of plastics compounds and their subsequent processing, these pigments can cause heavy wear due to metal abrasion at mixer walls and in high-speed screw and cylinder molding machines, slitter damage during film tape production, wear in spinnerets etc. This can be remedied by selecting other pigments, although it is generally necessary to compromise with regard to fastness requirements, or using pigment concentrates instead of powdered pigments, or finally by replacing machinery or mold parts especially subject to mechanical attack by parts with hardened surface.

10.9.2 Weather-resistant coloration of injection-molded parts [38]

Bottle and shipping crates made from HDPE and recently also from PP copolymer can be mentioned as a typical example of products for outdoor use. On the basis of many years' experience, high-grade inorganic pigments have largely managed to gain a hold on this market. The following can be mentioned as the most important of these products:

white:	aftertreated, stabilized titanium dioxide (rutile),
yellow:	cadmium yellow, nickel titanium yellow, chrome titanium yellow,
red:	cadmium red, iron oxide red,
blue:	cobalt blue, ultramarine blue,
green:	cobalt green, chrome oxide green,
brown:	iron oxide brown, chrome iron brown,
black:	carbon black (furnace).

Particular mention should be made of the outstanding weather resistance of cadmium yellow used alone. This statement must be qualified for pigment combinations, e.g. with cobalt blue or cobalt green. Moreover, in outdoor applications in countries with a hot, damp climate, a shift in shade towards blue can occur after a relatively short time as a result of oxidation of the sulfidic cadmium yellow pigment. Ultramarine pigments are less resistant to weathering.

The use of organic pigments is restricted to selected products, since, especially in blends with titanium dioxide, marked coloristic changes may take place within a few months, with the performance properties of polyethylene products being finally impaired to a considerable degree by degradation and embrittlement.

Although inorganic pigments protect polyolefins for an appreciably longer time against embrittlement, polyolefin articles used in outdoor applications should in principle be stabilized against the influence of light. The type and quantity of the relevant stabilizers are dependent on polymer type and intended usage, as well as on the climate of the area concerned (cf. chapter 3 on light stabilizers).

10.9.3 Extrusion

10.9.3.1 Hollow moldings

Almost all the known types of colorants are used for coloring hollow moldings:

a) pigment mixtures, matched to shade, also free-flowing,
b) pigment concentrates in the form of granules, grains, powder, suitable for automatic metering,
c) liquid concentrates dustfree, suitable for automatic metering.

When using powdered pigments or dustfree pigment preparations in powdered form, the addition of a maximum 0.5% of an adhesion promoter to the natural plastics granules can be recommended as a means of improving the quality of the coloration. Apart from important pigment properties such as heat stability and fastness to migration and light, chemical resistance is also of interest. In this connection, a knowledge of the product to be packaged in the container is of decisive importance if the coloration has to meet the requirement for physiological inertness. With organic pigments, particular attention must be paid to the resistance of the colorant to migration into products containing grease or oil. The purity of the colorants as regards heavy metal and aromatic amine content is also a very important factor. The coloration level of pigment preparations should be selected so as to ensure that the carrier material does not exceed the maximum permissible concentration in the end product (pay attention to the recommendations of the health authorities of the different countries and to the corresponding statutory regulations).

10.9.3.2 Plastic films, fibers, film tapes [39, 40]

The coloration of plastic films, fibers and film tapes takes place almost exclusively with concentrates in granule form, but can also be carried out to advantage with powdered pigments if the homogenizing capacity of the extruder is only moderate or only small quantities of colorant have to be dispersed with large quantities of natural material. Powdered pigments that have not been pre-dispersed are hardly suitable for direct coloration. The amount of concentrate to be added depends on depth of shade required, layer thickness and opacity and can vary between 1 and 10%. With fibers, the criteria to be examined, in addition to heat stability and lightfastness, are: wash fastness (DIN 54010, test 3), and rub fastness (DIN 54021). It should be borne in mind that rub fastness can be influenced by the type and quantity of the spin finish and softening agent.

Particularly when used in outdoor applications, polyolefin fibers and films are damaged by the u.v. content of sunlight, and pigments can increase or decrease the destructive action [41]. The addition of light and heat stabilizers in suitable combinations and concentrations enables any adverse effect by certain pigments to be prevented or reduced at least [42].

10.10 Coloration of styrene polymers

10.10.1 Coloration criteria

In previous years, the coloration of general purpose polystyrene and styrene copolymers was largely carried out by the resin manufacturer. This took place almost exclusively in the melt with single pigments using kneaders or extruders. Because of technical problems and particularly in view of improving the quality of coloration in the finished products, the processor carrying out his own coloration nowadays gives preference to pigment concentrates, occasionally to such in liquid form.

Direct coloration using readily dispersible colorants in powdered form is, of course, also possible. The latter are dry blended with the polystyrene pellets in a mixer at room temperature, a maximum 0.5% of an adhesion promoter or dispersing agent being added where necessary, and subsequently processed under relatively high back pressure on screw injection molding machines, with good mixing action, or on extruders.

When colorants are dry blended with polystyrene, account must be taken of the fact that transparent, general purpose polystyrene is a very hard, brittle thermoplastic. With pigments which are especially sensitive to abrasion, the sharp cutting edges and corners of the granules can cause shifts in shade. A coloristic change due to metal abrasion during mixing in metal drums is also possible. Plastics-coated mixer walls or blending in plastics vessels (tumbler) are suitable alternatives.

Impact-modified polystyrene (IPS) and polystyrene terpolymers (ABS) are not generally transparent and, ABS especially, possesses a more or less pronounced inherent yellow color. A relatively high level of pigmentation in the range from 3 to 5% is thus necessary. Coloration with powdered pigments can consequently only be carried out at by the polymer manufacturer with the aid of heated kneaders, since the high pigment concentration needed to compensate for the natural color cannot be taken up by the surface of the granules. This also applies to the more weather-resistant acrylonitrile-styrene-acrylic ester terpolymers (ASA).

Coloration with weather-resistant pigments possessing high heat stability is recommended.

The coloration of thermoformable, coextruded ABS/PMMA sheets for the sanitary engineering sector by the plastics processor constitutes a difficult problem. The high quality requirements with regard to dispersion and shade retention are fulfilled exceptionally well by pigment concentrates with all-round compatability, such as are supplied by BASF AG under the name "®Sicoversal pigments".

10.10.2 Colorants for polystyrene and SAN

The high transparency of these plastics permits the manufacture of colored glass-clear articles. The products best suited for this purpose are soluble dyes, which on account of their monomolecular solution in the plastics mentioned develop high tinctorial strength at the very low concentration of 0.03% and, in addition, possess very high dispersibility. As a result of the characteristically high tinctorial strength of soluble dyes, opaque colorations can be economically produced in combination with inorganic pigments. Whereas the transparent coloration, good to moderate lightfastness values are obtained, in reductions with white pigments there is a marked deterioration in this property. Because of the high glass transition temperatures of these polymers and their good dissolving power, there is only a small risk of migration when dyes are used. In the case of utility articles and toys, which are manufactured from impact-modified PS and ABS, extraction tests must be carried out without fail to demonstrate product safety.

Apart from soluble dyes, almost all organic and inorganic pigments stable at a processing temperature of 220 to 280 °C are suitable. Partly soluble organic pigments not only change the shade in an uncontrollable manner, but also their typical stability characteristics.

Pigment concentrates in granule or powdered form facilitate coloration by the plastics processor. Mention should also be made of liquid concentrates in which dyes and/or pigments are dissolved or pre-dispersed in suitable carriers.

Their use results in the following advantages:

- Volumetric metering of liquid concentrates on suitable automatic units,
- mold release agents are often no longer needed,
- no dust formation during processing,

- larger quantities of granules can be colored for stock using gravity or high-speed mixers,
- suitable mixers also permit the coloration of plastics in powdered form.

Since liquid carriers generally possess only limited compatibility with polymers, attention must be drawn to the restrictions on the quantities added, which usually amount to a maximum of 1.5%. Regarding the approval to use liquid concentrates for utility articles it is advisable to consult the manufacturers of such concentrates.

10.11 Polyamides, polycarbonate

Due to their chemical structure, polyamides and polycarbonate are hydrophilic and sensitive to moisture. A pre-condition for problem-free processing is, therefore, good pre-drying of the polymers themselves and of all admixed materials, i.e. including the pigments or pigment concentrates. Almost all inorganic pigments with a heat stability of up to 300 °C and above are suitable for polyamides. The selection of organic pigments and dyes is considerably restricted, since these colorants are quickly affected by the reducing influence of polyamide melts and may be destroyed at high processing temperatures. On the other hand, it has been observed that in packaging articles, in particular those made from polyamide 6, cadmium pigments, for example, can interact with acidic materials packed in these containers and stain them.

In the case of polyamides it is, therefore, always advisable to consult the manufacturers with regard to the fastness of colorants against specific products to be packaged. The polar nature of polyamides also permits a subsequent coloration of the surface by dyes in an aqueous liquor, as is customary in the textile industry.

Polycarbonate can not only cause difficulties during coloration due to the presence of moisture (hydrolysis), but wetting problems have also been encountered, which can generally be eliminated by small additions of magnesium stearate. The range of colorants (some organic pigments and dyes, almost all inorganic pigments) is likewise restricted due to the fact that the processing temperatures exceed 300 °C.

10.12 Polyurethane

PUR foams are colored almost exclusively with pigment pastes. This can be carried out by adding the color paste directly at the mixing head of the foaming machine or by stirring it into the polyol component. Because of their high tinctorial strength, preference is given to organic pigments, although, due to the possible presence of organic solvents as blowing agents, it is essential to check fastness to migration.

For the coloration of polyurethane coating and injection molding compounds, the market offers special pigment concentrates. These contain both high-strength organic pigments and inorganic pigments with high opacity. These pigment concentrates must be converted to suspension form by means of specific solvents and subsequently added in portions to the coating compound.

10.13 Engineering plastics

These include polysulfones, polyoxymethylenes, polytetrafluorethylene, polyethylene terephthalate etc. Because they are used chiefly for technical applications, such materials are only in specific instances colored with special pigment concentrates. Coloration is thus generally carried out by means of single pigments or pigment concentrates with all-round suitability.

10.14 General remarks

Since pigment properties are known to depend not only on the chemical structure of the pigment class, but also on particle size distribution, finish and surface coating of the pigments, as well as the manufacturing process, the reader is referred to the trade literature and the technical data of the pigment manufacturers when more detailed information is required. Because of the vast number of different pigments, the latter have been given a Color Index (C.I.) number and registered [8]. For the reasons listed above, however, pigments with the same Color Index number from different manufacturers can differ in their range of properties. Testing of the colorant in the plastic type for which it is intended and under conditions conforming to those in practice would, therefore, seem expedient.

10.15 Technological trend and future prospects

Plastics are materials of our time and have completely discarded the stigma of being regarded as "substitute materials". In close cooperation with the engineering industry, plastics manufacturers and processors have developed modern processing technologies that have opened up almost unlimited design opportunities with plastics. Moldings of modern design are also expected to generate a certain appeal, so that a stimulating effect is exerted on the person viewing them. This can be given additional emphasis through color. The majority of plastics articles are, after all, marketed in colored form. Color increases the attractiveness of the final product, color improves visual appeal.

The pigment industry supplies colorants in various forms. The trend away from high-dusting powdered pigments towards low-dusting or dust-free, free-flowing grades and, where necessary, liquid colorants is alone for reasons of industrial hygiene dictated by the demands of the time. Measures to protect the environment, as well as to increase efficiency and safety will inevitably entail cost burdens. The considerations given by the plastics processing industry to transferring its color compounding activities to the colorant supplier have for some time been of topical interest due to the need to reduce costs. Pigment concentrates matched to shade will thus continue to gain in importance.

An additional task is to furnish these granular colorants with additives such as antioxidants, light stabilizers, antistatic agents, flame retardants etc. in accordance with requirements. The diversity of plastics types on the market makes the requirement for pigment preparations with all-round suitability even more pressing. For economic reasons alone, it is difficult both for the processor of different thermoplastics and for the colorant manufacturers to maintain for each plastic a pigment concentrate in a wide range of shades and with a specific carrier. The capital tied up in inventory, including possibly dead inventory rises and the increase in delivery capability gained by carrying out one's own coloration is largely lost through the need to stock a large number of products.

In pigment research, the search for new structures is continuing, the price/performance factor coming increasingly to the forefront. The market potential for pigment preparations will continue to rise. The need to comply with certain statutory requirements has led to stagnation in a number of inorganic colored pigments in spite of the latter's good fastness properties and favorable price level. The demand for stable organic pigments and dyes, also for engineering plastics thus continues.

For the pigment researcher this opens up a wide field of activity and for the pigment industry a secure outlet for its products, since these materials will in the future continue to gain in importance.

10.16 Index of suppliers

This list does not claim to be complete.

A detailed list of manufacturers, commercial firms and trade names is contained in *Lückert:* Pigment- und Füllstofftabellen, *M.* and *O. Lükkert,* Laatz, 1977 as well as in Modern Plastics Encyclopedia, McGraw-Hill, Inc. New York, 1983/84.

American Cyanamid Co., Chemical Group
Wayne, NJ, USA

American Hoechst Corp., Industrial Chemicals Div.
Pigments Dept.
Coventry, RI, USA

BASF AG, D-6700 Ludwigshafen

BASF Lacke + Farben AG, Ressort Pigmente
D-7000 Stuttgart 30

Bayer AG
D-5090 Leverkusen

Blythe Colours Limited, Cresswell, Stoke on Trent
St 11 9rd/England

Bruchsaler Farbenfabrik
D-7520 Bruchsal

Cabot GmbH
D-6450 Hanau 9

Cappelle Frères Pigments
F-59433 Halluin Cedes

CdF (APC) Chimie S.A., Paris
F-92080 Paris la Defense 2

Ciba-Geigy AG
CH-4002 Basel

Ciba-Geigy Corp. Plastics & Additives Div.
Hawthorne, NY, USA

Colores Hispania SA
Barcelona, 19/Spain

Dainichi Seika Color & Chemicals Mfg. Co., Ltd.
Tokyo, Japan

Dainippon Ink and Chemicals Inc.
Tokyo, Japan

Degussa, Geschäftsbereich Pigmente
D-6000 Frankfurt/Main

Du Pont de Nemours, E.I., & Co.
Wilmington, DE, USA

Eckart-Werke
D-8510 Fürth

Ferro B.V.
NL-3002 AB Rotterdam

Harshaw Chemical Co.
Cleveland, OH, USA

Heubach, Goslarer Farbenwerke, Dr. Hans Heubach & Co.
D-3394 Langelsheim

Hoechst AG
D-6230 Frankfurt/Main 60

ICI Americas Inc.
Wilmington, DE, USA

Deutsche ICI GmbH
D-6000 Frankfurt/Main 71

James Brown
London W1X OBP/England

Køge Kemisk Vaerk Køge
DK-1414 Kopenhagen

E. Merck
D-6100 Darmstadt 2

Novosystems GmbH
D-2100 Hamburg 90

Nubiola S.A.
Barcelona 10/Spain

Reckitt's Colours S.A.
F-59560 Comines

Sachtleben Chemie GmbH
D-4100 Duisburg 17

Sandoz AG
CH-4002 Basel 13

Sandoz Colors & Chemicals
Hanover, NJ, USA

Silo SPA
I-10148 Torino

Sumitomo Kagaku K.K.
Osaka, Japan

Sun Chemical Corp., Pigments Div.
Cincinnati, OH, USA

Swada Ltd.
Stratford, London E 15 2QN/England

TG Kronos Titan GmbH
D-5090 Leverkusen 1

Tioxide International Ltd.
Billingham, Cleveland TS 23 1 PS/England

Toyo Ink Manufacturing Co., Ltd.
Tokyo, Japan

10.17 References

[1] *D. Hayer:* Einfärben von Kunststoffen, Carl Hanser Verlag, München 1962.
[2] *E. Herrmann:* Handbuch für den Pigmentverbraucher, 1. Aufl., Curt R. Vincentz-Verlag, Hannover 1962; 2. Aufl. Stuttgart 1982.
[3] *E. Herrmann:* Kunststoffeinfärbung, Zechner und Huttig-Verlag, Speyer am Rhein, 1976.
[4] *E. Hertel:* Farbproben zur Prüfung des Farbsinnes; Stilling'sche Tafeln, G. Thieme-Verlag, Leipzig, 1939.
[5] VDI-Gesellschaft Kunststofftechnik: Einfärben von Kunststoffen, VDI-Verlag, Düsseldorf 1975.
[6] DIN-Taschenbuch 16: Farbechtheitsnormen, Beuth-Verlag, Berlin/Köln/Frankfurt 1973.
[7] DIN-Taschenbuch 30: Anstrichstoffe, Pigmente, Beuth-Verlag, Berlin/Köln/Frankfurt 1973.
[8] Colour Index, 3. Aufl. und Ergänzungsbände, The Society of Dyers and Colouries-Verlag, Bradford 1971.
[9] *G. Kaufmann:* Plastverarbeiter 20 (1969) p. 457.

[10] *E. Herrmann:* defazet 29 (1975) p. 116.
[11] *G. Benzing:* Einfärben im Spritzguß in [5].
[12] *E. Herrmann:* VDI-Berichte 130 (1968) p. 19.
[13] *L. Gall:* Kunststoff-Rundschau 17 (1970) p. 282.
[14] *H.-J. Lenz:* Farbenlehre und Farbmessung in [5].
[15] *H.-J. Lenz:* Licht und Farbe in [5].
[16] *E. Herrmann:* Kunststoffe 62 (1972) p. 695.
[17] *E. Herrmann:* Plastverarbeiter 18 (1967) p. 897.
[18] *E. Herrmann:* Farbe + Lack 69 (1963) p. 601.
[19] *J. Richter:* Plastverarbeiter 19 (1968) p. 333.
[20] *E. Herrmann:* Farbe + Lack 72 (1966) pp. 966, 1055.
[21] *E. Herrmann:* Kunststoffe-Plastics 10 (1963) 3, p. 273.
[22] *H.-J. Beyerlin:* defazet 28 (1974) p. 302.
[23] *J. Koerner:* Lösliche Farbstoffe in der Kunststoffindustrie in [5].
[24] *K. Heinle:* Farbe + Lack 73 (1967) p. 735.
[25] *H.-J. Mair:* Kunststoffe 57 (1967) p. 930.
[26] *R. Rode:* Ruß als Pigment für Kunststoffe in [5].
[27] *K. Kaluza:* Grundlagen der Pigmentdispergierung in [5].
[28] *H. Rumpf:* Chem. Ing. Tech. 37 (1965) 3, p. 187.
[29] *A.W. Neumann, P.J. Sell:* Z. Phys. Chem. 227 (1964) p. 197.
[30] *B. Schindler:* Diss., Universität Stuttgart, 1973.
[31] *E.W. Washburn:* Phys. Rev. 17 (1921) p. 273.
[32] *W. Damm,* in: VDI-Gesellschaft Kunststofftechnik: Aufbereiten von PVC, VDI-Verlag, Düsseldorf 1976.
[33] Firmenschrift BASF Farben + Fasern AG: Einfärben von Kunststoffen mit ®Sicoplast-Farben, Anwendung, technische Hinweise, TI/p. 1023/4.79.
[34] *E. Paschke, E. Ullrich:* Spritzgießen von Polyolefinen, in: Kunststoff-Handbuch, Band IV Polyolefine, Carl Hanser-Verlag, München 1969, p. 475.
[35] *D.H. Zettler:* Typische Verfahrensweisen und Probleme beim Einfärben von Thermoplasten in [5].
[36] *W. Damm:* Der Abbau von PVC unter Einfluß von Sauerstoff und Licht, Fakultas-Verlag, Wien 1975, p. 216.
[37] *W. Damm:* Colouring of Polyolefines for highest Light- and Weatherfastness, Theoretical Conceptions about the Decomposition of PVC, Norweg. Plastverband, Kongsberg, 1977.
[38] *G. Benzing:* Plastverarbeiter 26 (1975) p. 65.
[39] Firmenschriften BASF Farben + Fasern AG: Technische Informationen über Pigmentpräparationen für die Einfärbung von Polyolefinen (®Euthylen-Pigmente TI/P 2650, ®Lufilen-Pigmente TI/P 2677d, ®Lufopril-Pigmente TI/P 264).
[40] *G. Biehler:* Chemiefasern Textilindustrie 29 (1979) pp. 848/853.
[41] *C. Ripke:* Chemiefasern Textilindustrie 30 (1980) p. 110.
[42] *C. Ripke:* Chemiefasern Textilindustrie 30 (1980) p. 30.

11 Flame Retardants for Thermoplastics

Dr. Herbert Jenkner, Chemische Fabrik Kalk GmbH, Köln,
Federal Republic of Germany

11.1 Introduction

Thermoplastics, being organic compounds of carbon and hydrogen, sometimes also containing oxygen and/or nitrogen, are more or less easily combustible. Efforts to develop flame-retarding plastics materials have been going along with the increasing use of thermoplastics. As a result, flame-retarding formulations are available today for all thermoplastics which strongly reduce the probability of their burning in the initiating phase of a fire. The possibility to make plastics flame-retardant secures the scope of utilization for thermoplastics and, in fact increases their range of application.

Because, however, for each field of application an optimum flame-retarding effect is required, a great number of processes and methods exist in order to meet these demands. The situation is complicated further, because almost every country has developed its own test methods and classification systems [22].

Although it can be assumed that the problems associated with the flammability of thermoplastics, with regard to initiating fires, have been largely solved with flame retardants, it must not be forgotten that in the case of fire the effectiveness of flame retardants depends on the period of time and intensity of such a fire. Even a product containing the most effective flame-retardants cannot resist a strong and long-lasting fire!

The demands on a flame retardant and thermoplastics formulated with such agents are manifold: The flame retardant should provide a durable flame-retarding effect by the addition of only small quantities of the additive; the manner of incorporation should be easy and it should be as cheap as possible; it should by no means cause corrosion of the processing equipment or cause plate-out. The mechanical and the physical properties of the base resin should not be affected. The additives should be non-toxic: neither fire effluents with increased toxicity, nor abundant smoke-nuisance should occur from the use of the flame retardant additives. Flame retardants for thermoplastics ought to show sufficiently high thermal stability and should not decompose at the processing temperatures required, nor undergo reactions with the thermoplastics. However, the fire retardant formulation must ensure that in case of fire, decomposition resp. evaporation of the flame retardant will take place. Finally, as to the end-product, great value is being laid on the resistance against weathering as well as on the stability against u.v. radiation. Besides, flame retardants should neither affect the dyeability, nor the polymerisation process when using compounds introduced during polymerisation. At any rate, they should not volatilize and smell; all in all the mechanical, optical and physical properties of the thermoplastics should be affected as little as possible.

It is understandable that these far-ranging demands cannot be satisfied by only one flame retardant and – over and above that – for all thermoplastics with their manifold applications. Regarding the fire retardant treatment demanded, one is indeed forced to seek the optimum flame retarding formulation for each thermoplastic and the specific application [1 to 4].

11.1.1 Economic significance

The market for flame retardants for plastics and textiles has already reached a remarkable volume. In 1977 180000 t of flame retardants have been produced in the USA, of which approx. 22000 t were organic bromine compounds.

In Europe the market for flame retardants is said to be approx. 52000 t:

bromine compounds	11000 t
chlorinated paraffins	5500 t
other chlorine compounds	6000 t
phosphoric acid esters	5000 t
antimony trioxide	10000 t
alumina trihydrate	15000 t
	52000 t

The consumption for 1985 is expected to be over 62000 t [19]. The Japanese market for bromine-containing flame retardants is estimated for 1980 at about 4000 to 5000 t.

Among the bromine-containing flame retardants the production of tetrabromobisphenol A (for epoxy resins and polycarbonates) is greatest, followed by brominated diphenyls and diphenyl ethers (for polystyrene and

ABS), hexabromocyclododecane (for polystyrene foam and for polystyrene) and brominated phenol- resp. bisphenol-A-ether (for polyolefins). The demand for bromine-containing flame retardants will expand probably at a greater rate than that of other fire retardant products, because

- requirements for fire safety, especially for the building trade and electrical appliances, but also for the field of public transportation, are going to become very strict;
- bromine compounds are particularly efficient, meeting even high performance demands with relatively small quantities of additives.

During the last years the sales of bromine-containing flame retardants increased only to a small extent. The reasons for this were the general economic situation and the discussions about the assumed toxicity and mutagenity of certain products. In spite of this, growth rates of 20 to 25% were anticipated in 1978 [5], that is to say that the consumption might increase by 1985 to:

- in the USA, from 22000 t to 70000–100000 t,
- in Europe, from 11000 t to 15000– 25000 t,
- in Japan, from 5000 t to 10000– 15000 t.

However, recent estimates are more cautious: An average growth rate only between 8 to 10% per annum is foreseen [19].

11.2 Theory of the mode of action of flame retardants

The combustion of thermoplastics is a succession of physical and chemical processes whereby substances, in reaction with oxygen, release heat and form the less energetic products water, carbon dioxide and carbon monoxide.

With a given energy supply the rate of temperature increase of the material depends on its specific heat, its density, thermal conductivity, latent heat and its heat of evaporation. This means, under the same conditions, it is more difficult to bring substances with high specific heat to a given temperature than those of lower specific heat.

Polystyrene, polyvinyl chloride and polyester have more or less the same thermal conductivity, that of polyethylene and polyamide-6 is approx. twice as high. The thermal conductivity of Pyrex glass is approx. three times as high as that of polyethylene, whereas copper conducts heat a thousand times better.

The combustion process is initiated by heating the plastics material up to its decomposition point through heat supplied by radiation, flame or convection. Thereby, besides non-combustible gases as carbon dioxide and water, combustible gases are formed, mainly hydrocarbons, hydrogen and carbon monoxide. If an adequate oxygen supply is available, the combustion process is started by pilot or self-ignition. With sufficient production of heat and heat reflection onto the material, the combustion proceeds automatically. On account of the rising temperature, the combustion rate increases.

In case of fire, the basic properties of thermoplastics can be classified by flammability and flame propagation which are closely connected. The composition of the plastics material, its form and density as well as its geometry influence the flammability particularly, whereas the flame propagation is mainly influenced by the amount of oxygen available.

11.2.1 Effect of halogen compounds

During combustion of thermoplastics, free radicals are formed by pyrolysis. The radicals combine with oxygen in a radical chain reaction. Heat release and further decomposition of the plastics material is caused thereby. For continued combustion it is necessary to have sufficient oxygen as well as combustible gaseous compounds. The combustion will be slowed down or interrupted if free radicals, which are evolved by pyrolysis, are blocked. It is presumed that the following reactions take place, when flame-retarded thermoplastics containing organobromine compounds as flame-retardants are used.

$HO^{\bullet} + CO = CO_2 + H^{\bullet}$	(1)	propagation, highly exothermic
$H^{\bullet} + O_2 = HO^{\bullet} + O^{\bullet}$	(2)	chain branching
$O^{\bullet} + HBr = HO^{\bullet} + Br^{\bullet}$	(3)	chain transfer
$HO^{\bullet} + HBr = H_2O + Br^{\bullet}$	(4)	chain termination

The radical chain reaction is interrupted when the highly reactive $HO^{\bullet}$-radical (spectroscopically proved [11]), which occupies a key position within the combustion process, is replaced by the less reactive $Br^{\bullet}$ radical (reaction 4). Experiments have shown, however, that halogen containing organic compounds may also react as undecomposed molecules [6, 7].

The superiority of bromine compounds is evident, i.e. in case of the combustion of a n-heptane/air mixture, the following minimum concentrations (in vol.-%) were found necessary for the extinction of the flame [3]:

dibromodifluoromethane	4.2%
dibromoethane	5.2%
ethyl bromide	6.2%
bromochloromethane	7.6%
hydrobromic acid	9.3%
carbon tetrachloride	11.5%
trichloromethane	17.5%
hydrochloric acid	25.5%

11.2.2 Effect of phosphorus compounds

Whereas halogen acids, which are released during the combustion of a flame-retarded thermoplastic containing organic halogen compounds, interrupt the chain reaction, the flame retarding action of phosphorus compounds is not yet completely understood. It is supposed, however, that in case of a fire, phosphorus compounds facilitate polymer decomposition, whereby the phosphoric acid formed reacts with the polymer and produces a char layer (protection of the surface) as well as water and nonflammable gases [3].

11.2.3 Halogen-antimony synergism

Antimony trioxide is an important component of flame-retarded thermoplastics containing halogen compounds. This compound serves as synergist and strengthens the effect of organic halogen compounds, although it itself is not a flame retardant. It is believed that as a result of using antimony trioxide and organobromine compounds, antimony tribromide is formed. It is decomposed to antimony oxybromide (SbOBr) and hydrobromic acid by the action of water released during the combustion. Hydrobromic acid is able to convert the highly reactive and chain-carrying hydroxy radical into the less reactive bromine radical (reaction 4).

At about 250 to 285 °C antimony oxybromide reacts to higher oxybromides and antimony tribromide (reaction 5):

$$5SbOBr = SbBr_3 + Sb_4O_5Br_2 \qquad (5).$$

In this way antimony oxybromide serves as reservoir for additional antimony tribromide. At temperatures around 500 °C the higher antimony oxybromides are converted back into antimony oxide by releasing again antimony tribromide (reaction 6):

$$3Sb_4O_5Br_2 = 2SbBr_3 + 5Sb_2O_3 \qquad (6).$$

Summarizing it can be said, that flame retardants exert three different modes of action:

- shielding of surface (charring),
- influence upon pyrolysis (lowering of decomposition temperature)
- influence on the combustion mechanism (chain termination).

With regard to thermoplastics, the third way is the most important one.

11.3 Testing methods

In view of the broad use of combustible materials including thermoplastics, especially in the transportation, building and electrical industries, the fire behavior of these materials must be evaluated adequately and so, besides security regulations, a series of laboratory-scale test specifications – mostly initiated by these industries – have been developed. Although these fire tests allow the comparison of one material with another, it is not possible to obtain absolute standards generally valid for all materials and under all conditions by means of these tests.

In order to avoid assessments on the burning behavior of materials by laboratory tests only, large companies like BASF AG, BAYER AG and HOECHST AG have provided test facilities, where the fire performance of materials can be tested on a larger scale and nearer to reality [8].

The fire performance of products does not only depend on the kind of material, but also on its shape, surface and mass. These material- and shape-bound influences, as well as the conditions of the product's application and the fire environment, are to be considered with respect to the fire hazard of the materials. Phenomena with increasing importance are smoke formation and toxicity of combustion gases.

Since the question related to the fire performance of materials (i.e. of thermoplastics) always includes the question of the test method used, specific test methods for the assessment of the fire hazard have to be applied which are directed to the intended application of the material.

For example, 15 different modes of testing exist according to the ASTM methods. They determine, for example, the ignition temperature, the extent of burning, the rate of flame spread, the smoke-density, flame-height and the loss in weight.

Out of the great number of laboratory test methods, a few which have some importance internationally should be briefly discussed.

ASTM D 635-77 (ISO R 1326-1970)

With this method plastics materials are tested especially with respect to extent of burning and burning time. At least 10 specimens ($12.5 \times 1.25 \times 0.3$ to 1.2 cm) with a reference mark 10 cm from the free end of the specimen are fixed horizontally in the longitudinal axis and at 45° in the transverse axis 1 cm above a wiregauge. The samples are ignited for 30 sec by means of a 2.5 cm high *Bunsen*-burner flame. If the reference mark is reached, the mean rate of burning is measured in cm/min; in the case of self-extinguishing before the 10 cm mark, the average time of burning (ATB) is recorded in seconds and the average extent of burning (AEB) in mm.

ASTM D 2863-77

This test method is also known as "oxygen index" or "limited oxygen index" (LOI) according to *Fenimore* and *Martin* [9]. With this procedure the concentration of oxygen in O_2/N_2 mixtures is determined at which the vertically mounted specimen – ignited at its upper end – just continues to burn. The size of the specimens is 0.65×0.3 cm with a length from 7 to 15 cm. The LOI-value is calculated according to the equation:

$$\mathrm{LOI} = \frac{[\mathrm{O_2}]}{[\mathrm{O_2} + \mathrm{N_2}]} \cdot 100$$

The LOI-values for a number of polymers are as follows:

polyoxymethylene	15.0
polypropylene	17.4
polyethylene	17.4
acrylonitrile/butadiene/styrene terpolymer	17.75
polystyrene	18.1
polycarbonate	22 to 24
rigid polyvinyl chloride	40
polytetrafluoroethylene	95

This easily reproducible test is particularly useful in the early state of developing new flame retardants, but conclusions as to the fire performance of a plastics material in case of an actual fire cannot be derived therefrom [10].

UL 94 (Jan. 1980) of Underwriters Laboratories Inc.

By this method the burning behavior of plastics materials is judged so that the dripping of the molten material is considered too. The specimen (size: 12.7 × 1.27 × 0.16 cm or 12.7 × 1.27 × 0.32 cm) is mounted vertically so that the lower end is placed 30.5 cm above a 0.6 cm thick surgical cotton layer. The specimen is ignited for 10 sec with a defined gas flame from below. If the specimen extinguishes after removing the ignition flame, it is ignited once more for 10 sec. The results of the test are classified as follows:

Class 94 V-0: Having a series of five specimens with ten ignitions, the sum of the afterflame times should not exceed 50 sec and no specimen should burn longer than 10 sec.
During ignition and combustion the cotton layer should not be set on fire by dripping material.

Class 94 V-1: Having a series of five specimens, which are ignited twice, the sum of afterflame times should not exceed 250 sec, no specimen should burn longer than 30 sec, the dripping material should not ignite the cotton layer.

Class 94 V-2: Like V-1, except that the cottonlayer is ignited by burning drops.

FMVSS 302 (Federal Motor Vehicle Safety Standard)

This test is used for the assessment of the burning behavior of materials used for vehicle interiors. A specimen of 35.5 × 10.1 × max. 1.3 cm is mounted horizontally and ignited for 15 sec. The rate of flame spread should not exceed 10 cm/min.

More severe conditions of motor vehicle producers demand that under these test conditions no flame spread over the whole length of the specimen takes place. It is to be ascertained that the length of burned specimen does not exceed certain maximum values.

DIN 4102, Part 1 (1981)

In the Federal Republic of Germany this standard (small-burner test and Brandschachttest) is used for testing and classification of the fire performance of building materials. Grades of fire performance, as laid down in this standard, are classified as (A) "non-combustible" and (B) "combustible" materials.

Thermoplastics as combustible building products belong principally to group B, which is divided again in the classes B 1, B 2 and B 3.

This means:

B 1 = "schwer entflammbar" (low flammability)
B 2 = "normal entflammbar" (moderately flammable e.g. like wood)
B 3 = "leicht entflammbar" (highly flammable)

Flame-retarded thermoplastics can at best be classified in class B 1. To achieve class B 1, the "Brandschacht" test has to be passed. This test is specified as follows:

Four sheets of 19 × 100 cm are placed at right angles to a horizontal standing chimney and are ignited from below for 10 min with a ring burner.

The test is passed, if the mean value of residual length of the specimen is at least 15 cm; if no specimen has a residual length of 0 cm, and if the mean smoke gas temperature does not exceed 200 °C.

In addition, the test report should indicate: after-glow and flaming droplets, flame height as well as time and maximum smoke gas temperature. Test certificates are supplied by state-operated testing laboratories; B 1 products need a special approval by the "Institut für Bautechnik" in Berlin.

The testing of B 2 building materials is done in the small-burner test ("Kleinbrennertest"). This test is used to differentiate the classes B 2 and B 3.

DIN 53438

The method is similar to the above mentioned "Kleinbrennertest".

NF P 92-501 to 92-507

Contained herein are the test methods as applied in France for assessing the fire performance of building and construction materials. Legislation and requirements for fire precaution in high rise buildings and buildings open to the public are described in a 700 page brochure entitled "Sécurité contre l'Incendie".

In NF P 92-507 building materials are classified as follows:

M 0 former description: non-combustible
M 1 former description: non-flammable
M 2 former description: low flammability
M 3 former description: moderately flammable
M 4 former description: high flammability
M 5 former description: extremely flammable

Test procedure: Four specimens of 30 × 40 cm are exposed with an inclination of 45°, to the radiation of a 475-Watt radiator for 20 min in order to judge the flammability of material according to the Epiradiateur test. M 1 is passed, if no flaming occurs on the exposed surface.

NF P 92-505 evaluates the formation of burning drops and
NF P 92-504 the flame propagation of materials.

ISO (International Organisation for Standardization)

Subdivided into Technical Committees (TC) ISO has worked out regulations for certain fields of application:

ISO/TC 92 is developing several reaction-to-fire tests for building materials and components. The testing of the combustibility or non-combustibility of those materials is performed according to ISO 1182–1979.

ISO/DP 5657 (draft proposal) relates to the ignitability of building materials.

ISO/TC 22 (road vehicles) recommends to test the burning behavior of vehicle interiors according to ISO 3795–1976 in analogy with FMVSS-test 302, but no classification is given.

ISO/TC 61 is concerned with plastics materials. ISO/DP 4598 corresponds to ASTM D 2863-77 (LOI-test).

ISO 181–1981 covers the determination of the flammability characteristics of rigid plastics in contact with an incandescent rod.

ISO 1210–1982 describes the determination of flammability of plastics in contact with a small flame applied to a horizontally positioned specimen.

British Standard (BS) 476

This British Standard deals with the fire performance of bulding materials and components. According to part 7 (surface spread of flame test), the test is performed with specimens with dimensions of 90 × 23 × max. 5 cm; the specimen is placed at an angle of 90 °C to the radiation source (duration of test = 10 min). Four classes are differentiated depending on the flame spread (mm) after 1.5 min and the final flame spread.

VDE-Standards (Verband Deutscher Elektrotechniker)

Electrotechnical products manufactured in Germany are specified by the VDE-regulations. These standards distinguish between the burning behavior of insulating materials and the fire characteristics of electrotechnical equipment and installations, their elements and their components:

VDE 0304,	part 3, glow bar test for rod-shaped insulating material, classification according to flame spread or burning rate;
VDE 0303,	arc-strength;
VDE 0471,	part 2, glow wire test at glow wire temperatures from 450 to 960 °C.
VDE 0471,	part 3, glowing contact test;
VDE 0471,	part 4, short circuit arc test.

Swiss Rules for use of combustible building materials in buildings (SNV* 520 183/2–1976)

The regulation differentiates the classes I to VI.

Class I:	extremely easily ignited and extremely rapidly burning,
Class II:	easily ignited and rapidly burning,
Class III:	highly combustible,
Class IV:	moderately combustible,
Class V:	low combustibility,
Class VI:	non-combustible.

Among the classes V and VI an additional class "quasi non-combustible" has been introduced. The fire behavior is evaluated by means of determining the ignitability and flame spread or by the extinction of a vertically oriented specimen. In addition, smoke emission is considered.

*) Schweizerische Normenvereinigung

11.4 Technically important flame retardants

11.4.1 Chlorine-containing compounds

Organic chlorine compounds are being used even today in considerable quantities as flame retardants for thermoplastics, mainly because of their favorable price in comparison to the more effective, but also more expensive bromine compounds.

Chlorinated paraffins with the highest possible chlorine content belong to the technically important chlorine compounds, whereas other organic chlorine compounds, such as chlorinated naphthalene, are scarcely used any more because of their poor effectiveness and the higher concentrations required.

The main disadvantage of chlorinated paraffins is their low thermal stability. They are applicable only in thermoplastics with processing temperatures below 200 °C. Even when adding PVC stabilizers it is not possible to increase the processing temperature substantially.

Derivates of hexachlorocyclopentadiene, obtained by *Diels-Alder*-reaction with maleic anhydride, show highly improved heat stability. Such chlorinated cycloaliphatics, like for example hexachloroendomethylene-tetrahydrophthalic acid (HET-Acid) with a melting point of 208–210 °C and a chlorine content of 54%, respectively its anhydride (melting point 238 °C), are used in thermoplastics. They are often used in unsaturated polyester resins, the same is true for tetrachlorophthalic anhydride.

Tetrachlorobisphenol A is being used in smaller quantities for the production of flame-retarded polycarbonates; chlorinated styrenes have been used as comonomers for styrene and ABS as flame retardants.

11.4.2 Bromine-containing compounds

Bromine-containing compounds are on a weight basis at least twice as effective as chlorine-containing ones. Substantial flame retardancy requirements can be achieved with bromine compounds. Because of the smaller quantities needed their use hardly influences the mechanical properties of the base resins and reduces markedly the hydrogen halide content of the combustion gases. They are readily incorporated and do not bloom. under even severe conditions. The reduced u.v. stability and the higher price are a disadvantage compared with organic chlorine compounds.

Aliphatic, cycloaliphatic as well as aromatic and aromatic-aliphatic bromine compounds are used as flame retardants nowadays. Dibromopentaerythritol, an aliphatic compound, is employed as a reactive flame retardant in linear polyesters and polyurethanes. The following cycloaliphatic compounds are suitable as flame retardants in polyolefins: hexabromocyclododecane and bromine-containing Diels-Alder-reaction products based on hexachlorocyclopentadiene.

Bromine compounds which contain aromatic-bound bromine – for instance octabromodiphenyl and decabromodiphenyl ether – possess outstanding heat stability; this means, that on account of the high processing temperatures required for linear polyesters or ABS, they are the preferred compounds to be used.

Hexabromobenzene, suggested as a flame retardant for polyamides, is a very stable aromatic bromine compound, showing however, a pronounced tendency to sublimation at the high processing temperatures required.

Trichloromethyltetrabromobenzene, pentabromomethylbenzene and pentabromophenyl benzoate have been proposed as flame retardants for ABS, polystyrene and polyesters.

Polymeric aromatic bromine compounds, such as poly(tribromostyrene), poly(pentabromobenzyl acrylate) or poly(dibromophenylene oxide) are suitable for application in polyamides, ABS and polyesters.

Aromatic-aliphatic bromine compounds, like the bis(dibromopropyl)-ether of tetrabromobisphenol A, or bromomethylated and as well, aromatic-ring-brominated compounds, such as 1,4-bis(bromomethyl)tetrabromobenzene, combine the high heat stability of aromatic-bound bromine with the outstanding flame retardant effect of aliphatic-bound bromine. They are therefore used mainly as flame retardants for polyolefins incl. polyolefin fibers [20].

Bromine compounds with exclusively aromatic-bound bromine are produced in large quantities. Occuping first place is tetrabromobisphenol A; it is employed as a reactive flame retardant in polycarbonates and epoxy resins. Brominated diphenyls, and especially brominated diphenyl ethers, have gained great significance.

Octabromobiphenyl, mainly consisting of nonabromodiphenyl and octabromodiphenyl, and decabromodiphenyl, with a content of approx. 92% decabromodiphenyl (remainder nonabromodiphenyl), are used for the manufacture of flame retarded ABS and polystyrenes.

Brominated diphenyl ethers, like pentabromodiphenyl ether (mixture, mainly consisting of pentabromodiphenyl ether, tetra- and hexabromodiphenyl ether), octabromodiphenyl ether (mixture, mainly nonabromodiphenyl ether and octabromodiphenyl ether) and decabromodiphenyl ether (approx. 92% decabromodiphenyl ether, remainder nonabromodiphenyl ether) are excellent flame retardants for those thermoplastics which have to be processed at high temperatures.

Up to now, tetrabromophthalic anhydride has been used mainly in the production of flame-retarded unsaturated polyester resins; beyond this application it is proposed as a flame retardant for saturated polyester resins.

Hydroxymethylated tetrabromobenzimidazolone is recommended as a reactive flame retardant for poly(butyleneterephthalate) [12]; poly(pentabromobenzene acrylates) and polymeric tetrabromoxylylene diacrylate have been developed for the same field of application.

Oligomeric bromine-containing ethers on the basis of tetrabromoxylylene dihalides and tetrabromobisphenol A (molecular weight between 3000

Table 1. Thermogravimetric analysis of aromatic bromine compounds

Product	Melting range (°C)	Weight loss (in %) at						
		100 °C	150 °C	200 °C	250 °C	300 °C	350 °C	400 °C
Pentabromo-diphenyl ether	–	0.12	0.42	1.33	3.6	18.2	–	–
Tetrabromo-bisphenol A	178	0	0	0.1	0.6	–	–	–
Tetrabromo-phthalic anhydride	269–271	0	0.1	0.3	2.0	14.0	77.2	–
Hexabromo-cyclododecane	170–200	0	0.45	7.0	–	–	–	–
Octabromo-diphenyl	200–290	0	0.35	0.5	0.6	1.3	5.5	–
Bis(dibromo-propyl)ether of tetra-bromobis-phenol A	85–110	0.35	0.35	0.45	1.6	–	–	–
Decabromo-diphenyl ether	290–310	0	0	0	0.1	0.4	2.8	–

and 7000) are useful flame retardants for polypropylene, polystyrene and ABS [13].

For the production of bromine-containing flame retardants see [14].

In table 1 the heat stability of selected aromatic bromine compounds is shown as determined by thermogravimetric analysis (heating rate 2 °C/min); The physical properties of bromine-containing flame retardants for thermoplastics are given in table 2.

Table 2. Physical properties of bromine-containing flame retardants for thermoplastics

Name	Molecular weight	Bromine content %	Melting point (°C)
Hexabromobenzene	551.6	86–86.5	325–336
Octabromodiphenyl [1]	785.5	81–82.5	200–290
Poly(tribromostyrene)	–	68–70	–
Pentabromodiphenyl ether [2]	564.6	67–71	–
Octabromodiphenyl ether [3]	801.5	79–80	200–235
Decabromodiphenyl ether	959.3	82–83	290–310
Tribromophenyl-allyl ether	370.9	63–65	74–77
Ethylene bis (tribromophenyl ether)	688	69–70	223
Bis(dibromopropyl)ether of tetrabromobisphenol A	943.7	65–67	90–110
Tetrabromobisphenol A	543.9	58–59	178
Tetrabromophthalic anhydride	463.8	67–68	269–271
Dibromoneopentylglycol	262.0	61.0	112
Hexabromocyclododecane	641.8	72–73	170–200
Poly(dibromophenylene oxide)	–	63–65.5	200–230

[1]) Technical product, consisting of highly brominated diphenyls; main constituents: octa-, and nonabromodiphenyl

[2]) Technical product consisting of tetra-, penta- and hexabromodiphenyl ether

[3]) Technical product, mainly octa- and nonabromodiphenyl ether

11.4.3 Phosphorus-containing compounds

Alkyl-substituted aryl phosphates are incorporated into plasticized PVC and modified PPO (Noryl) to a great extent, for other thermoplastics they are less important.

Quaternary phosphonium compounds are recommended as flame retardants for ABS and polyolefins. A phosphorus compound, which has been widely used in the past, namely tris(dibromopropyl)phosphate, is no longer in use now because of it's mutagenic and carcinogenic potential.

Phosphorus-nitrogen compounds, like aziridinylphosphine oxide, have been recommended for use as flame retardants for plastics materials, but they have not yet gained technical significance in the thermoplastics field.

Red phosphorus, sometimes in micro-encapsulated form, and alkali salts of phosphoric acids are used in polyamides as flame retardants.

Flame-retarded polyesters are obtained by employing oxaphospholane derivatives as comonomers in poly(alkyleneterephthalates) [21].

11.4.4 Synergists

Synergists, if used alone, are scarcely flame retardants, but they amplify the effect of other flame retardants; therefore the result is far above the effectiveness of the single components. In some formulations for thermoplastics it is possible to work without synergists, but this requires higher quantities of flame retardants. For economic reasons, however, this is mostly unacceptable.

In special cases, if i.e. high transparency is required, increased quantities of halogen-containing flame retardants in thermoplastics are tolerated. Commonly used synergists are: boron compounds, tin oxide, zinc oxide, zinc borate, aluminum trioxide and -trihydroxide as well as antimony silicates. Antimony trioxide, however, is the most widely used synergist with halogen-containing flame retardants.

In general, quantities between 3 and 8% Sb_2O_3 are sufficient; only in case of extremely high fire safety demands are quantities of up to 15% necessary.

For the purpose of a more homogenous distribution of antimony trioxide the importance of particle size has been pointed out in recent publications.

Antimony pentoxide seems to be even more effective [15]. Organic antimony compounds, which are of special interest for transparent thermoplastics, have obviously not yet achieved importance because of their hydrolytic instability and toxicity.

Organic peroxides and azo-compounds show synergistic effects in special formulations.

11.5 Smoke suppressants

Besides the combustibility of thermoplastics, the smoke development and toxicity of combustion gases are very important factors. The problem of smoke reduction has not yet been solved satisfactorily, because the demands on reduced flammability and smoke suppression are principally of antagonistic nature.

So far, most of the development work has been carried out in the field of PVC: sodium aluminum hydroxy carbonate is an effective smoke reducer and hydrochloric acid acceptor. In the meantime it has been removed from the market, because of its presumed carcinogenic potential. Highly dispersed aluminum trihydrate is today widely used as a flame retardant and smoke suppressant as well.

Antimony trioxide-metal borates (barium-, calcium- and zinc borate) are good smoke reducers. Zinc borate of the formula

$$2ZnO \cdot 3B_2O_3 \cdot 3{,}5H_2O$$

showed the best effect with regard to flame retardancy and reduced smoke development of PVC [16]. Occasionally, such additives are used as fillers and are, therefore, added in larger quantities; this results in a "dilution" of the PVC, which means a further reduction of smoke density. On the other hand, it is possible to diminish smoke development of rigid PVC to about 45% by adding rather small quantities of molybdic oxide. 2% of molybdic oxide reduce smoke density to about 35% already. Mixtures of molybdic oxide with other oxides, i.e. cupric oxide, reduce the smoke formation to about 54% (2,5% cupric oxide + 2,5% molybdic oxide). Thereby it is interesting to see that, with the addition of molybdic oxide as smoke reducer, the LOI-value of the rigid PVC will not decrease, but slightly increase [17].

Other smoke suppressants for PVC resp. polystyrene are iron dicyclopentadienyl(ferrocene) as well as iron, manganese, and chromium complexes of 8-hydroxyquinoline. In the case of polystyrene, the presence of halogen-containing flame retardants and smoke reducers imparts a negative influence on smoke-reduction. Phthalocyanine complexes of iron, copper, manganese, cobalt and vanadium do not have this unfavorable property.

The addition of 6% of tetraphenyl lead is said to reduce the smoke density of ABS by more than 50%.

Several methods have been proposed for the determination of the smoke density. Reproducible values are obtained with the NBS-chamber (laboratory method worked out by the National Bureau of Standards), whereby a specimen of 7.6×7.6 cm is heated by radiation. The smoke density is measured optically [18].

11.6 Applications for flame-retarded thermoplastics

The main fields of application for flame-retarded thermoplastics are the electrical, transportation, building and furniture industries. Some uses are mentioned in the following as examples: Flame retarded cable insulation (mostly made from plasticized PVC, but also from high and low density polyethylene) is used to a great extent. Important applications for flame-retarded styrenic polymers are cabinets for TV and phono-devices, for electronics and for telephone installations.

For vehicle construction and vehicle interiors, large amounts of thermoplastics are used. Special test requirements are hereby to be met. They are very strict for aircraft and ship construction, but of moderate severity for automotive construction and for automotive interiors.

Ducts, very often manufactured from flame-retarded polyolefins, are frequently used in the building industry, whereas flame-retarded PVC is preferred for floor and wall coverings. Whereas in Europe flame-retarded drain pipes are manufactured mainly with polypropylene, in the USA they are mostly produced with flame-retarded PVC. Roof-films of flame-retarded polyethylene are also used.

The use of appropriately flame-retarded polyolefins, high-impact polystyrenes and ABS is common for the production of flame-retarded furniture.

Further important applications for flame-retarded thermoplastics are: office machine housings, leisure time items, flame-retarded ABS containers, household dustbins, various injection-molded parts for electrical installations, life-saving devices, attachment parts, profiles and sheets of polyethylene and polypropylene, and plasticized PVC conveyor belts.

11.7 Examples for the formulation of flame-retarded thermoplastics

Flame-retarded thermoplastics may be prepared in several ways:

1. Thermoplastic powders are blended with the flame retardants and other additives in a fluid or turbine mixer, extruded by means of a twin screw extruder and pelletized.

2. Previously pelletized thermoplastics are mixed expediently with powdered flame retardants and other additives in heatable kneading machines (Banbury mixer). The homogenized mixture is transferred to a roll mill from which a sheet of desired thickness is removed continuously. Conversion to box-shaped pellets follows in a belt-granulator.

3. The addition of flame retardants can also take place via concentrates (masterbatches), whereby the concentration of the flame retardant can be between 50 to 80%. These flame retardant concentrates are premixed intensively with the base resin. During the subsequent extrusion or injection molding sufficient homogenization is obtained. In the following formulations the properties of flame-retarded thermoplastics are compared with those of the base resin, the properties of the latter are given in brackets.

The test methods used are:

Fire performance:

ASTM-D 635-77,	the value for ATB (=Average Time of Burning) is given in seconds, for AEB (=Average Extent of Burning) in millimeters,
ASTM 2863-77,	abbreviation: LOI-Test, specification in % O_2,
DIN 53438	similar to small-burner test in DIN 4102, part I
DIN 4102, part I	"Brandschacht" test,
UL 94	(specimen thickness in mm)

Physical properties:

Density according to DIN 53479 (ASTM D 792-75, D 1505-75), specified in kg/m^3;

Melt flow index (MFI) 190/5, according to DIN 53735, (ASTM D 1238-73), specified in g/10 min;

Notched impact strength, according to DIN 53453, (ASTM D 256-73, ISO R 179–1961), specified in kJ/m^2;

Impact strength, according to DIN 53453, specified in kJ/m^2;

Tensile strength, according to DIN 53455 (ASTM D 638–77, ISO/R 527–1966), specified in N/mm^2;

Elongation at break, according to DIN 53455 (ASTM D 638-77, ISO R 527–1966), specified in %;

Temperature of deflection under load (HDT) according to procedure B of DIN 53461 (ASTM D 648-61, ISO 75–1974), specified in K.

11.7.1 High density polyethylene (HDPE)

89.9%	high density polyethylene
5 %	octabromodiphenyl
1.6%	chlorinated paraffin (70% chlorine)
3.5%	antimony trioxide

are processed to a material of the following properties:

ASTM	ATB	<5	(210)
	AEB	20	(100)
UL 94 (1.6)		V 2	(burning)
LOI		25.75	(16.75)
density		1045	(946)
HDT		341	(344)

Melt flow index, notched impact strength, tensile strength and elongation at break are scarcely different from non flame-retarded material.

In order to reach a HDPE-classification of V 0 according to UL 94 with samples of 3.2 resp. 1.6 mm thickness, the following formulation is appropriate:

80%	high density polyethylene,		
10%	decabromodiphenyl,		
10%	antimony trioxide		
LOI		26.5	(16.75).

11.7.2 Low density polyethylene (LDPE)

90.5%	Low density polyethylene,
6.0%	chlorinated paraffin (70% chlorine)
3.5%	antimony trioxide

show the following behavior:

ASTM	ATB	<5	(160)
	AEB	18	(100)
UL 94 (3.2)		V 2	(not to classify)
LOI		23.75	(17.5)

Tensile strength, elongation at break and impact strength scarcely differ from normal LDPE.

Formulation with octabromodiphenyl:

89.9%	low density polyethylene,
4.9%	octabromodiphenyl,
1.6%	chlorinated paraffin (70% chlorine)
3.6%	antimony trioxide

results in a material, the fire-behavior of which is, according to ASTM as well as to UL 94, the same as in the recipe with chlorinated paraffin. The LOI-value increases from 25.5 to 26.75.

Notched impact strength, tensile strength, elongation at break and HDT are practically the same as with normal LDPE, whereas Shore hardness is slightly increased.

11.7.3 Polypropylene

94.5%	polypropylene
3.5%	bis-(dibromopropylether) of tetrabromobisphenol A
2.0%	antimony trioxide

result in a flame-retarded polypropylene with the following properties:

ASTM	ATB	5	(195)
	AEB	14	(100)
LOI		27.5	(18.25)

Polypropylene, with a V 1 rating according to UL 94 consists of

60%	polypropylene,
27%	Dechlorane® plus,
13%	antimony trioxide.

Properties of the compound:

UL 94 (1.6)	V 1	(burning)
LOI	25.5	(18.25)
tensile strength	25	(24)
elongation at break	0.7	(4.9)
HDT	370	(360)

11.7.4 Plasticized PVC

A colorless plasticized PVC-sheet which according to DIN 53438 extinguishes before reaching the mark (procedure K 1), has the following composition:

53.5%	suspension-PVC,
27.7%	dioctylphthalate,
4.9%	epoxidized oleic acid ester,
4.9%	pentabromodiphenyl ether,
8.3%	tricresyl phosphate,
0.5%	lubricant,
0.4%	stabilizer (butyltin carboxylate)

In comparison to the normal sheet there is no difference in low-temperature brittleness and elongation at break.

This formulation also passes the FMVSS 302 test.

11.7.5 Polystyrene

High-impact, impact and crystal polystyrene as well as polystyrene foam are flame-retarded according to the following formulations:

11.7.5.1 High-impact polystyrene

80%	high-impact polystyrene
15%	Dechlorane 602
5%	antimony trioxide

properties:

UL 94 (1.6)	V 1	(burning)
LOI	26	(16.5)
impact strength	53	(76)
elongation at break	0.25	(0.4)

Tensile strength, HDT and notched impact strength of the base resin and the flame-retardant resin are almost equal.

11.7.5.2 Impact polystyrene

80%	impact polystyrene
15%	octabromodiphenyl
5%	antimony trioxide

result in a flame-retarded polystyrene with the following properties:

UL 94 (1.6 resp. 3.2)		V0	(burning)
DIN 4102		B2	(B3)
ASTM D 635	ATB	<5	(150)
	AEB	17	(100)
LOI		25.0	(16.5)
melt flow index		10.4	(8.5)
impact strength		30	(42)
notched impact strength		4.4	(5)

Tensile strength, elongation at break and HDT vary scarcely in comparison to normal polystyrene.

11.7.5.3 Crystal polystyrene

97%	polystyrene
3%	hexabromocyclododecane

result in a material with the following burning behavior:

ASTM D 635	ATB	<5	(117)
	AEB	28	(100)
UL 94 (1.6)		V2	(burning)
DIN 4102		B2	(B3).

The physical properties are scarcely different from untreated crystal polystyrene.

11.7.5.4 Polystyrene foam

82%	polystyrene
13%	decabromodiphenyl ether
5%	antimony trioxide

gives a material substantially not different in its mechanical properties from normal polystyrene foam.

Burning behavior:

ASTM D 635	ATB	<5	(90)
	AEB	19	(100)
UL 94 (6.4)		V0	(burning)

11.7.6 ABS

ABS with a V1 rating according to UL 94 V-1, is composed as follows:

78% ABS,
17% bis-(2,4,6-tribromophenoxy)ethane,
5% antimony trioxide.

A flame-retarded ABS, fulfilling UL 94 V-0 (sample thickness: 1.6 mm) is formulated as follows:

82% ABS,
13% octabromodiphenyl ether,
5% antimony trioxide.

Burning behavior:

ASTM D 635	ATB	<5	(195)
	AEB	10	(100)
LOI		24	(18)
UL 94 (1.6)		V0	(burning)
density		1200	(1050)
notched impact strength		5	(8)

Tensile strength, surface resistance and electric strength (ASTM D 149-75) are practically identical with untreated ABS.

11.7.7 Polyamides

Polyamide 6 with or without fiberglass-reinforcement can be flame-retarded as follows.

11.7.7.1 Polyamide 6

84% polyamide 6,
10% decabromodiphenyl ether,
6% antimony trioxide

result in a material scarcely different from untreated material with regard to impact strength and notched impact strength, whereas tensile strength decreases from 88 to 67.

Burning behavior:

ASTM D 635	ATB	<5	(178)
	AEB	17	(60)
LOI		22.75	(21.75)
UL 94	(1.6)	V 1	(burning)
	(3.2)	V 0	(burning)

11.7.7.2 Fiberglass-reinforced polyamide 6

53% polyamide 6,
22% fiberglass,
18% Dechlorane 602
7% antimony trioxide

give a material with a UL 94 V-0 rating (sample thickness 1.6 mm).

LOI	32	(21)
tensile strength	135	(170)
elongation at break	0.02	(0.03)
impact strength	32	(40)
notched impact strength	5	(12)
HDT	438	(493)

11.7.7.3 Polyamide 12

85% polyamide 12,
10% decabromodiphenyl ether
5% antimony trioxide

result in a material of the following properties:

ASTM D 635	ATB	<5	(445)
	AEB	15	(100)
LOI		29	(20)
UL 94 (1.6)		V 0	(burning)
notched impact strength		9	(15)
tensile strength		47	(53)
elongation at break		2.1	(3)

11.7.8 Saturated Polyester

Polyester resins, with, as well as without, glassfiber reinforcement, can be flame-retarded as shown in the following sections:

11.7.8.1 Polyester

88%	poly(butylene terephthalate), (PBTP),
8%	octabromodiphenyl
4%	antimony trioxide

result in a material of the following properties:

ASTM D 635	ATB	<5	(240)
	AEB	16	(100)
LOI		28.5	(22.0)
UL 94 (1.6)		V 0	(burning)

11.7.8.2 Polyester with glassfiber reinforcement

87%	PBTP (30% glassfiber),
9%	decabromodiphenyl
4%	antimony trioxide

result in a material of the following properties:

UL 94 (1.6)	V 0	(burning)
LOI	33	(19)
tensile strength	95	(135)
elongation at break	0.02	(0.03)
notched impact strength	5.9	(10.5)
impact strength	15–16	(39)

11.7.9 Polycarbonates

Flame-retarded polycarbonates contain preferably tetrabromobisphenol A as flame retardant added before polycondensation.

Untreated polycarbonate can be processed to a flame-retarded polycarbonate as follows:

93.1%	polycarbonate,
5.0%	decabromodiphenyl ether
1.9%	antimony trioxide

Properties:

ASTM D 635	ATB	<5	(17)
	AEB	12	(24)
LOI		40	(25)
UL 94 (1.6)		V0	(V2)

11.8 Technological trend and outlook

Thermoplastics have so many outstanding material properties and are so economical that they are used in a steadily rising number of applications, replacing other materials like wood and metal.

Mainly in the building sector, in public transportation and in electrical equipment it is quite clear that government and industry will have to pay more attention to the increasing safety and more strict fire precaution demands. Flame-retarded thermoplastics, which do not evolve considerable amounts of smoke and toxic gases in case of a fire, are needed more and more.

As long as it is not possible to produce cheaper plastics materials, which have improved properties relative to fire safety on account of their chemical structure, the demand on additive flame retardants will increase further.

It is believed that bromine-containing flame retardants will gain a higher market share. Their economic advantages are clear, because they fulfill even high demands at relatively low concentrations, whereby the most important material properties are affected not at all or to a small extent only.

Attempts are still going on to make higher molecular weight or polymeric bromine compounds available as well as reactive bromine-containing monomers which can be introduced by copolymerization.

11.9 Index of suppliers of flame retardants

Flame retardants are produced by more than 50 companies. The following list does not claim to be complete. Additional addresses are found in Modern Plastics Encyclopedia.

Trade mark	Chemical class	Producer
Adine	Organic bromine compounds	M & T Chimie 92062 Paris la Défense, F
Amgard	Phosphoric acid ester	Albright & Wilson London SW1X 7QD, GB
Chlorowax	Chlorinated paraffins	Diamond Shamrock Plastics Div. Cleveland, Ohio 44114, USA
Cereclor	Chlorinated paraffins	ICI Millbank, London SW 1P3JF, GB
Dechlorane	Chlorine containing cycloaliphatics	Occidental Chemical Corp. Niagara Falls, NY 14302, USA
Disflamoll	Organic phosphates and halogenated organic phosphates	Bayer AG 5090 Leverkusen, D
Exolit	Chlorinated paraffins, organic phosphates	Hoechst AG, 6230 Frankfurt/M., D
Firebrake	Inorganic flame retardants	US Borax & Chem. Corp. Los Angeles, Calif. 90010 USA
Fire Fighters	Organic bromine compounds	Great Lakes Chem. Corp., West Lafayette, Ind. 47906 USA
Fireguard	Organic bromine compounds	Teijin Chem. Co., Nishishinbasi, Tokyo, Japan
Firemaster	Organic bromine compounds	Great Lakes Chem. Corp., West Lafayette, Ind. 47906 USA
Fire-Shield	Organic bromine compounds, Antimony trioxide	PPG Industries Inc., AMF O'Hare, Chicago, Ill. 60666, USA
Firex	Organic halogenated compound with antimony trioxide	Dr. Th. Böhme KG 8192 Geretsried, D
Flammastik	Flame retardant on the basis of antimony	Chem. Fabrik Grünau GmbH 7918 Illertissen, D

Trade mark	Chemical class	Producer
FR-	Organic bromine compounds	Dead Sea Bromine Co., Potash House, Beer-Sheba 84100, Israel
Hydral	Hydrated alumina	Aluminium Company of America, Pittsburgh, PA 15219, USA
Ixol	Organic bromine and chlorine compounds	Kali-Chemie AG, 3000 Hannover, D
Martinal	Hydrated alumina	Martinswerk GmbH, 5010 Bergheim, D
Nonnen	Organic bromine compounds	Marubeni Corp., Osaka 530-91, Japan
Nyacol	Flame retardant on the basis of antimony	Nyacol Products Inc. Ashland, MA 01721, USA
Phosflex	Organic phosphates	Stauffer Chem. Co., Specialty Chem. Div., Westport, CT 06880, USA
Proban	Phosphoric acid esters, quaternary phosphonium compounds	Albright & Wilson, London SW1X 7QD, GB
Pyro-Check	Organic bromine compounds	Ferro Corp., Chemical Division, Bedford, OH 44146, USA
Pyrovatex	Organic phosphates	Ciba-Geigy AG, 4002 Basel, CH
Sandoflam	Flame retardants based on organic bromine compounds	Sandoz AG, 4002 Basel, CH
Santicizer	Phosphoric acid esters, halogenated organic compounds	Monsanto Co., St. Louis, MO 63166, USA
Saytex	Organic bromine compounds	Ethyl Corp., Chemical Group, Baton Rouge, LA 70801, USA
Tardex	Organic bromine compounds	ISC-Chemicals Avonmouth, Bristol BS11 9HP, GB
Timonox	Antimony oxides	Cookson Ltd., Stoke-on-Trent ST3 7PX, GB
Triox	Antimony oxides	Société Nouvelle des Mines de la Lucette, 75008 Paris, F

Trade mark	Chemical class	Producer
Zerogen	Hydrated alumina	Solem Industries, Inc., Norcross, GA 30092, USA
–	Smoke-reducer based on molybdenum	Climax Molybdenum Co. Ltd London SW1X7DL, GB
–	Organic bromine compounds	Daichi Chem. Industries, Osaka, Japan
–	Organic phosphates	SKW Trostberg AG, 8223 Trostberg, D

11.10 References

[1] *H. Vogel:* Flammfestmachen von Kunststoffen, Hüthig-Verlag, Heidelberg, 1966.
[2] *C.J. Hilado:* Flammability Handbook for Plastics, Technomic, Stamford, Conn. 1969
[3] *J.W. Lyons:* The Chemistry and Uses of Fire Retardants, Wiley-Interscience, New York/London/Sydney/Toronto, 1970.
[4] *H. Jenkner:* Kunststoffe 62 (1972) p. 690
[5] Chemical Age; 050050 1978, p. 11.
[6] *E.R. Larsen:* Fire Retardants Chemistry I (1974), Vol. 1
[7] *E.R. Larsen:* Proc. Annu. Reinf. Plast. Compos. Inst. soc. Plast. Ind. 32., Washington, Febr. 1977, Sect. 17-B, p. 1/7.
[8] Kautschuk und Gummi – Kunststoffe 31 (1978), 6, p. 451 and Chemie-Ingenieur-Technik 47 (1965), p. A 452.
[9] *C.P. Fenimore* and *F.J. Martin:* "Modern Plastics, 44 (1966), p. 144.
[10] Kunststoffe 64 (1974), p. 153.
[11] *H. Hoffmann:* Kunststoffe 61 (1971), p. 813.
[12] *H. Batzer* and *J. Habermeier:* J. of Fire Retardants Chemistry 5 (Febr. 1978), p. 24.
[13] *R. Burkhardt, E.N. Petersen* and *N. Vollkommer:* Chemiker-Zeitung 102 (1978), I, p. 11.
[14] *H. Jenkner:* Plaste und Kautschuk 24 (1977), 6, p. 415.
[15] *R.F. Langere, O.E. Löffler* and *T.F. Santini:* USP 4, 026, 819.
[16] *J. Cowan* and *T.R. Manley:* The British Polymer Journal (1976), 6, p. 44.
[17] *F.W. Moore:* Climax Molybdenum Company of Michigan, 1977.
[18] *G.F. Lawson* and *E.L. Kay:* J. of Fire Retardant Chemistry 2 (Aug. 1975), p. 132.
[19] Chemical Age, May 22, 1981, p. 11.
[20] *H.W. Finck, G. Tscheulin:* Kunststoffe 71 (1981) 5, p. 320.
[21] *K. Weissermel, H.J. Kleiner, M. Finke:* Angewandte Chemie 93 (1981), p. 256.
[22] *J. Troitzsch:* International Plastics Flammability Handbook, Carl Hanser Verlag, D-8000-München, 1983.

12 Antistatic Agents

Dr. H.W. Finck, Sandoz Ltd., Basel, Switzerland

12.1 Problems of electrostatic charging

Antistatic agents occupy a special place among plastics additives. The use of such agents is necessitated in most cases not so much by legislative (fire prevention) or technological (stabilization) reasons, but primarily by aesthetic and, not least, safety considerations.

12.1.1 Forms of static charging

The phenomenon of static electricity was known in the ancient world (Thales), but in its mechanisms it has remained a mystery to many people up to the present day.

When two flat, insulated plates of copper and zinc are brought together and quickly separated, the different charges on the contact surfaces can be detected with an electroscope. This property is not confined to metals; on nonmetallic materials, however, vigorous frictional contact is necessary to generate static charges. On immobile objects, for example, static electricity may build up simply by friction with the ambient air.

The charge capacity of a solid depends on the condition of its surface, on the dielectric constant, the surface resistivity, and the relative humidity of the surrounding atmosphere. While the charge capacity is inversely proportional to the dielectric constant and relative humidity, it is directly proportional to surface resistivity. The sign of the charge is invariably from the material with the higher to the one with the lower dielectric constant.

By virtue of their chemical constitution most plastics are powerful insulators, a property which makes them indispensable materials for high frequency equipment such as radar. An associated disadvantage of the insulation property is the accumulation of static electricity, which is not discharged fast enough due to the low surface conductivity of most plastics – a difference between plastics and metals.

On the basis of their surface resistivity, the various classes of materials can be arranged on the scale in figure 1.

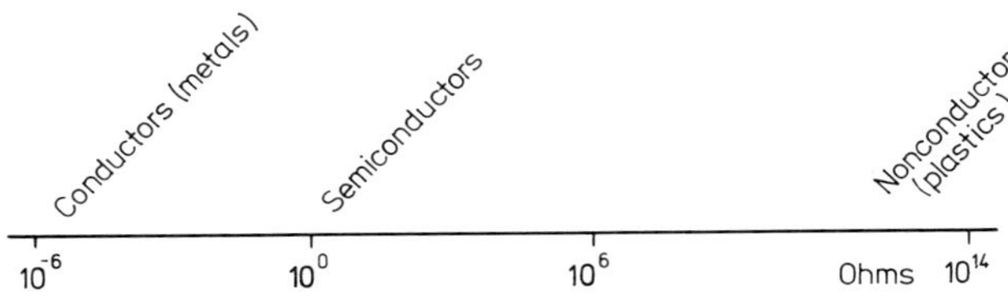

Fig. 1. Classification of materials by their surface conductivity.

In the service life of plastics products, static charging can give rise to troublesome effects and indeed serious and even dangerous situations, of which the following may be mentioned as most important:

- Heavy contamination of plastic surfaces,
- interference with the sound reproduction of records by dust particles,
- static buildup on people passing over synthetic fiber carpeting or plastics floorcoverings, with a subsequent "shock" as the charge flows off, usually when the person touches a door handle; this can be an unpleasant experience,
- production delays due to clinging of adjacent films or sheets,
- lump formation during pneumatic transportation of powders,
- spark formation through heavy charging with subsequent ignition of dust or solvent/air mixtures, which has been a cause of many destructive explosions.

12.1.2 Countermeasures to suppress static charging

Increase the relative humidity in the environment of the molded product: as the relative humidity rises, the surface conductivity increases allowing the charge to dissipate more rapidly. There is a reciprocal relation between the surface resistivity and the relative humidity (figure 2). The relation shown here was determined for polyolefins and does not apply to all polymers. With the water-absorbent polyamides, for example, at a relative humidity higher than 65% there is virtually no danger that problems with static electricity will occur. On the other hand, very low relative humidity values of the order of $\ll 20\%$ prevent charge equalization over the surface. In these conditions the only really effective systems are those that reduce the volume resistivity by means of a conductive matrix (see 12.3.2).

Increase air conductivity by ionizers working on the electrical or radioactive principle which enables the charge to dissipate into the surrounding air.

Increase the surface conductivity by chemical assistants (antistatic agents), which incorporated into the plastic or applied to the surface allow the static charge to be discharged.

12.2 Chemical structure of antistatic agents

Antistatic agents are substances which are added to plastics molding formulations or to the surface of molded articles in order to minimize the accumulation of static electricity. In general terms antistatic agents can be divided, according to the method of application, into internal and external agents.

12.2.1 Internal antistatic agents

Internal antistatic agents are incorporated in the polymer mass as surfactant additives, either before or during the molding process. They are all of interfacially active character, and via migration (see 12.3.1.1) accu-

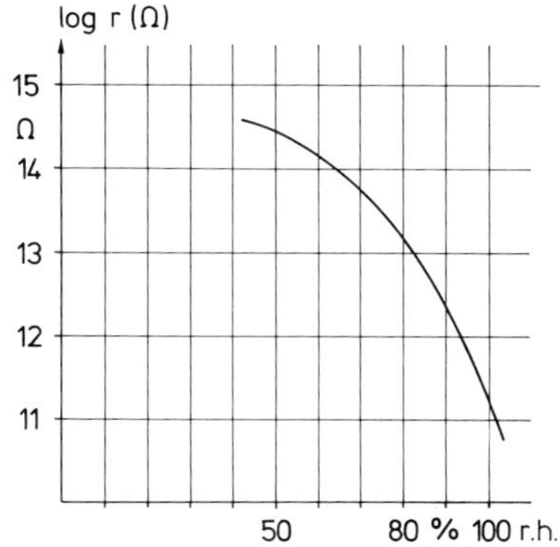

Fig. 2. Correlation between surface resistivity of polyolefins and relative humidity [26].

mulate on the surface of the molding. The molecule is composed of a hydrophilic and a hydrophobic component; the hydrophobic part confers a certain compatibility with the particular polymer and is responsible for anchoring the molecule on the surface, while the hydrophilic part takes care of the binding and exchange of water on the surface.

Most of the patented antistatic agents in use today, being interfacially active, can be divided fairly easily into cationic, anionic and nonionic compounds. There is a fourth class of amphoteric compounds, which has been developed primarily in Japan, but these have not achieved any importance in the European or American markets.

12.2.1.1 Cationic compounds

In these the active part of the molecule normally consists of a bulky cation, often containing a long alkyl radical, e.g. a quaternary ammonium, phosphonium or sulphonium salt, whose quaternary group may occur in a ring system, e.g. imidazoline. The anion is generally the one formed in the quaternization reaction, e.g. the chloride, methosulphate, nitrate etc. The quaternary ammonium salts have won a leading position among the commercial products of this class.

Cationic antistatic agents work best in polar substrates, e.g. PVC and styrenic polymers. Their application is however restricted by their adverse influence on the thermostability of certain polymers (see 12.4.1). Moreover, quaternary ammonium salts have not been granted FDA approval. Quaternary compounds have frequently been suggested for polyolefins in patents and published papers, but they have to be used at concentrations 5 to 10 times greater than those of nonionic compounds in order to develop comparable antistatic action.

12.2.1.2 Anionic compounds

In this class of compounds the active part of the molecule is the anion, which in most cases is an alkylsulphonate, sulphate or phosphate, a dithiocarbamate or a carboxylate. Quite often alkali metals, less frequently alkaline earth metals, are used as cations. Among industrial users, sodium alkylsulphonates have met with the widest acceptance, as they exhibit satisfactory antistatic action in both PVC and styrenic polymers. Their application in polyolefins is subject to the restriction referred to in 12.2.1.1.

12.2.1.3 Nonionic compounds

These compounds have an uncharged, interfacially active molecule of significantly lower polarity (polyethylene glycol esters or ethers; fatty acid esters or ethanolamides; mono- and di-glycerides; ethoxylated fatty amines; to name only the most important). They are supplied for the most part in liquid form or as waxes with a low softening region.

The low polarity of this class makes the members ideal internal antistatic agents for polyethylene and polypropylene, with which they are highly compatible. The differences between the various polyethylene and polypropylene types in density, crystallinity and molecular superstructure mean that the optimum structure for every given antistatic agent has to be arrived at by variation in the chain length of the alkyl radical and in the number of hydroxyl or polyether groups, so as to ensure that it will perform effectively in the intended application. For instance, typical agents for polypropylene are less effective in low density polyethylene, and vice versa.

For the sake of completeness, mention should be made of the slip additives and waxes which, by reducing the friction coefficient, exercise some slight antistatic action. Among the representatives of these more physically active substances are stearyl alcohol and stearylamine.

It is frequently found that by applying excessive amounts of a structurally unsuitable additive, e.g. a quaternary ammonium salt in a polyolefin polymer, certain antistatic effects are attainable, but in such cases one has to contend with undesirable side effects as well, such as surface smearing, molding difficulties due to lubricating action, thermal degradation and so forth (see 12.4.1).

A good synoptic view of numerous patented and industrially-employed compounds is provided by *Balbach* [1] for polyolefins, and by *Riethmayer* [2] for all plastics materials, notably PVC.

12.2.2 External antistatic agents

External agents are applied to the surface of the molded article from aqueous or alcoholic solution. Because of the method of application, the structural specificity mentioned in connection with internal antistatic agents is of no importance. More or less, all interfacially active compounds are effective, as well as numerous hygroscopic substances such as glycerin, polyols and polyglycols, in which the surface activity feature is absent. Their suitability is explained by the fact that compatibility with and controlled migration in the polymer play no part in their performance.

12.3 Mode of action of antistatic agents

The static charging of plastics can be minimized either by increasing the surface conductivity or the volume conductivity. The ideal situation is one in which the charges are dissipated more rapidly than they are generated, so that no accumulation takes place. As a rule, the charge on a plastics surface is dissipated more or less rapidly after interruption of its generation, depending on the nature of the antistatic finish.

12.3.1 Increase in surface conductivity

It is generally assumed that interfacially active molecules accumulate on the surface and are oriented with the hydrophobic part containing the paraffin chain extending into the plastic and the hydrophilic part pointing outwards where it is able to adsorb water on the surface. In consequence of their interfacially active character, antistatic agents decrease the phase boundary angle between water and plastic, thus allowing water to be uniformly distributed on the surface. A water film, whose thickness depends on the atmospheric humidity, forms on the plastics' surface, thus increasing the conductivity by means of an ion conduction process. This also explains why the surface conductivity and hence the antistatic action progressively decrease with decreasing atmospheric humidity (see figure 2).

More recently it has been postulated that charge exchange, in addition to the ion conduction mechanism, is effected through a constant exchange of water between the surface and the environment. This would be in accord with the mechanism of air ionization (see 12.1.2). In contrast to this, however, the antistatic agent on the surface acts as a contact point for the charge exchange.

In addition, charge transfer can be accomplished by a proton shift. Antistatic agents bearing OH or NH_2 groups are able to associate in chain form via hydrogen bonding, and display antistatic activity even at low atmospheric humidity, unlike compounds which are able to form only intramolecular hydrogen bonds [3].

Many antistatic agents also show hygroscopic properties, thereby intensifying the attraction of water to the surface. At constant atmospheric humidity a hygroscopic compound combines more water at the surface and so increases the antistatic effectiveness.

Application of antistatic agents in the normal concentration range (see 12.4.1) brings about a decrease in the surface resistivity from 10^{14}–10^{16} Ω to approximately 10^{8}–10^{10} Ω. A further decrease is attainable only by increasing the volume conductivity (see 12.3.2).

12.3.1.1 Migration

For an accumulation of molecules to build up on the surface, the antistatic agent has to migrate from the interior of the plastics product during or after molding. The progress of migration can be tracked indirectly, e.g. by continuous measurement of the surface resistivity (see 12.5.2). Depending on polymer type, the density, and the molding process, antistatic agents act, i.e. reduce the surface resistivity, more or less rapidly. The process can be understood by imagining island-like structures of antistatic agent being initially formed on the surface and then gradually expanding to create a continuous film of a thickness which, depending on the plastic and the agent itself, may consist of one to several molecular layers (see 12.3.1.2). This can be demonstrated on polyolefin blown film, for example, by continuous measurement of the surface resistivity at numerous different points on the surface. Depending on the draw ratio, it may take several days before a uniform resistivity value is reached on the entire surface. In other cases, the antistatic effect is detectable almost immediately after molding and later increases only minimally, e.g. in the application of paraffin sulphonates in rigid PVC.

Differences in crystallinity, and hence in density, have a critical effect on the migration of antistatic agents. In injection-molded polyolefin panels, onset of antistatic effectiveness in LDPE is accomplished almost at once after molding, whereas it takes much longer in HDPE and is strongly retarded in polypropylene. For example, the maximum effectiveness in polypropylene panels is not reached until 10 to 20 days after molding.

The buildup of effectiveness depends further on the molding process. In compression- or injection-molded polyethylene, for instance, the effect is measurable immediately after molding, whereas in drawn or blow molded film the antistatic agent takes considerably longer to develop its activity. In polypropylene stretched tapes the effectiveness is inversely proportional to the draw ratio, as the surface area/volume ratio increases with increasing draw ratio, so more substance is needed to cover a given area. At lower antistatic agent concentrations a correspondingly longer time has to be allowed until migration has brought about full coverage of the surface.

In addition to this simple, largely qualitative proof of migration, there are a number of other methods, all of which are more laborious but which supply quantitative data. The most accurate one is the tracer method with radioactive isotopes. When a radioactively labelled substance is compounded with the molding material, virtually every molecule washed off the surface can be determined [4, 5]. In other tests the plastics' surface is washed and the surface tension of the aqueous wash solution determined [3]. The metal ions of anionic compounds are determined by flame photometry [6], or by thin-layer chromatography of the solution [7]. By these diverse means, the diffusion coefficients of antistatic agents in polymers are determinable, which show good agreement in spite of the different methods employed.

These methods also enable interactions between different additives to be detected. For instance, low concentrations of polyethylene glycols or glycerin bring about an increase in the rate of diffusion of sodium alkyl sulphonates by a factor of 2 to 20 [6, 7]. In contrast the migration, and consequently the effectiveness of antistatic agents, is substantially reduced by certain other additives commonly found in plastics, two examples being carbon black and titanium dioxide [8]. Nickel-containing u.v. stabilizers as well are apt to retard the migration of many antistatic agents.

The long term effectiveness of antistatic agents is important. Periodic evaluations of samples containing internal antistatic agents over one year did not show any change in their activity. Diminished protection against static buildup occurs only if antistatic agent molecules are removed from the surface by evaporation or cleaning, but the deficiency is overcome over some length of time by continued migration from the internal deposits. An exception to this behavior is probable with molded articles stored at high temperature, as these conditions induce more extensive losses which may finally exceed the critical level. In the case of external antistatic agents the long term effectiveness is severely limited by evaporation from the surface and by migration into the interior of the molded product (see 12.4.2).

12.3.1.2 Surface concentration

The methods referred to in 12.3.1.1 are suitable for determining the surface concentration of antistatic agent. Here again, measurement of the surface resistivity of polyolefins enables the activity of the deposited agent to be determined in quantitative terms. When the surface of polyethylene or polypropylene containing an internal antistatic agent is thoroughly cleaned, e.g. by washing with a water/alcohol mixture, at first the substantially higher resistivity value of a sample without an antistatic finish is obtained, but after a few hours or days, the surface resistivity regains its original value, the time required depending on the density of the polymer. This test can be repeated more than ten times, or until almost all the available additive is exhausted.

In rigid PVC and styrene polymers, the regenerative capacity of the antistatic effect is much weaker. On surfaces which have been cleaned only once, the action does not return to a satisfactory level until a long period of time has elapsed, or indeed it may never be completely recovered.

What are the surface concentrations that suffice for an antistatic finish? In the literature 0.1 to 10 molecular films on various plastics surfaces are cited. A monomolecular film of dodecyl ethanolamine on high density polyethylene is said to give a half-life of 2.5 s (see 12.5.1) [3], which according to table 2 (see 12.5.3) is roughly equivalent to a surface resistivity of $10^{10}\ \Omega$. In agreement with this finding, a surface resistivity of $10^{10}\ \Omega$ has been measured on polyethylene and polystyrene carrying monomolecular films of lithium alkylsulphonates. In contrast to these are published data on ethoxylated fatty amines on polyethylene (0.2 molecular films $\approx 0.1\%$ of the total antistatic addition) [7]. After repeated surface cleaning, this concentration is regained exactly. On styrene polymers, sodium alkylsulphonate films build up to a thickness of 2 (crystal PS) to 8 layers (impact resistant PS), which represents 0.6–2.1% of the total added antistatic agent. Unlike polyethylene, after the surface has been washed only 0.4–0.5 molecular films form again, which is evidence of the aforementioned low capacity of regeneration in polystyrene [7].

12.3.2 Increase in the volume conductivity

To safeguard against explosions stemming from electrostatic charging, the surface resistivity of a plastics molding in a hazardous environment may not exceed a value of $10^6\ \Omega$ [9]. As this value is not reachable with the usual internal or external antistatic agents alone, the volume conductivity of the molded article has to be increased to facilitate sufficiently rapid equalization of the charge. In place of ion conduction, electron conduction takes over the mechanism of the antistatic action. The agent selected has to be so distributed in the polymer mass as to form a continuous conductor system uninterrupted by polymer components. Obviously, far greater amounts of additive are necessary to meet this requirement than are needed to increase the surface conductivity. On the other hand, the level of effectiveness is greater by a multiple of the power of ten.

Thus in polyolefins containing 20–30% carbon black a specific volume resistivity of $10^2\ \Omega \times \mathrm{cm}$ is reached, which approaches the conductivity of metals ($10^{-4}\ \Omega \times \mathrm{cm}$). Clearly the mechanical properties of the molded article will suffer at such high filler concentrations. For this reason, work is underway on modifying the carbon black types in order to reduce the necessary concentrations. Employing active types of large surface area and structured particle shape (conductive carbon blacks), it is now possible to attain the aforestated value of $10^2\ \Omega \times \mathrm{cm}$ with only 5–10% additions [10, 11]. Besides the structure of the carbon black particles, the realizable conductivity also depends on the degree of crystallinity of the polymer and the molding technique. Higher crystallinity increases the conductivity, while it is impaired by too strong or too weak shearing forces during the molding cycle.

Another way to increase the volume conductivity is to incorporate metal powders in the polymer mass, e.g. silver powder in polystyrene [12]. Cost is a big factor here, in addition to which the thermal stability of the polymer has to be kept in mind. As is known, numerous transition metals reduce the polymer stability as they catalyse the decomposition of peroxides into free radicals and thereby accelerate the degradation of the polymer.

Here it is relevant to note that at extremely low relative humidity values of $\ll 20\%$, this type of antistatic system is the only one capable of neutralizing electrostatic charges.

To round off this survey, the special field of electrically-conductive plastics should be mentioned. These are polymers which by reason of their structure (conjugated double bonds, metal complexes) possess high volume conductivity [13, 14]. All the named systems which increase the volume conductivity are however deeply colored due to the pronounced electron mobility in these substances.

12.4 Application

Antistatic agents are either blended with the polymer in the mass – the internal types – or applied from solution or emulsion to the surface of the molded article – the external types. A third method of application, the use of copolymerisable antistatic agents, has been described in several patent specifications but is not found in industrial practice. Overall it is too expensive and is not sufficiently adaptable to the many and varied types of polymers in a manufacturer's range.

12.4.1 Incorporation in the polymer mass

This method calls for additives that are stable in the temperature region of 250–300 °C so as to withstand the molding cycle. As previously stated (12.2.1), the chemical structure of a given agent has to be closely attuned to the polymer type for which it is intended. In spite of these stringent selection criteria, mass application has become accepted for plastics. Only molding processes conducted at extremely high temperatures are an exception, as no sufficiently heat-stable compounds for these have yet been found. The advantages of adding antistatic agent in the mass are the greater ease of handling and lower cost of application, plus excellent regenerative capacity of the antistatic action in certain polymers after surface cleaning.

The concentrations that are necessary to provide adequate protection against static buildup differ in a characteristic manner from one polymer to another; it is probable that the differences are closely connected with the morphological structure and the glass transition temperature (T_g).

Table 1. Application concentrations of antistatic agents in plastics.

Polymer	Antistatic agent (%)
LDPE	0.05–0.1
HDPE	0.2 –0.3
PP	0.5
Rigid PVC	0.5 –1.5 (dependent on the stabilizer system)
PS, crystal	2.0 –4.0
PS, impact-resistant	1.5 –2.5

Table 1 lists the normal concentrations for the principal polymers; for certain molding techniques, however, the necessary amounts are likely to be greater than these.

When determining the dosage rate for an antistatic agent it should be borne in mind that too high concentrations lead to the formation of tacky, smeared films on the surface. Not only do such films detract from the visual appeal of the product, they are also an obstruction to subsequent printing or lettering processes.

Antistatic agents are normally formulated into the molding material along with other additives or pigments. If a heatable internal mixer is in service, preliminary blending is unnecessary. If blending is carried out in an extruder, the additives and the polymer granules or powder are compounded in a mixer to coat the polymer particles with the additives prior to extrusion.

High antistatic agent concentrations may cause interference in molding since most of these products are viscous liquids and are apt to act as lubricants in the molding material and hinder its flow into the extruder. This difficulty can be overcome by the inclusion of a small amount of a filler which will increase the friction, e.g. SiO_2. Latterly, master batches of antistatic agents have been strongly promoted as a solution to such problems. This dosage method is of course advisable only on condition the required concentration in the polymer is not too high, which means chiefly in polyolefins (see table 1).

Antistatic agents are capable of reducing quite considerably the heat stability of heat-sensitive polymers. One consequence of this is that the use of the highly effective quaternary ammonium salts in rigid PVC is severely restricted. Sodium alkylsulphonates are superior to them in heat stability, but have the disadvantage of diminishing the transparency of the molding. Further, the use of sodium alkylsulphonates necessitates careful screening of the PVC heat stabilizers, since interactions with these too are a possibility (see 12.3.1.1).

In contrast to tin stabilizers, lead and certain barium/cadmium types, when used jointly with sodium alkylsulphonates, cause discoloration, although this can be suppressed by addition of phosphonites. Generally speaking, for antistatic agents, as for all additives, optimum blending components and concentrations should be determined by end-use tests of the finished product.

12.4.2 Surface application

Surface application is practiced most widely with textiles; it is rarer to find it used for plastics. The molded article is sprayed or wet with a solution of the antistatic agent, or dipped into the solution, and then dried. One example is in the finishing of PVC floorcoverings, where surface active agents are added to the rinsing water at the cleaning stage. The advantage of this method is freedom of choice, as any antistatic agent can be used without regard to the thermal stability of or compatibility with the polymer.

Along with the higher cost of this plant- and labor-intensive method, a big drawback of external application is the inadequate stability of the antistatic effect. On the one hand, the action is extinguished once the surface is cleaned with the aid of moisture, since there is no internal agent that can migrate to the surface to replace the loss, and on the other, the effect often wears off due to migration of the agent into the interior of the product [15].

Formerly it was thought that 100 mg antistatic agent per m^2 surface was necessary to produce a satisfactory effect. Recent experimental work has however shown that about 10 mg of a quaternary ammonium salt or sodium alkylsulphonate or 25 mg fatty acid monoglyceride per m^2 of the plastics surface is sufficient [15].

12.5 Measurement of the antistatic activity

All the methods used to demonstrate the antistatic activity of these agents are derived from two principles of measurement:

- measurement of the maximum charge and its decay (half-life)
- measurement of the surface resistivity.

These two quantities are interconnected, as is shown in some detail in 12.5.3. Scrupulous conditioning of the test sample before and during measurement is imperative if reproducible results are to be obtained, because the relative humidity and temperature play a big part in determining the dielectric properties of the molded article. Touching the surface for measurement with the fingers must be avoided at all costs, as impurities tend to increase the conductivity.

12.5.1 Measurement of the static charge

A solid can be charged by rubbing or spraying. The charge generated by friction depends on a number of factors and is difficult to reproduce.

A simple method yielding qualitative information is the cigarette ash test. The sample is charged by rubbing it with a wool cloth and is then held over finely powdered cigarette ash, which is attracted to the surface to a degree commensurate with the charge.

A second qualitative method which simulates practical conditions is the "carbon black chamber test" [16], in which the plastic sample is exposed to an air jet bearing carbon black particles and subsequently evaluated on the degree of contamination. Although the reproducibility of the rubbing method has been increased by means of special instruments [17], it is probably easier to charge the sample by spraying. The charge density is determined by measurement of the potential [18] or the field intensity [19]. Both methods are used in the plastics industry.

The charge density and the half-life of decay can be determined neatly without touching the sample in an instrument built in Japan, the "Honestometer" [20, 21], which is a further development of an original idea of *Shashoua* [22] and does credit to its name. The rapidly rotating sample (1 250 rpm) is charged by spraying from a needle electrode, the charge is conducted through the electrode, picked up by a voltmeter and amplified. After switching off the potential, discharge is plotted by a recording instrument to give a curve on which the half-life is read off.

12.5.2 Measurement of the surface resistivity

The surface resistivity of plastics is determined with spring or annular electrodes as specified in DIN 53482. It is an important criterion for evaluating the effectiveness of antistatic agents, since with declining surface resistivity the proneness of plastics to static charging decreases.

The method used to determine the volume resistivity cannot be employed for normal antistatic finishes that function on the surface. Whereas the surface resistivity is reduced by several powers of ten, the volume resistivity remains unchanged [3]. Only in polymers loaded with carbon black (12.3.2) does it begin to decline [10].

12.5.3 Correlation between half-life period and surface resistivity

The surface resistivity and the discharge time are interrelated as shown in the following formula, which describes the exponential discharge process of capacitors:

$$V_t = V_0 \cdot e^{-t/R \cdot C} \tag{1}$$

where V_0 is the initial voltage (V),
V_t the voltage at time t (V),
t the discharge time (s),
R the surface resistivity (Ω),
and C the capacitance (F).

The half-life period τ (*s*) is found by recording the point at which the voltage has fallen to one half of the starting value, i.e. when $V_\tau = V_0/2$. Equation (1) yields:

$$\tau \approx 0.7\, R \cdot C \tag{2}$$

Our experience has confirmed the correlation between τ and R. In table 2 are listed the values of τ and R obtained in a great number of measurements on different plastics.

Table 2. Correlation between surface resistivity and half-life period in plastics.

Surface resistivity (Ω) (DIN 53482)	Half-life period τ (s) (Honestometer)	Antistatic action
$<10^9$	0 (no charge)	excellent
10^9 to 10^{10}	1	very good
10^{10} to 10^{11}	2 to 10	good to satisfactory
10^{11} to 10^{12}	10 to 60	moderate
$>10^{12}$	>60	insufficient

Table 2 indicates that plastics with a surface resistivity of 10^9 Ω are no longer open to static charging. The limit value of the resistivity 10^6 Ω, given in 12.3.2 thus provides additional safety. Between 10^9 Ω and 10^{12} Ω the surface resistivity can be successfully used to evaluate static charges and charge decay; above 10^{12} Ω it appears to be less suitable for this purpose [17].

12.6 Areas of application for antistatic agents

Internal antistatic agents are now used in a limited number of polymers. By far the largest quantities go into polyethylene and polypropylene (film, sheet, injection-molded articles, and containers, especially packagings). The amounts used in the PVC sector, e.g. for records, are smaller; estimates of the volume of this application diverge [23].

Recent years have seen the growing use of antistatic compounds in styrenic polymers (polystyrene, impact-resistant polystyrene). In contrast to polyolefins, higher loadings of up to 4% antistatic agent are necessary in these substrates for a good level of activity (see table 1); consequently the cost factor prohibits their use except for speciality products such as video cassettes, instrument housings and the like.

In what are called "engineering plastics", which include polyethylene terephthalate, polycarbonate and polysulphone, and in polyacrylates, internal antistatic agents have not yet found any utility for two main reasons. Firstly, these materials are molded at high, in some cases very high temperatures, and secondly, some of them have important intrinsic properties, e.g. transparency in polymethacrylates and polycarbonates, which must not suffer damage under any circumstances.

For products to be used in environments where fire and explosion hazards exist, an increasing proportion of the molding material is loaded with carbon black to impart high conductivity. Examples are packs for explosives, conveyor belts for mines, and floorcoverings for operating theatres and chemical manufacturing plants.

Product safety is a further major requirement for packaging materials with an antistatic finish, i.e. non injurious to health. Since antistatic agent molecules are present on the plastics surface, contact with the packaged food is unavoidable. In polyethylene and polypropylene, no serious problems arise from the low content of antistatic agent that is required to conform to the specifications of the authorities that grant permission to additives under their national legislation. As is known, the limit values set by these bodies relate to the amount of migrated and analytically determinable product. In plastics such as styrenic polymers which require a higher content of antistatic agent (see table 1), these limits are occasionally exceeded. It is quite possible that one antistatic agent, e.g. a sodium alkylsulphonate type, would be accepted in a particular plastic, e.g. crystal polystyrene, in view of its low migration capacity, while in another, e.g. impact-resistant polystyrene, it might well be rejected on the ground of excessive migration [7].

12.7 Technological trend and outlook

Over the past few years antistatic agents have shown a steady annual growth rate of around 10%, except in 1974–75, the years of the oil crisis when business generally was slack [24].

In the USA it was expected that new specifications for the antistatic finishing of military electronics packagings would be issued in 1980, which would have sparked off a 30% upswing in the sales of internal antistatic agents, but as it happened the formulation of the new specifications under the authority of the Navy Bureau of Ships has been delayed [25].

In antistatic agents for polyolefins, the foremost area of use, no radical innovations are expected in the next few years. However, the aggregate state of the products has been modified, with some products being launched in powder form that offer easier blending with the polymer, as well as a growing number of concentrates.

The many years of searching for antistatic compounds suitable for styrenic polymers have yielded some successes, although the high concentrations that are necessary have so far imposed strict limitations on the use of these agents.

In spite of intensive research, an antistatic agent for rigid PVC which will not impair the transparency or the thermostability remains a desideratum.

With the availability of highly active carbon black grades of high surface area, the former application concentrations of 20–30% have been cut to 5–10%, an important advance for retention of the polymer's mechanical properties. The future will show whether further economics are realizable here.

12.8 Trade names and manufacturers of antistatic agents

The following list of trade names and manufacturers does not claim to be complete. In the USA updated reference books, notably the Modern Plastics Encyclopedia (MPE), are published at regular intervals, but there are many gaps in the information provided by European sources. The details of products of European manufacture not included in the MPE were supplied mainly by the manufacturers in response to direct inquiries.

The products named are in the main, internal antistatic agents, but they can, if necessary, be applied to the surface of molded articles.

Trade name	Manufacturer or supplier
Antistatic Agent	Hexcel Corp., Lodi, NJ 07644, USA
Armostat	Akzo Chemie GmbH, 5160 Düren, D
Atmer	ICI Ltd., Millbank, London SW1P 3SF, GB
Bärostat	Chem. Werke Otto Bärlocher GmbH, 8000 München 50, D
Berkstat	Steetley Co. Ltd., Gateford Hill, Workshop, Notts., S81 8AF, GB
Cyasta	American Cyanamid Co., Bound Brook, NJ 08805, USA
Dehydat	Henkel & Cie. GmbH, 4000 Düsseldorf, D
Drewplast	Harwick Chem., Corp., Akron, OH 44305, USA
Electrosol	Alframine Corp., Paterson, NJ 07524, USA
Escostat	East Coast Chem. Co., Cedar Grove, NJ 07009, USA
Gafstat	GAF Corp., New York, NY 10020, USA
Hostastat	Hoechst AG, 6230 Frankfurt (Main), D
Irgastat	Ciba-Geigy AG, 4002 Basel, CH
Lankrostat	Lankro Chem. Ltd., Eccles, Manchester M30 0BH, GB
Markstat	Argus Chem. Corp., Brooklyn, NY 11231, USA
Michel XO	Michel & Co., Inc., New York, NY 10004, USA
Nopcostat	Diamond Shamrock Corp., Morristown, NJ 97960, USA
Pegosperse	Glyco Chem. Inc., Greenwich, CT 06830, USA
Polystat	Jos. Cosfield & Sons Ltd., Warrington, Cheshire WAS 1AB, GB
Statexan	Bayer AG, Bayerwerk, 5090 Leverkusen, D
Tebestat	Th. Böhme KG GmbH & Co., 8192 Geretsried, D

12.9 References

[1] *G. Balbach:* Kunststoffe 67 (1977) pp. 154–159.
[2] *S. Riethmayer:* Gummi, Asbest, Kunststoffe 26 (1973) pp. 76–88, 182–184, 298–308, 419–429, 507–512.
[3] *W. Biedermann, G. Langhammer & W. Reicherdt:* Plaste u. Kautschuk 19 (1972) pp. 119–126.
[4] *A.E. Sherr & E.A. Vitalis:* Ind. & Eng. Chem. Product Research & Development 2 (1963) p. 97.
[5] *K. Figge:* Kunststoffe 61 (1971) pp. 832–838.
[6] *K. Rombusch & G. Maahs:* Angew. Makromolekulare Chemie 34 (1973) pp. 50–70.
[7] *K. Woggon & W.J. Udhe:* Die Nahrung 21 (1977) pp. 235–245.
[8] *K. Othmer:* Encycl. of Chem. Techn. Vol. II, p. 666, New York 1963.
[9] Berufsgenossenschaften der chem. Industrie: Richtlinien zur Vermeidung von Zündgefahren infolge elektrostatischer Aufladung. Verlag Chemie, Weinheim, 1972.
[10] *R. Gilg:* Kunststoffberater 22 (1977) pp. 262–265, 312–321.
[11] *W.F. Verheist:* Kunststoffe 66 (1976) pp. 701–703.
[12] *R. Matsushita, M. Senna & H. Kuno:* J. Mat. Sc. 12 (1977) pp. 509–516.
[13] *G. Kossmehl & M. Rohde:* Makromolekulare Chemie 178 (1977) pp. 715–722.
[14] *R. Reichherzer:* Kunststoff-Rundschau 18 (1971) pp. 431–434.
[15] *K. Polzhofer & H.D. Lehmann:* Angew. Makromolekulare Chemie 36 (1974) pp. 145–151.
[16] *R.E. Schanzle:* Modern Packaging 37 (1964) p. 129.
[17] *G. Heyl & G. Luttgens:* Kunststoffe 56 (1966) pp. 51–54.
[18] *J.L. Rogers:* SPE Journal 29 (1973) pp. 28–34, 52–55.
[19] *H. Kramer & D. Messmer:* Kunststoffe 54 (1964) pp. 696–699 and Kunststoff-Rundschau 18 (1971) pp. 265–271.
[20] *H. Marumo:* Advances in Static Electricity (Proceedings of the 1st Congress on Static Electricity), Vienna, (1970) pp. 445–458.
[21] *P.E. Slade:* Modern Textiles Mag. 3 (1973) pp. 68–70.
[22] *V.E. Shashoua:* J. Polymer Sc. 33 (1958) p. 65.
[23] Modern Plastics 47 (1970) pp. 9, 104.
[24] Modern Plastics 54 (1977) pp. 9, 64.
[25] Modern Plastics 57 (1980) pp. 9, 53.
[26] Sandoz research findings.

13 Fluorescent Whitening Agents

Kurt Berger, Ciba-Geigy AG, Basel, Switzerland

13.1 Introduction

For a long time fluorescent whitening agents have been used to improve the appearance of finished articles in the paper, soap, and textile industries. Along with the development of thermoplastics, fluorescent whitening agents have penetrated this field too.

The objectives of using fluorescent whitening agents in plastics are the same as in paper and textiles:

- Improvement of the initial color of plastics, which are inherently often slightly yellowish,
- production of brilliant white end-use articles (such as fashion goods of imitation leather, packaging materials etc.),
- increase of the brilliancy of colored, and black-pigmented articles.

In principle, fluorescent whitening agents can be used in all plastics.

A technically suitable fluorescent whitening agent shows high whitening effectiveness at low concentrations. In practice, concentrations of 100 to 500 ppm are used in thermoplastics. Only special applications require concentrations exceeding 1 000 ppm.

13.2 Definition of a fluorescent whitening agent

Most thermoplastics absorb light in the blue spectral range of natural daylight ("blue defect"), hence causing a more or less pronounced yellow appearance (figure 1 a). There are two ways to compensate for this deficiency, i.e. to increase the degree of whiteness:

- Compensation of the blue deficit or
- increase of reflection.

Already decades ago brightening of textiles was achieved by adding very small amounts of a blue pigment, an operation known as "bluing".

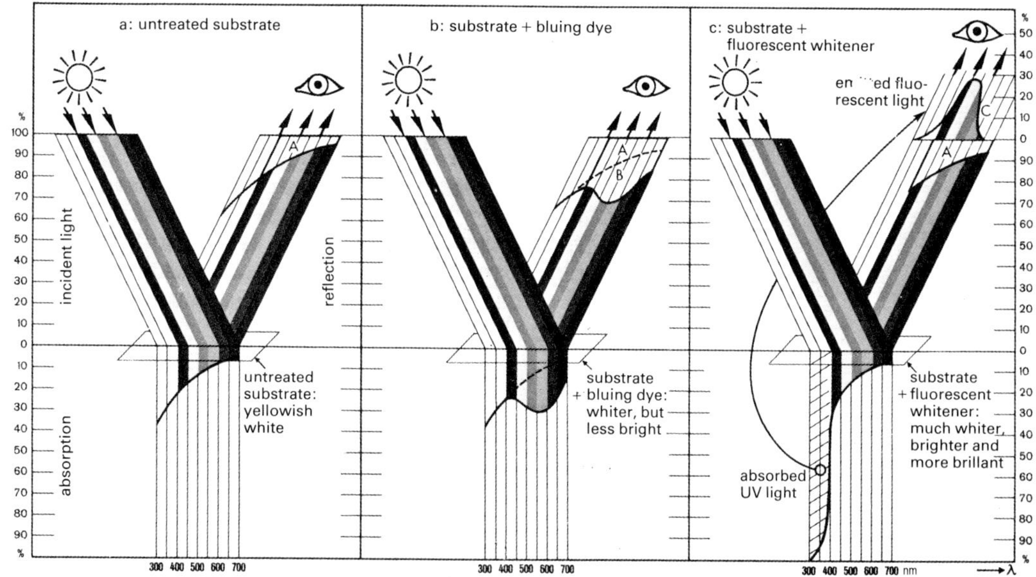

Fig. 1. Absorption and reflection of daylight on white surfaces.
a: Untreated substrate absorbs mainly blue light (A) → yellow cast (blue defect).
b: Blued substrate: compensation of yellow cast (A) by additional absorption of green-yellow light (B) → luminosity loss.
c: Fluorescent whitened substrate: remission + fluorescence → yellow cast compensated + excess of blue light (C).

In this way, the yellowish cast of the substrate is compensated by increasing the reflected blue portion. The substrate appears whiter, but the total reflected light is reduced (figure 1 b), in consequence it appears duller [1].

In contrast fluorescent whitening agents are capable of absorbing invisible u.v. radiation (in the wavelength range of about 360 to 380 nm) converting it to longer wavelength and reemitting it as visible blue or violet light. Thereby, the unwanted yellowish cast of the substrate is compensated and, in addition, more visible light in the range of 400 to 600 nm is reflected than was originally incident; hence the article appears whiter, brighter and more brilliant (figure 1 c).

The physical process in the fluorescent whitener agent molecule follows the pattern shown in figure 2: By absorption of photons (energy) from ultraviolet light, the whitener molecule raises its energy from the ground state S_0 to the excited electron state S_2. Owing to non-radiating deactivation it loses part of its energy and drops to the level S_1 and finally, because of fluorescent emittance, back to the ground state S_0.

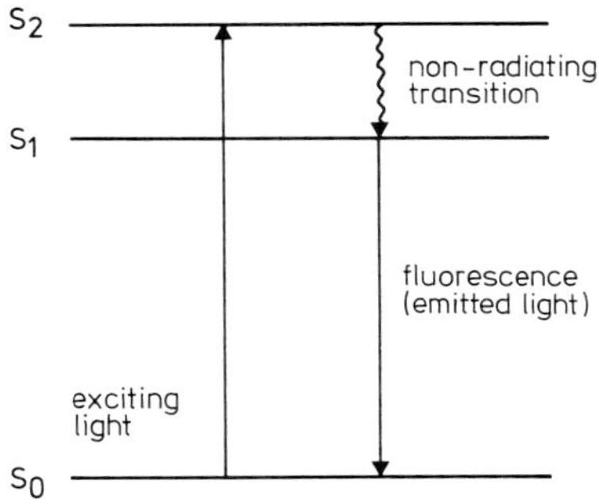

Fig. 2. Simplified functional diagram of physical processes in an excited whitener molecule.

Since a certain portion of the received energy is reemitted as non-radiating energy, the emitted photons are at a lower energy level, i.e. the radiation is of a longer wavelength. Small amounts of foreign matter (such as catalysts, impurities) are able to partially or completely quench the fluorescence, due to absorption of excitation energy [2].

13.3 Requirements

An essential criterion for the technical suitability of a fluorescent whitening agent is its lightfastness in the substrate under consideration. Fluorescent whiteners are usually more or less light sensitive in the wavelength range needed to excite the molecule to fluoresce. Consequently, the lightfastness of whiteners is limited, and is considerably lower than the lightfastness of pigments and the light stability of most plastics. Colored structures, which may lead to additional yellowing must not be generated at the end of the photochemical degradation of a fluorescent whitener.

The lightfastness of a fluorescent whitener is considerably influenced by the substrate (figure 3). In general we find the following ranking of lightfastness:

Crystal polystyrene > polyvinyl chloride > polypropylene
polyethylene > polyurethane

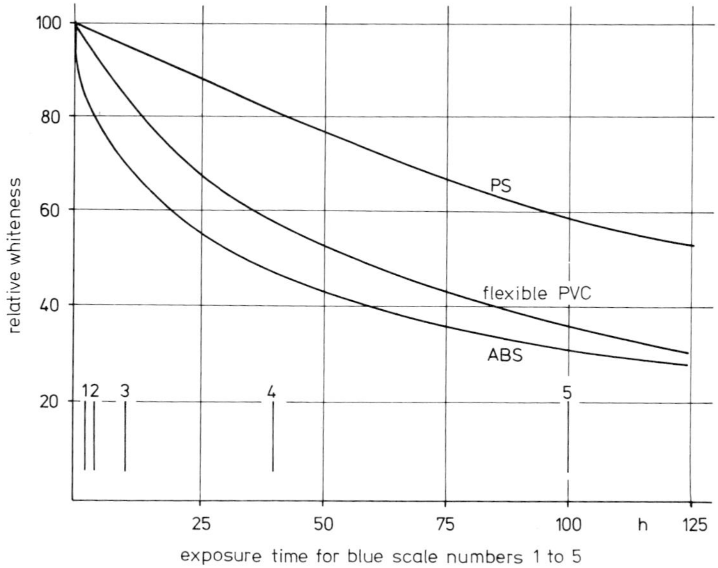

Fig. 3. Lightfastness of a fluorescent whitener (bis-benzoxazole) in different plastics. Relative whiteness as a function of exposure time in a Xenotest 150 with filtered xenon radiation (corresponding to daylight under window glass).

To achieve maximum effectiveness with a fluorescent whitener it is imperative that the whitener is fully dissolved and homogeneously distributed in the finished article, i.e. the distribution must be monomolecular. Sufficient compatibility is necessary to avoid exudation. The requirements regarding thermal stability and low volatility of fluorescent whiteners for thermoplastics are higher, than e.g. for whiteners for use in textiles, which are applied by wet dyeing.

Migration stability has to be taken into consideration particularly in flexible polyvinyl chloride and polyurethane coatings. In flexible polyvinyl chloride, an insufficiently compatible whitener may be transported to the surface by plasticizer migration. Consequently partial loss of whitener results; furthermore, the surface may become discolored in spots.

The evaluation of a whitener is also influenced to a certain extent by its hue. In general, neutral blue to blue-greenish hues are preferred.

13.4 Chemical classes

Only few of the known patented chemical classes of fluorescent whitening agents possess the properties required for mass whitening [2 to 6].

The most important chemical classes are:

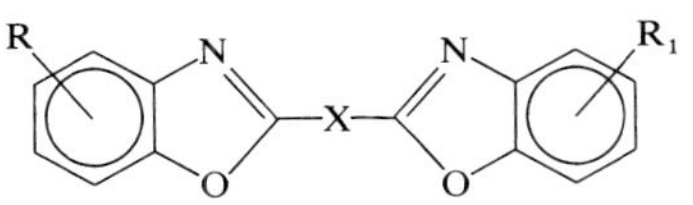

(1)

bis-benzoxazoles

triazine-phenylcoumarins
benzotriazole-phenylcoumarins
naphtotriazole-phenylcoumarins

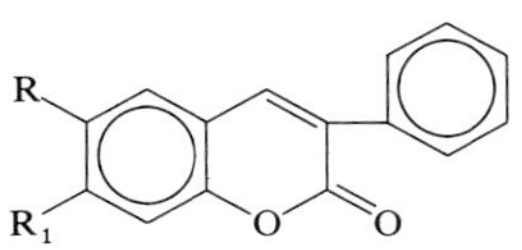

(2)

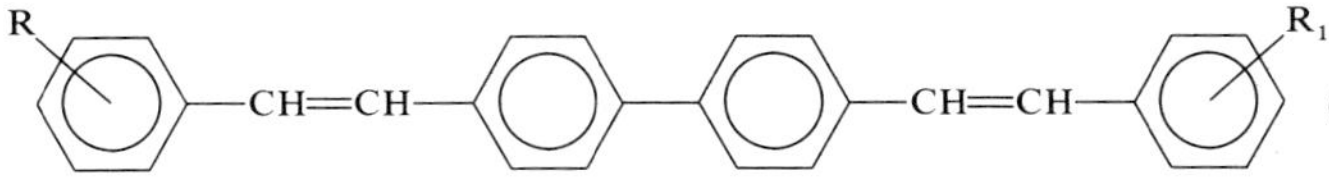

(3)

bis-(styryl)biphenyls

In principle, fluorescent whitening agents belonging to these chemical classes can be used in all thermoplastics.

13.5 Incorporation and product forms

The incorporation of fluorescent whiteners in thermoplastics is carried out exclusively into the mass and before the forming process; i.e. the whitener is not applied subsequently by means of solutions, hence the term "mass whitening".

Usually, fluorescent whiteners are introduced by dry blending with the plastics material (powder or pellets). If necessary, an adhesion promoter as e.g. butyl stearate is added. In flexible polyvinyl chloride, fluorescent whiteners are frequently applied as a solution or dispersion in a plasticizer.

To achieve distribution of improved homogeneity at low concentrations, the fluorescent whiteners are often added as a concentrate or as a master blend, e.g. 10% active whitener blended with chalk or a plasticizer.

13.6 Selection

The determining factors for the selection of the most suitable fluorescent whitener for a given plastics material are:

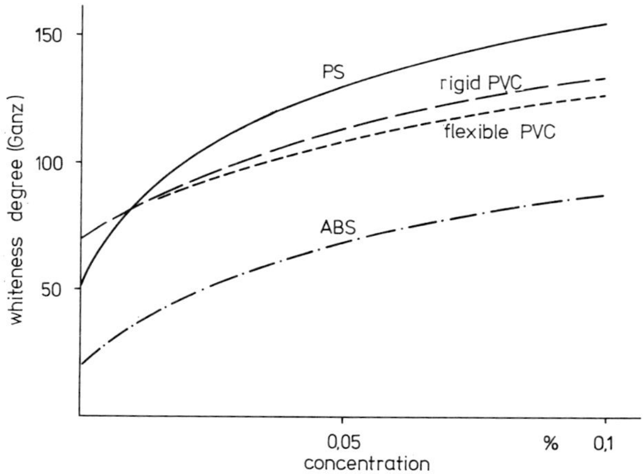

Fig. 4. Whitening effect of a bis-benzoxazole whitener in different pigmented (titanium dioxide, rutile) plastics as a function of concentration.

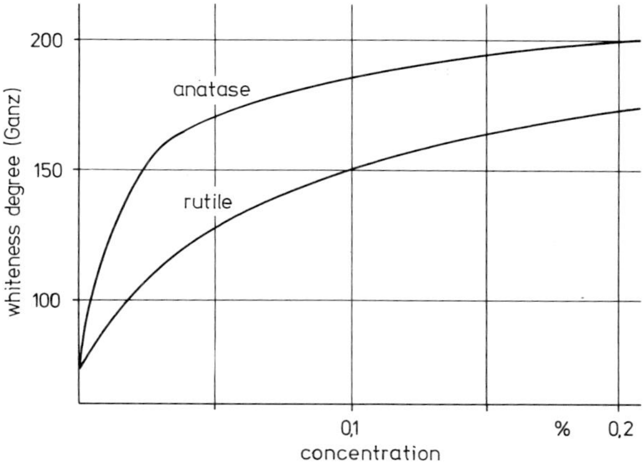

Fig. 5. Whitening effect of a bis-benzoxazole whitener in flexible PVC (5% titanium dioxide) as a function of concentration. Comparison anatase/rutile.

- Achievable whitening effect,
- lightfastness,
- compatibility.

Lightfastness and achievable whiteness degree of a given fluorescent whitener depend on the substrate (figure 4). In addition, however, the whiteness degree may be influenced by processing conditions. Furthermore, possible interactions with other components in the formulation such as e.g. white pigments and u.v. absorbers have to be taken into account. Titanium dioxide white pigments absorb in the long-wave u.v. range, which is essential for exciting fluorescent whiteners, thus generating lower whiteness degrees. Anatase pigments absorb at 380 nm about 40%, rutile pigments on the other hand about 90% of the incident radiation (figure 5).

U.V. absorbers too, absorb radiation in the fluorescent whitener exciting range, and may therefore lead to reduced whiteness. However, this whiteness decrease is small since the fluorescent whitener on the surface receives sufficient energy to become excited.

13.7 Examples of application

The most important applications of fluorescent whiteners in thermoplastics are:

Plastics	Processing/end use
rigid and flexible PVC	films, extrusion, injection molding, plastisols, imitation leather
crystal polystyrene (PS) impact polystyrene (IPS) styrene-acrylonitrile copolymers (SAN) styrene-butadiene-acrylonitrile terpolymers (ABS)	extrusion, injection molding
polyurethane (PUR)	coating, injection molding
polycarbonate (PC)	extrusion, injection molding
polypropylene (PP)	films, injection molding
polyethylene (PE)	films, injection molding, paper coating

13.7.1 Polyvinyl chloride

Usually bis-benzoxazoles or phenylcoumarins are used in flexible and rigid polyvinyl chloride, their lightfastness is 3 to 4 on the blue scale.

The incorporation of fluorescent whiteners does not present particular difficulties. By dry blending with PVC powder together with the commonly used heat stabilizers, while compounding or on the mill, a sufficiently homogeneous distribution is obtained. In flexible polyvinyl chloride, the fluorescent whitener is often predissolved or dispersed in the plasticizer. Nevertheless, it is frequently preferable to add the fluorescent whitener in the form of a concentrate or a blend. For transparent formulations, concentrates with plasticizer or PVC are used. For pigmented formulations, blends with chalk are also employed.

To improve the basic whiteness of polyvinyl chloride, extremely small amounts of a fluorescent whitener are sufficient. A concentration of

50 to 100 ppm, or even less, is usually adequate to compensate a yellowish hue. Transparent formulations achieve thereby a water-clear appearance. For ultra-white-pigmented PVC articles, fluorescent whitener concentrations of about 500 ppm are required; for rutile-pigmented up to 1000 ppm. Overdosage must be avoided because certain fluorescent whiteners tend to greening at higher concentrations, i.e. a less pleasing whiteness hue will be obtained and, furthermore, the whiteness degree may be reduced.

In general, the heat stabilizer formulation has only a minimal influence on the whitening effect.

The processing temperatures of polyvinyl chloride are usually in the range of 160 to 200 °C. Thus, the requirements on the heat stability of fluorescent whiteners are not too severe. On the other hand, solubility and migration stability (especially in flexible PVC) are of great importance.

13.7.2 Polystyrene and styrene copolymers

The same bis-benzoxazole and phenylcoumarin types can be used with these plastics as with polyvinyl chloride. The lightfastness in crystal polystyrene is one half to one point (blue scale) better than in polyvinyl chloride, in other styrene polymers, however, somewhat worse.

Bis-(styryl)-biphenyls also yield excellent whitening effects but their lightfastness is somewhat less than that of the chemical classes mentioned above.

Crystal polystyrene exhibits slight inherent fluorescence, which may be quenched by traces of yellow-colored degradation products caused by thermal stress during processing (extrusion, injection molding). However, as with polyvinyl chloride, it is possible to compensate a yellowish hue with minimal concentrations (less than 50 ppm) of a fluorescent whitener. In white-pigmented crystal polystyrene (mostly with rutile titanium dioxide) ultra-white articles are achieved with about 500 ppm of a fluorescent whitener.

Styrene copolymers such as impact polystyrene (IPS), styrene-acrylonitrile (SAN), acrylonitrile-styrene-acrylate (ASA), acrylonitrile-butadiene-styrene (ABS), styrene-butadiene-styrene (SBS) exhibit on occasion a distinct yellowish cast, which can be compensated by the addition of a fluorescent whitener. To that end, higher concentrations of fluorescent whiteners are needed than for crystal polystyrene. To obtain adequate whitening effects in pigmented systems, about 1000 ppm and even more have to be added. The whitening effects are usually not as pronounced as in crystal polystyrene and often a white with a greenish cast is obtained. To ascertain the best fluorescent whitener and the optimal concentration, tests are recommended.

The incorporation of fluorescent whiteners in styrene polymers is usually effected by dry blending of the whitener with the polymer pellets prior to processing.

13.7.3 Polycarbonate

Polycarbonates are processed at temperatures of about 300 °C and higher, whereby yellowing may occasionally occur. These high processing temperatures obviously impose restrictive requirements on the thermal stability and volatility of a fluorescent whitener.

With amounts of about 50 ppm of the same chemical classes as for polyvinyl chloride and styrene polymers it is possible to compensate yellowish discolorations of unpigmented polycarbonate. For ultra-white (pigmented) articles concentrations of about 500 ppm are needed.

The addition of fluorescent whiteners to polycarbonate is the same as in styrenics: by dry blending with the dried pellets prior to processing. For improved adhesion of the whitener to the polycarbonate pellets the use of an adhesion promoter is recommended, such as butyl stearate in amounts of 50 to 100 ppm.

13.7.4 Polyurethane

In the polyurethane field, only the one-component systems are usually fluorescent whitened. They are used for films and coatings (solution coating, thermoplastic coating). The fluorescent whitener is directly added to the polyurethane solution or dry blended with the pellets.

Special attention has to be paid to solubility and migration stability with cast films because during the drying process, there is the possibility of transporting the fluorescent whitener to the film surface by vaporizing solvent. In that way random yellow spots of exuded optical whitener may appear.

The lightfastness of fluorescent whiteners in polyurethane is in general distinctly poorer than in other plastics. For ultra-white articles concentrations in the order of 500 ppm are required.

The selection of the most suitable fluorescent whitener should be made by means of undertaking tests in the given polyurethane formulation and by taking into account recommendations of the fluorescent whitener producers.

13.7.5 Polyolefins

Polyolefins are occasionally also mass-whitened, especially e.g. polyethylene films for paper coating. An essential criterion for the choice of an adequate product is its compatibility with the substrate. The compatibility with polyolefins is generally in the following decreasing order:

Polypropylene > high density polyethylene > low density polyethylene

Lightfastness in polyolefins is generally poor (blue scale number 1 to 2).

13.8 Testing of fluorescent whiteners

13.8.1 Whitening effect

The evaluation of the whitening effect of a fluorescent whitener may be carried out visually or by means of colorimetric measurements.

For visual evaluation so-called "white scales" are used for comparison, e.g. the plastic white scale developed by Ciba-Geigy. This consists of a melamine molding resin and is graded with yellow pigments, or fluorescent whiteners into 12 different whiteness degrees. The calibration point is physically ideal white, approximately represented by magnesium oxide or barium sulfate. This calibration point represents by definition the whiteness degree 100.

The result of any visual evaluation depends considerably on luminosity and spectral energy distribution of the incident light, as well as on physiological influences and the color sensitivity of the observer. The assessment should be carried out in daylight (behind window glass, or even better, outdoors) with adequate brightness between 10 am and 3 pm facing north in the northern hemisphere and south in the southern hemisphere.

The visual evaluation is not satisfactory because of the numerous influences which are hard to keep constant. For that reason in recent years the instrumental assessment of whiteness has been increasingly adopted [7].

White, as a physiological impression, is not directly measurable, for that reason it has to be calculated from colorimetric data. On the basis of physical measurements, objective characteristic values are obtained, free of subjective influences. A specified light source in the color measuring device is a prerequisite for reproducible results independent of time, location or operator.

The literature describes many formulas for the measurement of whiteness degrees. Mostly they are based on the tristimulus values X, Y, Z and the chromaticity coordinates x, y and z according to CIE-standard (Commission Internationale de l'Eclairage) [8]. These tristimulus values permit not only the expression of the whiteness impression in *one* single number, but also furnish another value (dominant wavelength) for the assessment of the color hue. In the evaluation of different whiteness degree formulas, it must be borne in mind that all such formulas are empirical statements, since color (also white) is a physiological impression and not a physical value. The tristimulus values can be measured by means of spectrophotometers or by three-filter instruments (e.g. Elrepho or Hunter).

Details of the colorimetrical and mathematical problems associated with instrumental whiteness assessment are beyond the scope of this chapter. Reference is made to the specialized literature [9 to 13].

Here we list only some of the most common formulas:

According to *Taube* $W = 4B - 3G$
according to *Berger* $W = 3B + G - 3A$

where A: amber (red), B: blue, and G: green, values which can be calculated by tristimulus values or directly measured with three-filter instruments;

according to *Ganz**) $W = D \cdot Y + P \cdot x + Q \cdot y + C$

D, P, Q and C are parameters of the formula which permit adjustment of the whiteness degree values to any white scale and takes into consideration the type of light source in the measuring instrument.

Whiteness degree formulas on the basis of Hunter coordinates:

According to *Hunter* $W = L - 3b$
according to *Stensby* $W = L + 3a + 3b$

here, L indicates the position of the color point on the axis of luminosity, a on the red-green axis and b on the yellow-blue axis (figure 6).

---

*) The whiteness degree values in this contribution have been assessed according to the *Ganz* formula.

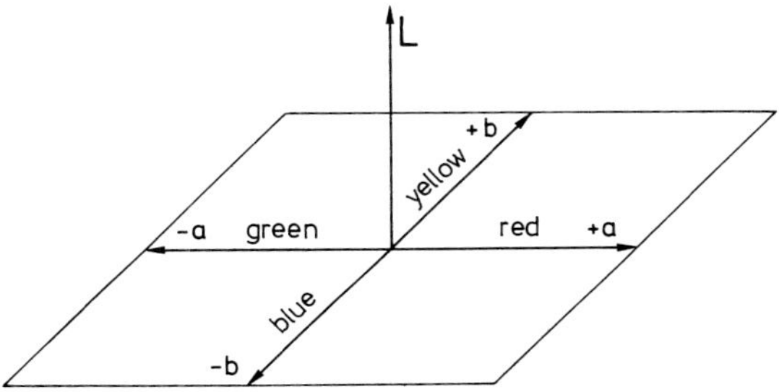

Fig. 6. Color coordinates according to *Hunter*. L: luminosity; a: red-green; b: yellow-blue.

13.8.2 Lightfastness

For the time being there are no international standards for the assessment of the lightfastness of fluorescent whiteners in plastics. Therefore, testing is conducted mostly with reference to DIN 54004, "Bestimmung der

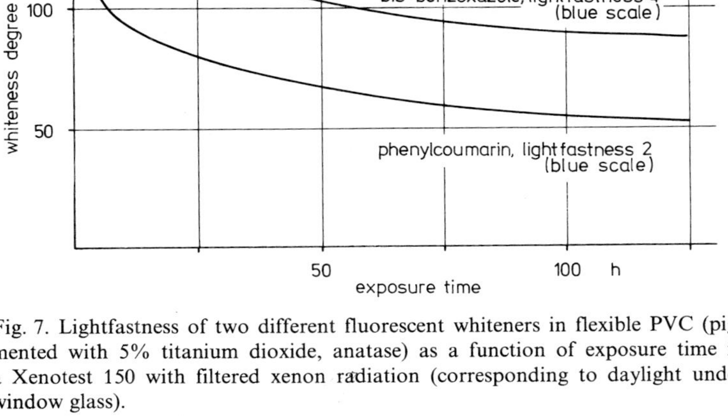

Fig. 7. Lightfastness of two different fluorescent whiteners in flexible PVC (pigmented with 5% titanium dioxide, anatase) as a function of exposure time in a Xenotest 150 with filtered xenon radiation (corresponding to daylight under window glass).

Lichtechtheit von Färbungen und Drucken mit künstlichem Tageslicht (gefiltertem Xenonbogenlicht)" or BS 4321.

The light exposure of the samples is carried out in a light exposure device with filtered xenon radiation, corresponding to the global radiation behind window glass (e.g. Atlas Weatherometer, Xenontest).

As a yardstick for the incident radiation an ISO blue scale [14] is used as a simple actinometer. It consists of eight standardized wool fabrics dyed with blue dyestuffs having different lightfastness. As end point, i.e. lightfastness number (blue scale number) the blue scale grade is taken at which the test sample and also the blue scale show the first color change compared with the unexposed control.

A better and more objective assessment can be obtained by measuring the change of whiteness degree as a function of exposure time (example for flexible polyvinyl chloride in figure 7).

13.8.3 Migration and exudation

Examination for migration and exudation effects is carried out best in the light of an u.v. lamp. Migration and exudation effects result in random yellow spots, which fluoresce distinctly brighter than the surrounding area.

13.9 List of trade names and producers of fluorescent whitening agents

The following list of trade names and producers is not claimed to be comprehensive.

Trade name	Chemical class	Producer
Blankophor	benzotriazole-phenylcoumarins	Bayer AG, Leverkusen, D
Eastman	bis-benzoxazoles	Eastman Chemical, Kingsport (Tn), USA
Fluolite	pyrene-titriazines	ICI, Manchester, GB
Hostalux	bis-benzoxazoles	Hoechst AG, Frankfurt, D
Leukopur	naphtotriazole-coumarins	Sandoz AG, Basel, CH
Uvitex	bis-benzoxazoles bis-(staryl)biphenyls	Ciba-Geigy AG, Basel, CH
Whitefluor	styryl-bis-benzoxazoles	Sumitomo Chem. Co. Ltd., Osaka, Japan

13.10 References

[1] Ciba-Geigy Rundschau (1973) 1, p. 5.
[2] *A. Brockes:* Fluorescent Whitening Agents, EQS Environmental Quality and Safety (Suppl. Vol. IV), Georg Thieme Verlag, Stuttgart, 1975, pp. 19/24.
[3] *G. Broadhurst, A. Wieber:* SPE J. 29 (1973) pp. 36/39.
[4] *K. Eschle:* Plastverarbeiter 21 (1970) pp. 629/632.
[5] *R. Zweidler, H. Häusermann:* in *Kirk-Othmer,* Encyclopedia of Chemical Technology, Vol. 3, 2nd Ed., 1964, pp. 737/750.
[6] *R. Zweidler:* Textilveredlung 4 (1969) pp. 75/87.
[7] *G. Anders:* Fluorescent Whitening Agents. EQS Environmental Quality and Safety (Suppl. Vol. IV), Georg Thieme Verlag, Stuttgart, 1975, pp. 83/93.
[8] DIN 5033, ASTM D 307-44.
[9] *R. Griesser:* Ciba-Geigy Rundschau (1973) 1, pp. 2/3; pp. 14/15.
[10] *A. Berger:* Farbe 8 (1959) pp. 187/202.
[11] *E. Ganz:* Appl. Opt. 15 (1976) 9, pp. 2039/2058.
[12] *R.S. Hunter:* J. Opt. Soc. Amer. 48 (1958) p. 597.
[13] *P.S. Stensby:* Soap. Chem. Soc. 43 (1967) pp. 80/88.
[14] ISO/R 105/V (1969), DIN 54004, BS 4321.

14 Biostabilizers

Dr. Joachim Lorenz, Ciba-Geigy Marienberg GmbH, Bensheim,
Federal Republic of Germany

14.1 Biodeterioration

14.1.1 Fundamentals

Natural polymers, such as cellulose and starch, biodegrade readily. On the other hand, most commercial synthetic polymers are resistant to biodegradation apparently because microorganisms are not able to assimilate them directly. Moreover, if man-made polymer compositions biodegrade, it usually is not the polymer itself, but other constituents of the formulations, most commonly plasticizers, which are degraded.

Microbiological attack of plastics received a great deal of attention beginning with the early days of World War II. In 1964, *Wessel* [1] published a comprehensive review on this topic. Nowadays biodeterioration (sometimes also called biological corrosion) of plastics is recognized as a problem, especially with plasticized PVC. It is not the PVC itself but the plasticizer which is attacked. Degradation of the plasticizer, which usually comprises a rather high portion of any plasticized PVC, results in strong reduction of the product's useful properties.

Several publications deal with the susceptibility of various plasticizers to microorganisms (see [1] and [2] for more literature references).

From these, one might prepare the following groupings:

None to low susceptibility:

- Phthalates,
- phosphates,
- toluene sulfonates.

Medium susceptibility:

- Adipates,
- epoxidized fatty acids.

High susceptibility:

- Oleates,
- stearates,
- polyesters.

But a table like this has to be used carefully. The initiation of the biodegradation of plasticizers, mostly esters by nature, is presumably via enzymatic cleavage of ester-bonds [3]. This means that any plasticizer can be attacked by microorganisms. Therefore, it is more a difference in time than in character which gives the impression of different susceptibilities to microorganisms.

Lubricants and related auxiliaries like slip and antiblocking agents exhibit microbiological susceptibilities different from PVC stabilizers, see e.g. [4]. Their use concentrations are generally much lower than those of plasticizers.

All these results were obtained with individual materials and do not necessarily reflect the behavior of finished articles of PVC since, within plasticized PVC, compatibility, diffusion equilibrium, and the like may influence microbial susceptibility to a high degree.

Studies on PVC articles proved that all kinds of microorganisms – bacteria, fungi, yeasts, sometimes algae – can produce deterioration. On plasticized PVC practically all types of microorganisms were found (see [2]).

14.1.2 Manifestation of biodeterioration

Staining:

Staining is likely caused by indelible dyes which are metabolic products of microorganisms. Besides the widely occurring pink stain, yellow, gray or black discoloration can also be experienced. Furthermore, staining may stem from microbiological degradation products by the influence of light and/or oxygen. Reduction of physicomechanical performance does not necessarily accompany color development.

Odor problems:

Metabolites of mold and mildew are probable sources of odor, as well as degradation products caused by microorganisms. Again, mechanical properties do not necessarily decrease with odor development, especially since odor generation needs minimum concentrations of substances.

Effect on electrical properties:

Deterioration of electrical properties, especially insulating power, may be caused by microorganisms. To produce such effects, microorganisms need only to settle on the surface of materials without any biological damage.

Enhanced dirt uptake:

Dirt pickup results from surface roughness and stickiness. Degradation by light or plate-out of plasticizer due to inappropriate compounding may be the reason. For plasticized PVC, this effect is often an indication of the beginning degradation of the plasticizer by microorganisms.

Permeability of solvents and gases:

Permeability can be influenced by microorganisms in two ways: degradation of the plasticizer itself or changes in the solvent resistance by degradation products of the plasticizer (e.g. alcohols) as well as metabolites of the microorganisms (e.g. acids).

Deterioration of mechanical properties:

This is the most serious consequence of microbiological degradation. Brittleness and shrinkage indicate the loss of plasticizer. Moreover, the degradation products may even function as accelerators for the degradation of the PVC polymer by light and/or oxygen.

Hygienic effects:

The growth of microorganisms on plastics' surfaces not only gives rise to poor appearance but it also creates the danger of supporting and dispersing even pathogenic strains. The possibility of transmitting infections is widely underestimated.

14.1.3 Fields of application of biostabilizers

Microorganisms in general need humidity (relative humidity 63 to 99%) and optimum temperature conditions ranging from 10 to 40 °C [5] for growth. If these environmental conditions exist, and plasticizers or other additives are available as nutrient, it becomes quite clear that the PVC products listed are endangered:

Building and construction:

- Inflatable buildings,
- barrier films (solid waste disposal sites),
- roofing membranes,
- electrical cord,
- vinyl hoses,
- wire jacketing.

House and garden, recreation:

- Floor covering,
- wall covering,
- appliance gaskets,
- shower curtains,
- bath mats,
- water beds,
- shoes and boots,
- awnings,
- tents,
- pool and pond liners,
- lawn furniture,
- coated fabrics,
- agricultural film.

Transportation:

- Automobile tops,
- truck covers,
- vinyl side molding,
- sight shields,
- tarpaulins,
- marine upholstery.

When considering the biostabilization of plasticized PVC it is necessary to take into account the conditions of installation. Proper subsurface preparation is necessary especially for floor and wall coverings, roofing membranes, swimming pool liners and the like. Since PVC sheets function as moisture barriers, growth conditions for microorganisms under those layers may become excellent, if adequate chemical and biological preparation were neglected. Contamination by underground molds may give rise to discoloration of well protected plastics material due to colored metabolites produced by the fungi underlaying the material. Odor problems are also often caused by this condition.

14.2 Biostabilizers

14.2.1 Definition

Light stabilizers and heat stabilizers (thermostabilizers) are products which prevent deterioration due to the influences of light, oxygen and heat. It seems logical to call biostabilizers all those products which ensure the protection of materials against the negative influence of microorganisms. This expression has the advantage of overcoming the confusion which is caused by using terms like fungicide, bactericide, bacteriostat, microbiocide etc., which sometimes are even misused or misunderstood.

14.2.2 Requirements

An effective biostabilizer should meet the following requirements:

- Broad spectrum of antimicrobial activity,
- effective at low concentrations,
- no negative influence on the material to be protected (e.g. no discoloration, no negative influence on heat stability or light stability),
- compatibility with plasticizers, processing aids, and other additives (stabilizers, lubricants, antioxidants, u.v. absorbers, etc.),
- no negative side effects (e.g. corrosion, skin irritation, odor and the like) of the end product,
- thermal stability, and none to low volatility under processing conditions,
- storage stability,
- easy application,
- long lasting effectiveness,
- low toxicity to man and environment,
- registration by authorities (e.g. EPA) in several countries.

14.2.3 Products

As usual, a great number of biologically active products are claimed in patents or proposed in publications. But only a few products or groups of products have achieved significant market volumes:

- 10,10′-oxy-bis-phenoxarsin (I),
- N-(trihalogenomethylthio)-phthalimide (II) and the corresponding tetrahydro derivative,
- diphenylstibine-2-ethylhexanoate (III),
- copper-bis-(8-hydroxyquinoline) (IV),
- tributyltin oxide and its derivatives (V).

The above-mentioned active ingredients are sold by several producers under different trade names. One may find them as 100% active materials

I 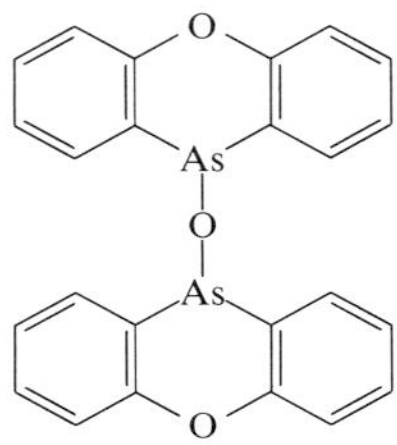

II 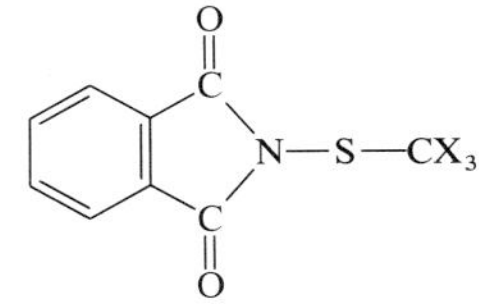

X = F, Cl

the corresponding tetrahydro derivative is also commercially available

III 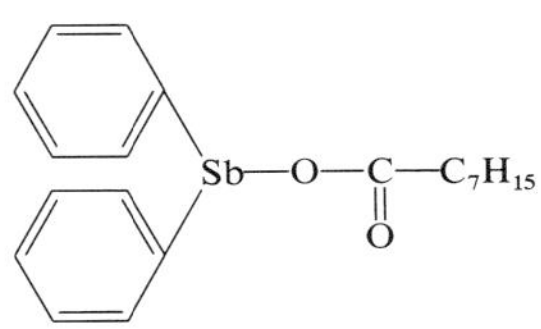

IV 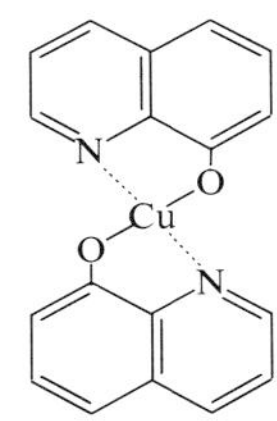

V $(H_9C_4)_3Sn{-}O{-}Sn(C_4H_9)_3$ and $(H_9C_4)_3Sn{-}O{-}C(=O){-}R$

as well as constituents of formulations. Formulations may contain plasticizers, costabilizers like epoxidized soybean oil, or solvents as vehicles. Accordingly, the recommended use concentration varies from 0.3% (pure active ingredient) to 5% (formulation). It is also dependent on the field of application.

14.2.4 Selection and testing of biostabilizers

Producers of biostabilizers usually offer a survey on the spectrum of activity of their products against microorganisms. They also provide physical and chemical properties and handling safety information for the compounds. Thus, a well prepared data sheet or brochure is an important source of information to help choose the appropriate product for a given application.

The "spectrum of activity" sometimes is provided as a list of microorganisms against which activity has been demonstrated. In many cases, a more detailed description includes the activity against single strains by defining the minimal inhibition concentration (MIC). This concentration is given in ppm (parts per million) or in mg/l (milligram per liter which corresponds to ppm). Both these units are preferable to MIC values expressed as percentage. MIC indicates the concentration of the active ingredient (or of the formulation) which prevents growth of a certain strain of a microorganism under the test conditions used. It gives no information whether the product kills that strain or stops its further growth only. The several strains are defined by their Latin names, sometimes letters and numbers are added (e.g. pseudomonas aeruginosa ATCC 13388) indicating the culture collection (ATCC = The American Type Culture Collection [6]) and the number (13388) under which this particular strain is available.

The spectrum of activity indicates strains against which a biostabilizer is highly effective (lowest MIC figure), strains against which a higher concentration of active ingredient is necessary (high MIC figure), and strains against which possibly no activity may be expected (marked with > for the highest concentration tested).

A comprehensive assessment of the all-round effectiveness of the biostabilizer in the envisaged plasticized PVC article cannot be drawn, however, from the knowledge of the spectrum of activity alone. The type and amount of plasticizer may influence the compatibility as well as the long time effectiveness of the biostabilizer. Thermostabilizers, fillers, lubricants, and pigments also may have an influence on the actual activity [7]. With some experience in this field, these influences are partially predictable but nevertheless the complexity of the total system requires more or less extensive testing.

The selection of the most suitable method of testing is of great importance and must take into account the specific application. From the historical point of view, many of these test methods derive from tests performed in the wood or textile industry. Know-how already existed for these natural polymers for the correlation of microbiological tests with practical results.

Some test methods have been published as standards (see chapter "Standards"), others may be found in the literature as special publications [8] or more general reviews [2, 9].

14.2.4.1 Test methods

14.2.4.1.1 Microbiological attack

There are three groups of test methods to evaluate the biodegradation of plastic specimens:

a) The test organisms are contacted only with the material to be tested. The test specimens in horizontal or upright positions are inoculated by bacteria or fungi.

b) In addition to the test specimen a complete nutrient is offered to the inoculum.

c) The resistance of the test material is investigated by a soil burial test or under the influence of humidity or water.

For groups a) and b) either commonly known defined strains of bacteria and fungi or strains isolated from deteriorated material (they are sometimes only roughly defined) are used. The test is performed with single strains or a "mixed culture". A "mixed culture" is obtained by mixing together several defined single strains or by taking that mixture which was found on deteriorated products. Tests with pure, defined strains are carried out with a few representatives strains, preferably with:

Bacteria:

- pseudomonas aeruginosa,
- staphylococcus aureus,
- escherichia coli.

Fungi:

- aspergillus niger,
- penicillium funiculosum,
- chaetomium globosum,
- trichoderma viride.

Yeast:

- candida albicans.

This list is not to be taken as a complete one. Literature, data sheets or special reports often use additional strains.

As a rule, the listed test organisms make it possible to evaluate the behavior of a biostabilizer in the test material. The behavioral survey, in addition to providing knowledge about the more complete spectrum of activity of the biostabilizer itself, allows estimation of the protection which could be attained.

Group c) of the methods is quite different insofar as the natural microflora (soil) determine the type and concentration of microorganisms. On the other hand, some additional parameters like humic acids from the soil, the content of oxygen or CO_2 in the water make this test more rigorous. Nevertheless, they cannot be recommended if reproducibility of results is important, especially at different sites and at different times.

14.2.4.1.2 Evaluation

The evaluation methods of the results of biodegradation can be classified in three groups:

a) Visual observation of the test specimens like
   - sample overgrown,
   - zone of no growth around the sample,
   - color changes,
   - changes in surface characteristics.

b) Physical tests like
   - weight changes,
   - tensile strength,
   - electrical properties.

c) Control of metabolites of microorganisms by
   - oxygen uptake.

14.2.4.2 Problems of interpretation

An ideal situation exists where technicians closely cooperate with microbiologists. The microbiologist should establish whether the results obtained from the tests are relevantly interpreted by the technician or demand further testing. Instead of generalizing descriptions, an example might give a better explanation:

Four samples of plasticized PVC had to be tested, one containing no biostabilizer, the others the biostabilizers A, B, and C, respectively. They were tested in a Petri-dish test. For this test, the specimen is placed on the top of nutrient agar in a Petri-dish and then inoculated with the test organism. If no active substance is available, the test organism will overgrow all the agar including the test sample (O in figure 1). Any substance controlling growth becomes obvious by protecting the sample from growth or even producing a zone of no growth around the specimen (halo), see A, B, and C in figure 1.

Figures 1 and 2 show the results of such a dish test. *Aspergillus niger* was used in the first test (figure 1), *penicillium funiculosum* in the second (figure 2). Based on the diameters of the halos around the test specimens,

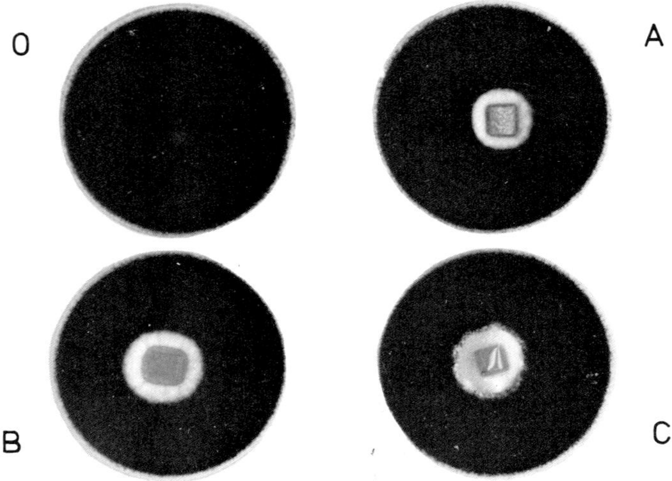

Fig. 1. Dish-test, test organism: *aspergillus niger*.

O: sample without biostabilizer,
A: sample containing biostabilizer A,
B: sample containing biostabilizer B,
C: sample containing biostabilizer C.

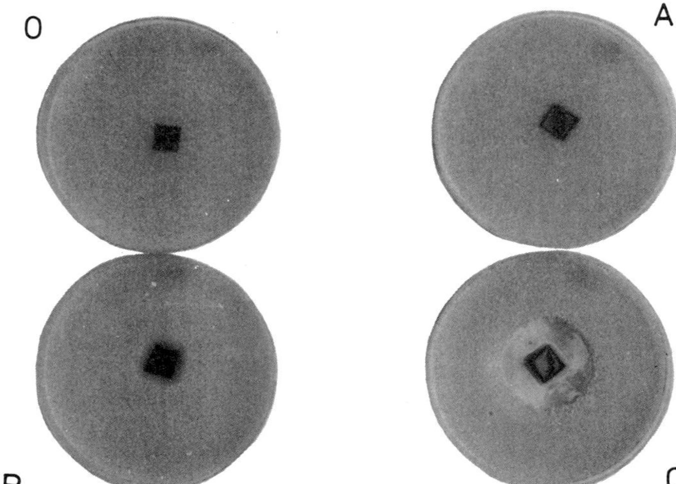

Fig. 2. Dish-test, test organism: *penicillium funiculosum*.

O: sample without biostabilizer,
A: sample containing biostabilizer A,
B: sample containing biostabilizer B,
C: sample containing biostabilizer C.

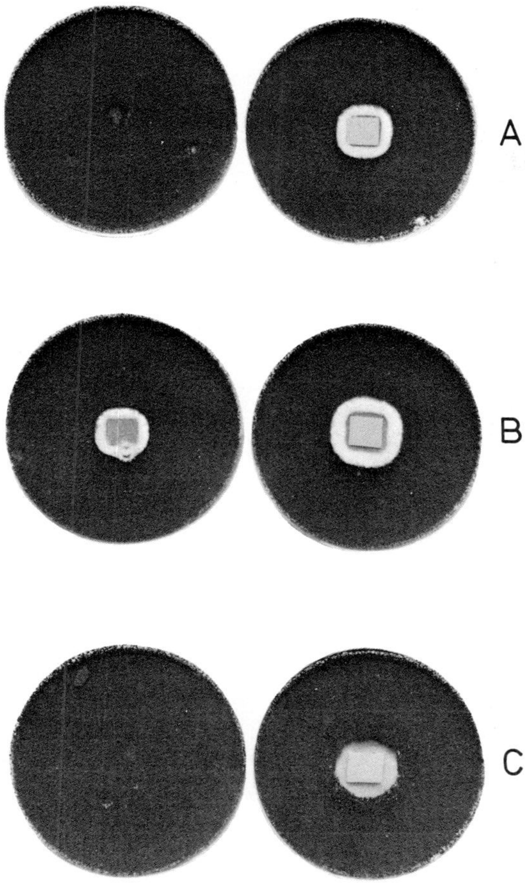

Fig. 3. Dish-test, test organism: *aspergillus niger*.

left column:	samples after 1 year outdoor exposure,
right column:	samples after 1 year storage in lab,
upper row:	samples containing biostabilizer A,
middle row:	samples containing biostabilizer B,
bottom row:	samples containing biostabilizer C.

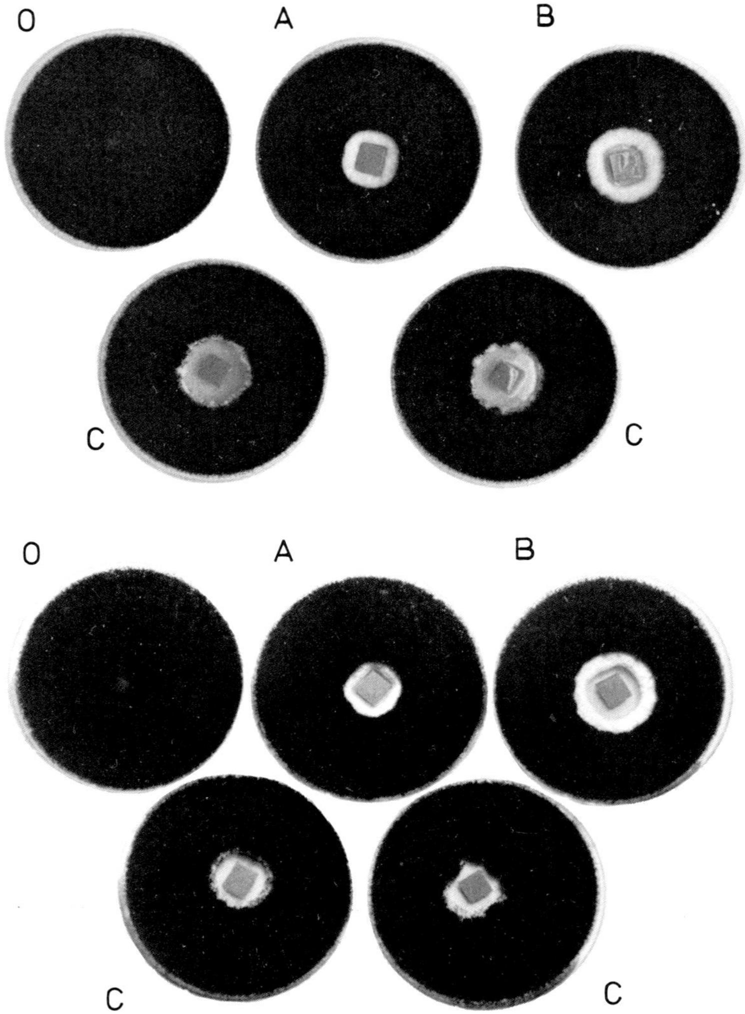

Fig. 4. Dish-test, test organism: *aspergillus niger*.

O: sample without biostabilizer,
A: sample containing biostabilizer A,
B: sample containing biostabilizer B,
C: sample containing biostabilizer C,

top group: samples without watering,
bottom group: samples after four weeks storage under running water.

product C was the most active of the three biostabilizers. Its superiority is clearly demonstrated against both the organisms tested.

Part of the samples were subjected to outdoor exposure for one year (45° south, free hanging) and then tested again in comparison to the samples which have been stored in the lab for the same period. Figure 3 shows the results of the test using *aspergillus niger*. The dishes on the left side contain the samples subjected to outdoor exposure, those on the right side contain the corresponding specimens stored in the lab (the blank was not repeated). Now product B will be accepted spontaneously as the best one.

This example clearly shows that the widest halo does not necessarily identify the best product. The existence of a halo indicates activity at a low concentration. It also demonstrates the mobility of the product and raises the issues of potentially easy depletion from the test specimen, water solubility (since agar solution has a high water content), and extractability by water.

Some other influences besides water can cause biostabilizer ineffectiveness, e.g. inadequate light stability or incompatibility with the PVC compound. In the reported case, a comparison of figure 1 with the right row of figure 3 (after storage in the lab) shows no differences. Thus, it can be concluded that no depletion on account of blooming took place during storage for over one year.

Additional testing was carried out to evaluate the resistance of the biostabilizer against extraction by water and to generate more complete information: test specimens not exposed to water (upper part of figure 4) and after four weeks' watering in a water tank in the laboratory (bottom part of figure 4) were tested against *aspergillus niger*. There is practically no difference in the growth-free zone using biostabilizers A and B. In contrast, the samples containing biostabilizer C show a much smaller halo diameter after watering. This means that product C is depleted by water from the PVC sheet.

A summary of the test results is presented in the following table:

Bio-stabilizer	Blooming	Water resistant	Light resistant	For outdoor use	For indoor use	
					wet	dry
A	none	yes	no	no	yes	yes
B	none	yes	yes	yes	yes	yes
C	none	no	?	no	no	yes

This example from an actual testing program but necessarily shortened to present the results with two test organisms only, should not be taken as a general recommendation. It was chosen to demonstrate two facts:

1) A test result (halo diameter in this case) allows different interpretations and
2) microbiological testing to define usefulness of a certain product gives a clear answer only if it is part of a well-conceived full testing program.

Often it will be impossible to delay the decision for the selection of an appropriate biostabilizer by waiting for the results of one year's outdoor weathering. Earlier decisions are made possible by comparing microbiological test results after artificial weathering or after storage in humidity cabinets and the like.

Kaplan and coworkers [8] published some years ago a report on testing 32 biological active compounds as biostabilizers in plasticized PVC. The flexibility of several sheets was tested after a soil burial test for 1, 2, 3, and 4 weeks, respectively. The test specimens had been buried in the original state, after watering, and after 100 hours exposure in a Weatherometer. With a few samples the test program was completed by conducting the soil burial test on specimens which had been weathered outdoors for one and three months.

The information in the literature [2, 9] coupled with other references found therein, should provide a stimulating source of ideas for the formulation of imaginative test procedures. A general rule for testing programs cannot be proposed. Knowledge of the testing art will permit selecting the best combination of tests which answers a specific question fast, cheaply and reliably.

14.3 Toxicity

14.3.1 Toxicity of biostabilizers

By definition, these products are toxic against microorganisms. In many instances they also have some toxicity for man. In this matter, one has to differentiate between toxicity of the active ingredient and toxicity of the formulation in which the active ingredient is offered.

Information on toxicity and necessary handling precautions are given in the safety data sheets supplied by the producers.

14.3.2 Toxicity of products

Usually, the concentration of biostabilizers in final articles is low enough to avoid potential harm to the user. If direct and permanent contact with the human skin or with drinking water or food is to be expected, biostabilizers should not be used or only if the special situation makes it indispensable. This is true also for articles which are destined for or assumed to be used by children, especially infants.

14.4 Technological trend and outlook

14.4.1 New biostabilizers

The extent to which new biostabilizers, if any, will enter the market cannot be predicted. In several current applications of biologically active products, environmental risks require new products. This does not refer to plastics where the objective of biostabilizers to protect the final article for a long time rules out any strong depletion into the environment.

The biostabilizers in the market have reached a high standard of performance. There seems to be no real need for new products. Nevertheless, optimization of existing products will take place.

A driving force for further development might be the general need to create products with better handling safety or enhanced occupational health conditions. But the real opportunities should not be overvalued; the costs of marketing new products may be high. There are not only the traditional costs of development but in several countries the additional high cost and time required to receive registration.

14.4.2 Market growth

The consumption of biostabilizers is expected to grow faster than the sales of articles in which they are used. Some developments (roofing membranes, solid waste disposal barrier films, etc.) started in the past with inadequate recognition of possible biodeterioration. Furthermore, in some fields of application a rather good market growth is expected, even if it is generally not as much as the predicted market doubling for single ply roofing membranes in the USA.

Last but not least, the need to save raw materials and energy by producing longer lasting materials is nowadays much better appreciated. This also calls for products which are more effective against biodegradation.

14.5 Standards

ISO 846-1978 (E): Plastics – Determination of Behavior under the Action of Fungi and Bacteria.

ASTM G 21-70 (D 1924-70): Determining Resistance of Synthetic Polymeric Materials to Fungi.

ASTM G 22-76 (Reapproved 1980) (D 2676-67T): Determining Resistance of Plastics to Bacteria.

ASTM D 876-71 (Edition 1975): Nonrigid Vinyl Chloride, Polymer Tubing used for Electrical Insulation (paragraph 63 to 70 Fungus Susceptibility).

ANSI[1])/ASTM E 154-58 (Reapproved 1975): Materials for use as Vapor Barriers under Concrete Slabs and as Ground Cover in Crawl Spaces (paragraph 10 to 13 Soil burial test).

AFNOR[2]) NF X 41-513: Protection des matières plastiques: Méthode d'essai de résistance des constituants aux microorganismes.

AFNOR NF X 41-515: Protection des matières plastiques: Méthode d'essai de résistance des matériels et appareillages aux microorganismes.

DIN 40046, Blatt 10, 1970: Klimatische und mechanische Prüfung für elektrische Bauelemente und Geräte der Nachrichtentechnik – Prüfung J: Schimmelwachstum.

DIN 53931–1969: Bestimmung der Widerstandsfähigkeit von Textilien gegen Schimmelpilze – Bewuchsversuche.

Draft DIN 53739 (1982): Einfluß von Pilzen und Bakterien – based on ISO 846-1978 – contains soil burial test based on EMPA[3]) VII/11 (1974).

MIL[4])-STD-810 B, method 508: Fungus.

---

[1]) ANSI = American National Standard Institute.

[2]) AFNOR = Association Française de Normalisation.

[3]) EMPA = Eidgenössische Materialprüfanstalt, St. Gallen, Switzerland.

[4]) MIL = Military Standards, Superintendent of Documents, U.S. Government Printing Office, Washington DC 20402.

MIL-E-5272C: Environmental testing, aeronautical and associated equipment, chapter 4.8 Fungus resistance tests.

Burial tests: DIN 53933 did not become standard due to problems of reproducibility, but see ASTM 684, AFNOR NF X 41-503, AFNOR NF X 41-514 and EMPA[1]) VII/11, 1974.

Further information on standards are included in *Pantke* [8] as well as in ref. [2] and [9].

14.6 Index of trade names and suppliers of biostabilizers

This list will not claim to be complete.

Trade name	Chemical name	Supplier
bioMet	diphenylstibine-2-ethyl-hexanoate	M & T Chemicals B.V. 4380 AB Vlissingen, NL M & T Chemicals Rahway, NJ 07065, USA
Cunilate	copper-bis-(8-hydroxy-quinoline)	Morton Thiokol Inc. Ventron Division Danvers, MASS 01923, USA; 1050 Bruxelles, B
Estabex	10,10′-oxy-bis-phenoxarsin	AKZO Chemie B.V. 3818 LW Amersfoort, NL
Fungitrol	N-(trichloromethyl-thio)-phthalimide	Tenneco Chemicals Piscataway, NJ 08854, USA
Intercide	tributyltin derivative	Interstab Chem. New Brunswick, NJ 08903, USA

Trade name	Chemical name	Supplier
Metatin	tributyltin oxide and derivatives	ACIMA 9470 Buchs/SG, CH
Preventol	N-(dichlorofluoromethyl-thio)-phthalimide	Bayer AG 5090 Leverkusen, D
Vancide	N-trichloromethyl-mercapto-4-cyclohexene-1,2-dicarboximide	R.T. Vanderbilt Norwalk, CT 06855, USA
Vinyzene	10,10′-oxy-bis-phenoxarsin	Morton Thiokol Inc. Ventron Division Danvers, MASS 01923, USA; 1050 Bruxelles, B

14.7 References

[1] *C.J. Wessel:* SPE Transactions 4 (1964) p. 193.

[2] *H. Becker, H. Gross:* Material und Organismen 9 (1974) p. 81.

[3] *H. Kühlwein, F. Demmer:* Kunststoffe 57 (1967) p. 183.

[4] *M. Pantke:* Material und Organismen 8 (1973) p. 249.

[5] *L.E. Hawker, A.H. Linton:* Micro-organisms, Edward Arnold (Publishers) Ltd., 1971, p. 155.

[6] The American Type Culture Collection, 12301 Parklawn Drive, Rockville, Maryland 20852. The 'Catalogue of Strains' contains a list of abbreviations for culture collections other than ATCC.

[7] *E.E. Cadmus, J.F. Brophy:* Kunststoffe 72 (1982) p. 233.

[8] *A.M. Kaplan, M. Greenberger, T.M. Wendt:* Polym. Eng. Sci. 10 (1970) p. 241.

[9] *M. Pantke* and *J.L. Osmon, R.E. Klausmeier, A.H. Walters* in: Biodeterioration Investigation Techniques, Applied Science Publishers Ltd., London (1977), p. 51, p. 77.

[10] *B. Doležel:* Die Beständigkeit von Kunststoffen und Gummi, Carl Hanser Verlag, München/Wien, 1978.

15 Chemical Blowing Agents

Dipl.-Ing. (FH) H. Hurnik, Bayer AG, Leverkusen,
Federal Republic of Germany

15.1 Technical and economic considerations

15.1.1 Introduction

Since the production of foams first began, more than a thousand compounds have been identified as chemical blowing agents with most of them described in patents. The majority of these compounds never progressed beyond the laboratory stage. Some of them were used on an economic basis for a short time and were later replaced by superior products. Today, there are no more than ten chemically different blowing agents of commercial significance worldwide.

15.1.2 Chemical blowing agents

15.1.2.1 Definition

The term "blowing agent" in the broadest sense denotes an inorganic or organic substance used in polymeric materials to produce a foam structure. Chemical blowing agents are products which decompose at high temperature, at least one of the decomposition products being a gas, which then expands the plastics material to give a foam structure. The decomposition process is generally exothermic and irreversible and takes place within a specific, narrow temperature range.

15.1.2.2 Required properties

The properties an ideal chemical blowing agent should possess were described in detail by *Reed* [1] in 1955. These requirements are still valid today.

The primary criterion governing the choice of a blowing agent for expanding a particular plastics material is its decomposition temperature. The blowing gas must be evolved within a defined temperature range, which should not be too large, and related to the processing temperature of the plastics material. The blowing agent must not decompose too spontaneously, as in the case of an explosive, otherwise heat buildup and combustion could occur due to the heat generated by the exothermic reaction. The gas formed should, if possible, consist of nitrogen and contain no explosive constituents. The blowing agent should be easy to incorporate and to disperse evenly in the plastics material concerned.

Neither the blowing agent nor its decomposition products should present any health hazard, adversely affect the thermal stability or mechanical strength of the plastic or have a corrosive effect. The decomposition residues should be compatible with the plastic and not exude or cause discoloration. Finally, the blowing agent should have a high gas yield and be economical to use. In view of all these conditions, it is obvious that no chemical blowing agent is ideal in every respect. Certain of these demands, however, are not necessarily relevant to the manufacture of every type of foam.

15.1.2.3 Incorporation into the compound

Chemical blowing agents usually take the form of powders of small particle size, though occasionally they are liquids. Uniform dispersion of the blowing agent in the plastics material to be expanded is crucial to the quality of the resulting foam, and special attention must be given to this stage in the process. Blowing agents for the expansion of PVC plastisols are normally added in the form of a pre-dispersed paste (blowing agent batch). This masterbatch often contains an additional amount of dispersing agent, apart from the plasticizer. This ensures faster, more even dispersion when the batch is stirred into the PVC plastisol than if the pure substance is added in powder form. In addition, directly dispersible blowing agent powders have recently been introduced; these make the formulation of a masterbatch unnecessary.

The production of cellular moldings on injection molding machines, known as thermoplastic foam molding (TSG), as well as the extrusion of molding compounds containing a blowing agent, known as thermoplastic foam extrusion (TSE), usually start from plastics granules [2]. The blowing agent may be added in a number of ways, the simplest being by application to the surface of the granules to be processed; this may be done in an eccentric tumbling mixer or on similar equipment. If the granules are to be transported pneumatically, prior application of a bonding agent such as paraffin oil or butyl stearate is recommended.

Blowing agents may also be incorporated using automatic metering and mixing devices mounted directly on the injection molding machine or extruder in place of the hopper. Such devices ensure reliable metering of blowing agents in the form of either powders or concentrates.

The latest development in the metering of blowing agents is based on liquid blowing agent dispersions which are injected straight into the barrel of the injection molding machine or the extruder via a measuring and pumping system. The action of the pump is linked electrically to the screw speed so that the blowing agent dispersion is metered only during the plastification stage of the injection molding process. In extrusion, however, the metering pump operates continuously [3 to 5].

15.1.2.4 Processing and applications

The processing methods for plastics compounds containing blowing agents are basically the same as for the corresponding compact materials. The oldest, and still the largest volume application for chemical blowing agents is in the PVC industry: foam coatings for handbags, suitcases etc., upholstery, clothing; foam backings for carpets; foamed floorcoverings (cushion vinyls), wallpapers, bath mats etc. In addition, the extrusion of pipe and profile in rigid PVC foam is gaining in importance [2, 6 to 8]. The main factor here is weight saving, which may be up to 50%. The majority of new developments in process technology, machine design and tooling have occurred in the injection molding of expandable thermoplastics other than PVC [2, 9 to 13]. Thermoplastic integral-skin foams are used for television sets and audio equipment, the automotive and transportation industries and in the construction and sports equipment industries. Chemical blowing agents are also employed in the high-pressure compression molding process with subsequent expansion, calendering with subsequent expansion, rotational foaming, mold foaming, sintering and extrusion blow molding.

15.1.2.5 Properties of foams

Cellular plastics possess a wide range of outstanding properties which make them superior to non-expanded materials in a number of applications.

The following properties are of particular significance:

- Lower density and therefore material saving;
- higher specific rigidity with integral-skin foams, taking into account the rigidity/weight ratio, the reason for this is that the third power of the wall thickness affects the rigidity of a molding;

- improved dielectric properties for cable insulations;
- improved heat and sound insulation;
- better acoustic behavior of housings for television sets and audio equipment;
- greater absorption of impact energy (packaging).

The addition of small amounts of blowing agent (on the order of 0.05 to 0.1%) eliminates sink marks in thick-walled injection-molded parts.

15.1.2.6 Economic importance

The chemical blowing agents in industrial use are azo compounds, hydrazine derivatives, semicarbazides, tetrazoles and benzoxazines.

The importance of the individual classes of products varies greatly. Azodicarbonamides and their modifications account for about 95% of all chemical blowing agents for cellular plastics. Of the remaining 5%, approx. 2% are sulfohydrazides, 2% are what are known as high-temperature blowing agents and 1% are other compounds. Approx. 85% of total blowing agent consumption is processed with polyvinyl chloride, the remaining 15% with polyethylene, polypropylene, polystyrene, ABS, modified polyphenylene oxide (Noryl), polycarbonate, polyamides and thermoplastic polyesters.

15.2 Product classes and modes of action

A knowledge of the chemistry of chemical blowing agents and an understanding of the decomposition process are important when choosing a blowing agent for a particular field of application. In this way, problems which may arise during their use are more readily understood and ideas more easily conceived for their solution.

15.2.1 Azo compounds

15.2.1.1 Azodicarbonamide

$$H_2N{-}CO{-}N{=}N{-}CO{-}NH_2$$

Azodicarbonamide is manufactured by a two-step process. In the first step (1) hydrazodicarbonamide is formed from hydrazine and urea; in the second step (2) this hydrazodicarbonamide is converted to azodicarbonamide by oxidation.

$$H_2N{-}NH_2 \;+\; 2\,H_2N{-}CO{-}NH_2 \xrightarrow[(H^+)]{} H_2N{-}CO{-}HN{-}NH{-}CO{-}NH_2 \;+\; 2\,NH_3 \quad (1)$$

$$H_2N{-}CO{-}HN{-}NH{-}CO{-}NH_2 \xrightarrow[(O)]{} H_2N{-}CO{-}N{=}N{-}CO{-}NH_2 \;+\; H_2O \quad (2)$$

The oxidizing agents used are chlorine, chlorates, hydrogen peroxide or nitric acid. Oxidation with chromium-VI compounds – previously a common method – has declined in importance for ecological reasons.

Depending on the particle size, azodicarbonamide is an orange to pale yellow, crystalline powder with a density of 1.65 g/cm^3 and a decomposition temperature in air of 205 to 215 °C. With its gas yield of approx. 220 ml/g, it is the most economical of all commercial blowing agents. It is virtually insoluble in most organic solvents and plasticizers. Its solubility in dimethyl sulfoxide at room temperature is approx. 5 g/100 g solvent; its solubility in dimethylformamide is somewhat less.

Azodicarbonamide conforms to BGA (West German Federal Health Office) recommendations and FDA regulations (175.300; 177.1210; 177.2600) for the manufacture of articles for food contact applications. It has excellent storage stability. Unlike a number of other chemical blowing agents, it is not self-igniting in contact with a naked flame and does not support combustion.

Azodicarbonamide has an LD_{50} p.o. for rats of >6.8 g/kg body weight; the corresponding figure for the decomposition residue is >6.4 g/kg body weight [14].

The thermal decomposition of dry azodicarbonamide in air gives rise to the following proportions of decomposition products [15, 16]:

gaseous products:	32% by wt
solid residues:	41% by wt
sublimate:	27% by wt

The gaseous products consist of:

nitrogen (N_2):	65% by vol
carbon monoxide (CO):	32% by vol
carbon dioxide (CO_2):	3% by vol

The solid residues and the sublimate are composed mainly of urazole (39%), cyanuric acid (26%), hydrazodicarbonamide (2%) and cyamelide (1%). The percentages are calculated on the quantity of azodicarbonamide used.

The decomposition of azodicarbonamide [15] appears to follow one of two reaction mechanisms (3, 4), depending on the formulation of the plastics compound and the temperature.

$$\begin{aligned} H_2N{-}CO{-}N{=}N{-}CO{-}NH_2 &\longrightarrow N_2 + CO \\ &\qquad\quad + H_2N{-}CO{-}NH_2 \\ H_2N{-}CO{-}NH_2 &\longrightarrow NH_3 + HNCO \end{aligned} \tag{3}$$

$$\begin{aligned} &2\,H_2N{-}CO{-}N{=}N{-}CO{-}NH_2 \\ &\longrightarrow H_2N{-}CO{-}HN{-}NH{-}CO{-}NH_2 + N_2 + 2\,HNCO \end{aligned} \tag{4}$$

The hydrazodicarbonamide formed is the starting product for urazole (5).

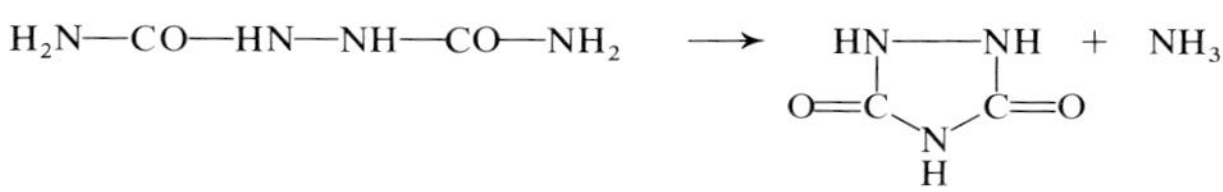
(5)

Cyanuric acid is formed from the intermediate product cyanic acid by trimerization (6). Both are also present in their tautomeric forms as isocyanuric and isocyanic acid respectively.

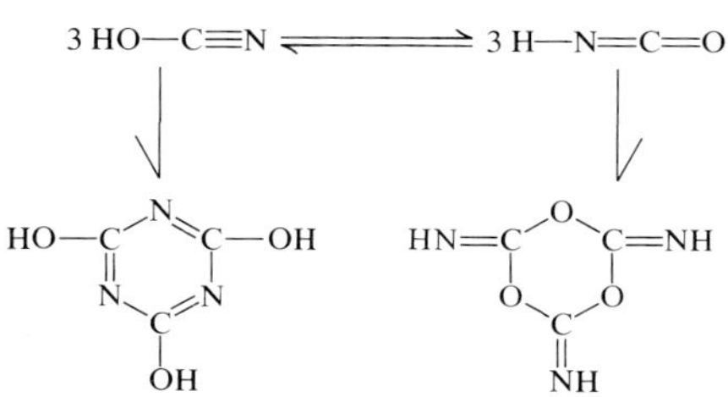
(6)

Another solid polymerization product, cyamelide, is also formed.

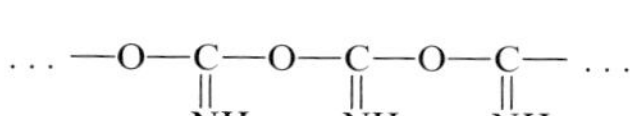

Moisture affects the decomposition of azodicarbonamide. In the presence of acids or bases at high temperatures it may hydrolyze to form hydrazodicarbonamide, nitrogen, carbon dioxide and ammonia (7).

$$\begin{aligned} &2\ H_2N{-}CO{-}N{=}N{-}CO{-}NH_2 \ + \ 2\,H_2O \\ &\longrightarrow \ H_2N{-}CO{-}HN{-}NH{-}CO{-}NH_2 \\ &\qquad + \ N_2 \ + \ 2\,CO_2 \ + \ 2\,NH_3 \end{aligned} \tag{7}$$

In the presence of a cold, aqueous alkaline solution, azodicarbonamide can be reacted to give the corresponding alkali salts (8).

If the solution is heated, the alkali azodicarboxylates decompose to form hydrazine, alkali carbonate, carbon dioxide and nitrogen (9).

$$\begin{aligned} &H_2N{-}CO{-}N{=}N{-}CO{-}NH_2 \ + \ 2\,NaOH \\ &\longrightarrow \ NaOCO{-}N{=}N{-}OCONa \ + \ 2\,NH_3 \end{aligned} \tag{8}$$

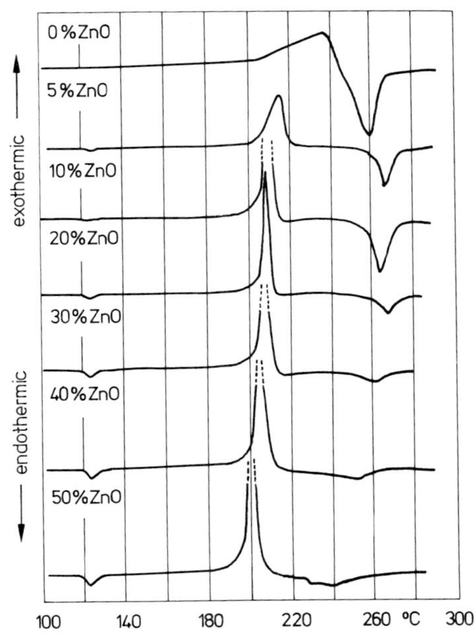

Fig. 1. Decomposition of azodicarbonamide with different proportions of zinc oxide kicker (DTA).

$$2\,NaOCO{-}N{=}N{-}OCONa \;+\; 2\,H_2O$$
$$\longrightarrow\; H_2N{-}NH_2 \;+\; Na_2CO_3 \;+\; 2\,CO_2 \;+\; N_2 \qquad (9)$$

The various possibilities that exist for influencing the way azodicarbonamide decomposes are of considerable importance in foam manufacture. Although the decomposition temperature of pure azodicarbonamide is very high, it can be reduced by means of a number of additives also used in the manufacture and compounding of plastics. The chemical structure of these active substances – known as "kickers" – may vary widely. Effective kickers include polyols, urea, amines, certain organic acids and bases and also a large number of metallic compounds, particularly basic lead, zinc, and/or cadmium compounds used mainly as stabilizers for PVC; certain fillers and pigments are also suitable.

With the aid of differential thermal analysis (DTA), it is possible to investigate how chemical blowing agents decompose and how kickers affect the rate and temperature of decomposition. Figure 1 shows the effect of increasing amounts of zinc oxide as kicker on the decomposition of azodicarbonamide. Increasing proportions of this kicker give rise to a spontaneous reaction at lower temperatures [7].

The lowering of the decomposition temperature in the presence of basic metallic compounds is attributed to the thermal instability of the metal azodicarboxylates formed as an intermediate stage in the alkaline medium (10).

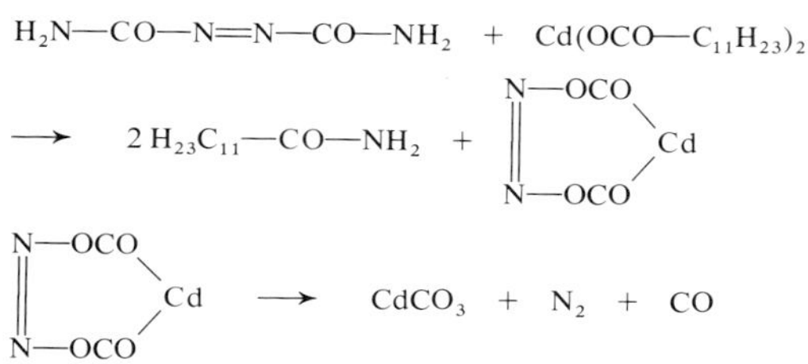

(10)

No kicker effect occurs in the presence of barium compounds, since the barium azodicarboxylate formed is thermally more stable than azodicarbonamide.

The rate at which azodicarbonamide decomposes, however, can be influenced not only by various kickers, but also depends on the particle size of the blowing agent. Since activation takes place in a heterogeneous system, the specific surface, and therefore also the particle size, obviously have a decisive effect on the decomposition rate. Azodicarbonamide of small particle size is more easily "kicked" than coarser material. Figure 2

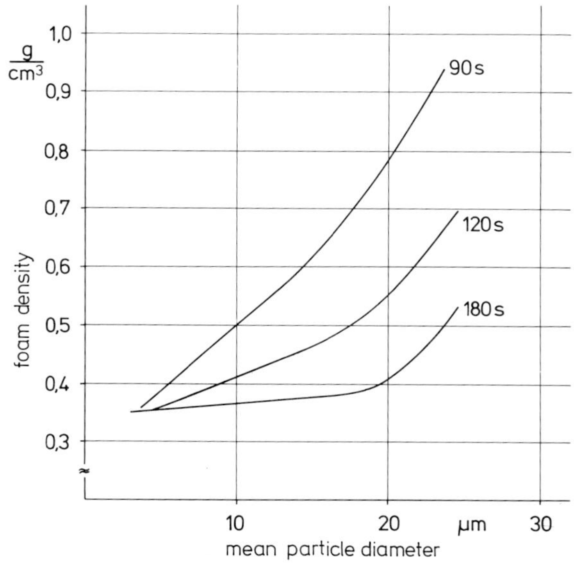

Fig. 2. Effect of azodicarbonamide particle size on the density of plasticized PVC foam.

Proportion of azodicarbonamide: 1.0%,
kicker/stabilizer: 3.0% dibasic lead phthalate,
residence time: 90 s, 120 s, 180 s at 190 °C.

shows the effect of particle size on the decomposition rate of azodicarbonamide in a PVC plastisol. These relationships are particularly important for the free-blowing expansion of plasticized PVC. With dynamic foaming methods such as injection molding or extrusion, the effect of particle size is of only secondary importance.

15.2.1.2 Modified azodicarbonamide

Azodicarbonamide-containing additives of any kind must be regarded as modified azodicarbonamide. The simplest and oldest form of modification is azodicarbonamide paste. In addition to the plasticizer, which forms the liquid phase, such a paste may also contain dispersing agents and kickers. The principal field of application is the expansion of PVC plastisols. The additives facilitate the dispersion of the blowing agent when it is stirred into the PVC plastisol, and kickers lower the decomposition temperature.

Another group of modified azodicarbonamides has been developed for the manufacture of integral-skin foams by extrusion and injection molding. The main purpose of the additives in this case is to modify the usual decomposition process of azodicarbonamide and suppress the formation of cyanuric acid, which is the cause of plate-out on the surfaces of molds, dies and screws. Although the coating thus formed can be easily removed, it interferes with production. The additives used here are zinc oxide and/or silicic acid with a low water content. Instead of cyanuric acid, however, a considerable quantity of ammonia is evolved. The additives also act as nucleating agents, giving a cell structure of uniform fineness.

Another possible method of modifying azodicarbonamide is to mix it with other chemical blowing agents, e.g. from the sulfohydrazide group. These "auxiliary blowing agents" decompose at lower temperatures than azodicarbonamide. The decomposition products formed, as well as the water split off when sulfohydrazides decompose, accelerate the decomposition of azodicarbonamide. Here, too, the formation of cyanuric acid is suppressed.

15.2.2 Hydrazine derivatives

15.2.2.1 4,4′-Oxybis(benzenesulfohydrazide)

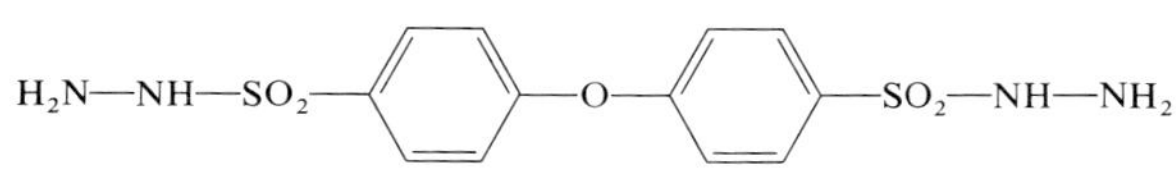

4,4′-Oxybis(benzenesulfohydrazide) is produced by the chlorosulfonation of diphenyl ether and subsequent reaction with hydrazine in an alkaline medium.

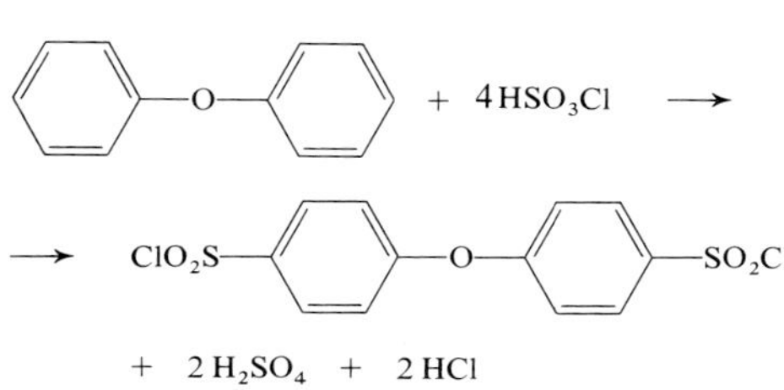

(11)

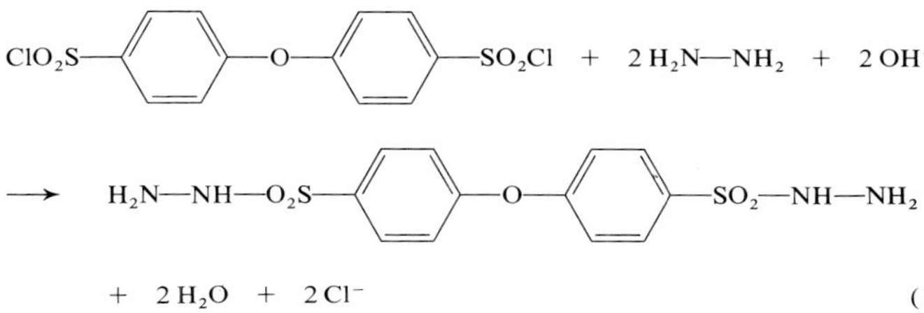

(12)

4,4′-Oxybis(benzenesulfohydrazide) is a white, crystalline powder with a decomposition temperature of 157 to 160 °C. The gas yield of 125 mg/g is equivalent to the theoretical yield. In the presence of plastics and rubbers the decomposition temperature is normally reduced. Effective decomposition accelerators for all sulfohydrazides include oxidizing agents as well as solvents containing hydroxyl groups, e. g. aliphatic alcohols or water in the presence of small amounts of alkali.

4,4′-Oxybis(benzenesulfohydrazide) is insoluble in most organic solvents and in cold water. It is slightly soluble in hot water or ethanol. In cold, aqueous alkaline solutions it dissolves and decomposes. With ketones, sulfohydrazones are formed.

FDA regulations "175.300; 177.1210; 177.2600" permit the use of 4,4′-oxybis(benzenesulfohydrazide) for the manufacture of articles for food contact applications. 4,4′-Oxybis(benzenesulfohydrazide) has an LD_{50} p.o. for rats of >5.2 g/kg body weight, the corresponding figure for the decomposition residue being >16.0 g/kg body weight [14].

The thermal decomposition of sulfohydrazides involves an intramolecular redox reaction which is partly of an endothermic and partly of an exothermic nature; the total amount of heat drops sharply upon decomposition.

By analogy with the decomposition of monosulfohydrazides, that of 4,4′-oxybis(benzenesulfohydrazide) may be represented as follows:

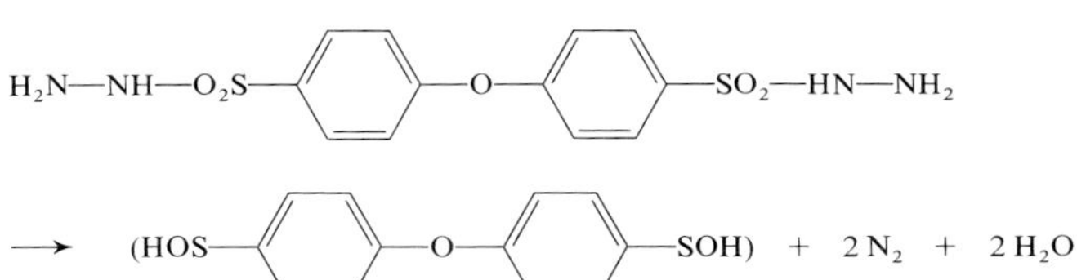

(13)

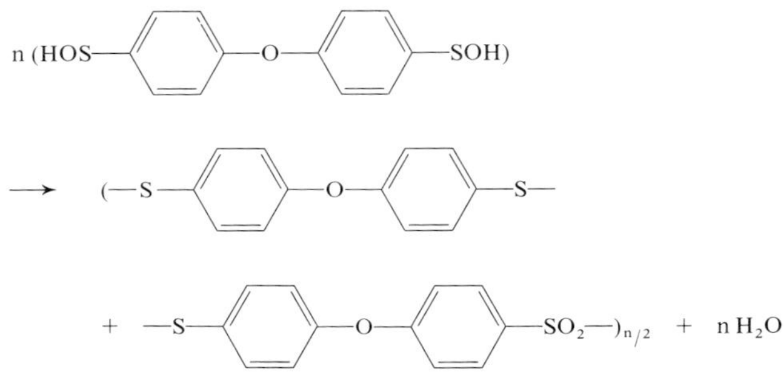

(14)

The sulfohydrazide initially gives rise to an unstable sulfenic acid, which stabilizes by disproportionation to give a disulfide and a polymeric thiosulfonate. These polymeric decomposition residues are virtually odorless. When the blowing agent is incorporated into a plastics material, the decomposition reaction is more controlled and the exothermic heat of reaction can be evenly dissipated. The decomposition residue therefore remains colorless.

15.2.2.2 Diphenylsulfone-3,3′-disulfohydrazide

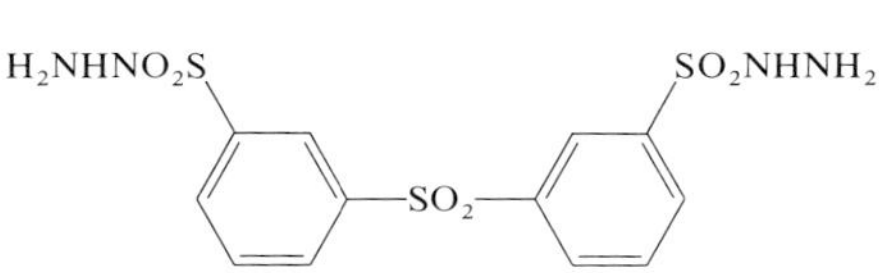

Diphenylsulfone-3,3′-disulfohydrazide is produced by the chlorosulfonation of diphenylsulfone and subsequent reaction with hydrazine in an alkaline medium. Diphenylsulfone-3,3′-disulfohydrazide is a white, crystalline powder with a decomposition temperature of approx. 155 °C and a gas yield of 110 ml/g.

Diphenylsulfone-3,3′-disulfohydrazide is virtually insoluble in most organic solvents and in cold water, but soluble in dimethylformamide, dimethyl sulfoxide, tetrahydrofuran, cyclohexanone and methyl ethyl ketone. Diphenylsulfone-3,3′-disulfohydrazide has an LD_{50} p.o. for rats of >5 g/kg body weight [18]. The decomposition of diphenylsulfone-3,3′-disulfohydrazide may be represented in the same way as that of the other sulfohydrazides.

15.2.2.3 Diphenylene oxide-4,4′-disulfohydrazide

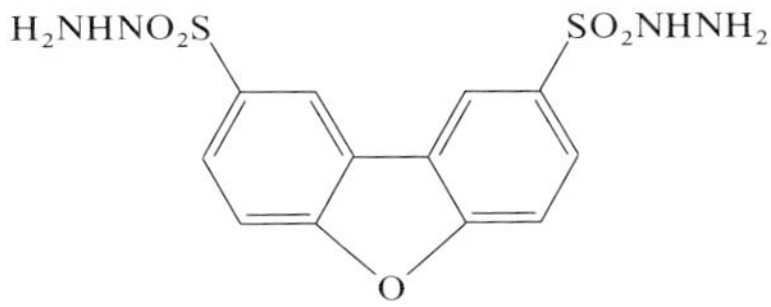

Diphenylene oxide-4,4′-disulfohydrazide is a white, crystalline powder with a decomposition temperature of 175 to 180 °C and a gas yield of 120 ml/g. It is virtually insoluble in most organic solvents and in cold water, but soluble in dimethyl sulfoxide and dimethylformamide.

Diphenylene oxide-4,4′-disulfohydrazide has an LD_{50} p.o. for rats of >2.5 g/kg body weight [18].

The decomposition of diphenylene oxide-4,4′-disulfohydrazide may be represented in the same way as that of the other sulfohydrazides.

15.2.2.4 Trihydrazinotriazine

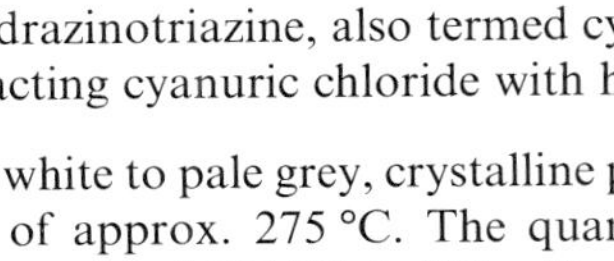

Trihydrazinotriazine, also termed cyanuric trihydrazide, is manufactured by reacting cyanuric chloride with hydrazine.

It is a white to pale grey, crystalline powder with a decomposition temperature of approx. 275 °C. The quantity of gas split off, measured at a temperature of 320 °C, is 225 ml/g; this consists mainly of nitrogen and ammonia. Melamine remains as a solid residue.

15.2.3 Semicarbazides

15.2.3.1 p-Toluenesulfonyl semicarbazide

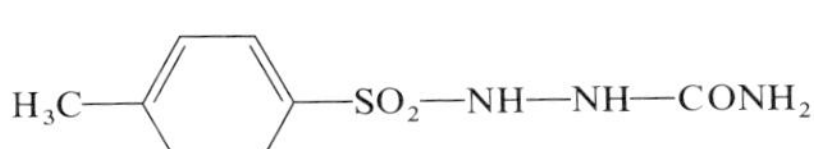

p-Toluenesulfonyl semicarbazide is manufactured by reacting p-toluenesulfohydrazide with sodium cyanate in an acid medium.

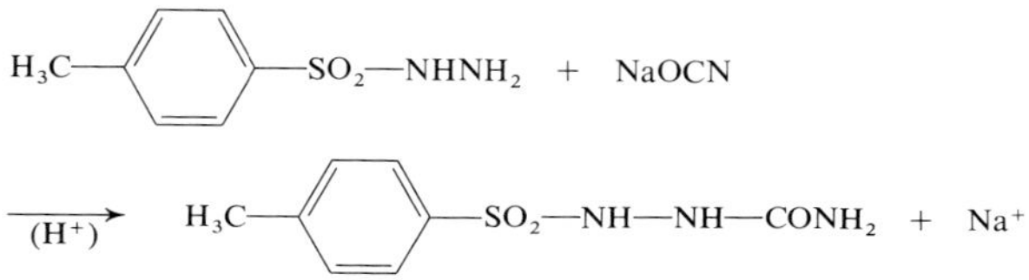

(15)

It is a white, crystalline powder with a decomposition temperature in air of 228 to 235 °C and a gas yield of 140 ml/g. When incorporated into plastics it decomposes at 213 to 225 °C. The following activators are effective in lowering the decomposition temperature: glycols, zinc compounds, stearic acid, urea, dibasic lead phthalate and phosphite, also azodicarbonamide and 4,4′-oxybis(benzenesulfohydrazide) as auxiliary blowing agents.

p-Toluenesulfonyl semicarbazide is insoluble in most organic solvents and in water. It dissolves in dimethyl sulfoxide, dimethylformamide, dimethylacetamide and in aqueous alkaline solutions.

p-Toluenesulfonyl semicarbazide has an LD_{50} p.o. for rats of >10 g/kg body weight, the corresponding figure for the decomposition residue being >4.3 g/kg body weight [14].

The dry decomposition of p-toluenesulfonyl semicarbazide gives rise to the formation of the following detectable gaseous and solid products [14]:

Gases:

55% N_2, 37% CO_2, 2% CO, 3% NH_3

Solid decomposition residue:

ditolyl disulfide and p-tolueneammonium sulfonate

The mechanism of decomposition is presumed to be the following:

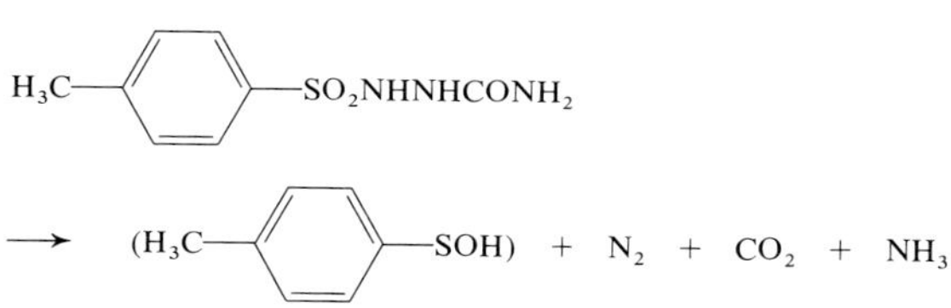

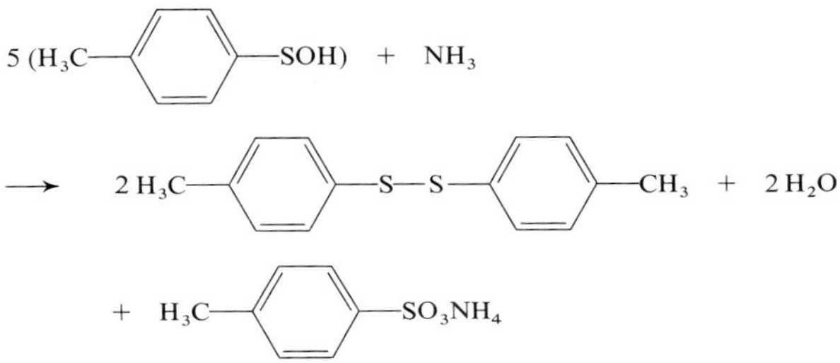

$$3\,NH_3 + 2\,CO_2 + H_2O \longrightarrow (NH_4)HCO_3 + NH_2COONH_4 \quad (16)$$

The toluenesulfenic acid formed initially is unstable and becomes rearranged to give ditolyl disulfide and p-tolueneammonium sulfonate. The sublimate consists of ammonium bicarbonate and ammonium carbamate.

15.2.4 Tetrazoles

15.2.4.1 5-Phenyltetrazole

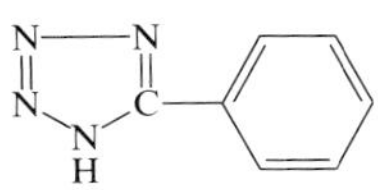

5-Phenyltetrazole is a white, crystalline powder with a decomposition temperature in air of 240 to 250 °C. The quantity of gas split off at a temperature of 260 to 265 °C is 210 ml/g. The gas consists entirely of nitrogen.

5-Phenyltetrazole is readily soluble in ethanol and other common solvents.

5-Phenyltetrazole has an LD_{50} p.o. for mice of 840 to 946 mg/kg body weight, the corresponding figure for the decomposition residue being 2175 mg/kg body weight [19].

Most of the residue from the thermal decomposition of 5-phenyltetrazole consists of three substituted, heterocyclic compounds. These are: aminodiphenyltriazole, 3,5-diphenyl-1,2,4-triazole and triphenyl-s-triazine.

15.2.5 Benzoxazines

15.2.5.1 Isatoic anhydride

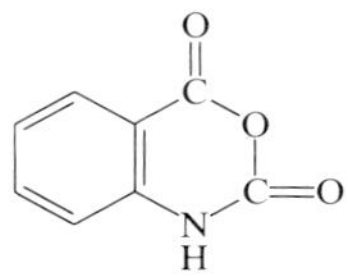

Isatoic anhydride [20] is a light beige to brownish, crystalline powder having a decomposition temperature in air of 210 to 225 °C. The gas yield measured at 250 °C is 115 ml/g, the blowing gas consisting of carbon dioxide. The decomposition temperature may be lowered by adding small amounts of acid or alkali. Isatoic anhydride is virtually insoluble

Table 1. Common chemical blowing agents for plastics: decomposition range, gas yield, principal uses.

Chemical description	Decomposition range in air (°C)	Gas yield (ml/g)	Used mainly for
Azodicarbonamide	205 to 215	220	PVC, PE, PP, PS, ABS, PA
Modified azodicarbonamide	155 to 220	150 to 220	PVC, PE, PP, EVA, PS, ABS
4,4′-Oxybis(benzene-sulfohydrazide)	150 to 160	125	PE, PVC, EVA
Diphenylsulfone-3,3′-disulfohydrazide	155	110	PVC, PE, EVA
Diphenylene oxide-4,4′-disulfohydrazide	175 to 180	120	PE, PVC, EVA
Trihydrazinotriazine	275	225	ABS, PE, PP, PA
p-Toluenesulfonyl semicarbazide	228 to 235	140	ABS, PE, PP, PA, PS
5-Phenyltetrazole	240 to 250	190	ABS, PPO, PC, PA, PBTP
Isatoic anhydride	210 to 225	115	PS, ABS, PA, PPO, PBTP, PC

in water and in most common organic solvents, but soluble in dimethylformamide and dimethyl sulfoxide.

Isatoic anhydride and its decomposition residue have an LD_{50} p.o. for rats of 2.5 g/kg body weight [18]. The principal decomposition products formed on the thermal decomposition of isatoic anhydride are aromatic polyamides of high molecular weight. In the presence of water, small quantities of anthranilic acid, benzoxazinone derivatives and benzamides are also formed.

15.2.6 Important properties of common chemical blowing agents

A list of the common chemical blowing agents for plastics together with their decomposition ranges, gas yields and main uses is given in table 1.

15.2.7 Testing of chemical blowing agents

In order to obtain a complete picture of the properties of a blowing agent and its suitability for various individual applications, physical, chemical, technological and toxicological tests are generally carried out. The following characteristics are investigated:

- Appearance and odor,
- degree of purity,
- impurities,
- bulk density and free-flowing properties,
- electrostatic buildup,
- explosion risk,
- particle size and particle size distribution,
- gas yield,
- mode and temperature of decomposition,
- other calorific data,
- decomposition residues,
- ash content and spectral analysis of the ash,
- practical, technological properties,
- toxicology and skin tolerance.

All these characteristics are determined by the raw material manufacturer; with a few exceptions they are the subject of continuous production control. This, however, does not release the processor from the obligation to carry out a systematic check on incoming goods. In particular, he should verify the properties of importance for the application concerned.

15.3 Technological trends and outlook

In recent years, developments in the field of chemical blowing agents have taken place parallel to those in the thermoplastics field. Market demand for engineering plastics with a high heat deflection temperature has led to what are known as high-temperature blowing agents. Isatoic anhydride and isophthalic acid bis (carbonic acid ethyl ester anhydride) [21, 22], for example, have satisfied the demand for products that split off gases other than nitrogen, in these cases carbon dioxide. It is difficult to assess at present how important these products will eventually become within the whole range of blowing agents.

The choice of blowing agent is governed increasingly by economic considerations, in addition to the purely technical aspects. The principal blowing agent will therefore continue to be the "commodity" product azodicarbonamide which, with its many possible modifications, can be adapted to different fields of application and processing methods.

Virtually all known thermoplastics can be expanded using the chemical blowing agents currently available; for this reason the development of new types of blowing agents in the near future is unlikely.

As with the raw materials, the development of machinery and processes for the manufacture of cellular plastics is largely complete. Thus all the technical prerequisites now exist for the importance of foams to increase still further through the opening up of new fields of application.

15.4 List of trade names and manufacturers of chemical blowing agents (1987)

This list of trade names and manufacturers makes no claim to completeness.

The classes of products are denoted as follows: (A) = azodicarbonamide, (B) = modified azodicarbonamide, (C) = 4,4′-oxybis(benzenesulfohydrazide), (D) = diphenylsulfone-3,3′-disulfohydrazide, (E) = diphenylene oxide-4,4′-disulfohydrazide, (F) = trihydrazinotriazine, (G) = p-toluenesulfonyl semicarbazide, (H) = 5-phenyltetrazole, (I) = isatoic anhydride.

Trade name	Product class	Manufacturer
Azobis	(A), (B)	Toyo Hydrazine Ind. Co., Tokyo, Japan
Azocel	(A)	Fairmount Chemical Co, Inc., Newark, N.J., USA
Cellcorn	(A), (B), (C)	Kum Yang, Seoul, Republik of Korea
Cellmic	(A), (C)	Sankyo Kasei Co., Ltd., Osaka, Japan
Celogen	(A), (B), (C), (G)	Uniroyal Chemical, Company, Inc. Middlebury, CT, USA
Espon	(A), (C)	Quimica Heterociclica Mexicana, S.A., Ecateped, Edo de Mexico
Evipor	(A)	EVM, Budapest, Hungary
Expandex	(H)	Polychem Products, Olin Corp., Stamford, CT, USA
Ficel	(A), (B), (F)	Schering Industrial Chemicals, Hauxton, Cambridge, GB
Genitron	(A), (B), (C), (F)	Schering Industrial Chemicals, Hauxton, Cambridge, GB
Kempore	(A), (B)	Polychem Products, Olin Corp., Stamford, CT, USA
Kem Tec	(I)	Sherwin-Williams Chemicals, Cleveland, OH., USA
Neocellborn	(C)	Eiwa Chemical Ind., Co., Ltd., Kyoto, Japan
Nitropore	(B), (C)	Polychem Products, Olin Corp., Stamford, CT, USA
Porofor	(A), (B), (I)	Bayer AG, Leverkusen, D
Unicell	(A), (B), (C), (G)	Dong Jin Chemical Ind. Co., Ltd., Seoul, Republic of Korea

Trade name	Product class	Manufacturer
Unifoam	(A), (B)	Otsuka Chemical Co., Ltd., Osaka, Japan
Vinstab	(A), (B)	Hebron S.A., Barcelona, E
Vinyfor	(A), (B)	Eiwa Chemical Ind. Co., Ltd., Kyoto, Japan
–	(A), (B), (C)	Société Française d'Organo-Synthèse, Gennevilliers, F

15.5 References

[1] *R.A. Reed:* Plastics Progress 1955, London, 1955, p. 51.
[2] *H. Piechota, H. Röhr:* Integralschaumstoffe, Carl Hanser Verlag, Munich/Vienna, 1975.
[3] Clark-Cooper Corp., Cinnaminson, N.J., USA, Technical Bulletin MP-442 "Chemair".
[4] European Plastics News 4 (1977) 7, pp. 17/18.
[5] Kunststoffe 66 (1976) p. 404.
[6] *H. Barth:* Kunststoffe 64 (1974) pp. 646/653.
[7] *H. Barth:* Kunststoffe 64 (1974) pp. 2/7.
[8] *H. Barth:* Kunststoffe 67 (1977) pp. 130/135.
[9] Kunststoffe 67 (1977) pp. 503/505.
[10] *W. Fillmann:* Kunststoffe 67 (1977) pp. 370/373.
[11] *H. Eckardt:* Kunststoffe-Plastics 23 (1976) 10, pp. 28/35.
[12] *A. Splichal:* Kunststoffe-Plastics 23 (1976) 8, pp. 13/17.
[13] *K. Mörwald:* Plastverarbeiter 28 (1977) pp. 305/310, 7, pp. 354/356 and 8, pp. 405/408.
[14] Celogen Blowing Agents, Bulletin No. ASP-3917, Uniroyal Chemical, Div. of Uniroyal, Inc., Naugatuck, Conn./USA.
[15] *R.A. Reed:* British Plastics 33 (1960) p. 469.
[16] *H.R. Lasman:* Modern Plastics Encyclopedia (1966) pp. 394/402.
[17] *P. Geisbüsch:* Experimental thesis at the IKV, archive no. SA 74/49 (1974), Aachen Technical University.
[18] Unpublished research by Bayer AG, Leverkusen.
[19] Technical Bulletin OOR-03-0274, Stepan Chemical Co., National Polychemicals Div., Inc., Wilmington, Mass./USA.
[20] US Pat. 3725321, Can. Pat. 987450.
[21] German Pat. 1252891.
[22] *W. Cohnen:* Kunststoff-J. 9 (1975) pp. 34/35.

16 Organic Peroxides as Cross-Linking Agents

Drs. A.L. Berg, Akzo Chemie BV, Amersfoort, The Netherlands

16.1 Introduction

The use of an organic peroxide as cross-linking agent was reported for the first time by *Ostromislenski* in 1915. In this experimental work, dibenzoyl peroxide, which at that time was employed for bleaching flour, was used for the vulcanization of natural rubber.

Since around 1950, interest in the industrial use of peroxides as cross-linking agents has been increasing, partly as a result of the introduction of saturated rubbers such as EPM and silicone rubber, which cannot be vulcanized with the usual sulfur systems.

From this time on, detailed scientific and technical work have been carried out in this field; a good overall survey of the publications that appeared up to about 1967 is provided by the synopsis compiled by *Loan* [1]. The primary intention of the present study is to deal with more recent developments.

Parallel to their application in elastomers, interest was being shown in the use of peroxides for the cross-linking of thermoplastics, the aim being to improve the latter's dimensional stability at elevated temperatures and their chemical resistance. Up to now, however, the cross-linking of thermoplastics for technical applications has been mainly limited to polyethylene and its copolymers.

Apart from the use of peroxides, it is also possible for polymers to be cross-linked by means of high-energy radiation. An advantage in this regard is that no decomposition products of peroxides are formed. A disadvantage, however, is the high level of capital expenditure and the cost of the necessary protective measures against radiation damage.

A further possibility is to allow water to diffuse into a siliconized polyethylene copolymer, Si—O—Si bonds being formed by hydrolysis and subsequent condensation and thus causing cross-linking to take place [2]. An alternative method derived from this is the grafting of vinyl siloxanes on polyethylene, which is initiated by peroxides, and subsequent cross-linking through the action of water. This technique is lately being employed in particular for pipes, containers and cable insulation in the low-voltage range.

16.2 Requirements

If the known organic peroxides are arranged according to their decomposition temperature, the range commences at a temperature below 0 °C and finishes at approx. 200 °C. This means that a number of peroxides can only be stored and handled under refrigeration conditions. This is probably the reason why peroxides are occasionally still regarded with a certain amount of reserve. There is no cause for this as regards peroxide cross-linking agents, since only peroxides with sufficient thermal stability can be incorporated in a polymer without undesirable, premature cross-linking taking place.

In order to be suitable for technical applications, a peroxide cross-linking agent must fulfil a number of conditions:

- It must be safe to handle during transportation, storage and processing.
- It must not be too volatile, so as to avoid losses during mixing.
- It must be sufficiently compatible in elastomers and plastics.
- The nature of its decomposition products must be such that rapid cross-linking takes place at the desired temperature, without a tendency to premature cross-linking (prevulcanization, scorch).
- It should react in such a way that cross-linking is the only modification to the polymer that occurs.
- It must also be active in the presence of fillers and reinforcements.
- Neither the peroxide nor its thermal decomposition products must accelerate aging.
- The peroxide and its decomposition products must be non-toxic and meet the requirements of industrial hygiene.

16.3 Fields of application

Cross-linking peroxides are used for:

- Wire and cable insulation and cable sheathing (e.g. made from PE, EPM, EPDM, CM),
- profiles continuously cross-linked by salt bath (LCM) or steam (PE, EPDM, CR),
- pipes and hoses free from extractable components, e.g. for car radiators (EPDM, NBR),
- shoe soles (foamed and cross-linked EVA, PE, SBR, etc.),
- roller coverings (EPDM, NBR, SI),
- rotational moldings (PE, EVA),
- medical supplies, e.g. tubes (SI),
- foamed articles with closed cell structure (PE, EVA, CM).

Polymers which are cross-linkable and not cross-linkable with peroxides are listed in table 1.

Unlike sulfur vulcanization, peroxides exhibit no appreciable differences in their rate of reaction in the presence of two different polymers. This offers the possibility for covulcanization of blends of saturated and unsaturated copolymers as a means of optimizing the chemical and physical properties of the final product.

By blending SBR and EPDM, for example, in a ratio of about 70:30, it is possible to obtain a vulcanizate with the mechanical properties of SBR and the ozone resistance of EPDM. In this and similar polymer blends, the vulcanization rates of the blend components differ to such an extent when a sulfur-based system is used, that the first polymer would already be completely vulcanized before vulcanization of the second had even started (in this case EPDM). The result would be an inhomogeneous vulcanizate. However, since peroxide cross-linking is not dependent on the presence of double bonds in the substrates, a peroxide cross-links both components simultaneously. A homogeneously cross-linked product is thus obtained. In a similar way, the ozone resistance of natural rubber can be markedly improved by blending it with EPDM.

In certain applications, EPDM is blended for cost reasons with ethylene-vinyl acetate copolymers (EVA), the well-known resistance of EPDM to ozone and oils being retained in full. Because of the presence of EVA, blends of this kind can only be cross-linked by peroxides or radiation.

Table 1. Peroxide cross-linking of thermoplastics and elastomers.

Cross-linking	
possible	not possible
ABS – acrylonitrile-butadiene-styrene copolymers	PP – polypropylene
CPE – chlorinated polyethylene	ACM – acryl ester-2-chlorovinyl ether rubber
CSM – chlorosulfonated polyethylene	CIIR – chlorinated butyl rubber
EVA – ethylene-vinyl acetate copolymers	ECO – copolymer of epichlorohydrin and ethylene oxide
PB – polybutene	IIR – butyl rubber
PE – polyethylene	
PVC – polyvinyl chloride	
ANM – acryl ester acrylonitrile rubber	
BR – polybutadiene	
CO – polyepichlorohydrin	
CR – polychloroprene	
EPM – ethylene propylene rubber	
EPDM – ethylene-propylene-diene rubber	
FKM – fluoro polymers	
IR – polyisoprene rubber	
NBR – butadiene-acrylonitrile rubber	
NR – natural rubber	
PTE – polysulfide	
SBR – butadiene-styrene rubber	
SI – silicone rubber	
IPS – impact polystyrene	

Blends of EPDM and polyethylene are also frequently used. When polyethylene is the predominant polymer (80% to 90% of the blend), EPDM improves the surface quality and dimensional stability during extrusion and at the same time makes it possible to operate at a higher speed and temperature. If EPDM predominates, however, the addition of polyethylene results in an increase in hardness.

16.4 Classification of peroxides

Characteristic of an organic peroxide is the peroxy group —O—O—, which through homolytic cleavage of its electron-pair bond can decompose to form two radicals. The general formula for such compounds is R_1—O—O—R_2, whereby R_1 and R_2 either symbolize organic radicals or an organic radical and a hydrogen atom. When both radicals R_1 and R_2 are a hydrogen atom, the simplest form of peroxide is obtained, H—O—O—H: hydrogen peroxide.

Organic peroxides can be subdivided into the following classes:

16.4.1 Hydroperoxides: R—O—O—H

Hydroperoxides are generally unsuitable for cross-linking reactions, since the scale of possible side reactions, e.g. ionic decomposition, is too great.

Although in a number of patents, hydroperoxides are claimed to be effective cross-linking agents at room temperature, this type of peroxides has up to now been scarcely employed in commercial practice.

16.4.2 Alkyl peroxides: R_1—O—O—R_2

They contain two organic radicals, wholly or partly aliphatic by nature.

The following subgroups in this class should be mentioned:

Dialkyl peroxides, e.g. ditert.butyl peroxide:

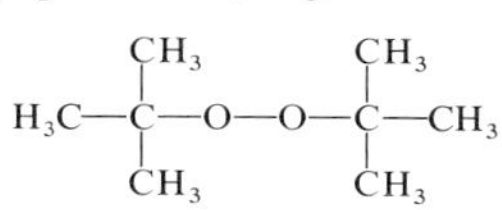

Physical form at 23 °C: liquid,
decomposition range:
approx. 190 °C,
usage: e.g. direct addition to cable insulation and sheathing compounds.

Diaralkyl peroxides, e.g. dicumyl peroxide:

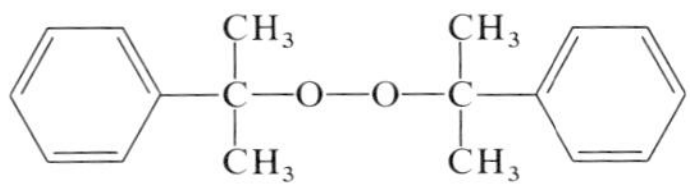

Physical form at 23 °C: crystalline,
decomposition range: approx. 170 °C,
usage: wide range of applications.

Alkylaralkyl peroxides, e.g. tert.butylcumyl peroxide:

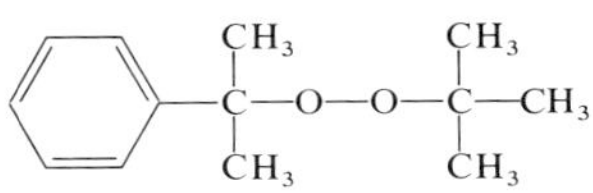

Physical form at 23 °C: liquid,
decomposition range: approx. 180 °C,
usage: e.g. direct addition to cable sheathing compounds.

Bis(tert.butylperoxyisopropyl)benzene:

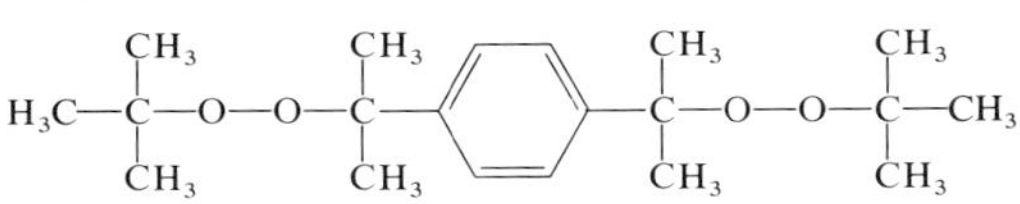

(predominantly in para position)

Physical form at 23 °C: solid,
decomposition range: approx. 180 °C,
usage: wide range of applications, very low odor nuisance.

16.4.3 Peroxyesters: $R_1{-}\overset{\overset{\displaystyle O}{\|}}{C}{-}O{-}O{-}R_2$

These compounds are derived from peroxy acids, $R{-}\overset{\overset{\displaystyle O}{\|}}{C}{-}OOH$, whose acidic hydrogen atom has been replaced by an aliphatic radical, usually the tert.butyl radical. The best known product in this group is tert.butylperoxybenzoate:

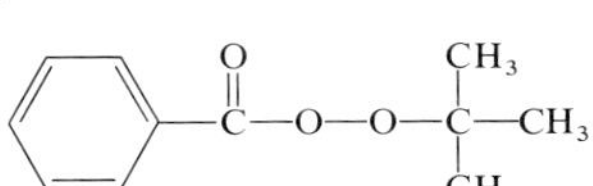

Physical form at 23 °C: liquid,
decomposition range: approx. 140 °C,
usage: e.g. silicone rubber, EVA, EPDM at low cross-linking temperatures.

16.4.4 Diacyl peroxides: $R_1—\overset{O}{\overset{\|}{C}}—O—O—\overset{O}{\overset{\|}{C}}—R_2$

Depending on the composition of the organic groups R_1 and R_2, these products can be subdivided as follows:

Dialkanoyl peroxides, R_1 and R_2 representing aliphatic groups.

Alkanoyl-aroyl peroxides. This group is characterized by the fact that one radical is aliphatic by nature and the other aromatic.

Diaroyl peroxides. Here the organic radicals consist solely of aromatic groups. The diaroyl peroxides include bis(2,4-dichlorobenzoyl)peroxide and the oldest cross-linking peroxide, dibenzoyl peroxide:

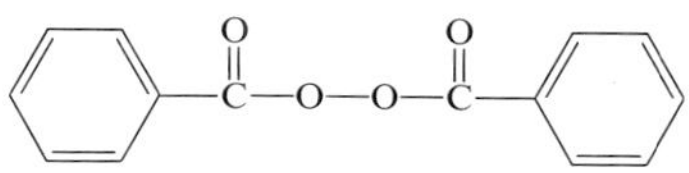

Physical form at 23 °C: crystalline,
decomposition range: approx. 120 °C,
usage: e.g. silicone rubber.

16.4.5 Peroxyketals

They can be regarded as derivatives of the corresponding ketals, in which the oxygen atoms of the ether bond have been replaced by peroxy groups. The ketal peroxides thus possess the following general structure:

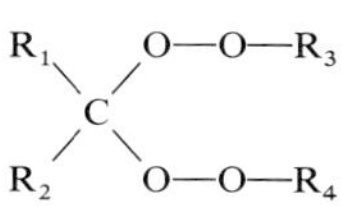

A peroxyketal known in commercial practice is 1,1-ditert.butylperoxy-3,3,5-trimethylcyclohexane:

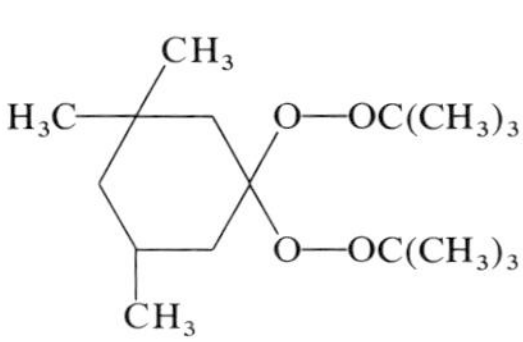

Physical form at 23 °C: liquid (for safety reasons only obtainable on carrier material),
decomposition range: approx. 150 °C,
usage: wide range of applications.

Mention should also be made of the fact that the chemical designation of many peroxides is often simplified in daily usage. Dibenzoyl peroxide is thus frequently referred by the simplified name of benzoyl peroxide.

16.5 Important characteristic values of peroxides

16.5.1 Active oxygen content

Each organic peroxide contains a certain amount of active oxygen, which in commercial products is generally between 1 and 15%. However, the active oxygen content is not a measure of its effectiveness as a cross-linking agent. The degree of cross-linking obtained at equal levels of active oxygen can differ in a given substrate. Determination of active oxygen thus only enables the peroxide content to be established.

There are a number of known analytical methods for determining the active oxygen content. The most commonly used method is iodometry. In order to prevent autooxidation of the iodide to form iodine, it is expedient to exclude atmospheric oxygen while carrying out the analysis. Reduction of the peroxide takes place in accordance with the following equation:

$$R_1{-}O{-}O{-}R_2 \;+\; 2\,HI \;\longrightarrow\; R_1OH \;+\; R_2OH \;+\; I_2$$

The iodine eliminated is titrated as usual with a 0.1 n sodium thiosulfate solution. Since the strength of the oxygen-oxygen bond depends on the substituents R_1 and R_2, different reaction conditions must be selected according to the structure of the peroxide. Hence, there is no generally applicable iodometric method of determining the active oxygen content. As regards solvent, temperature and reaction time, for example, the procedure must be adapted to the peroxide in question.

16.5.2 Determination of half-life time

Thermal stability, i.e. the rate of thermal decomposition of a peroxide bond, is influenced in particular by the groups in the direct vicinity. The term half-life time ($t_{1/2}$) has become generally established as a measure of the rate of decomposition. Half-life time is defined as the time

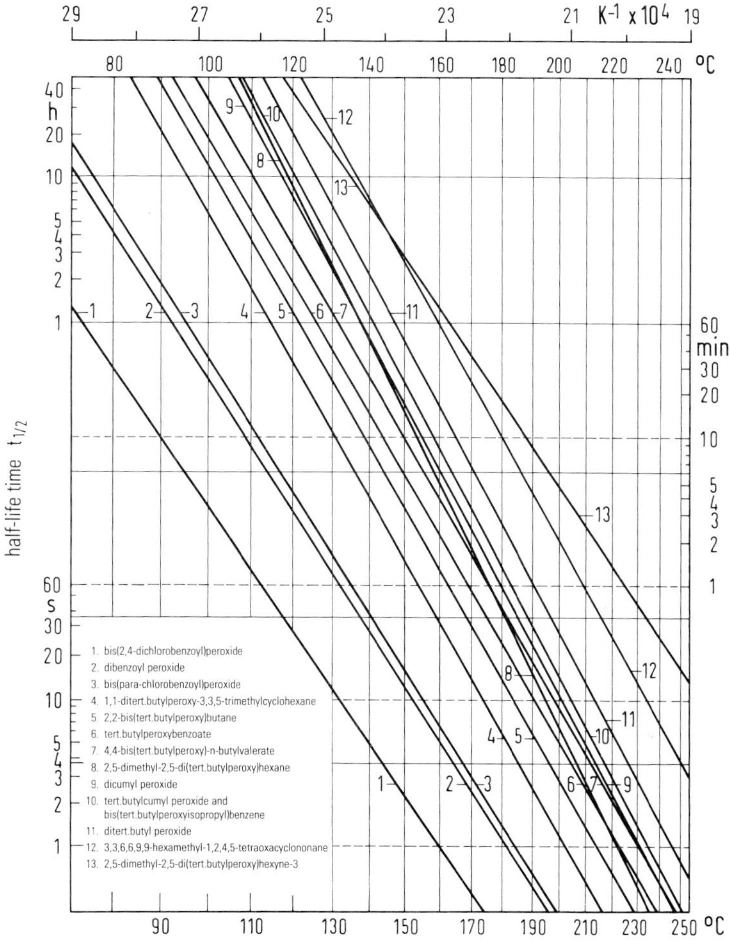

Fig. 1. Temperature/half-life time relationships for decomposition of peroxides.

in which at a constant temperature a given amount of peroxide decreases by half.

The peroxide dissolved in an inert solvent is tested at regular intervals and at constant temperature to determine the peroxide concentrations still present. Similar tests are carried out at other temperatures. From the data obtained it is possible by the usual method to formulate the temperature dependence of the half-life time value of the peroxide used (cf. figure 1).

Since decomposition of peroxides takes place in accordance with a first-order reaction, half-life time is primarily dependent on temperature, irrespective of the amount of peroxide used. 50% of the peroxide originally present therefore decomposes during the first half-life time, a further 50% of the remaining peroxide during the second half-life time etc. In theory, 100% decomposition is thus never attained. In fact, the amount of peroxide remaining after seven half-life times is so small that the effect of any additional cross-linking is negligible. Table 2 shows the relationship between the number of half-life times and the amount of peroxide decomposed.

Table 2. Relationship between number of half-life times and the amount of peroxide decomposed.

Number of half-life times	Amount of peroxide decomposed [%]
1	50.0
2	75.0
3	87.5
4	93.75
5	96.9
6	98.4
7	99.2
8	99.6
9	99.8
10	99.9

The half-life time values determined in an inert solvent (figure 1) are not readily applicable to the conditions prevailing during commercial usage. Half-life time values should thus only be regarded as a relative means of comparison.

16.6 Selection of peroxides

Selection of a peroxide is primarily determined by the polymer to be cross-linked and its processing conditions such as residence time and temperature and by the requirements that have to be met by the finished article, e.g. freedom from odor. A further factor is that the degree of cross-linking is a function of the peroxide quantity, whereas the rate of the cross-linking reaction is determined by the temperature.

The amount of peroxide required depends on the structure of the peroxide, the polymer to be cross-linked and the presence and quantity of other additives. Data on the amounts of peroxide required for the cross-linking of important polymers are listed in table 3.

Satisfactory mechanical properties can already be obtained with the lowest quantity of peroxide. However, permanent set (foamed articles) does not improve until fairly high quantities of peroxide are used. The mechanical properties generally deteriorate when the maximum quantities are exceeded. Higher quantities than the minimum are likewise necessary if the compound contains substances that can scavenge free radicals, e.g. antioxidants or extender oils. The negative influence of these types of additives can be partly compensated for by the use of coagents (cf. section 16.8.5).

Table 3. Amount of peroxide needed for cross-linking of polymers.

Polymer	Equivalents of peroxide per 100 kg polymer	Parts per weight peroxide per 100 parts polymer		
		A	B	C
EPM, EPDM	9.0 to 15.0	6.8 to 11.3	6.1 to 10.1	3.8 to 6.3
NBR	3.5 to 6.0	2.6 to 4.5	2.4 to 4.1	1.5 to 2.5
PE	2.0 to 10.0	1.5 to 7.6	1.4 to 6.8	0.8 to 4.2
EVA	3.5 to 7.0	2.6 to 5.3	2.4 to 4.7	1.5 to 3.0

A: 1,1-ditert.butylperoxy-3,3,5-trimethylcyclohexane (40% active material)
B: Dicumyl peroxide (40% active material)
C: Bis(tert.butylperoxyisopropyl)benzene (40% active material)

16.7 Radical reactions of peroxides and plastics

The function of the radicals formed from the decomposition of peroxides is to produce polymer carbon radicals by means of hydrogen abstraction which subsequently react with each other forming C—C bonds. When this process is repeated several times, a three dimensional network is ultimately obtained. However, the primary radicals formed from the peroxides and the substrate radicals P· produced with their assistance can undergo other reactions that do not necessarily contribute to cross-linking.

16.7.1 Mechanisms of radical reactions from decomposition of peroxides

16.7.1.1 Dicumyl peroxide

The reactions possible during radical decomposition are indicated by means of the commercially important product dicumyl peroxide.

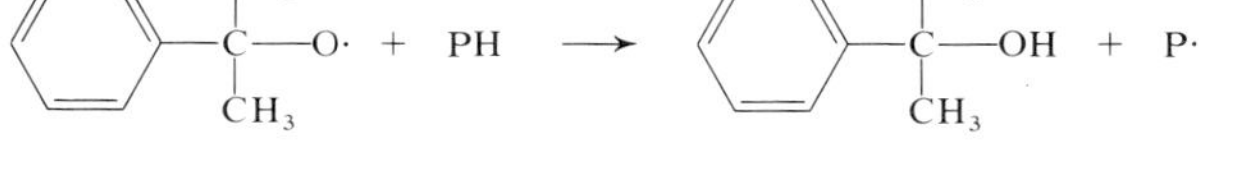

(I)

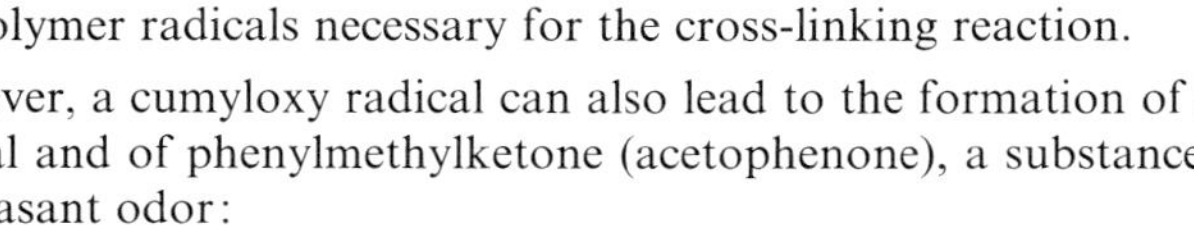

(II)

The substrate radicals formed in accordance with reaction II represent the polymer radicals necessary for the cross-linking reaction.

However, a cumyloxy radical can also lead to the formation of a methyl radical and of phenylmethylketone (acetophenone), a substance with an unpleasant odor:

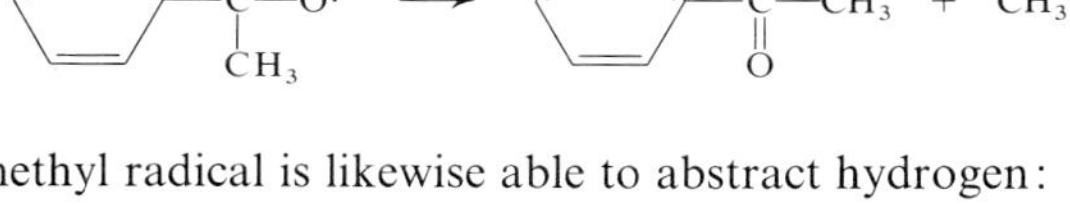

(III)

The methyl radical is likewise able to abstract hydrogen:

$$CH_3\cdot + PH \longrightarrow CH_4 + P\cdot \quad \text{(IV)}$$

or recombine – although on an insignificant scale:

$$2\,CH_3\cdot \longrightarrow C_2H_6 \quad \text{(V)}$$

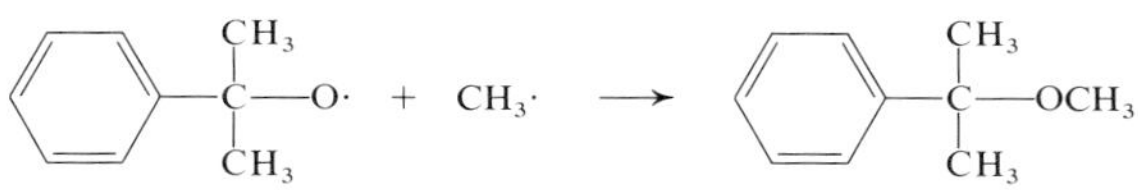

(VI)

Similar reactions take place during decomposition of ditert.butyl peroxide, with tert.butanol and acetone as the main products of decomposition.

16.7.1.2 Bis(tert.butylperoxyisopropyl)benzene

The reactions taking place during the thermal decomposition of bis(tert.-butylperoxyisopropyl)benzene are shown in figure 2.

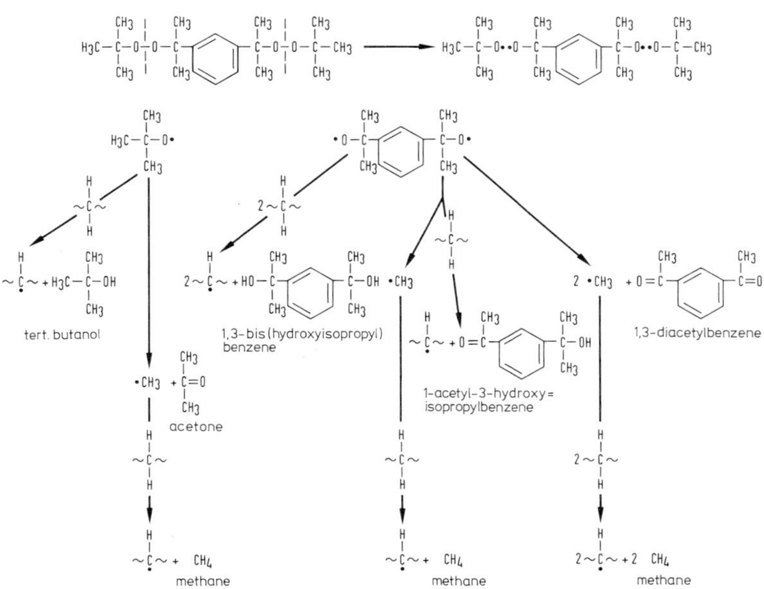

Fig. 2. Reaction scheme for the thermal decomposition of bis(tert.butylperoxyisopropyl)benzene.

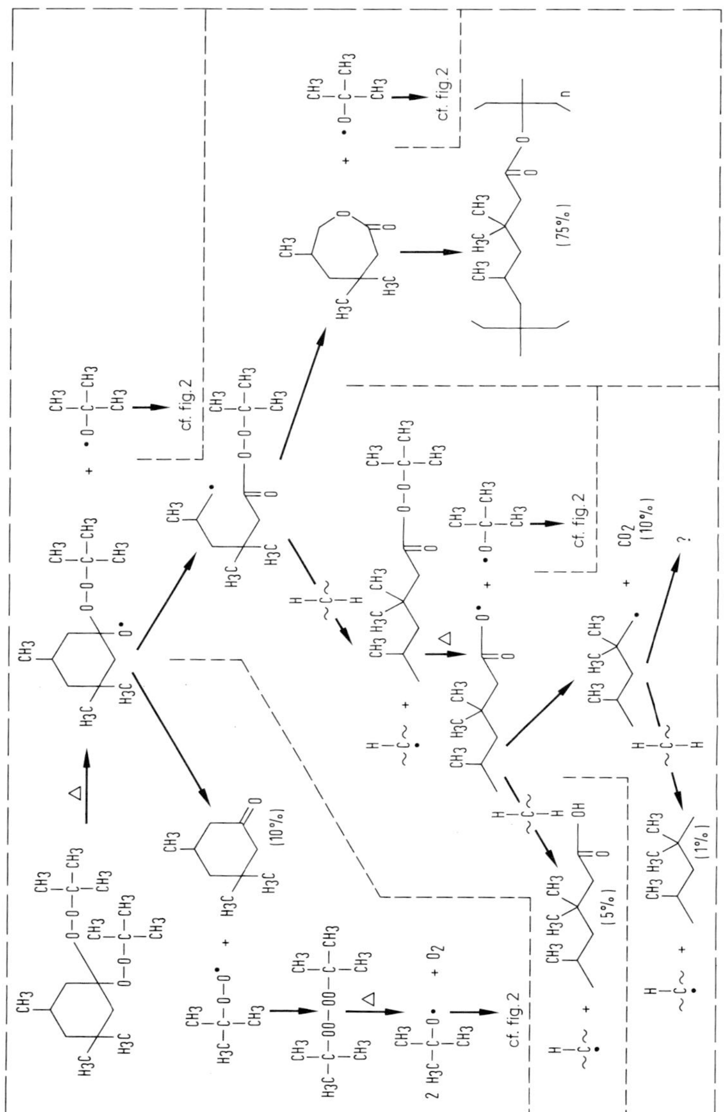

Fig. 3. Reaction scheme for the thermal decomposition of 1,1-ditert.butylperoxy-3,3,5-trimethylcyclohexane

Here, a tert.butoxy radical is formed, whose possible secondary reactions are similar to those already described for the cumyloxy radical. Mono- and dialcohols, as well as ketones, are formed. When the mixture has cooled they are present in solid form and, if not sufficiently compatible with the substrate (excessive amounts added), can cause blooming.

16.7.1.3 1,1-Ditert.butylperoxy-3,3,5-trimethylcyclohexane

It is assumed that the thermal decomposition of this product takes place as shown in figure 3 [3]. Although the compound contains two peroxy groups, only one contributes to cross-linking.

16.7.1.4 Bis(2,4-dichlorobenzoyl)peroxide

The principal decomposition products of this peroxide, which has been investigated as a model substance in pentadecane, are 2,4-dichloroben-

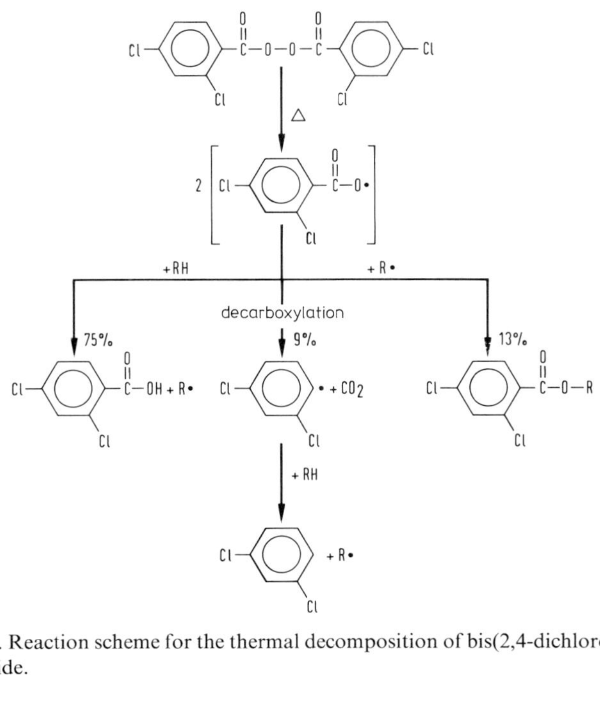

Fig. 4. Reaction scheme for the thermal decomposition of bis(2,4-dichlorobenzoyl)-peroxide.

zoic acid, 2,4-dichlorobenzene, carbon dioxide and pentadecyl 2,4-dichlorobenzoate (figure 4). The decomposition of benzoyl peroxide results in similar but chlorine-free reaction products, although the individual reaction products vary in their proportions.

16.7.2 Possible reactions undergone by the polymer radicals

The substrate radicals resulting from hydrogen abstraction can in addition to the main reaction leading to the desired cross-linking.

$$2\,P\cdot \longrightarrow P{-}P$$

also undergo a number of side reactions; the most important are

chain cleavage

disproportionation

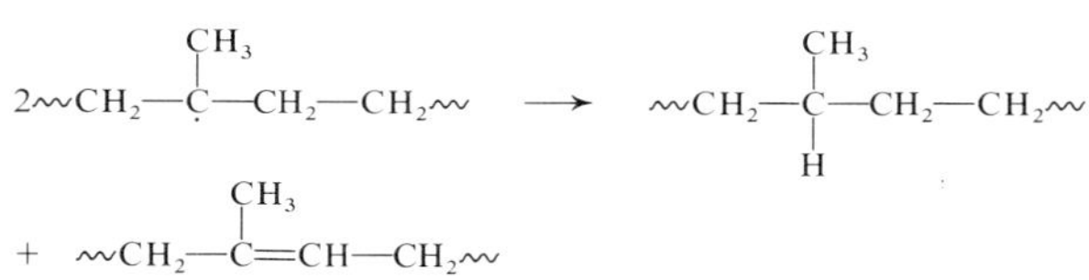

intramolecular recombination (ring-formation)

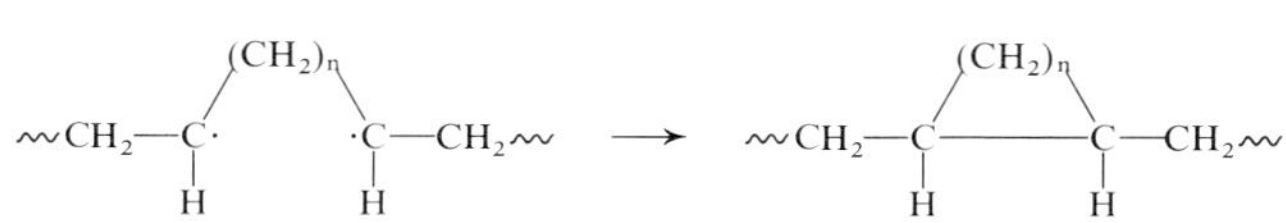

recombination with radicals originating from peroxide decomposition

$$P\cdot + CH_3\cdot \longrightarrow P{-}CH_3$$

$$P\cdot + RO\cdot \longrightarrow P{-}OR$$

Various groups of workers have been concerned with the investigations of the competing reactions, which have been studied either in a polymeric substrate [1, 4, 5] or in low molecular weight compounds such as dodecane or pentadecane [3, 6]. Particular attention was paid to the influence of radical concentrations, temperature and type of substrate. It was found that when the temperature is raised, there is an increase in chain cleavage and disproportionation reactions.

16.7.2.1 Studies carried out in model systems

In order to avoid the analytical problems that arise when working with polymeric substrates, *van Drumpt* and *Oosterwijk* [6] have used chemically defined, liquid n-alkanes as substrates. As a result of the short substrate chains and low concentrations in which the peroxides were used, the reactions of the substrate radicals formed resulted in dimeric and trimeric reaction products that could be quantitatively determined by means of known analytical techniques.

16.7.2.2 Analytical methods for investigating cross-linked systems

The methods most frequently used for determining the degree of cross-linking are comparatively simple to carry out. The first comprises the extraction of a sample of the cross-linked product with boiling xylene for a minimum period of 12 hrs.

The degree of cross-linking is expressed by the ratio:

$$\frac{\text{weight after extraction}}{\text{weight before extraction}} \cdot 100$$

The second method utilizes torque measurement in a rheometer (e.g. Monsanto rheometer, oscillating disc rheometer, ODR). For the investigation of peroxide-containing compounds, an oscillation velocity of between 3 and 100 oscillations per minute with an amplitude of $\pm 3°$ should be set. Higher amplitudes and frequencies cause excessive heat buildup in the sample and thus lead to distorted results. Special instruments are available for measurements on granular starting materials.

As an example, figure 5 shows the variation in torque observed as a function of time at 180 °C for low density polyethylene with 3 parts dicumyl peroxide per 100 gram LDPE [7]. At the start, the polymer melts and the torque decreases to a minimum value M_L; as cross-linking commences, the torque increases to a maximum value M_H.

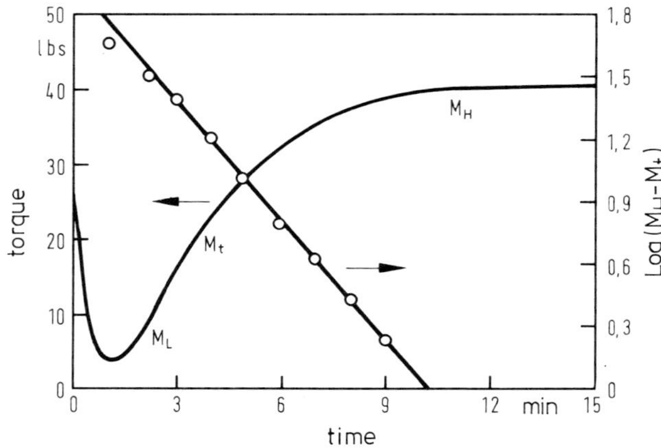

Fig. 5. Rheometer curve for the cross-linking of low density polyethylene with dicumyl peroxide at 180 °C (Monsanto rheometer).

In figure 5, the logarithm of $(M_H - M_t)$ is at the same time plotted as a function of time, M_t signifying the torque at time t. The path of the curve is almost linear, i.e. the cross-linking reaction is 1st order and comprises the main reaction. Within the experimental tolerances, its velocity constant shows, as expected, a close conformity with measurements made at other peroxide concentrations.

In similar experiments, *Harpell* and *Walrod* [7] were able to show that the peroxides investigated by them closely obey the law of a 1st order reaction over a wide temperature range. Plotting the logarithm of the velocity constant k as a function of $1/T$ demonstrates that the cross-linking reaction conforms to the *Arrhenius* law.

The investigation of the cross-linking reaction in low and high density polyethylene with 2,5-dimethyl-2,5-di(tert.butylperoxy)hexyne-3 showed identical velocity constants [7]. This means that the low concentration of the tertiary CH bonds present in low density polyethylene does not significantly influence the kinetics of cross-linking. Other, generally more elaborate methods and processes have been described by *Hummel* and co-workers [8].

16.7.3 Inhibition of the cross-linking reaction by atmospheric oxygen

During peroxide cross-linking, the reaction can be wholly or partly inhibited, in particular at the surface, by the admission of oxygen. This effect is based on the fact that oxygen reacts extremely rapidly with the substrate radicals P•, peroxy radicals POO• being formed. The latter react comparatively slowly, with the result that no cross-linking occurs at these places. The consequence of this is that the surface is inadequately cross-linked and in the case of elastomers remains tacky.

16.8 Influence of blend components on the cross-linking reaction

16.8.1 Fillers

Insufficient cross-linking in the presence of a number of acidic fillers, in particular silicic acids, results – at least in part – from the catalyzed ionic decomposition of the peroxides. There is, however, also the possibility that the adsorption of peroxides on the surface of certain fillers such as talc and silicates leads to inadequate cross-linking. In such cases, this can be remedied by increasing the peroxide quantity and/or by the addition of coagents (cf. section 16.8.5).

16.8.2 Plasticizers, extender oils

With the exception of plasticized PVC, cross-linked thermoplastics do not generally contain plasticizers or extender oils. In rubbers and a number of special applications in which these auxiliaries are used, these additives reduce the cross-linking effect to a greater or lesser extent.

In this respect, aromatic oils have the most adverse effect, whereas naphthenic and paraffinic oils exhibit only a slight effect and plasticizers such as dioctyl phthalate and alkylbenzenes have virtually no influence.

16.8.3 Antioxidants

Most antioxidants reduce the effectiveness of peroxides. According to *Harpell* and *Walrod* [7], a normal level of poly-2,2,4-trimethyl-1,2-dihydroquinoline diminishes the efficacy of peroxides by approx. 10% and causes a slight discoloration. In the presence of antioxidants such as alkyl substituted phenylenediamines and sterically hindered phenols, cross-linking is inhibited to a greater or lesser extent, depending on the structure of the peroxide and the antioxidants.

16.8.4 Blowing agents

In the case of peroxide cross-linking, the blowing agent most frequently used is azodicarbonamide in combination with an activator (kicker). This activator serves to lower the decomposition temperature of the blowing agent and enables the peroxide and blowing agent decomposition rate to be adapted to one another. Zinc oxide/stearic acid is often employed as a kicker. This combination has no adverse influence on the thermal stability of peroxides and their effectiveness during cross-linking. It is thus possible to produce articles with a homogeneous, fine-cell and closed cell structure.

16.8.5 Coagents

Coagents are sometimes used to increase the cross-linking density or cross-linking yield. The most effective coagents include compounds with several allyl functions. A typical representative of this class is triallyl cyanurate (TAC), a product frequently used in commercial practice.

Other coagents are:

ethylene glycol dimethylacrylate	(EDMA)
diallyl terephthalate	(DATP)
triallyl isocyanurate	(TAIC)
trimethylol propane trimethacrylate	(TRIM)
m-phenylene-dimaleimide	
1,2-cis-polybutadiene	

Ahne, Wiedenmann and *Kleeberg* [9] have found that triazines containing allyl groups make possible a higher degree of cross-linking and, with an efficiency comparable to that of TAIC, improved adaptation of polymer and coagent to one another and thus also impart better storage stability and processability of the blends. According to [9], the use of triazines of this type is particularly advantageous in the peroxide cross-linking of polyolefins containing fillers, especially in the presence of acidic fillers. In addition, it is stated that improved stability towards thermooxidative degradation is achieved.

16.9 Peroxides as coupling agents

Specifically suited compounds or systems such as cobalt naphthenate, isocyanates or resorcinol/formaldehyde resins have been used for many years for improving the adhesion of elastomers to metals and synthetic fibers. Studies by *Fan* and *Shaw* [10] have shown that with certain silyl peroxides, chemical bonds can be formed between a polymer and a second substrate.
Berg and *Endstra* [11] have investigated this, taking as an example vinyl tris(tert.butylperoxy)silane (VTBS). In combination with various peroxide cross-linking agents they obtained excellent adhesion of polymers such as EPDM, PE, silicone etc. to glass, metal, wood and synthetic fibers. It was found, however, that the compound used for the study is too toxic to be suitable as a commercial product. Silyl peroxides with lower toxicity are at present being tested for the same application.

16.10 Trends in peroxide cross-linked polyethylene (XLPE) and ethylene-vinyl acetate copolymers

In recent years there has been a marked increase in the use of cross-linked polyethylene, in particular for cable insulation. Other fields of application that are becoming increasingly important are hot water pipes, as well as foamed and cross-linked polyethylene for use as insulating and packaging material. Although by β-radiation cross-linking it is possible to obtain articles free from the products of peroxide decomposition, this process – as already mentioned – requires a high level of capital expenditure and can only be used up to a certain polymer thickness. Polyethylene cross-linking is thus predominantly carried out with the aid of peroxides. This also applies to ethylene-vinyl acetate copolymers used for cable and wire insulation and for shoe soles (also foamed), as well as in polymer blends.

Although dicumyl peroxide and bis(tert.butylperoxyisopropyl)benzene are the peroxides at present most frequently used in the cable industry, in recent years growing interest can be observed in peroxides with low volatility, which are liquid at or close to room temperature and can be metered direct into the extruder during cable manufacture. Compounding and energy costs can thus be reduced. Ditert.butyl peroxide is in many cases too volatile, although from the economic aspect it is by all means of interest. Tert.butylcumyl peroxide and its blends with other peroxides are at present being investigated. Peroxide masterbatches in pellet form, which instead of an inorganic substance as carrier material contain a polymer compatible with the final mixture, are recently gaining in importance.

Growing interest is lately being shown in peroxides that are capable of being excited in a microwave field and thus also make possible the cross-linking of nonpolar blends. As the initial field of application, polyethylene cable insulating material cross-linked with ditert.butylperoxy terephthalate is at present in the development stage. The advantages mentioned are: considerable savings in energy cost in comparison with steam and dry cross-linking, appreciably smaller overall length of the installation in question [12].

Recently a new class of cross-linking agents has been introduced to the market [13]. Compounds like ethyl-O-benzoyl laurohydroximate, preferably used together with a coagent, allow cross-linking of polyethylene in the temperature range above 220 °C without premature cross-linking (scorch) at processing temperatures up to 180 °C.

16.11 Safety aspects

16.11.1 Stabilization

Organic peroxides are known to be products that are basically unstable. It should be noted, however, that individual peroxides differ very greatly with regard to their stability.

Strictly speaking, each organic peroxide possesses its own specific hazard features and would thus require special precautionary and safety measures. Fortunately, however, it is not necessary for each organic peroxide on the market to be stored and handled under different conditions. For this purpose it is possible to provide general guidelines, since the manufacturers of organic peroxides deliver their products in a form that ensures that they possess adequate impact and shock resistance, as well as storage stability. Higher stability is achieved by means of stabilization, i.e. mixing with substances which under the conditions encountered in commercial practice are inactive vis à vis the peroxide in question and facilitate its handling. The stabilizer can be in solid or liquid form, e.g. chalk or a plasticizer such as dibutyl phthalate or a mineral oil.

Some blends of organic peroxides contain stabilizers not for safety reasons but to improve their performance properties: in such cases, the stabilizer serves only as a diluent.

16.11.2 Thermal stability

It is a specific characteristic of organic peroxides that they already decompose at relatively low temperatures by exothermic reaction. The temperature at which exothermic decomposition commences varies for each organic peroxide. For safety reasons it is thus necessary to know the temperatures at which this process takes place to such a degree that it could give rise to a dangerous situation.

Only relatively stable peroxides can naturally be used for cross-linking reactions, since they are generally incorporated at fairly high temperatures.

There are several methods of determining thermal stability. One of the best known is the spontaneous heating test in a Dewar vessel, a method developed by the Federal Institute for Materials Testing in Berlin. This involves a process known as heat buildup on storage [14]. In this test, 400 ml of the substance being tested are placed in a Dewar vessel that is not firmly sealed and has been insulated so as to comply with the containers normally used. The Dewar vessel containing the sample is stored in a circulating air oven at constant temperature, the temperature of the oven and the specimen being recorded. The test procedure employed is to raise the oven temperature in stages until a temperature is reached at which during a maximum test period of 7 days a noticeable increase occurs in the temperature of the sample.

This test provides an indication of the ambient temperature at which no appreciable decomposition of the peroxide takes place. However, it supplies no information on the temperatures permissible during processing.

With bis(tert.butylperoxyisopropyl)benzene, for example, no self-accelerating decomposition is observed at an oven temperature of 50 °C. Based on the recommendations of the United Nations regarding the transportation of dangerous goods, this substance thus requires no special temperature control. 40% dicumyl peroxide can also be regarded as relatively stable. It is accepted by the international transport organizations – IMCO (sea transport), RID (rail transport), ADR (road transport) – as a safe product.

In addition to the above-mentioned spontaneous heating test, there are a number of other refined test methods for measuring thermal sensitivity. The so-called isotherm test provides information on heat evolution by a peroxide as a function of time under conditions in which the heat of reaction produced is removed and measured. In the adiabatic storage test, the exothermic heat of reaction is retained in the specimen, thus resulting in an increase in temperature. The conditions encountered in commercial practice lie between these two extremes.

The advantage of these new methods lies in the fact that the data calculated from the measurements are not dependent on peroxide quantity, whereas the results obtained in the spontaneous heating test in a Dewar vessel apply only to a commercial container holding 25 to 50 kg peroxide.

Since slow decomposition of organic peroxides can already take place at relatively low temperatures, it is advisable for quality reasons to protect peroxides against heat. *It is important, therefore, not to store cross-linking peroxides for longer than one week at temperatures above 30 °C.*

Even after a period of several months, cross-linking peroxides stored at a temperature of $\leqq$30 °C exhibit virtually no reduction in their active oxygen content.

16.11.3 Flammability and fire behavior

When peroxides are rated according to their fire behavior, it is possible to encounter products with low flammability, as well as those which after ignition burn very rapidly, even without being in contact with air (deflagrate). A familiar method of determining the flammability of solvents and other liquids is to measure their flash point by means of a standard technique. These tests are based on the determination of the lowest temperature at which a flammable vapor/air mixture is formed. A disadvantage of flash point determination when applied to peroxides is that the results obtained can lead to misunderstandings due to the fact that many organic peroxides already exhibit considerable exothermic decomposition at temperatures below their flash point. For this reason, the flash point test is generally not a suitable means of rating the flammability of peroxides. Organic peroxides and their blends should in principle be regarded as readily flammable materials and handled accordingly.

16.11.4 Influence of impurities

Numerous organic peroxides are highly sensitive to contamination by impurities such as rust, dust, paper, wood etc. Contact with heavy metals and their salts generally causes decomposition to accelerate.

16.11.5 Physiological properties

A discussion of peroxide cross-linking agents would be incomplete without attention being drawn to the possible health hazards that may arise during handling and usage of these products. Based on the classification of acute toxicity, peroxide cross-linking agents can be rated as "slightly toxic" to "virtually nontoxic" [15]. Contact with eyes, skin and mucous membranes should be avoided.

For further information, the reader is referred to the manufacturers' leaflets and to [16].

16.12 List of trade names of organic peroxide producers

For this list of products and trade names completeness is not claimed.
Major available cross-linking peroxides on the market (situation mid 1983).

Chemical structure	Composition	Producers and trade names	
		Akzo Chemie BV Netherlands	Hercules USA
2,5-dimethyl-2,5-di-(tert.butylperoxy)-hexyne-3	90%, liquid	Trigonox 145	
	45%, on carrier		
	25%, masterbatch		
2,5-dimethyl-2,5-di-(tert.butylperoxy)-hexane	90%, liquid	Trigonox 101	
	45%, on carrier	Trigonox 101-50 D, 45 B	
Ditert.butyl peroxide	95%, liquid	Trigonox B	
	50/74%, liquid formulation	Trigonox B-50 E	
Bis(tert.butyl-peroxyisopropyl) benzene	90–100%, solid	Perkadox 14-90, 14 S	Vulcup R
	40%, on carrier	Perkadox 14-40 B, 14-40 K	Vulcup 40 C, 40 KE
	40%, masterbatch	Perkadox 14-40 MB, 14-40 A, 14-40 P	Vulcup 40 MB
Tert.butylcumyl peroxide	95%, liquid	Trigonox T	
	40–50%, on carrier	Trigonox T-50 D	
Dicumyl peroxide	90–100%, solid	Perkadox BC/BCff	Dicup R, T
	40%, on carrier	Perkadox BC-40 B, BC-40 K	Dicup 40 C, 40 KE
	40%, masterbatch	Perkadox BC-40 MB, BC-40 A	Dicup 40 MB

Luperox GmbH W.-Germany	Montefluos S.p.A. Italy	N.O.F. Japan	Peroxid Chemie GmbH (Interox) W.-Germany	Witco Chemical USA
Luperox 130		Perhexyn 25 B		
Luperco 130XL				
Luperco 130 EVA-25				
Luperco 101		Perhexan 25 B	Interox DHBP	
Luperco 101 XL			Interox DHBP-45-IC	
Luperox Di	Peroximon DE	Perbutyl D	Interox DTBP	DTBP
Luperox Diop 31				
	Peroximon F	Perbutyl P	Interox DIPP	
	Peroximon F 40	Perbutyl P-40	Interox DIPP-40-IC, DIPP-40-G	
Luperco 802 G	Peroximon F-40 MG			
Luperco 801		Perbutyl C	Interox BCUP	
		Perbutyl C-40	Interox BCUP-50-IC	
Luperco 500 R, 500 T	Peroximon DC	Percumyl D	Interox DCUP	
Luperco 540 C, 540-KE, 540-G	Peroximon DC 40, DC 40 K		Interox DCUP-40-IC,	
Luperco 540 PE 540 LMB	Peroximon DC 40 MG			

Chemical structure	Composition	Producers and trade names	
		Akzo Chemie BV Netherlands	Hercules USA
4,4-ditert.butylperoxy-n-butylvalerate	40%, on carrier	Trigonox 17-40 B	
	40%, masterbatch	Trigonox 17-40 MB	
1,1-ditert.butyl-peroxycyclohexane	50%, liquid formulation	Trigonox 22-B-50	
1,1-ditert.butyl-peroxy-3,3,5-tri-methylcyclohexane	40%, on carrier	Trigonox 29-40 B	
	40%, masterbatch	Trigonox 29-40 MB	
Tert.butylperoxy benzoate	100%	Trigonox C	
	50%, on carrier	Trigonox C-50 D	
Dibenzoyl peroxide	50%, paste	Lucidol S-50 S	
Bis(2,4-dichloro-benzoyl)peroxide	50%, paste	Perkadox PD-50 S PD-50 F	
Bis(p-chloro-benzoyl)peroxide	50%, paste	Perkadox PC-50 S	
2,2-bis(tert.butyl-peroxy)butane	40/50%, liquid formulation	Trigonox DB-50	
Ethyl-3,3-bis(tert.-butylperoxy)butyrate	40%, on carrier		

Luperox GmbH W.-Germany	Montedison S.p.A. Italy	N.O.F. Japan	Peroxid Chemie GmbH (Interox) W.-Germany	Witco Chemical USA
Luperco 230 XL		Perhexa V	Interox NBV-40-IC	
Luperco 331		Interox	USP 400 P CH-50 AL, FT	
Luperco 231 XL		Perhexa 3 M	Interox TMCH-40-IC	
Luperco 231 LMB				
Luperco P-XL		Perbuzyl Z	Interox TBPB-50-IC	Esperox 10 XL, KXL
Luperco AST		Nypex BS	Interox BP-50-P-SI	
Luperco CST		Nypex CS	Interox DCLBP-50-P-SI	
			Interox CLBP-50-P-SI	
Luperox 220		Perhexa 22	Interox BU-40-IC	
Luperco 233 XL				USP 333 XL, 333 KXL

16.12.1 Index of major peroxide producers

1. Akzo Chemie BV
   3800 AE Amersfoort, The Netherlands
2. Hercules Incorporated
   International Department Hercules Power
   Wilmington, Delaware 19899, USA
3. Luperox GmbH
   8870 Günzburg, West Germany
4. Montefluos S.p.A.
   Milano, Italy
5. Nippon Oil & Fats Co. Ltd.
   Chiyoda-ku, Tokyo, Japan
6. Peroxid Chemie GmbH
   8023 Höllriegelskreuth, West Germany
7. Witco Chemical (U.S. Peroxygen Division)
   Richmond, California 94804, USA

16.13 References

[1] *L.D. Loan:* Rubber Chem. Technol. 40 (1967) p. 149.
[2] Technical information Dow Corning: Sioplas E, a new crosslinkable polyethylene.
[3] *J.D. van Drumpt:* Paper being prepared for publication in J. Polym. Sci.
[4] *L.D. Loan:* Pure Appl. Chem. 30 (1972) p. 173.
[5] *F.P. Baldwin, G. Ver Strate:* Rubber Chem. Techn. 45 (1972) p. 709.
[6] *J.D. van Drumpt, H.H.J. Oosterwijk:* J. Polym. Sci. 14 (1976) p. 1495.
[7] *G.A. Harpell, D.H. Walrod:* Rubber Chem. Technol. 46 (1973) p. 1007.
[8] *K. Hummel:* Kautsch. Gummi Kunstst. 30 (1977) p. 807.
[9] *H. Ahne, R. Wiedenmann, W. Kleeberg:* Kautsch. Gummi Kunstst. 28 (1975) p. 135.
[10] *J.L. Fan, R.G. Shaw:* J. Adhes. 4 (1972) p. 261.
[11] *A.L. Berg, W.C. Endstra:* Gummi Asbest Kunstst. 30 (1977) p. 280.
[12] *W.C. Endstra:* Revue Générale des Caoutchoucs et Plastiques 60 (1983) p. 35.

[13] Preliminary data sheet: Development product D 501, AKZO Chemie BV, Amersfoort, The Netherlands.
[14] Verordnung über die Beförderung gefährlicher Güter auf der Strasse. Rn 3152/1.
[15] *W.L. Sutton:* J. Occupat. Med. 5 (1963) 4, pp. 212/217.
[16] *R.S. Waritz:* Toxicology of organic peroxide crosslinking agents for elastomers, Meeting Rubber Division, Am. Chem. Soc., Cleveland, Ohio, October 1977.

17 Nucleating Agents for Partly Crystalline Polymers

Ir. Joost Jansen, DSM, Geleen, The Netherlands

17.1 Crystallization of polymers

Polymers crystallize from a melt under the following conditions:

- The molecular structure of the polymer must permit crystalline ordering: insufficient mobility of the main chain, irregularly sited side groups, and branching or bulky side chains can prevent crystallization.
- The crystallization temperature must be below the melting point, but not too close to the glass point, so that the molecule chains have the mobility necessary for crystallization.
- Nuclei must be present to initiate the crystallization, from which crystallites are then formed, the latter arranging themselves into a superstructure, the spherulites.
- The rate of crystallization must be sufficiently high.

The overall rate of crystallization can be determined from the nucleus density and the growth rate of the spherulites [1, 2].

Figure 1 shows in general form for the case of thermodynamically determined, *homogeneous* and *thermal* nucleus formation, the dependence of the rates of nucleation and crystallization on the temperature of a supercooled melt [1]. It can be seen that in the event of strong undercooling (or rapid cooling, as is the case in practice) many spherulites begin to grow at the same time, whereas at smaller undercooling (or in the case of slow cooling) only a few spherulites are formed.

Table 1 shows the maximum attainable degrees of crystallinity x_c(max), the maximum crystal growth rate v_{max}, and the melting point T_m for a number of polymers [2].

Polymers with a medium crystal growth rate, such as polyamide 6, polypropylene, and polyethylene terephthalate, can obviously be more strongly undercooled than the rapidly crystallizing polyethylene, and for

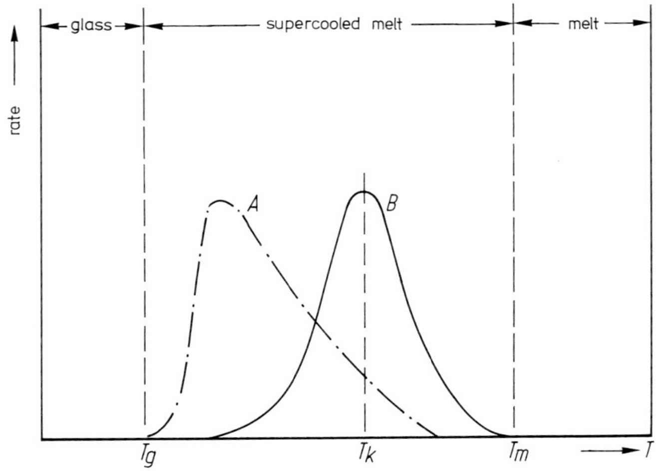

Fig. 1. Crystal growth (A) and nucleation (B) in dependence on the degree of supercooling of a polymer melt.

T_g: glass transition point,
T_k: crystallization temperature,
T_m: melting point.

Table 1. Characteristic parameters of the crystallization of polymers.

Polymer	Maximum degree of crystallinity x_c(max)	Maximum crystal growth rate v_{max} (μm/s)	Crystalline melting point T_m (°C)
High-density polyethylene	0.80	33	141
Polyamide 6.6	0.70	20	267
Polyamide 6	0.35	3.3	229
Isotactic polypropylene	0.63	$3.3 \cdot 10^{-1}$	183
Polyethylene	0.50	$1.2 \cdot 10^{-1}$	270
Isotactic polystyrene	0.32	$4.2 \cdot 10^{-3}$	240
Polycarbonate	0.25	$1.7 \cdot 10^{-4}$	267

this reason respond very well to *heterogeneous* and *athermal* nucleation with the aid of foreign substances – the *nucleating agents*. Conventional processing of polymers such as polystyrene and polycarbonate, which have a very low crystal growth rate, always results in an amorphous form.

The number and size of the spherulites formed are determined by whether the nucleus formation is homogeneous or heterogeneous. Since very many more nuclei are formed in the presence of a nucleating agent, under the same cooling conditions the resulting spherulites will be considerably smaller than in material to which no nucleating agents have been added. Nucleated polymers therefore have a finer grain structure than unnucleated ones, and this is reflected in their physicochemical characteristics; thus, coarse-spherulitic plastics are more brittle and less transparent or translucent than ones with the same crystalline fraction but a fine-spherulitic structure.

The so-called *spontaneous* nucleation, i.e. nucleation without an intentional introduction of a nucleating agent, is generally believed to be due to foreign substances such as catalyst residues, regranulate, oxidatively degraded polymer, or other, processing-determined impurities about whose physicochemical nature very little is known. Spontaneous nucleation should be distinguished from *autonucleation*, in which the crystallization is triggered by incompletely melted down crystallites of the polymer.

17.2 Requirements for nucleating agents, substance classes

True nucleating agents were discovered empirically. The more precise details relating to the connections between the type of the polymer and the chemical and physical character of the nucleating agent and its action have not yet been completely clarified. However, in the case of organically based nucleating agents a few general observations can be made on what makes a substance capable of nucleation:

- the nucleating agent should be wetted or absorbed by the polymer;
- it should be insoluble in the polymer;
- its melting point should be higher than that of the polymer;
- it should be homogeneously dispersable in the polymer melt, in as fine a form as possible (1 to 10 µm).

Nucleating agents can be roughly classified as follows:

- inorganic additives such as talc, silica, kaolin;
- organic compounds such as salts of mono- or polycarboxylic acids, certain pigments;
- polymers such as ethylene/acrylic ester copolymers.

In practice, concentrations of up to 0.5% are used, concentrations higher than this not, as a rule, producing any further enhancement of the nucleation effect. The nucleating agents are incorporated as powder/powder mixtures, as suspensions or solutions, or in the form of a master batch. Whichever method is used, good pre-dispersion of the nucleating agent is necessary if an optimal effect is to be achieved.

17.3 Characterization of the activity of nucleating agents

The method used most frequently to study the crystallization behaviour of a polymer melt and to determine the activity of a nucleating agent is differential thermal analysis (DTA, DSC) [3].

In this procedure the crystallization of the melt is followed at a constant rate of cooling and the temperature at which the crystallization rate reaches its maximum is determined. This temperature, at which the heat released by the crystallizing melt climbs to a peak at a constant rate of cooling, is generally known as the crystallization temperature. It increases with increasing nucleus density, i.e. with increasing activity of

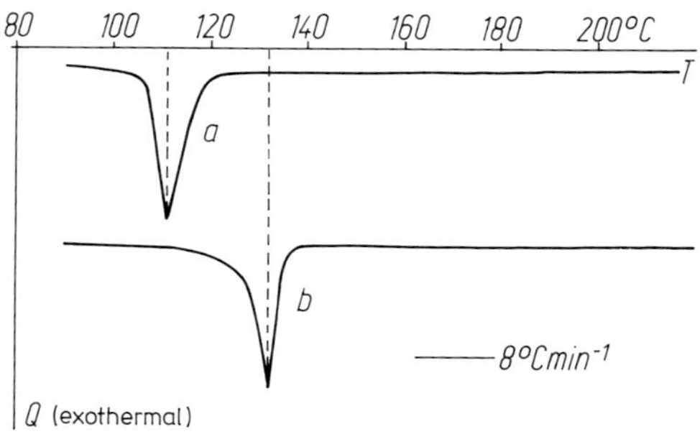

Fig. 2. DSC crystallization diagrams for polypropylene, cooling rate 8 °C/min. (a) without nucleating agent, (b) with 1% sodium p-tert.butylbenzoate.

without nucleating agent

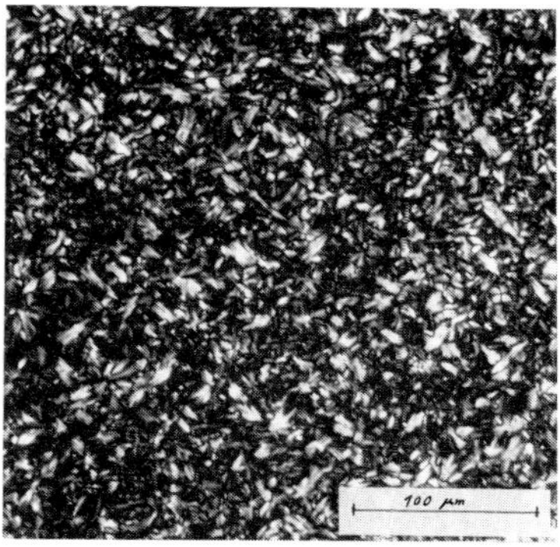

with nucleating agent

Fig. 3. Influence of a nucleating agent on spherulite size of polypropylene.

the nucleating agent. Figure 2 shows the DSC crystallization curves of polypropylene with and without a nucleating agent [4], while figure 3 demonstrates the influence of a nucleating agent on the size of the spherulites [5]. Faster rates of cooling lower both the crystallization temperature and the height of the peak.

The size of the spherulites is normally measured with a polarizing microscope. In the case of not too strongly nucleated samples the spherulite density or the nucleus density can be determined by counting the spherulites.

17.4 Influence of nucleating agents on characteristics of the polymer

Most industrial nucleating agents produce high degrees of crystallinity resulting in increased hardness, elasticity modulus, tensile strength, and yield point compared with the unnucleated material. In addition, the fine-grain spherulite structure of nucleated polymers improves optical properties such as transparency or translucency, the elongation at break, and the impact strength, since as a result of this structure a more uniform stress distribution is produced on mechanical stress.

Shortening of the cycle times in the injection molding of nucleated molding compositions as a result of the higher crystallization temperature and faster overall rate of crystallization resulting from the higher nucleus density is of economic importance; the crystallization process then sets in at smaller supercooling and is completed after a shorter cooling time [6].

Because of the increased crystallization tendency of nucleated materials, after-crystallization is often prevented. The latter effect is able to modify the mechanical properties and dimensions of finished articles (after-shrinkage) on prolonged storage.

In practice, nucleation effects are used to enhance processing and finished-product characteristics, especially of polyethylene terephthalate, polypropylene, and polyamide 6; in the case of polyethylene and polybutene-1 the technological significance of these effects is far smaller.

17.5 Nucleating agents for polyethylene terephthalate (PET)

The low crystallization rate (see table 1) and the slow nucleus formation [2] of PET are particularly advantageous for the manufacture of highly transparent, partly crystalline PET bottles, but they constitute a serious handicap in the injection molding of PET. Injection into heated molds (e.g. to about 150 °C) – to accelerate crystallization and to shorten cycle times – yields finished articles of low crystallinity that are difficult if not impossible to remove from the mold and too soft; if very long and uneconomical cycle times are used the resultant articles are too brittle (large spherulites). These problems can be avoided by the use of nucleating agents [1], and good material properties are then obtained (many small spherulites) with economically acceptable cycle times.

Table 2 gives an idea of the wide variety of nucleating agents for PET put forward in the patent literature [7].

In practice, insoluble, inorganic nucleating agents such as metal oxides, metal salts, pigments, and certain minerals [8] with particle sizes of 3 µm are preferred and are used in concentrations of about 0.5%. They can be added before, during, or after the polycondensation, in the form of dry, fine powders or in suspension; tumbling with the granulate in a drum is another possibility.

Figure 4 shows the degree of crystallization of PET as a function of the nucleating agent used and the crystallization time at 100 °C, the substantial reduction of the crystallization time of nucleated PET is clearly evident [9].

Table 2. Nucleating agents for polyethylene terephthalate.

Inert, insoluble substances
Mineral fillers such as chalk, gypsum, clay, kaolin, mica, talc, silicates
Pyrophyllite
Pigments such as cadmium red, cobalt yellow, chromium oxide
Metals: metal oxides such as titanium dioxide, magnesium oxide, antimony trioxide; phosphates
Carbonates and sulfates, preferably of the alkaline-earth metals
Boron nitride
Sodium fluoride
Carbon black
Organic compounds, on their own or with inert solids
Salts of monocarboxylic or polycarboxylic acids
Montan wax and montanic ester salts
Diphenylamine, tetrachloroethane
Acetone, nitromethane, benzene, toluene
Halogenated alkanes, such as tetrachloroethane
Benzophenone, tetralin
Aromatic alcohols and amines
Alkali aralkylsulfonates
Epoxides
Polymers, on their own or with inert solids
Polyolefins: PE, PP, poly-4-methylpentene-1, poly-3-methylbutene-1
Copolymers of ethylene and unsaturated carboxylic esters
Ionic copolymers of ethylene and salts of unsaturated carboxylic acids
Copolymers of styrene derivatives and conjugated dienes

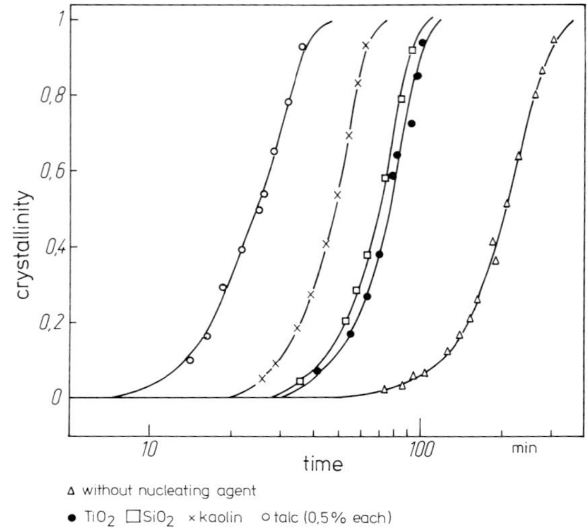

Fig. 4. Crystallization isotherms of nucleated polyethylene terephthalate at 100 °C.

17.6 Nucleating agents for polyamides

Nucleation of polyamides 6, 6.6, and 6.10 results in the following modifications of the properties [10, 11]:

- higher crystallinity,
- increased bending and tensile moduli,
- increased surface hardness,
- increased yield point,
- elevated deflection temperature under load,
- improved abrasion resistance,
- lower elongation at break and impact strength,
- reduced water absorption.

Transparent injection-molded articles with greater wall thicknesses (>2 mm) cannot be made even from nucleated material, since the spherulite diameter exceeds 5 µm; the reduction of the cycle times in injection molding is between 3 and 30% [12].

Highly disperse silica [12] is generally used as the nucleating agent, in concentrations of about 0.1%. The other possible additives are molybdenum disulfide, iron sulfide, titanium dioxide, talc, and sodium phenyl-

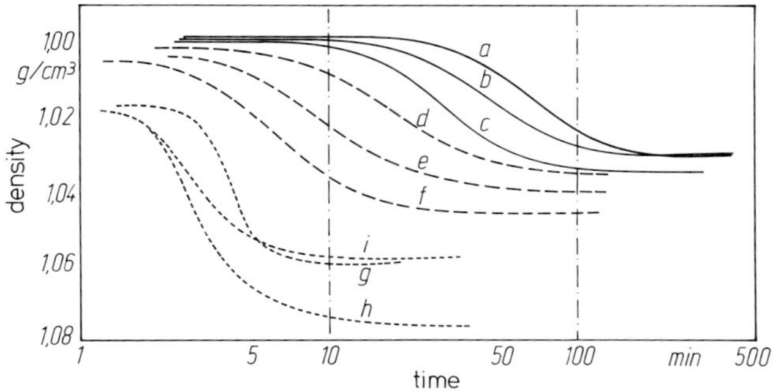

Fig. 5. Time dependence of the density of polyamide 6 melts during isothermal crystallization at various crystallization temperatures T_k.

a: $T_k = 205\ °C$, d: $T_k = 200\ °C$, g: $T_k = 180\ °C$: without nucleating agent,
b: $T_k = 205\ °C$, e: $T_k = 200\ °C$, h: $T_k = 180\ °C$: with 1% polyamide 6.6 powder,
c: $T_k = 205\ °C$, f: $T_k = 200\ °C$, i: $T_k = 180\ °C$: with 1% PET powder.

phosphinate. A list of nucleating agents for polyamide 6, mostly suggested in the patent literature will be found in [13].

Higher-melting polymers such as polyamide 6.6 or polyethylene terephthalate are also used in polyamide 6. Figure 4 gives a clear picture of the increase in density, i.e. crystallinity, as a function of time with falling isothermal crystallization temperature, and of the crystallization rate accelerated by the nucleating agent and leading to higher degrees of crystallinity [13].

17.7 Nucleating agents for polypropylene (PP)

In spite of the availability of comprehensive literature on the nucleation of polypropylene, its commercial application is relatively recent. Nucleated PP molding compositions are frequently used for the manufacture of thin-walled (< 0.4 mm) injection-molded articles, in the case of PP with a high melt index, the addition of a nucleating agent allowing a 30% shortening of the cycle time. The nucleation of PP is also important for improving the optical properties of films, nucleated films having lower haze, higher gloss, and improved clarity (see table 3).

Table 3. Optical properties of polypropylene films.

Film type	Blown film			Cast film		
Film materials	Un-nucleated poly-propylene	Nucleated polypropylene		Un-nucleated poly-propylene	Nucleated poly-propylene	Oriented poly-propylene
Mass temperature	220 °C	220 °C	260 °C	280 °C	280 °C	
Haze (%)	36	20	17	4	7	1
Gloss	19	28	40	71	65	90
Clarity*)	39	29	10	1	2	2

*) Measured with an EEL clarity meter
Note that lower clarity figures correspond to better optical properties

Injection-molded parts made from nucleated PP have better strength characteristics in comparison with unnucleated PP, as shown in table 4 [5].

In addition to the improvement of the mechanical properties, the tendency towards delamination, excessive flash formation and that towards sinking and voiding is reduced [14].

The clarity of PP molded articles is considerably improved by nucleation, this effect being particularly discernible in random PP copolymers.

The inorganic additives [4] such as talc, various silicas, carbon black, and kaolin have a poor nucleating effect. Salts of aliphatic monobasic or dibasic acids or arylalkyl acids such as sodium succinate, sodium glutarate, sodium caproate, sodium 4-methylvalerate, aluminum phenylacetate, and sodium cinnamate have a medium nucleating activity. Particularly active nucleating agents are the alkali metal or aluminum salts of aromatic or alicyclic carboxylic acids such as aluminum benzoate, sodium or potassium benzoate, sodium β-naphthoate, lithium benzoate, and aluminum tert.butylbenzoate. Bis-benzylidene sorbitol has recently become available as a nucleating agent particularly suitable for PP [15].

Beck [3, 16] and *Binsbergen* [17] have given a general synopsis of the activity of nucleating agents for PP. The free acids of the above-mentioned salts do not have any useful nucleating action in ash-free PP, but do exert a certain effect in PP with a sufficiently high catalyst residue (approx. >5 ppm Al) [17].

Table 4. Influence of nucleating agents on mechanical properties of polypropylene (injection-molded ASTM specimens).

Logarithmic viscosity number (dl/g)	2.6		2.5					2.4	
Nucleating agent	none	nucleating agent A	none	nucleating agent A	nucleating agent E, increasing concentration →			none	nucleating agent A
Tensile test, 0.2 in/min									
– yield point σ_y, psi	4680	4830	4590	4800	4860	4820	4750	4500	4640
– elongation at yield, %	9.4	8.9	9.2	9.6	9.0	9.1	12.1	12.5	11.0
– tensile modulus, 10^6, E, psi	2.1	2.5	2.3	2.4	2.7	2.6	2.2	1.7	2.0
Tensile test, 2 in/min									
– yield point σ_y, psi	5290	5480	5220	5440	5500	5380	5640	5630	5730
– elongation at yield, %	6	9	5	6	8	8	5	–	–
– break stress, psi	3470	3360	3280	3430	3080	3550	3450	2370	2560
Bending modulus, 10^5 psi	2.31	2.36	2.30	2.48	2.60	2.64	2.51	–	–
IZOD notch impact strength*) at 23 °C, ft.lb/in of notch	0.70	0.74	0.64	0.72	0.90	0.81	0.99	0.58	0.63
Hardness									
– Shore D	71	72	73	75	74	74	74	–	–
– Rockwell M	26.5	29.0	26.0	30.0	32.5	32.5	32.0	–	–

*) Standard notch radius 0.01″

The organic nucleating agents are preferably milled down to 5 μm in a ball mill and are normally mixed dry with PP powder in concentrations of up to 0.5%. Addition in the form of solutions with subsequent drying [18] or in the form of a master batch seems to be less effective.

17.8 Nucleating agents for polyethylene (PE)

Since polyethylene is a rapidly crystallizing polymer (see table 1), it is not surprising that comparatively little has been published on its nucleating agents. In "uncontaminated" high density PE a nucleating agent such as 1% potassium stearate (not very effective in PP) seemingly has a comparatively large effect [19], the size of the spherulites being reduced in this way from 47 μm (unnucleated) to 13 μm (nucleated). The addition of nucleated PE or of a higher polyolefin [20] is believed to lead to similar effects in low density PE. In the injection molding of high density polyethylene, during which pronounced row orientation generally occurs in the flow direction, certain organic pigments can give rise to nucleation, resulting in the buildup of high internal stresses and undesired distortion phenomena [21].

17.9 Nucleating agents for polybutylene (PB)

Addition of adipic or p-aminobenzoic acid raises the crystallization temperature by some 15 °C compared to unnucleated PB. The lattice transformation characteristic of polybutylene, from the tetragonal form II into the rhombohedral form I, is not accelerated by nucleating agents [22].

17.10 References

[1] *D.W. van Krevelen:* Int. Congress Amsterdam on "Processing Polymers to Products" (1963) p. 167.
[2] *D.W. van Krevelen:* Properties of Polymers, Elsevier, 1976.
[3] *H.N. Beck, and H.D. Ledbetter:* J. Appl. Polym. Sci. 9 (1965) p. 2132.
[4] *K. Binder:* Chem. Kunst-Aktuell 29 (1975) pp. 168/175.
[5] *C.J. Kuhre:* SPE Journal 20 (1964) p. 1113.

[6] *U. Johnsen* and *G. Spilgies:* Kolloid-Z. u. Z. Polymere 250 (1972) pp. 1174/1181.
[7] *R. Vieweg* and *L. Goerden* (eds.): Kunststoffe Handbuch [Polymer Handbook], Vol. VIII, Polyester, Hanser Verlag München, 1973, p. 701.
[8] *L.B. Morgan:* J. Appl. Chem. 4 (1954) p. 160.
[9] *G. Groeninchx et al.:* J. Appl. Polym. Sci. (Polym. Phys. Ed.) 12 (1974) pp. 303/316.
[10] *M.I. Kohan:* Nylon Plastics. Wiley & Sons, New York, 1973, p. 424.
[11] *G.A. Toelcke:* Mod. Plast. 45 (1968) pp. 117/126.
[12] *L.L. Scheiner:* Plastics Techn. 13 (1967) 4, p. 37.
[13] *R. Vieweg* and *A. Müller* (eds.): Kunststoffe Handbuch [Polymer Handbook], vol. VI, Polyamide, Hanser Verlag, München, 1966 pp. 227 and 245.
[14] *J.E. Lane:* Br. Plastics 39 (1966) p. 528.
[15] Product Data Sheet MILLAD 3905. Milliken Chemicals, Spartanburg. SC/USA.
[16] *H.N. Beck:* J. Appl. Polym. Sic. 11 (1967) pp. 673/685.
[17] *F.L. Binsbergen:* Polymer 11 (1970) pp. 253/267 and 309/326.
[18] *U. Johnsen:* Kolloid-Z. u. Z. Polymer 240 (1970) pp. 762/765.
[19] *K.A. Narh, J.A. Odell, A. Keller* and *G.V. Fraser:* J. Mat. Sci. 15 (1980) pp. 2001/2009.
[20] German Patent 1095427 (1966).
[21] *A. Turturro et al.:* Br. Polym. J. 5 (1973) pp. 129/139
[22] *M. Cortazar:* Polymer Bulletin 5 (1981) pp. 635/641.

18 Aspects of Plastics Additives Relative to Industrial Hygiene and Food Quality

Dr. Roland Leimgruber, Research and Consulting Company (RCC), Itingen, Switzerland

18.1 Introduction

Health risks must be estimated in all cases where human beings are in contact with chemicals. This applies to any type of chemically active substance and, therefore, also plastics additives (subsequently called additives). The following phases are important in the life of an additive:

- Production (raw materials, their impurities and metabolites),
- transportation,
- incorporation of the additive, and transformation of the plastics material,
- use of the final object,
- disposal of an object (metabolism of the structure during burning, during exposure to environmental influences etc.).

The problems relating to the incorporation of the additives into the plastic (industrial hygiene) and to the use of the final object (e.g. food packaging materials, children's toys, textiles etc.) will be discussed. Questions of industrial hygiene during the production of the additives are the responsibility of the additive manufacturer and are not included in our discussion.

Because there are no problems of industrial hygiene which concern additives exclusively, industrial hygiene itself will be discussed in general terms. In contrast, problems concerning additives during the use of the final object will be discussed specifically.

The toxicological risk of a chemical substance depends on the doses as well as the chemo-biological mechanism of the substance in question. Many different chemicals are used as additives and their respective chemical structures vary greatly. As a consequence, the biological properties of the substances may rank from inactive up to highly active, which means, from practically not toxic to toxic.

To illustrate, antioxidants such as BHT (2,6-ditert.butyl-4-methylphenol), because of their low toxicity, are used in some countries as direct food additives. On the other hand, additives are known which are so toxic that their use in plastics is banned completely (e.g. o-tricresylphosphate, which has been employed in early times as plasticizer).

Because of the large number of additives and their biologically different individual activities, it will not be possible to discuss their toxic properties. Rather, the basic problems will be covered using some typical examples and, where necessary, references to the literature will be provided. We would like to emphasize at this point that the examples mentioned have been randomly selected and the selection was influenced by the importance of the substance, the typical toxicological pattern, and the availability of reliable information.

18.2 Industrial hygiene

18.2.1 Introduction

In handling plastics additives, any possible contamination of man must be minimized; this is a valid principle not only for additives but for all other chemicals as well. Before the first contact with an additive, the possible toxic hazard for man has to be estimated to establish the necessary safety precautions. In practice, the highest potential for the contamination of man can usually be found in filling facilities or in places where drums or other packages are opened and emptied. As soon as the additives are incorporated in the polymer the direct exposure of man and the toxic risks are very small. Indeed, during polymer conversion highly volatile additives may contaminate the atmosphere, vapors and gases may lead to a toxic effect for humans, but appropriate protection measures can prevent this.

18.2.2 Basic problems of exposure to chemicals

Chemicals can be introduced into the human body via three routes:

- by ingestion,
- by inhalation,
- by absorption through the skin.

Introduction by ingestion is rarely a risk in industry. Absorption of dusts and vapors by inhalation and through the skin are the main routes of human contamination. Although additive manufacturers are supplying additives more and more in non- or low-dusting forms, dust exposure can cause skin or mucous membranes irritation, and occasionally sensitization may occur. As mentioned before, these problems happen usually only on opening additive packages and during incorporation into the polymer. By using effective local ventilation and by wearing goggles, gloves and dust masks, reliable protection can be provided.

Toxic effects are graded as acute, medium or long-term (chronic toxicity). In industry, the main intoxications are of acute nature. This concerns hazards to man which are caused by single, double or threefold exposure – mostly in high doses. In contrast, toxic effects in the organism can be found also by long-term exposure in small doses, as e.g. to organic and inorganic compounds of heavy metals, which often are used as stabilizers.

Today, diseases based on chronic exposure to minute doses have become more and more important. The most dangerous effect is cancer. So far, no carcinogenic effects have been ascertained for the main additives in use, but this does not mean that no such substances exist. It is important that all unnecessary chronic exposure of man to any chemicals should be avoided.

Since an increasing number of test methods are discussed in brochures, safety data sheets, and in regulations etc. they will be discussed in the following sections as far as plastics additives are concerned.

18.2.3 Acute toxicity tests

The acute hazard of a product is usually characterized by means of the following six tests. Depending on the intended use and the physical and chemical properties of the substance only the necessary tests are performed. Handling precautions and safety measures are usually established on the basis of these tests.

18.2.3.1 Acute oral toxicity

The acute oral toxicity is the toxicity of a substance after single application by mouth (per os). This value is called LD_{50} (Lethal Dose) or DL_{50} (Dosis Letalis) and means the amount of substance per body weight of the test animal (mg/kg), which kills 50% of the animals after a single oral application. These values are relative ones and can only be classified in comparison to other substances (cf. table 1) to give useful indications. To estimate the degree of toxicity of a substance based on the LD_{50} value in [6] a classification is proposed (cf. table 2).

Table 1. Acute oral toxicity in rats for different plastics additives

Chemical structure	Use	LD_{50} p.o. rats
Dimethoxyethylphthalate [1]	plasticizer	2750 mg/kg
4-methyl-2,6-ditert.butylphenol [2]	antioxidant	1700 mg/kg
2(2′-hydroxy-3′-tert.butyl-5′-methyl-phenyl)5-chlorobenzotriazole [3]	u.v. absorber	>5000 mg/kg
Azodicarbonamide [4]	blowing agent	>6000 mg/kg
Triphenyllead acetate [5]	biocide	230 mg/kg

Table 2. Classification of the acute oral LD_{50} in rats [6]

LD_{50} by single application in rats	Designation in use
1 mg/kg or less	extremely toxic
1 to 50 mg/kg	highly toxic
50 to 500 mg/kg	moderately toxic
500 to 5000 mg/kg	lightly toxic
5000 to 15000 mg/kg	practically nontoxic
>15000 mg/kg	relatively harmless

18.2.3.2 Acute dermal or acute percutaneous toxicity

The resorptive toxicity of a substance is tested by a single application on a defined area of depilated skin. Thereby, the amount of the test substance which kills 50% of the test animals is measured in mg/kg body weight. Therefore, that value is also called acute dermal LD_{50} (cf. table 3).

18.2.3.3 Acute eye irritation test

The irritation of the eye is measured after application of the test substance into the lachrymatory sack. Rabbits are used generally for this type of test. After a single application, the interaction of the chemical substance with the cornea, the lens, the iris, the conjunctiva and the eyelid is assessed. This test is also known as the primary eye irritation test (cf. table 3).

Table 3. Acute toxicity of a light stabilizer and of an antioxidant [7]

Test	Nickel complex of 3,5-ditert.butyl-4-hydroxy-benzylmono-ethyl-phosphonate (light stabilizer)	n-Octadecyl-β(4'-hydroxy-3',5'-ditert.-butylphenyl)propionate (antioxidant)
Acute LD_{50} p.o., rats	2000 mg/kg	>10000 mg/kg
Acute LD_{50} dermal, rats	>3100 mg/kg	> 2000 mg/kg
Primary skin irritation, rabbits	minimally irritating	nonirritant
Sensitization, guinea pigs	no sensitization	no sensitization
Acute dust inhalation, rats, 4 h (LC_{50})	4800 mg/m^3 air	1800 mg/m^3 air

18.2.3.4 Acute skin irritation test

The irritation of the skin after a single or repeated application of the test substance onto the depilated skin of the animals is judged (cf. table 3). This test is also called the primary skin irritation test.

18.2.3.5 Acute inhalation toxicity

The test animals are exposed for two to four hours to a dust, an aerosol, a vapor or a gas. The acute LC_{50} (Lethal Concentration) is measured, i.e. the concentration in mg/m^3 air, which kills 50% of the test animals (cf. table 3).

18.2.3.6 Sensitization

This type of test, which is usually performed with guinea pigs, tests the possible allergenic reaction of a substance (allergy) (cf. table 3).

For technical reasons, it is not possible very often to apply the test substance in doses higher than a defined quantity. In such cases the value is expressed as "greater than" or ">".

18.2.4 Investigations of long-term toxicity

Improvements in industrial hygiene make long-term exposures to *high doses* practically of no importance, even though intoxications with high doses can happen by accident. On the other hand, long-term exposure to *small doses* is still essential. The contamination of man by highly volatile and highly dusty substances can not be avoided completely.

Chronic intoxication may often produce different toxic effects than those from the single exposure of large doses. For example, vinyl chloride monomer shows a relatively small acute oral toxicity, whereas it produces cancer in small doses over a long period of time. Mainly mutagenic and carcinogenic effects are known from chronic exposure to chemicals.

The possibilities for the evaluation of the potential carcinogenicity of a chemical by animal trials are limited. Only exposure over the whole animal life may show significant results. For such tests, relatively short-lived animals as e.g. mice or rats have to be used. Even for a dog with its higher life expectancy, it would mean that after 12 years or more, the first results would be available.

In addition, lifetime studies in animals have the disadvantage that the results can be transferred to man only with reservations. For example there are substances known which show carcinogenic effects in certain species while such effects have never been found in man. On the other hand, there are substances known to be carcinogenic to man, but similar effects have never been found in animals. Nevertheless, all chemicals showing carcinogenic effects in animal trials should be regarded as suspected carcinogens in man. In handling such substances, special handling precautions are necessary. For different reasons (e.g. lack of laboratory capacity, high cost, long-term tests, weak significance etc.), it is only exceptionally possible to test industrial chemicals in a carcinogenicity study, usually a lifetime study.

Investigators have long been looking for timesaving and relevant test methods to evaluate carcinogenic properties of substances. It seems that initial steps have been found which test the mutagenic effect of chemicals in bacteria and cells of mammalians. The so-called "*Ames*-test" is well known (developed by *B. Ames*). With this test the mutagenic effect of a substance on different strains of bacteria can be tested. Because there are some parallels between mutagenic properties and carcinogenic effects, it is hoped that some indications of the potential carcinogenic effect will be obtained by the results of this test. Since many substances show mutagenic properties by the *Ames*-test but no carcinogenic effect in mammalians, this type of test should never be regarded as a test for carcinogenicity in man. Therefore, a full range of additional short-term tests in bacteria as well as in mammalian cells have been developed. But each of those tests is relevant to a single aspect in the initiation of cancer by chemicals. Therefore, only competent toxicologists should give interpretation on results of such tests.

18.3 Additives as a component of articles for every day use

18.3.1 Introduction

When plastics materials are used there is always a slight possibility that incorporated additives may migrate in small quantities to the surface of the article and into the surrounding medium and so directly or indirectly into the human body. This can happen with food packaging materials, children's toys, textiles, packaging materials for pharmaceuticals etc. Very often this will result in a subchronic or chronic exposure of man to small doses. Therefore, the health authorities of the different countries have become aware of the problems connected with the approval of additives for plastics food packaging materials; the same or similar regulations are enforced for the use in textiles and children's toys. Such regulations are also the basis for the approval of pharmaceutical packaging materials, whereby of course, there are additional requirements in regard to product-specific interactions between packaging material and packaged goods. In the following we shall discuss only the problem of "food packaging materials" because we can cover over 90% of all requirements.

Although only about 15% of all plastics products will be in direct contact with food, for practical reasons, much more of the production of the large volume plastics as polyolefins, PVC, polystyrene etc. comply with the requirements established for food packaging materials.

Before the common use of plastics as food packaging materials, regulations existed for the conventional packaging materials. Already around 1900, hygienists realized that lead and other toxic metals may migrate from the coatings of porcelain and enamel into food. But not before the large use of plastics as food packaging materials did the health authorities of the different countries start to reflect on the transition of non-intentional substances from packaging into the packaged material.

As a matter of fact, the oldest known regulations for packaging materials concerned mainly the limitation of the content of lead, arsenic, antimony, mercury, tin and zinc. But none of these older regulations gave any explanation about the toxicity of materials and under which conditions and in which quantities they would be toxic.

The quick development of plastics as packaging materials made it necessary to specify the regulations for their use. Packaging materials such as porcelain, glass, metals, paper, paperboard etc. have been used for a long time. Plastics appeared suddenly on the market and were suspect because of their novelty. This is the main reason that today much more is known about plastics than about other conventional packaging materials.

It is interesting to mention that the first appropriate regulations concerned exclusively the migration of additives into food. Only recently has the transfer of residual monomers, oligomers etc. come into discussion.

18.3.2 Basic considerations concerning the use of additives in plastics food packaging materials

"The expression *insoluble* is only a relative one. No substance is totally insoluble and it does not meet the reality if someone states that a plastic should be totally insoluble." This statement was made by *Lehmann* of the FDA (Food and Drug Administration), already in 1956 on the basis of widespread investigations; at least certain minute quantities of additives may migrate from the packaging material into the packaged food. This depends mainly on the following parameters:

Type of plastic, diffusion properties of the additive in the plastic, nature of the packaged food, storage temperature, surface condition of the plastic, contact time between plastic and packaged food, as well as interaction between the different additives in respect to their migration properties.

All these parameters may influence the migration behavior of an additive. The migrated quantity of an additive on the one hand and its toxicity on the other hand determine whether it is suitable for use in plastics food packaging materials. To test suitability for such applications, two basic principals have been formulated on the occasion of the 5th session of the Plastics Commission of the Federal Health Office (Federal Republic of Germany) in 1971 by its chairman *R. Franck:* "For guaranteeing the purity of our food, we have to prove that:

1. the harmless composition of a packaging material is guaranteed;
2. every unavoidable transition of a nonfood ingredient into the food should be as small as possible."

The food packaging regulations in the different countries have been established on the basis of similar considerations. We shall discuss them later. First, the test methods will be discussed from which we get the information about the suitability of an additive for food packaging materials.

18.3.3 Migration behavior of an additive

As already mentioned, the producer of a packaging material has to show which quantities of an additive migrate into the packaged food. He has to carry out the necessary tests. Because it is impossible to test a packaging material under all conceivable conditions it was necessary to establish some test models. The appropriate test conditions are standardized in most countries and they differ only slightly. It is mandatory to chose conditions as temperature, contact time, food simulant etc. in accordance with conditions in practice. The test conditions as they are proposed by the Federal Health Office (FRG) are summarized in table 4.

When the conditions in use differ substantially from the conditions mentioned in table 4, the test conditions have to be modified in accordance to practice. This is valid for plastics exclusively used in contact with frozen food. It has to be mentioned here that in the regulations of the FDA (USA), a much more detailed list of test conditions for the different applications is available than in any other country.

The tests listed in table 4 are so-called static migration tests. For the time being, these conditions are accepted in all countries with the exception of the USA. Here, for storage conditions near room temperature during several months, the measurement of the dynamic migration property is required, i.e. that at a temperature of 49 °C (120 °F) samples

Table 4. Test conditions for migration studies as recommended by the BGA (Bundesgesundheitsamt, Federal Republic of Germany) [8]

Intended use	Test temperature	Duration of test
Contact at room temperature up to several months (including hot filling)	40 °C	10 days (10 × 24 hrs)
Short-term contact up to 70 °C	70 °C	2 hrs
Short-term contact between 70 and 100 °C	100 °C	1 hr
Contact at 121 °C (sterilization), followed by storage at room temperature	121 °C	30 min
Contact at refrigeration conditions up to several months	10 °C	10 days (10 × 24 hrs)

Table 5. Test conditions for migration studies as recommended by the Food and Drug Administration (FDA, USA) [9]

Simulant	Pre-conditioning: temperature/time	Migration study: temperature/time
Dist. water	121 °C / 250 °F } during 2 hrs	cool to 49 °C (120 °F) and continue until no change in the slope of the migrant concentration/time curve (cf. figure 1)
3% by wt. acetic acid	100 °C / 212 °F } during 2 hrs	
8 or 50% by vol. ethanol	71 °C / 150 °F } during 2 hrs	
n-Heptane, 96% by vol. ethanol, fat or synthet. triglyceride [1])	66 °C / 150 °F } during 2 hrs	

[1]) e.g. HB 307, NATEC, Gesellschaft für naturwissenschaftlich-technische Dienste m.b.H., Behringstr. 154, 2000 Hamburg 50, Federal Republic of Germany.

of the simulant are taken periodically and the migrated quantity of additive is determined until there is practically no change of slope in the concentration/time curve; in addition a thermal pretreatment of the test specimens is prescribed (cf. table 5). The typical shape of such a curve for a light stabilizer is shown in figure 1.

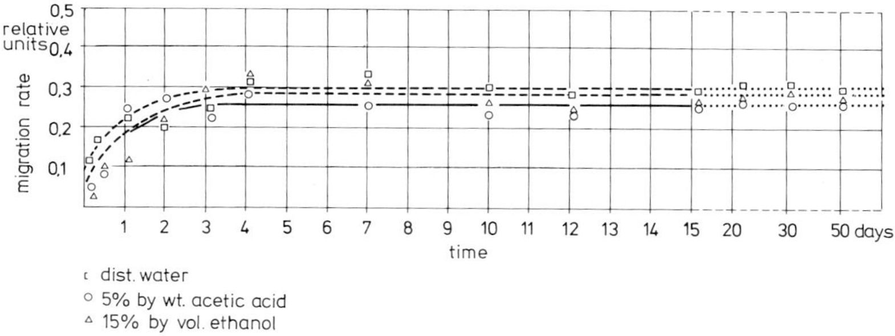

Fig. 1. Migration rates of 3,5-ditert.butyl-4-hydroxybenzoic acid 2,4-ditert.butyl-phenyl ester as a function of time at 49 °C (120 °F). Polypropylene plaques (contact on both sides) in different aqueous food simulants [9]

Very often, using the dynamic method, it can be shown after a few days already whether equilibrium is reached and what would be the magnitude of migrant in the food.

It is problematic whether extractions at 40 or 49 °C better simulate storage at room temperature. Indeed, articles intended exclusively for use at room temperature may be exposed to peak temperatures up to 30 °C. But test temperatures of 40 or 49 °C may give a wrong picture of the actual migration behavior in practice. Near 40 °C, the difference of a few degrees in test temperature may influence significantly the migration rate of an additive. This can be shown by the migration curve of an antioxidant in polypropylene as in figure 2.

For analytical and other reasons, it is preferred worldwide to determine the migration properties of an additive using food simulants. Today the following four types of simulants are used:

- distilled water,
- aqueous acetic acid,
- aqueous ethanol,
- natural fats or oils or a standardized synthetic mixture of triglycerides.

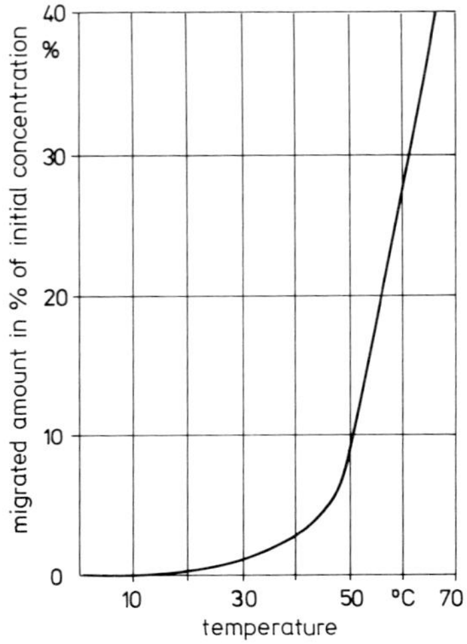

Fig. 2. Influence of temperature on the migration rate of an antioxidant from polypropylene [10]

Food simulants to be used following the recommendations by FDA and BGA are specified in the tables 6 and 7.

Depending on the final use, one or another simulant may be dropped. The severest problem in such investigations is the migration of additives into fats and oils, because substances are known which can be tolerated

Table 6. Food simulants as recommended by FDA (USA)

Type of food		Food simulant
Aqueous:	fat <5% ethanol <0.5% pH >4.6	distilled water
Acidic:	pH <4.6	3% by wt. acetic acid
Fatty:	fat >5%	natural fats or synthetic triglyceride [11, 12]
Alcoholic:	ethanol >0.5%	8 or 50% by vol. ethanol

Table 7. Food simulants as recommended by BGA (Federal Republic of Germany)

Food simulant	Type of food
1. Dist. or deionized water	aqueous food, pH $\geqq 4.5$
2. 3% by wt. acetic acid	acidic aqueous food, ph $\leqq 4.5$
3. 10% by vol. ethanol	aqueous alcoholic food
4. Standardized synthetic mixture of triglycerides or natural fats and oils	fatty food in direct contact with packaging material

Table 8. Test conditions for migration studies as recommended by the European Council [time (t), temperature (T)]

Contact at intended use	Migration study: time/temperature
1. Contact time: $t > 24$ hrs	
1.1. $T \leqq 5$ °C	10 days at 5 °C
1.2. 5 °C $< T \leqq 40$ °C [1])	10 days at 40 °C
2. Contact time: 2 hrs $\leqq t \leqq$ 24 hrs	
2.1. $T \leqq 5$ °C	24 hrs at 5 °C
2.2. 5 °C $< T \leqq 40$ °C	24 hrs at 40 °C
2.3. $T > 40$ °C	following country-specific regulations
3. Contact time: $t \leqq 2$ hrs	
3.1. $T \leqq 5$ °C	2 hrs at 5 °C
3.2. 5 °C $< T \leqq 4$ °C	2 hrs at 40 °C
3.3. 40 °C $< T \leqq 70$ °C	2 hrs at 70 °C
3.4. 70 °C $< T \leqq 100$ °C	1 hrs at 100 °C
3.5. 100 °C $< T \leqq 121$ °C	30 min at 121 °C
3.6. $T > 121$ °C	following country-specific regulations

[1]) For plastics materials in contact with food which on account of the labelling or on account of law have to be stored at <20 °C, the following conditions for migration studies have to be used: 10 days at 20 °C.

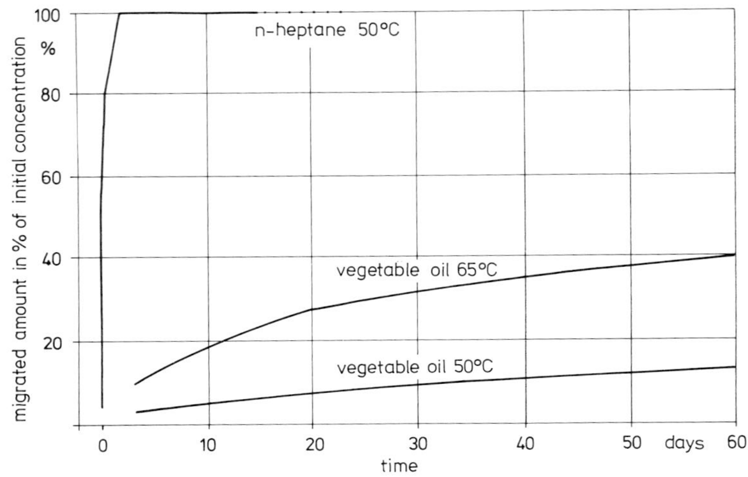

Fig. 3. Migration behavior of an additive in polypropylene [13]

in food only in the area of 10, 50 or 100 ppb (parts per billion, one part per 10^9 parts). But the detection of such small quantities of additives in fats is often only possible by using radioactive labelling. In some countries, e. g. USA and Switzerland, it is intended to solve the problem by using n-heptane or 96% by vol. ethanol (USA) and n-hexane or ethyl acetate (Switzerland) as fat simulants. With n-heptane and n-hexane, the extraction results may be divided by a factor of 5. As a matter of fact, these simulants have to be used carefully since they do not affect *all* polymers in the same way. As shown in figure 3, n-heptane as well as n-hexane may lead to absolutely wrong results for polyolefins. In the Swiss regulations [14], therefore, ethyl acetate is proposed as fat simulant for polyolefins and, since about 1981, the US Food and Drug Administration (FDA) accepts this also. Since such "alternate simulants" are always more demanding in terms of migration resistance of an additive than natural fats, it is desirable that other countries in addition to the USA and Switzerland would accept such alternatives.

Recently, migration cells have been proposed by different authors where there is only a one-side-contact of the test sample with the simulant.

Our investigations have shown that under these conditions, the test sample may be strained in such a way that the resulting migration rates may be up to a factor of ten higher than the real migration rates.

The migrant quantities are usually expressed in mg/dm^2 or in $\mu g/dm^2$ (in the USA $mg/inch^2$ or $\mu g/inch^2$).

18.3.4 Toxicological investigations

To prove the safety of substances migrating from packaging materials into the foodstuff, these substances must be tested in animals. Impurities, meaning migrants in the foodstuff, are usually very small quantities on the order of ppm or ppb. Such contamination of packaged food can be found not only with plastics but practically with all packaging materials. The assumption is made that a person daily eats food packaged in the same packaging material, i.e. having the same contaminants. Therefore, the appropriate test model for studies in animals will also be carried out over long periods with small doses.

The duration and the scope of these studies depend mainly on the quantity which migrates into food. Apart from small differences the tests required are very similar in most countries. The general situation at the present time is shown in table 9.

Table 9. Approximate scope of feeding studies required with additives for plastics food packaging materials

Migrated amount in food		Feeding studies
	≦50 ppb	acute toxicity
USA:	50 ppb to ca. 1 ppm	90 days rat and 90 days dog
Europe:	50 ppb to max. 60 ppm	90 days rat
USA:	>ca. 1 ppm	2 years rat, possibly 18 months mouse or hamster
Europe:	depending on NEL[1]) based on a 90-day feeding test	mutagenicity teratogenicity

[1]) explanation cf. 18.3.4.1.2

Depending on the migrant quantity, three areas can be distinguished: i.e. a first area where the quantity of a migrated substance is below 50 ppb. So far in most countries only acute toxicity tests are required if no highly suspected substances are concerned (as for example potential carcinogens, mutagens etc.).

In the area between 50 ppb and a few ppm's, one 90-day feeding test in one species (mostly rats) is required. The only exception in the matter is the USA where an additional feeding study of a second species (non-rodent) is mandatory.

In the third range above a few ppm, these food contaminants are treated in the same way as an intentional food additive. In this case, in addition to the 90-day feeding test, one or two chronic tests and sometimes multi-generation tests, teratogenicity and mutagenicity tests, are needed. It has to be emphasized at this point that more than 90% of all additives approved for plastics food packaging materials belong to the first or the second area of the above scheme.

18.3.4.1 Feeding trials

Since we are concerned with the introduction of migrated substances into man via food, such tests have to be carried out orally. If possible the test substance has to be mixed together with the feed; only exceptionally, if for example the substance is not stable in the feed (hydrolysis etc.), it is permissable to give the test substance via tube to the animal.

18.3.4.1.1 Range-finding study

To find the appropriate dose level for the 90-day study, a screening test or so-called range-finding test has to be performed. Thus it can be found if the substance will be accepted by the animal and whether toxic effects can be expected. This is done with a 10- to 28-day feeding test. The timing and the expenses for this test are shown in table 10. Before carrying out a range-finding test, acute oral data should be available.

18.3.4.1.2 90-day subchronic feeding

The 90-day feeding test is a basic test for estimating the subchronic toxicity of a substance. The species preferred for this test are rats. Mice are used only occasionally. In the United States, where a second species (non-rodent) is required for the approval of a plastics additive, the second test is usually carried out with dogs. During the tests, the substance is applied in three to four dose levels, whereby the highest concentration is determined where no toxic effects in the animals are found. This dose level is called the "No Effect Level" (NEL). It is expressed in mg applied substance per kg animal per day (mg/kg). As shown later, the amount which is believed not to produce any toxic effect in man over a longer period will be estimated from this NEL (for time and cost estimates cf. table 10).

Table 10. Approximate cost and time needed for subchronic and chronic feeding studies

Type of feeding study	Approximate time period for feeding, evaluation and reporting	Approximate cost in DM (1983)
28 days, rat (range-finding) at three dose levels with one control group	3 to 4 months	$22 \cdot 10^3$ to $30 \cdot 10^3$, depending on scope
90 days, rat with complete pathology	ca. 6 months	$100 \cdot 10^3$ to $180 \cdot 10^3$, depending on scope
90 days, dog with complete pathology	ca. 6 months	$120 \cdot 10^3$ to $180 \cdot 10^3$, depending on scope
Lifetime feeding, rat (incl. carcinogenicity)	ca. 3 to 4 years	$800 \cdot 10^3$ to $1000 \cdot 10^3$, depending on scope
Lifetime feeding, mice (incl. carcinogenicity)	ca. 2 to 3 years	$800 \cdot 10^3$ to $1000 \cdot 10^3$, depending on scope

18.3.4.1.3 Chronic tests, i.e. application over a lifetime period

If the quantity of contamination found by migration tests exceeds a certain level (cf. table 9), or, if the substance shows a toxicologically suspicious structure, feeding tests over a lifetime period are required in addition to teratogenicity and mutagenicity tests. Nowadays, these studies are carried out mainly with respect to potential carcinogenicity. Preferably, rats or mice are used since other species have too long a lifetime. These tests are costly and time-consuming (cf. table 10).

18.3.5 Basis for the suitability of additives for food packaging materials

18.3.5.1 ADI and PADI

ADI value (=Acceptable Daily Intake) is the amount of an additive which it is believed will cause no toxic effects in man over prolonged oral application. The ADI value is calculated on the basis of the NEL found in the *subchronic,* resp., *chronic* feeding test in animals, which is expressed in mg/kg animal and day. By dividing the above NEL by a security factor of 100 to 1000, the NEL for man is calculated. By multiplying the NEL for man with the average body weight of 60 kg, the ADI value for man is obtained.

The expression PADI (=Packaging Acceptable Daily Intake) was created specially for additives in food *packaging materials.* The ADI value for *direct* food additives is calculated on the basis of a NEL value found in *chronic* animal studies (lifetime studies). As mentioned in section 18.3.4.1.2, the basis for *indirect* food additives, i.e. migration from the packaging material into the food, is a 90-day *subchronic* feeding test.

In order to be able to distinguish how a given ADI value was obtained, it was agreed that a ADI value has to be based on a chronic feeding study. In all other cases where such values are calculated on the basis of a 90-day feeding test the value is called PADI. Both values, the ADI and PADI are different also with regard to the security factors which are used for their calculation. When a chronic feeding test is available a security factor of 100 to 200 is applied, whereas for 90-day feeding tests a factor of 200 to 1000 is used. Since these values for plastics

Table 11. Example for the calculation of the PADI value of an additive

The 90-day feeding study in rats resulted in a "no effect level" (NEL) of 250 mg/kg/day.	
NEL	250 mg/kg/day
Security factor	500
NEL for man	$\frac{250\ \text{mg/kg/day}}{500} = 0.5\ \text{mg/kg/day}$
PADI value for man	0.5 mg/kg/day × 60 kg
	PADI = 30 mg/day

additives are mostly based on subchronic tests we shall speak in the following only about PADI.

If the calculated PADI value (table 11) is greater than the migrated amount found in the migration test, usually the substance will be approved. Thereby the assumption is accepted that man eats at a maximum 1 kg of the same foodstuff a day over a longer period. In some cases, for example fatty food, where the daily intake is below 1 kg, correction factors may be used.

18.3.5.2 Conditions for the approval of an additive

It was shown in section 18.3.3 how the migration into food is measured under standard conditions. If the measured migration is below the calculated PADI, the additive is suitable for food packaging materials. The inequality of figure 4 has to be shown.

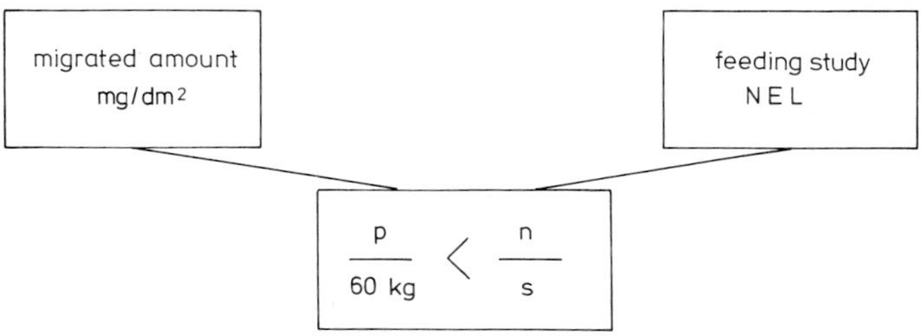

n = NEL in mg/kg/day resulting from a feeding study in animals
p = quantity incorporated by man and day
s = security factor

Fig. 4. Approval conditions for plastics additives

18.3.5.3 Estimate for the daily intake of indirect food additives (FDA proposal)

Different eating habits (for example daily intake of fat, vinegar, alcohol etc.) as well as the frequency of different plastics materials as food packaging so far have been taken into account too infrequently (cf. section 18.3.5.2) or have been completely neglected. Recent proposals from the FDA [15] concerning those factors will be discussed here, since these procedures will be integrated in the revised guidelines for preparing indirect food additive petitions for FDA.

The expression "TDD" (=Total Daily Diet) is newly introduced: this is the estimated level of each of the migrating species that can be expected in the total daily diet.

In addition, the expression "Consumption Factor" (CF) has been defined. The Consumption Factor is the portion of the diet likely to contact specific packaging materials. This term reflects the ratio of the weight of food contacting a specific packaging material to the weight of all food packaged.

The "food type distribution value" (f_T) reflects the nature of the food that will typically contact the food packaging material. To account for the very nature of food contacting a specific packaging material, the food type distribution values were calculated for each packaging material to indicate the fraction of aqueous, acidic, alcoholic and fatty food.

In the FDA proposal it is assumed:

About 41% of all packaged food are packed in plastics materials (CF= 0.41). From that, about 49% are in contact with aqueous food (f_T=0.49), 16% with acidic food (f_T=0.16), less than 0.01% with alcohol (f_T=less than 0.01) and 34% with fat (f_T=0.34). The CF and f_T values for different plastics are listed in table 12.

Table 12. Consumption Factor CF and food type distribution value f_T for different plastics

Plastics material	CF	f_T			
		aqueous	acidic	alcoholic	fatty
Polyacrylonitrile	≦0.001	≦0.01	≦0.01	≦0.01	0.97
Polyacrylates Phenolic resins etc.	0.147	0.17	0.40	0.31	0.12
EVA	0.021	0.30	0.28	0.28	0.14
Ionomers	0.004	≦0.01	≦0.01	≦0.01	0.97
Polycarbonate	≦0.001	0.97	≦0.01	≦0.01	≦0.01
Polyester	0.043	0.01	0.97	≦0.01	≦0.01
Polyolefins	0.328	0.67	≦0.01	≦0.01	0.31
Polystyrenes	0.075	0.67	≦0.01	≦0.01	0.31
PVC	0.110	≦0.01	0.23	0.27	0.49
PVDC	0.035	≦0.01	≦0.01	≦0.01	0.97
Waxes	0.053	0.47	≦0.01	≦0.01	0.51
Cellulose	0.011	0.05	≦0.01	≦0.01	0.93

The amount M of an additive in the food which contacts a packaging material is found by multiplying the appropriate f_T value with the migrated amount in the respective food simulants:

$$M = f_{\text{aqueous}} \cdot (M_{H_2O}) + f_{\text{acidic}} \cdot (M_{\text{HOAc}}) + f_{\text{alcoholic}} \cdot (M_{\text{EtOH}}) + f_{\text{fatty}} \cdot (M_{\text{heptane}}) \cdot [\text{ppm}].$$

The concentration C in the daily diet is calculated by multiplying M with the consumption factor CF. In addition, it is assumed that the total daily intake of food for man is 3 kg (total). The probable Estimated Daily Intake (EDI) for the additive is then determined using the following equation:

$$\text{EDI} = 3000\ \text{g/day} \cdot M \cdot \text{CF}.$$

For special cases, it is possible to make additional corrections.

As mentioned before, those are FDA recommendations, but it is known that these describe FDA's current approach for estimating exposure to indirect food additives and describe the type of information that should be supplied to help estimate probable exposure. This proposal is so important because, for the first time, different eating practices and packaging material types are taken into account.

18.3.6 Petition types for additives intended for food contact

The object of any food law is the protection of the end user from health injuries resulting from eating food. It must be emphasized that there is not, and there will never be, a 100% perfect regulation, but every law together with its rules is a compromise between theory and practice, or better, between perfection and practicability. When the conditions for an approval (cf. section 18.3.5.2) are fulfilled, there will be an approval granted following two different principles whereby in some countries both principles are accepted together.

18.3.6.1 Principle I: Limitation of the additive in the plastics material

In the case that the conditions discussed in section 18.3.5.2 are fulfilled, approval will be granted. This approval is usually published in the so-called positive lists of a country whereby a maximum amount of an additive in a plastics material is defined. The manufacturer of a plastics food packaging material may find in the appropriate list, what amount of an additive is allowed to be incorporated in a compound, complying with the hygienic regulations of a certain country. The costly tests for obtaining such an approval are usually initiated by the additive or plastics manufacturer. It is then the responsibility of the authorities to check each case and determine whether the found concentration of an additive in the final object exceeds the maximum allowed concentration.

This first principle shows a lot of advantages for the processor of plastics but it also shows some disadvantages especially for the control authorities. Since the analytical tests (migration tests) are not always carried out on the final object, the direct relation to practice may not be given:

- By other processing conditions and for a number of other reasons, the final object may show in practice other properties than were found with the model type.
- The control authorities are usually not able to carry out the necessary tests on the final object because that is only possible with relatively expensive equipment and highly specialized experts.

18.3.6.2 Principle II: Limitation of the migrants resp. impurities in the food

To avoid the disadvantages of principle I, a second principle has been discussed for about 10 years, that is the limitation of the amount of migrants in the food, the so-called *global migration.* The present standard value for the global migration is 60 ppm resp. 10 mg/dm^2. Several authorities have incorporated this principle in their latest edition of regulations. Thus not only known, but also unknown impurities are quantitatively limited. Since the known impurities are substances which are added to the formulation, as for example additives, pigments etc. the unknown substances are those which may be formed during production and conversion of the packaging material and about which neither quantitative nor qualitative values with respect to their toxicological properties are available. The global migration shows the advantage that it can be measured in the known food simultants rather easily (the only exception being fat).

Following this principle, approved additives are no longer registered in the positive list by their maximum approved concentration but with their PADI or ADI value. This value reflects the maximum amount migrated in 1 kg food which is consumed in one day. The Council of Europe has listed the additives in two groups: in group A are those additives which have a PADI or ADI greater than 60 mg, and in group B those which have a PADI or ADI smaller than 60 mg.

This procedure has an advantage: if the global migration is below 60 mg/kg food, only the additives of the list B have to be tested in respect to their *specific* migration. By this second principle the manufacturer of the final object is obliged to run analytical tests on every new packaging material. But this is difficult to realize in practice because:

- the majority of converters are technically unable to run the very difficult specific migration tests themselves;
- the necessary test capacity of institutions able to carry out all necessary tests is presently too small for the demand.

18.3.6.3 Present situation

The use of principle I shows the disadvantage that the authorities are not able to control it fully, under principle II, the authorities have the possibility to check the final object with a reasonable effort, whereas the necessary testing is too costly for most processors and can not be realized in practice.

For that reason, different countries (for example The Netherlands, Italy, Switzerland, Spain etc.) try to work out a compromise for both principles:

1. Additives intended for food contact have to be considered by toxicologists. Taking into account a certain defined standard (for the time being 60 mg/day/man), the products are contained either in list A or B.
2. Lists which have been established by toxicologists on the basis of scientific facts have to be checked by the appropriate authorities.
3. For all products of list B, the limitation in the plastics material will be defined in addition to the limitation of the maximum allowable migration into the food, whereby this value may be different for different food types.

When a final object complies with one of the two conditions – limitation in the plastic or limitation in the food – that is sufficient and the other limitation is not valid, that means one or the other condition has to be fulfilled. In table 13, an example is given for a light stabilizer.

Table 13. Light stabilizer approved on basis of principle I respectively II

Chemical structure	Limitations in food	Limitations in the plastics
2(2′-hydroxy-3′-tert.butyl-5′-methylphenyl)5-chlorobenzo-triazole	10 ppm in food or 1.66 mg/dm^2 contact area	PP max. 0.5% PE max. 0.5% PVC max. 0.3% PS max. 0.6%

18.3.7 Procedures for approval in different countries

18.3.7.1 Europe

Most European countries have food regulations where the use of additives for plastics food packaging materials is regulated. For the EC (European Community) a new directive is proposed which may be enforced within the next few years, with the consequence that the different legislations in the member states will disappear. For the time being, the situation can be summarized as follows:

Federal Republic of Germany: Positive list following principle I based on recommendations.

France: Positive list following principle I.

Italy: Positive list following principle I and II.

Belgium: Positive list following principle I and II.

The Netherlands: Positive list following principle I and II.

Spain: Positive list following principle I and II.

Austria: Positive list following principle I.

Switzerland: Approval of an additive by the Federal Health Authorities following principle I or II and in addition positive list of the Council of Europe.

Scandinavia: So far no positive list is published. The Scandinavian countries follow the recom-

	mendations of the Federal Republic of Germany or the positive list of the FDA (USA).
German Democratic Republic:	Positive list following principle I.
Czechoslovakia:	Positive list following principle I.
Remaining Eastern European Countries:	So far no positive lists published. The situation is unclear.
United Kingdom:	Positive list following principle I which is published by the British Plastics Federation as a code of practice.
Greece:	Positive list following principle I.

18.3.7.2 Other countries

USA:	Positive list following principle I.
Canada:	Approval following principle I for the single final object.
Brazil:	Positive list following principle I.
Japan:	Positive list for different polymers following principle I.
Australia:	Approval following principle I.

18.3.8 Cost

This chapter should not be concluded without giving an idea about cost and time needed for an international approval. Not only are the toxicological investigations very expensive – in addition the legislation in the different countries varies in such a way that certain investigations have sometimes to be carried out repeatedly under country-specific conditions.

18.3.8.1 Migration test

The requirements for the detection limit of traces of a migrated additive in food become more severe. The greatest difficulties arise in the specific detection and determination of trace amounts of additives in fats and oils.

If it is not possible, based on the structure of an additive, to detect the substance by a specific method within the required detection limit radioactive labelling has to be used. Nevertheless, a method for the specific detection of an additive in the plastics material and in the food has to be elaborated. The cost for working out these methods and the total migration behavior of an additive ranges today between US $ 25000 and 50000.

18.3.8.2 Toxicological investigations

As mentioned before, an approval for an additive is granted in most countries only under the conditions that a 90-day feeding test in rats has been carried out. In table 10 it is shown that the cost – depending on the size of the test – amounts to US $ 50000–90000.

In addition, an analytical method for the specific detection of the substance in the feed has to be elaborated (in the range of ppm). This feed analysis has to be conducted periodically during the toxicity test. The cost for such an analysis varies between US $ 7000 and 15000. Therefore, the total amount for an up-to-date feeding test, including the necessary analytical work is between US $ 60000 and 100000. If a petition for an additive is also filed in the United States, an additional 90-day feeding in a non-rodent species (usually dog) is required, with the cost of US $ 60000 to 100000. In some countries it seems that, in

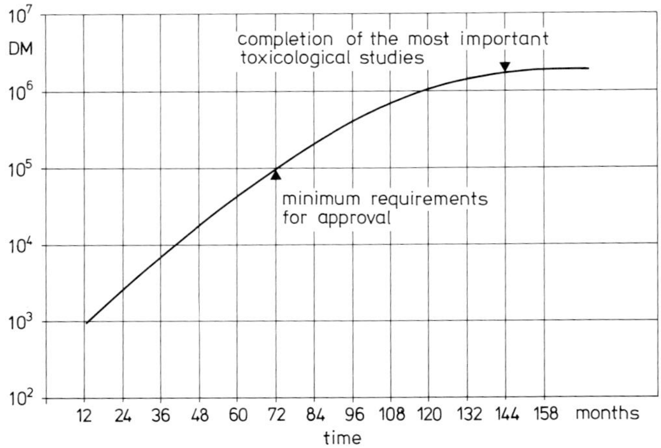

Fig. 5. Approximate cost and time needed for the approval of an additive in plastics packaging materials in contact with food

addition to the subchronic tests, one or two lifetime studies will be required in the future.

Since most of these tests cannot be conducted in parallel, the testing needs a great deal of time. From the beginning of the first tests until the first approval is granted, 2 to 3 years are usually needed. For an approval in the main industrial countries, another 3 to 4 years are necessary.

In figure 5 the time and cost estimates for approval of an additive in plastics food packaging materials are shown.

It is expected that the requirements for approval of an additive will get more demanding in the future. This fact will tremendously influence the cost for the development of an additive. As a consequence, it will be more and more difficult to get a new additive for food packaging materials on the market within reasonable time limits.

18.4 Definition of the most common toxicological expressions

ADI (Acceptable Daily Intake):	The acceptable daily intake of a substance which is established on the basis of chronic animal studies.
Acute dermal toxicity:	Percutaneous toxicity of a substance by one single exposure to the skin.
Acute oral toxicity:	Toxicity of a substance after one single oral application (per os).
Acute toxicity:	The toxic effect appears after one single application of the toxic substance (acute animal study, one single application).
Acute inhalation toxicity:	Reaction of a test animal after short-term inhalation of the test substance.
Allergens:	Substances which provoke allergenic reactions of an organism.
Eye irritation test:	Test of the injury of a substance after one single application into the lachrymatory sack of the animal.
Carcinogenicity:	Toxic effects showing malignant tumors.

Chronic toxicity:	The injury will happen after a long-term application (chronic animal study: application over a period of for example 2 years with rats. Today most applications are made over a lifetime period.).
Consumption Factor (CF):	Portions of the diet likely to contact specific packaging materials.
Dermal, cutaneous:	Concerns the outside skin.
Estimated probable Daily Intake (EDI):	Estimated quantity of the daily intake by man of an additive through food.
Food type distribution value (f_T):	The amount of food type contacting each packaging material.
Lethal Dose (LD):	The dose of a substance which leads to the death of the individual. Also called DL (Dosis Letalis).
LD 50 (LD_{50}):	The applied dose level in animal studies which kills 50% of the animals within a few days; also called DL 50 (DL_{50}).
MAC value (= Maximum Allowable Concentration):	The concentration of a substance in the air of which it is believed that the influence within a working week of daily 8 to 9 hours will not do any harm to the workers in the work environment.
Mutagenicity:	Toxic effect on the gene substance which may lead to the injury of the organism.
No Effect Level (NEL):	Maximum tested dose level in a chronic resp. subchronic animal study where no toxic effect in the test animal was found.
No Observable Effect Level (NOEL):	see NEL
PADI (= Packaging Acceptable Daily Intake):	The acceptable daily intake for man of a substance migrating from packaging material into the food which was established by a 90-day animal study.
Percutaneous:	Penetrating the skin.
Peroral (per os):	By route of the mouth.
Primary irritation:	Reaction of skin tissue by one single application of the substance.

Skin irritation test:	Injuries of the skin after one single or different application of the test substance on the skin.
Subchronic toxicity (subacute toxicity):	The toxic effect is shown after a few applications, for example subchronic tests: daily application over a period of 90 days.
Teratogenicity:	Toxic effects consisting of malformation of the embryo by treating the mother with toxic substance during pregnancy.
Total Daily Diet (TDD):	The estimated total amount of a migrated substance in the food.

18.5 References

[1] *F.A. Patty:* Industrial Hygiene and Toxicology, Vol. II, Interscience, New York, 1967, p. 1904/1906, table 19.
[2] Registry of Toxic Effects of Chemical Substances, Vol. II, U.S. Dep. of Health, Education and Welfare, Cincinnati, 1977.
[3] Ciba-Geigy: unpublished results.
[4] *R.R. Joiner, F.D. Vidal, H.C. Marks:* Cereal Chem. 40 (1963) p. 439.
[5] Ciba-Geigy: unpublished results.
[6] *H.W. Gerarde:* Toxicology and Biochemistry of Aromatic Hydrocarbons, Elsevier Monographs (1960) p. 128.
[7] Ciba-Geigy: unpublished results.
[8] 46th communication of the BGA, Bundesgesundheitsblatt 24 (1981) 4/5, p. 81.
[9] Ciba-Geigy: unpublished results.
[10] Ciba-Geigy: unpublished results.
[11] *K. Figge, S. Eder, H. Piater:* Dtsch. Lebensmittelrdsch. 68 (1972) p. 359.
[12] *K. Figge:* Food and Cosmetics Toxicology 10 (1972) p. 815.
[13] Ciba-Geigy: unpublished results.
[14] Schweizerisches Lebensmittelbuch, chapter 48: Gebrauchsgegenstände aus Kunststoff, eidgen. Drucksachen- und Materialzentrale, Bern, 1982.
[15] Food Chemical News 23 (1982) p. 34.

19 Compound Development and Compounding of Thermoplastics

Dr. Gerhard Menzel, Chemische Werke Hüls AG, Marl,
Federal Republic of Germany

Part I: Compound Development

19.1 Introduction

The preceding chapters describe the most important additives and their effects in various polymers. They are used for the most part in thermoplastics whose commercial range comprises more than 50 structurally different plastics which in turn are available in a number of grades tailored for specific applications (figure 1). The prerequisite for such a classification is a careful investigation of the actions of the particular additives in the plastic or plastic type to be stabilized, so that an optimum additive formulation can be developed for the case in question.

19.2 Classification of additives

In a simplified scheme, additives can be divided into two principal groups (figure 2).

Process additives:

These are additives which facilitate the processing of a plastic or, as in many cases, make it possible at all. For example, rigid PVC cannot be processed in the absence of stabilizers and lubricants. Polyolefins, styrenic polymers and polyamides, too, could not be processed economically, if antioxidants and other additives were not used.

Functional additives:

These impart properties to the plastic or enhance properties already present to a limited extent. For example, polypropylene has low light and weather resistance, and after a few months' outdoor exposure becomes

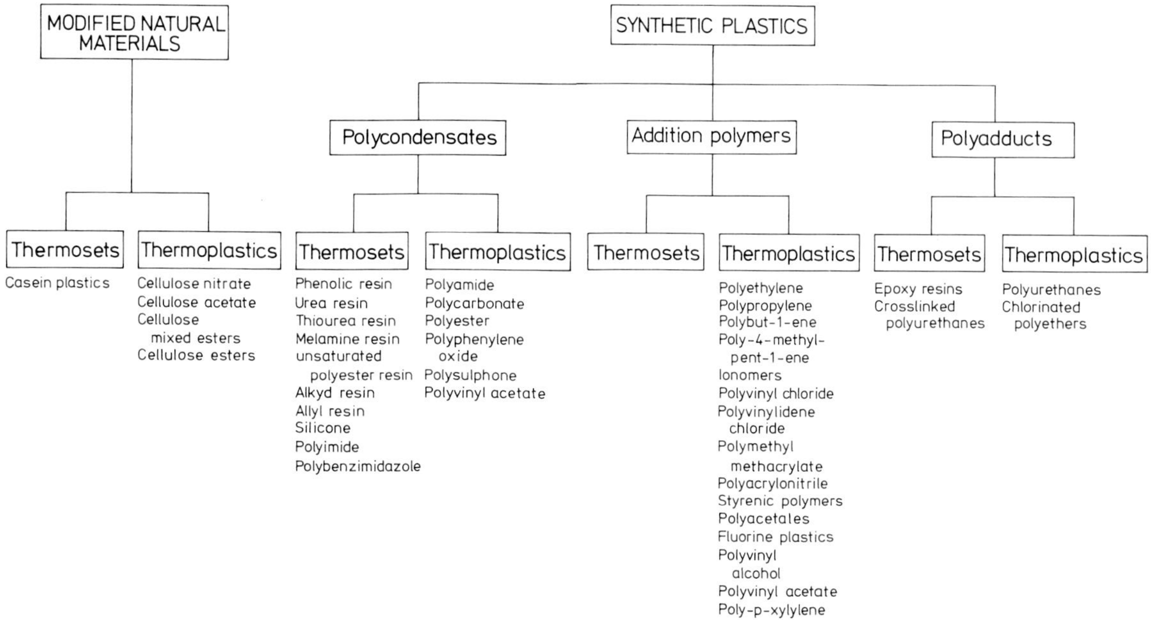

Fig. 1. Classification of plastics according to preparation, reaction and structure

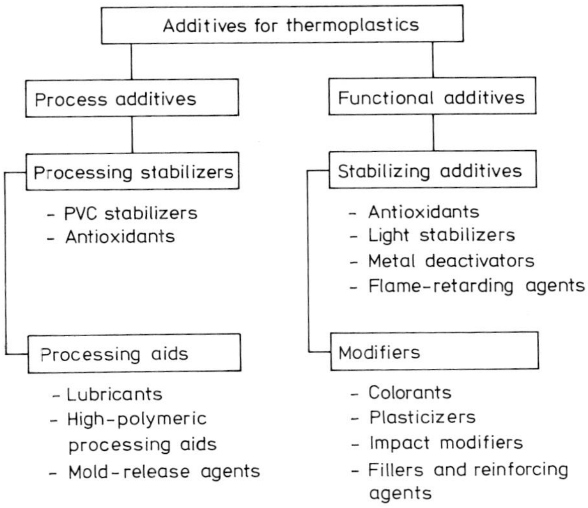

Fig. 2. Classification of plastics additives

brittle and can no longer be used. However, certain light stabilizers render polypropylene highly light-stable and weather-resistant, and have thus opened up new fields of application for this material.

19.3 Compounding

Plastics raw materials are generally already combined with additives by the manufacturer, before being supplied to the processor. Because plastics are very versatile, there are economical and also technical reasons for adding the additives, not during polymerisation, but in a downstream process known as *compounding* (cf. figure 3).

The field of compounding is based on the development of a practical formulation. By simulation of the main process conditions using process-related laboratory measuring equipment and processing units, it is possible to arrive step by step at the formulation of a plastics compound, initially the so-called *control formulation*.

The optimum composition of a *practical formulation* is determined by the processing method used, the spectrum of properties required and the fields of use of the finished plastics articles.

19.4 Systems analysis of processing procedures [1]

Even today, compound development or design of a practical formulation are frequently carried out in an empirical manner. Questions and the solutions to them are not treated systematically but are associated with the particular problem in hand, the technical equipment available and the knowledge of the requirements which the plastics article has to satisfy. In this case, the effects of the additives in the particular plastic are tested and interpreted on the basis of isolated investigations. Systems analysis of industrial processes, on the other hand, is aimed at working out relationships, as far as possible in quantitative form, and hence providing the preconditions for carrying out selective development and make it possible to control a system so as to achieve the desired parameters. Investigation of individual process steps is inadequate for this purpose; rather, it is necessary to consider the entire system, taking into account the specific properties of additives and plastics (see figure 4).

The system relevant to a plastics processing procedure comprises the following stages (cf. figure 3):

- polymerization (P),
- formulation and compound development (F),

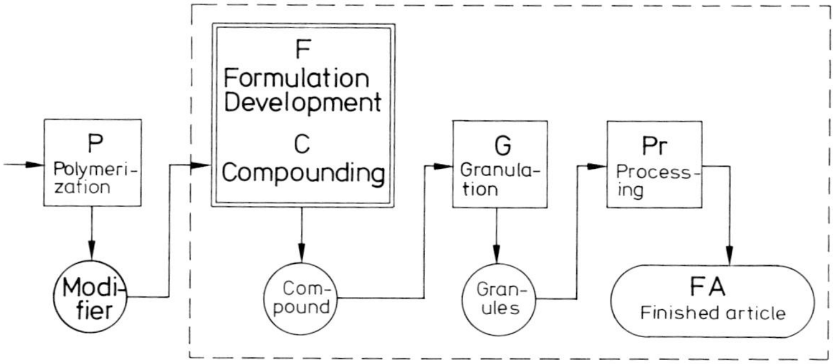

Fig. 3. Plastics production and processing system

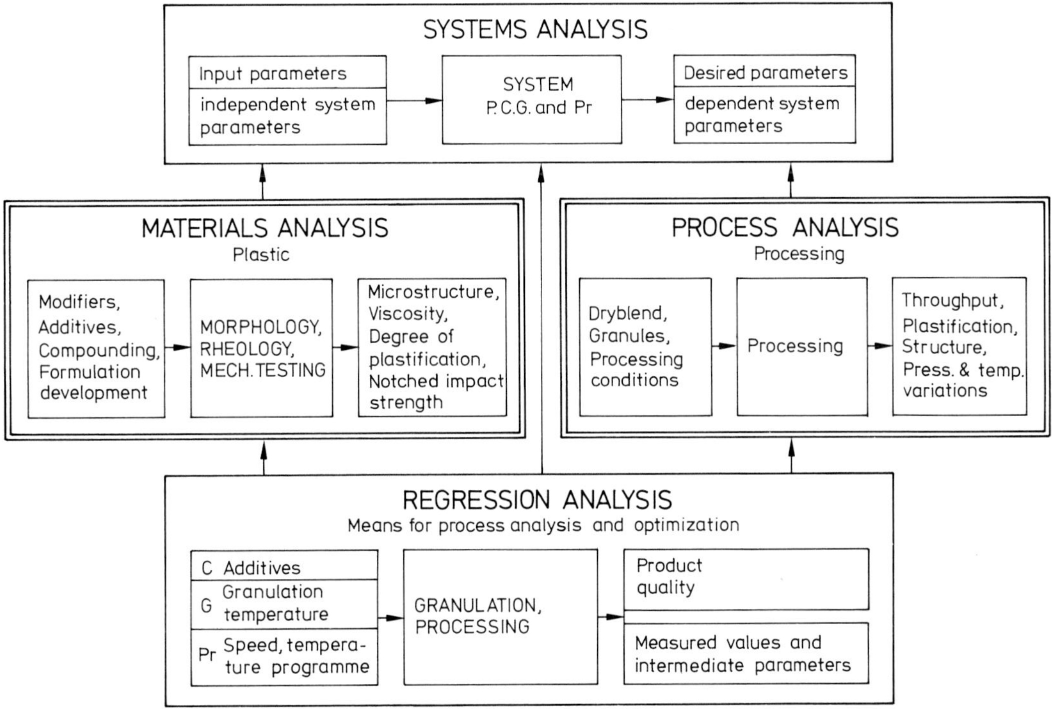

Fig. 4. Systematic model of plastics production and processing (survey of selected system parameters)

- compounding (C),
- granulation (G),
- processing (Pr),
- finished article (FA).

The first step in a systems analysis consists in collecting the parameters relevant to the individual stages of the system and arranging them methodically in a system matrix (cf. figure 5). For the individual system stages, classification according to

- starting material (initial properties),
- process conditions,
- product properties

may prove to be advantageous.

19.4.1 System stages [1]

Before a reduced system matrix is considered, a brief description of the stages belonging to the system will first be given.

Polymerization

The polymerization conditions prevailing result in important properties of the polymers, e.g. thermal stability, polymer shape (particle size and shape and particle size distribution) and plasticization behavior (the latter applying in particular to PVC), being established as early as at the polymerization stage.

Formulation development/compounding

Compounding is intended to mean formulation design and production of a processable molding material.

Granulation

This procedure, in which a powder mixture of a plastics raw material with additives is converted to processable granules, may itself be regarded as a processing step.

Processing – thermoplastic forming to give a finished article

For an optimum compound formulation, the spectrum of properties of a plastics article is perhaps most strongly influenced by the processing procedure.

Independent system parameters, Input parameters / Dependent system parameters, Desired values		Polymerization: Starting materials	Polymerization: Process conditions	Compounding: Formulation	Compounding: Mixing conditions	Granulation	Processing (Extrusion): Material	Processing (Extrusion): Extrusion conditions
		VC (vinyl chloride) VE (deionised water Initiators Emulsifiers etc. Modifiers (HI)	Polymerization method Type and size of apparatus Devolatilization Dewatering Working up Stirrer speed Temperature programme Procedure Reaction time	Type of PVC Stabilizers Lubricants Fillers Pigments (Modifiers)	Mixing method Size of mixer Temperature programme Mixing and cooling times Component feed	Granulation method Extrudate and size Temperature programme Speed Cooling conditions	Granules Dry blend	Method Extruder (type, size) Screw Barrel Die Downstream units Inlet grooves Metal surfaces Speed(s) Temperature programme Take up conditions Cooling conditions
P	Yield Deposits in the kettle Reaction time Productivity							
Crude PVC	K-value Volatile constituents Particle size distribution Particle shape Mean particle size Pourability Bulk density Thermal stability Fleckiness Electrical properties Plasticizer uptake							
C	Throughput Homogeneity (Time of mixing cycle)							
Dry blend	Pourability Bulk density Color Thermal stability Electrical properties Mechanical properties Rheological properties Performance characteristics							
G	Throughput Productivity							
Granules	Size of granules Shape of granules Degree of plasticization Thermal stability Chlorine/chloride content							
Pr	Throughput Productivity Transport behavior (pulsation) Plasticization behavior Ejection behavior Cooling behavior Power Mass temperature Melt homogeneity							
Profile	Geometrical constancy Mechanical properties Electrical properties Optical properties Performance characteristics Notched impact strength Gloss Roughness Grooves Stria Flecks etc.							

Fig. 5. System matrix for PVC

19.4.2 Reduced system matrix [1]

So-called process analyses relate to the analysis of an individual system stage (process) comprising various process sections. However, working through the entire system matrix would be beyond the economically acceptable limits for any problem and any scheme of investigation. For this reason, in practice only a few relevant system parameters are selected from the total matrix and are represented in the form of a reduced

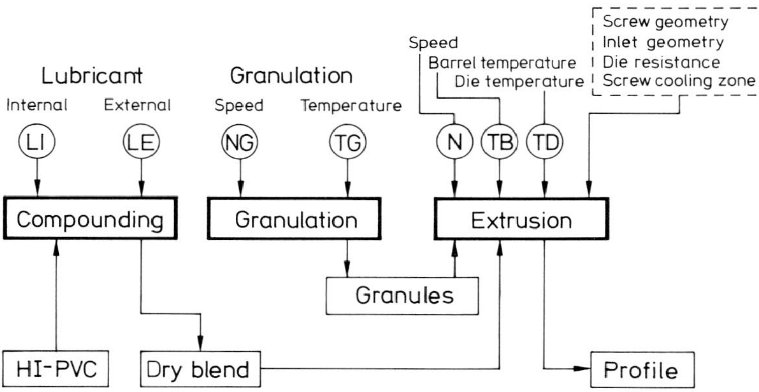

Fig. 6. Reduced system matrix for the example of the extrusion of impact resistant PVC

system matrix (cf. figure 6). The investigations to be carried out are then generally concentrated on the effects of:

- the nature of the additives and their concentrations,
- the granulation conditions,
- the processing conditions and behavior, and
- the properties of the finished article.

While the *dependent system parameters* are predetermined by the problem and the object and do not require more detailed explanation, the independent system parameters have to be discussed briefly.

The *independent system parameters,* i.e. the process additives, alter the rheological behavior and the morphology of a plastic and as a result have a particularly lasting effect on the processing procedure and quality of the finished article. The investigative work required in this connection

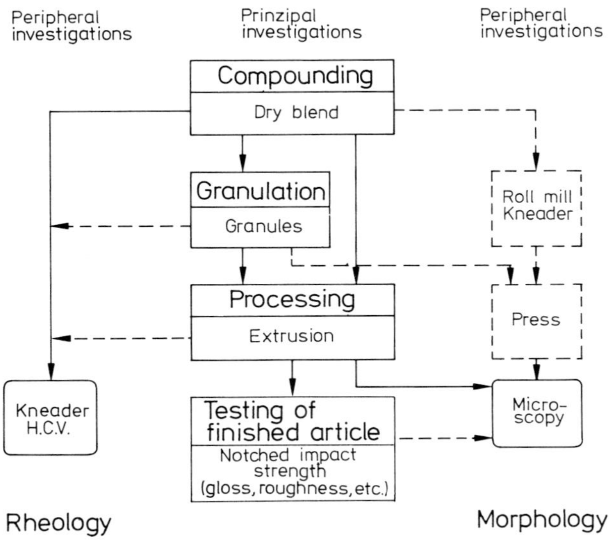

Fig. 7. Principal and peripheral investigations
H.C.V. = high pressure capillary rheometer

may be divided into principal investigations and peripheral investigations; a prerequisite of such investigations is an extensive system of experimental, test and analytical techniques (cf. figure 7).

The systems analysis outlined above makes no claim to completeness, since it merely relates, by way of example, to a narrowly restricted, reduced model. However, any system once worked out is constantly subject to alterations as a result of increased demands on the material and in connection with further technical development (rationalization). Consequently, new parameters are often introduced, and their effects both on the individual process and on the system have to be checked.

The design of a comprehensive model is the basic precondition for controlling all processes taking place in a system and for rapid adaptability to constantly changing problems. In practice, a system as complex as plastics development, production and processing can be handled efficiently only with the aid of electronic data processing (cf. figure 8).

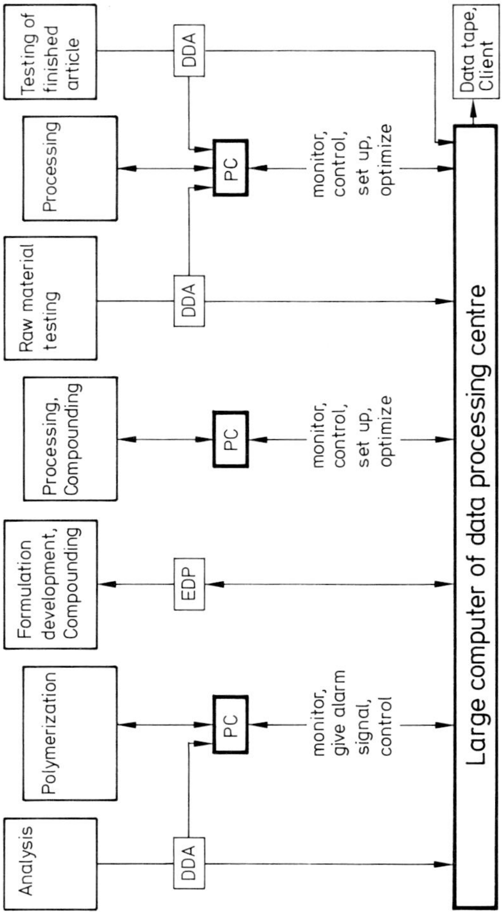

Fig. 8. System automation concept

DDA: Digital data acquisition, *PC*: Process control computer, *EDP*: Electronic data processing

19.4.3 Optimization

A frequent problem of application technology is the determination of the optimum amount of additives for a plastics compound with respect to processability, properties of the finished article and cost of formulation. Optimization calculations can be usefully applied to such cases; if the regression equations are extended by a price equation, the additives represent the variables. If the predetermined technical conditions and qualities are kept constant, the optimization calculation gives the most economical price for the formulation. The computed optimization results should be confirmed by practical experiments, and the additional data obtained can, if necessary, be used to improve the optimization calculation.

19.5 Effect of the morphology of PVC in the compounding [2]

In view of the wide-ranging industrial and economical importance of PVC and its specific properties with regard to production, stabilization and processing, the system PVC/additives should be discussed in somewhat more detail.

The large number of good properties of PVC can only be utilized if its limited thermal stability is increased by means of suitable stabilizer systems and if the technical difficulties in processing, which are predominantly due to rheological factors, are reduced or eliminated by the use of process additives (processing aids). In contrast to the completely amorphous polymers, such as commercial standard polystyrene and polymethyl methacrylates, PVC contains ordered structures which may also be regarded as crystalline moieties; however, the amounts of these are so small that no exact and convincing description of these structures exists to date. On the other hand, it appears certain that the specific structure of PVC makes possible more intensive interactions with added low molecular weight and high molecular weight molecules than is the case in mixtures of other amorphous polymers containing comparable components. For example, the mode of action of plasticizers in PVC has been described in detail on the basis of the special structure of polyvinyl chloride. It is well known that plasticization with low molecular weight compounds was an important stage of development for the thermoplastic processing of PVC on an industrial scale.

The supermolecular structure of PVC is established already during polymerization and depends on the polymerization conditions (emulsion polymerization, suspension polymerization or mass polymerization). Compounding of these various types of PVC is therefore carried out by different procedures since the absorptivity for additives is heavily influenced by the morphology of the particular type of PVC.

Because of its particulate structure, the *PVC powder particle* contains voids into which the additives can penetrate during the compounding process (cf. figure 9). As the largest morphological unit it remains unchanged during production of the compound, so that the primary particle structure, too, is retained. It may be assumed that at least some of the additives act as a "lubricant" between the *primary particles* (= globules = microparticles) (cf. figure 10). Uniform distribution of the additives on the primary particles is achieved only when processing is carried out in the melt.

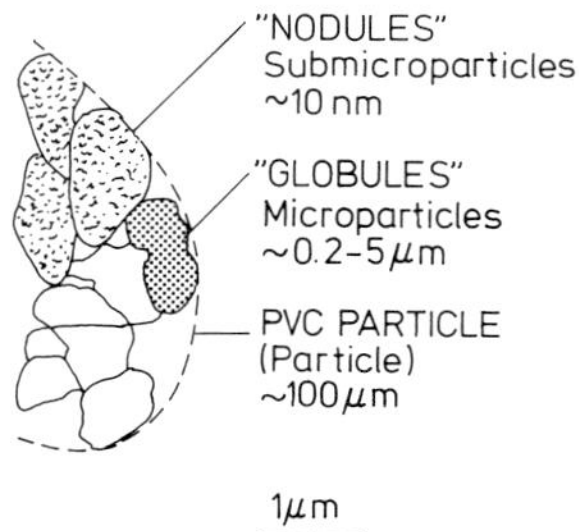

Fig. 9. Microstructure of PVC

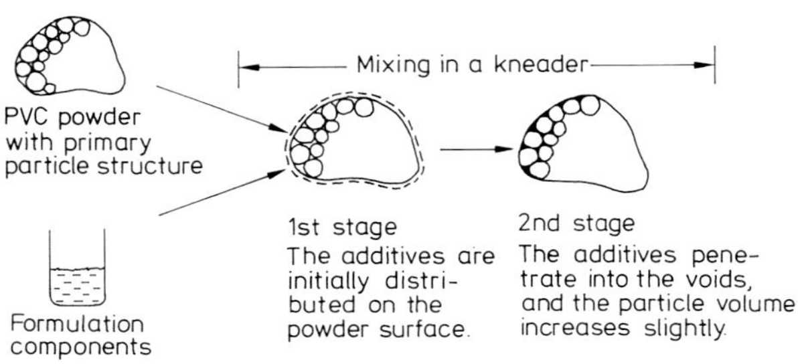

Fig. 10. Schematic representation of the processing of PVC

Lubricants and other additives not only make it possible for the PVC particles to slide away from one another but also reduce the adhesion of the polymer to the walls of the processing equipment. The pronounced effect of even small amounts of additives is therefore obvious.

The viscosity of a PVC melt is a function of the primary particle size; under otherwise identical conditions (formulation, temperature, compressive stress and shearing stress), the PVC containing the smaller primary particles has the higher melt viscosity.

19.5.1 Processes during the melting of PVC [2]

A precondition for a high degree of homogenization of a PVC melt is very uniform distribution of the additives on the PVC powder particles. During the plasticization process, the powder particle structure breaks up from about 160 °C, and the primary particles (globules) may then be regarded as the largest structural units. The distribution of the additives on the primary particles which takes place in this step has a strong effect on the properties of the finished article. Fluid additives or additives which melt at $\lesssim$160 °C probably diffuse into the globules in small amounts, the major part, however, remains in the voids.

A further plasticization stage begins in the temperature range from about 190 °C. The primary particles now begin to melt; if the additives are sufficiently compatible, they form a molecular dispersion, but if they are incompatible they are distributed in the PVC in the form of discrete particles.

At still higher processing temperatures, the morphology of the finished article is characterized by the occurrence of HCl bubbles; because of their notch effect, they have an adverse influence on the mechanical properties of the finished article, and the tensile strength and elongation at break decrease sharply. The temperature range in which the thermal decomposition of PVC has a marked effect on the mechanical properties depends of course on the type and amount of stabilizers. Since it may be assumed that relatively high stabilizer concentrations are present especially at the surfaces of the morphological units (powder particle or primary particle) it is possible, using effective stabilizers and lubricants to process PVC into satisfactory end use articles without substantial damage to the polymer. Thus, the "structuring" of PVC has considerable technical consequences.

19.5.2 Distribution of additives in rigid PVC [2]

It is probable that there is an inhomogeneous distribution of additives, such as stabilizers and lubricants, even after the PVC powder particle structure has broken up. In the most advantageous case, they are uniformly distributed over the primary particles; owing to the customarily short residence times of the PVC melt at temperatures above 160 °C, complete diffusion into the globules becomes unlikely.

With the aid of scanning electron microscopy (SEM), *N. Berndtsen* [2] investigated a lead-stabilized PVC extrudate containing calcium stearate as a lubricant. When the samples are irradiated in the SEM, in addition to the secondary electrons, which are used to produce the picture, X-rays are also formed. The energy of the X-rays permits qualitative and quantitative determination of the elements present at any particular point in the sample.

In the analysis, only chlorine could be detected in the interior of the particle, i.e. only PVC is present (cf. figure 11). In contrast, at the particle boundary the chlorine content is only one third of the content in the interior of the particle (i.e. the particle boundary contains only about

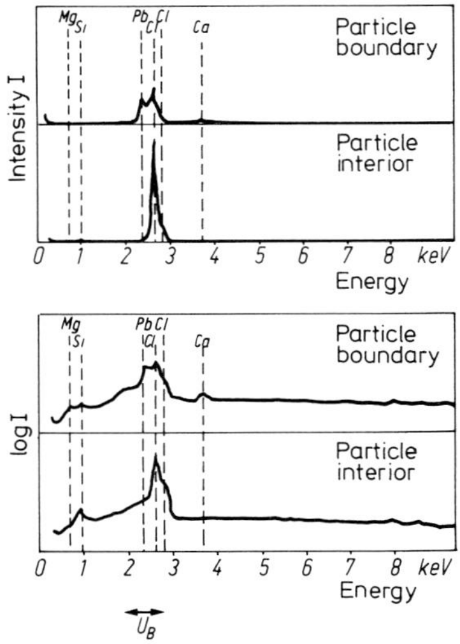

Fig. 11. Analysis of a rigid PVC extrudate obtained at low processing temperature

30% PVC), while lead and calcium were clearly detectable. Taking into account the lead content (50% Pb) of the stabilizer employed and the calcium content (8% Ca) of the calcium stearate used as lubricant, the composition at the particle boundary is approximately as follows: 30% PVC, 25% lead stabilizer and 20% lubricant. On the other hand, the interior of the particle consists of pure PVC, as mentioned above. Viewed in this way, it is clear that in *the area of the powder particle* there is a inhomogeneous distribution of the additives. The additives which occur preferentially at the particle boundary are certainly partly responsible for the good adhesion of the particles to one another and hence have a decisive effect on the quality of the finished article.

Part II: Compounding of Thermoplastics with Additives

19.6 Introduction

Thermoplastics are predominantly obtained as powders with particle sizes of about 1 μm to >1 000 μm. Exceptions to this include, for example, low density polyethylenes obtained by the high-pressure process, polyamides, and polyesters, which as a rule are discharged from the reaction vessel as a melt and immediately thereafter granulated. In order to obtain processable molding materials, the plastics raw materials, which are generally obtained in an unstabilized form, have to be mixed with additives, i.e. compounded. This compounding procedure has to be matched to the performance characteristics and product-specific properties of the particular plastics and additives. The aggregate state and the form of the additives frequently play an important role in this process.

19.7 Methods of incorporating additives [3]

In principle, there are four possible methods of incorporating additives (cf. Fig. 12):

1. Additives are added to the monomers, i.e. before or during the production of the polymer.
2. Additives are added during the working up of the polymer.

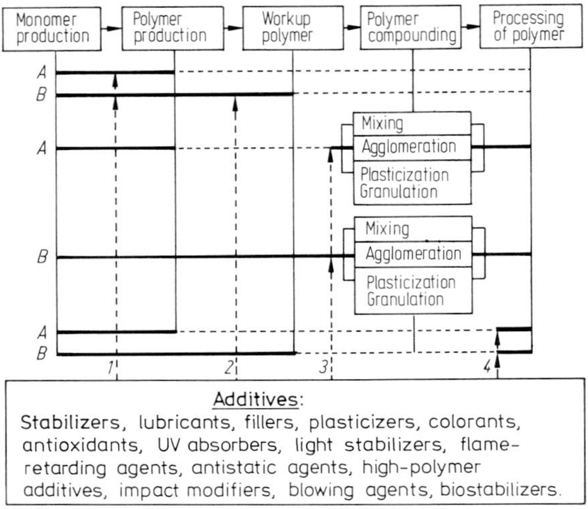

Fig. 12. Possible methods of incorporating additives

3. Additives are added in a separate manufacturing step (polymer compounding).
4. Additives are incorporated directly before or during molding of a plastic to a semi-finished product or finished article.

19.7.1 Addition of additives before or during a polymerization process

Additives which participate in the reaction are to be distinguished from those which are inert.

The method can of course only be used by the producer of the base resins; it is assumed that additives are sufficiently stable under polymerization conditions and do not have an adverse effect on the kinetics of polymerization and hence on the polymer properties. Typical examples of this procedure are the stabilization of impact-resistant polystyrene and the pre-stabilization of PVC produced by suspension polymerization; in both cases, suitable phenolic antioxidants are added to the monomers.

19.7.2 Addition of additives during the working up of the polymer

As in the case of method *1*, this method too has to be carried out by the polymer manufacturer. The additives used must satisfy specific requirements determined by the process, such as emulsifiability, solubility in certain solvents, low volatility or stability to hydrolysis. Polymers obtained in dispersed form (emulsion, suspension or slurry processes) are frequently provided with some or all of the required additives before being isolated. This applies, for example, to modified styrene polymers, such as IPS and ABS. Here in particular, antioxidants in emulsified form are added to the ABS latex (see Chapter 1.8.9.1). As a result of coagulation at an elevated temperature, the additives are taken up by the relatively large secondary particles in such a manner that they do not separate out subsequently. Polyolefins prepared by solution, slurry or gas-phase processes are frequently provided with antioxidants at this stage in order to keep oxidative damage during isolation and drying at a low level.

19.7.3 Addition of additives in a separate manufacturing step: polymer compounding

As an independent process, this method provides the greatest range of variation and has also become the most important method. It is used equally by the polymer manufacturer, the compounder and the final processor. The following versions are distinguished:

- *Powder mixing:* mixing below the softening point of the particular thermoplastic;
- *Paste production:* mixing a paste-forming solid with relatively high amounts of a liquid, e.g. a plasticizer (virtually restricted to PVC);
- *Agglomerate production:* mixing and heating at *above* the softening point but *below* the plasticization temperature;
- *Production of granules:* mixing, plasticization and comminuting.

Powder mixtures, agglomerates and granules constitute alternative forms for PVC. On the other hand, polyolefins, styrene polymers, polyamides, etc. are predominantly converted to granules (pellets), since additives exhibit too little adhesion to more or less non-polar polymer powders, with the result that separation would occur. This is undesirable and has adverse consequences.

19.7.4 Addition of additives before or during processing

This procedure is principally used by processors, for incorporating pigments or pigment formulations (self-coloration), additive systems in master batch form which have been developed for specific applications, blowing agents, fillers or reinforcing agents. For metering the additives it is desirable for them to be free-flowing and have a low dust content and a narrow particle size distribution, as is true for example, with granules. These can be metered easily without great expense. The advantages of granules are:

- in contrast to powder mixtures, granules are dust-free,
- no separation of plastic and additive takes place,
- granules are free-flowing and possess a high bulk density,
- the processing machine is not subject to such high demands, since the homogenization (carried out during the production of granules) and the molding of the plastics material to semi-finished products or finished articles are carried out independently of one another, in two separate process steps,
- granules are safer to process.

The choice of the compounding method and of the possible versions is governed on the one hand by the criteria which apply to the particular additional processing steps for the plastic to be compounded and on the other hand of course by the cost-efficiency in production and processing.

19.7.5 General operations in the compounding of plastics

A flow diagram for the compounding of plastics is shown in figure 13. The process includes the following steps:

a) storage of starting materials in silos (plastics raw materials and additives),
b) metering in the starting materials,
c) mixing, devolatilization and, if appropriate, cooling,
d) feeding in, homogenization, devolatilization, plasticization and molding,
e) comminuting,
f) cooling and, if appropriate, devolatilizing, and
g) storage in bunkers and packaging.

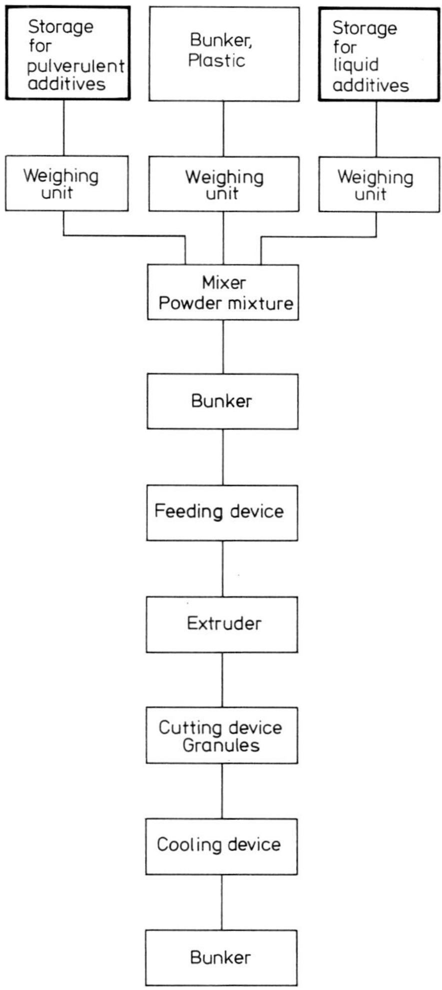

Fig. 13. Flow diagram for the production of plastics granules, agglomerates or PVC pastes.

If processes *d* and *e* are omitted, the plant is a continuously or batchwise-operated plant for the production of powder mixtures. Depending on the design of the units for the processes listed under *c*, it is also possible to produce PVC agglomerates or PVC pastes.

19.7.5.1 Storage of starting materials in silos

Plastics powders (=plastics raw material) and pulverulent and liquid additives are generally stored in outdoor silos, which are charged at the site of manufacture of the raw material directly from the polymerization plant or from railway cars or tanker trucks.

Pulverulent stabilizers, lubricants and colorants (mostly comprising several individual components predispersed in a carrier) are advantageously packed in transportable containers by the producer and, using suitable transport, are introduced upstream from the appropriate metering station. Such a procedure is economical and at the same time advisable from the point of view of safety, since it avoids transfer of material between containers and the risk of dust formation (dust explosions!).

19.7.5.2 Metering of starting materials

An essential component of an automatic compounding plant is a multi-component metering unit for pulverulent and for fluid components. A distinction is made between batchwise and continuous metering and between volumetric and gravimetric metering. The choice between *batchwise* and *continuous* metering depends on the construction and the mode of operation of the downstream units; the required precision of metering determines whether *volumetric* or *gravimetric* metering is preferable. In particular, the continuous metering of small amounts with the necessary precision presents difficulties, so that the requirements are reduced if the additives are metered as premixes in the form of, for example, powder/powder concentrates, master batches, etc.

19.7.5.3 Mixing

Two types of mixing processes are distinguished:

- mechanical mixing and
- pneumatic mixing.

Apart from a few special cases, plastics compounding is carried out using mechanical mixing. Mechanical mixers are characterised by rotating components: either mixing elements of some type rotate in a station-

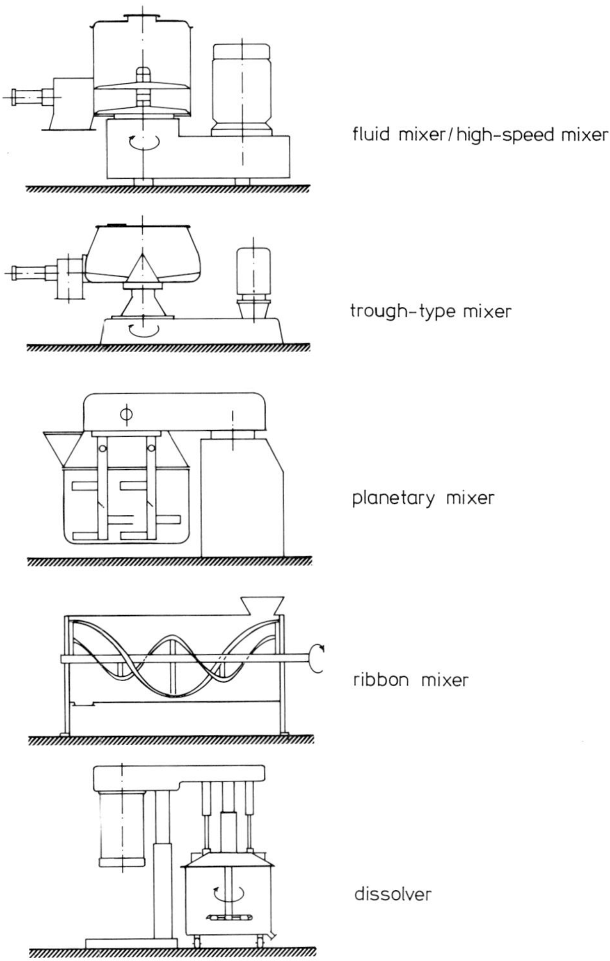

Fig. 14. Industrially important mixing units

ary housing or the housings are themselves designed to be rotatable. The large number of existing designs are classified on the basis of the parameters characteristic of the mixing process (cf. figure 14).

The cost efficiency of large units, progress in the construction of weighing belts and the development of premixed stabilizer concentrates and colorant concentrates will promote the development of continuously-operated plants. In special cases, where further processing is carried out directly, it should be possible to dispense with premixing as a separate compounding step.

19.7.5.4 Agglomeration

The production of agglomerates is used particularly in compounding PVC, and consists of two phases:

- the mixing and heating phase and the
- mixing and agglomeration phase.

If PVC powder is mixed with additives and is heated to *above* its softening point, but only as high as its *lower* range of plasticization with the aid of high-speed rotors, compacted, partially plasticized, agglomerated secondary particles of homogeneous composition are formed. With respect to production processes using powder mixtures and to the processing properties, agglomerates are similar to granules in many respects. This applies in particular to semi-rigid PVC (about 20 parts of plasticizer per 80 parts of PVC).

19.7.5.5 Granulation

Melt granulation, which comprises as process steps, mixing, compacting, homogenization, devolatilization, plasticization and comminuting to give granules of uniform particle size, is a compounding method which for most fields of use permits processing of a plastic to give a product of the required quality.

According to the sequence of the process steps, granulation processes are classified as follows:

- Hot granulation processes: Extrusion – cutting – cooling
- Cold granulation processes: Extrusion – cooling – cutting.

Granulators in which the granules are cut and cooled in air are used for the granulation of rigid and soft PVC, since these materials in the plastic state have only a slight tendency for the particles to stick together.

Hot granulators in which the granules are cooled in water are employed as a rule for polyolefins, styrene polymers and polyamides.

The granulation process is in general as follows:

The polymer powder is premixed with additives and the mixture is fed, via a feeding device, to the plasticization unit (e.g. extruder or kneader). Resins with a low bulk density are deaerated or precompacted in order to achieve high transport efficiency in the plasticization units. In the plasticization unit, the mixture is transported, heated, plasticized with an increase in pressure, devolatilized, and homogenized in the plastic state. The plasticized material is forced through a die and converted to extrudates of uniform diameter. These extrudates are cut into pellets of very similar size by means of granulation devices (rotating blade) arranged concentrically or eccentrically with respect to the die. The pellets are cooled to below 40 °C by means of cooling devices so that subsequent damage to the product does not occur during storage.

The plasticization units used for the granulation process are machines of various construction: roll mills, kneaders and in particular twin-screw extruders.

Apart from hot-granulation devices, in special cases strand and band granulators are employed for cutting cooled plastics strands or bands (cold-granulation devices).

19.7.5.6 Producing pastes

Pastes can be produced from emulsion-polymerized PVC and from microsuspension-polymerized PVC in conjunction with plasticizers. This is done using mixers with slowly or rapidly rotating mixing elements which can be mounted horizontally or vertically. For further homogenization and dispersing of pastes together with colorants and fillers, frictional mills have been used for decades. The paste is subjected to friction by exactly defined slots (roll mills and disc mills).

Mixers with rapidly rotating mixing elements which are mostly arranged vertically and are in the form of a disc canted at the circumference are referred to as dissolvers. The particular advantage of such units is that paste formation, homogenization and dispersing are carried out in one operation.

Paste mixers are generally operated at a reduced pressure of up to about 2.67 mbar, which ensures effective and rapid removal of the air enclosed when the paste is formed as well as vaporization of the residual moisture.

The machines described operate batchwise or continuously. Production of pastes in automated closed plants from the silo station for the starting materials to storage of the ready-prepared paste in bunkers constitutes the state of the art.

19.8 Possible mistakes in the compounding procedure [4]

In the course of *formulation development* for plastics molding materials, trials on the processing machines subsequently to be used e. g. extruders, injection-molding machines, calenders, are unavoidable. However, for *routine testing* of plastics molding materials compounded batchwise or continuously in accordance with established formulations, the stated procedure is not absolutely necessary.

Apart from a few special cases, a processing test using an appropriately set up test-extruder, followed by an assessment of the extrudate in the form of a sheet or a profile, proves to be a relatively simple and inexpensive test method. It provides very reliable information on the materials quality, even in the case of those plastics molding materials which are not processed with extruders.

The extrudate is generally assessed visually, the defects being evaluated qualitatively and quantitatively. A great deal of practical experience is required in order to be able to recognize defects correctly and classify them reliably, particularly since these defects sometimes occur in transition forms and mixed forms. If it is not possible to make a reliable identification with the naked eye, or with the aid of a magnifying glass (magnification about 10 times) or a microscope, physical or chemical methods of investigation will have to be relied upon. The most commonly occurring phenomena which appear as defects in the extrudate are listed below, together with their causes. Errors attributable to the compounding stage (with the exception of gross formulation errors) are given special attention.

1. Relatively small, hard, predominantly transparent particles (up to about 500 μm) — flecks, also called fish eyes

Generally result from the raw material, probably crosslinked, horny particles, which cannot be eliminated in a standard compounding process. Plasticizers, modifiers or inadequately dispersed additives are also possible causes.

2. Relatively large agglomerated particles (nodules) or horny, hard particles — solid agglomerates, partially thermally decomposed

These are generally raw material particles, which are thrown upwards during the mixing process, or tacky additives, which have become deposited in the upper part of the mixing unit. When these particles fall down towards the end of the mixing process and are not sufficiently homogeneously dispersed, inhomogeneities can result.

3. Brownish to black, frequently hard particles — thermally decomposed or "burnt" particles

Possible causes include insufficient stabilization and inhomogenous distribution of stabilizers. Such particles frequently behave like foreign matter in the extrudate, since they do not form any bond with the surrounding material.

Dark stripes in the extrusion direction in otherwise colorless molding materials likewise indicate "burnt" materials.

4. Blisters or bubbles in the extrudate — moisture, increased amounts of volatile components and decomposition products

These can be caused by overheating the molding materials during compounding, by enclosure of air or by partial decomposition of the additives.

5. Deposits and plate-out — incompatible additives

Indicates that homogenization of the formulation components may be inadequate.

6. Pale, powdery pockets — solid additives, fillers and other pale, pulverulent formulation components

Insufficient distribution during the mixing process. Crumbling leads to the formation of craters or holes in the finished article.

7. Color flecks, color stripes or color stria — incomplete distribution or dispersion of colorants (dyes, pigments)

Color flecks are pigment agglomerates which result in the formation of holes when dissolved away.

Color stripes are formed from pigment agglomerates or from inadequately dispersed colorant concentrates during final processing under conditions which no longer permit adequate pigment dispersion.

Color stria (or color clouds) are an indication of non-uniform dispersion or metering of the colorant.

Where a suitable colorant has been chosen, the above defects generally result from a mixing process which has not been properly carried out.

If, after the quality has been tested by the extrusion test there is, in an exceptional case, still doubt with regard to the suitability of the tested material for the intended purpose, it is advisable to carry out an additional process-specific test (e.g. injection-molding test, roll mill test, coating test).

19.9 Prospects and technical trends

The problems relating to the use of additives in plastics have long become a well-defined field, especially since they have developed on the one hand from specific empirical technical concepts and on the other hand from the study of aging reactions in synthetic high polymers.

The development of plastics today is concerned in the main less with novel synthetic methods or novel starting materials than with opening up new fields of use. An important aim in this respect is the improvement of the properties of plastics or the extension of their spectrum of properties. This is primarily achieved by means of additives; today, these are not just processing assistants, but are modifiers, imparting important technical advantages and performance characteristics to plastics articles. This means that the level of performance and the further development of plastics technology depend substantially on the plastics additives.

In the use of additives, too, problems concerning the rational use in automated processing plants are arising to an increasing extent with regard to environmental protection and the improvement of product safety. Additive mixtures, also referred to as "additive systems", can contribute to the solution of such problems, since they contain stabilizers, processing aids and to an increasing extent also pigments, fillers and reinforcing agents.

Additive systems in the form of pre-metered packs (one-packs), flakes, granules, agglomerates and sintered powders simplify the compounding process and the processing procedure. They reduce the cost of metering and mixing and rationalize the storage. Mixing risks are virtually excluded and industrial hygiene problems are avoided.

Additive systems are more than the sum of the individual additives; they encompass the results of developments in application technology and practical experience in plastics. This know-how is expressed both in terms of the composition and in terms of the forms supplied and the specific synergistic actions of additive systems.

Further questions concerning the development and use of additives arise in the field of multi-phase plastics (polyblends, such as PP/EPDM, or PC/ABS; see chapter 8). Solutions to a whole range of problems associated with the use and development of additives, and the effects of these solutions on the overall system of plastics production, compounding, processing and application will only be worked out through more intensive use of electronic data processing.

19.10 References

[1] *J. Warmuth:* Analyse des Verarbeitungsprozesses von schlagzähem PVC. Thesis, TH Aachen, 1978.

[2] *N. Berndtsen:* Einfluß der Verarbeitung auf Struktur und Eigenschaften von Polyvinylchlorid. Thesis, TH Aachen, 1978.

[3] *H. Schiffers* (Editor): Grundlagen der Aufbereitung von Polyvinylchlorid, VDI-Verlag GmbH, Düsseldorf, 1978.

[4] *A. Jansen* in: Aufbereitung von PVC. VDI-Verlag GmbH, Düsseldorf, 1976, page 419.

Index

Advertisement

PARALOID MODIFIERS FOR PVC

ROHM AND HAAS

THE MOST COMPLETE RANGE OF MODIFIERS AVAILABLE TO SERVE THE PVC INDUSTRY

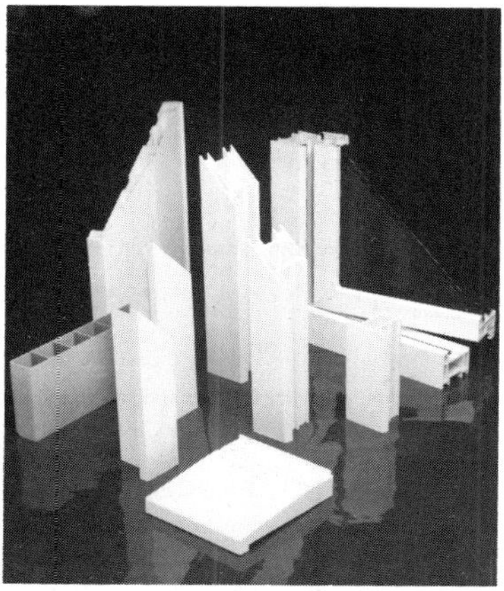

- **Processing Aids**
  PARALOID K 120 N
  K 125

- **Lubricating Processing Aid**
  PARALOID K 175

- **MBS Impact Modifiers**
  PARALOID KM 611
  BTA III F
  BTA III FFI
  BTA III OR
  BTA III N2
  BTA 702
  BTA 753

- **Acrylic Impact Modifiers**
  PARALOID KM 334
  KM 323 B

- **PVC Foam Modifier**
  PARALOID KM 318 F

- **Gloss Control and Anti-Blocking Agent**
  PARALOID KF 710

PARALOID PROCESSING AIDS

Rigid PVC is more difficult to process than other large volume thermoplastic materials such as polyethylene or polystyrene. Its non-homogeneous morphology results in difficult fusion and low melt strength. It also suffers from a narrow processing temperature range between fusion and degradation and shows a tendency to stick to processing equipment.

For rigid PVC formulations these difficulties have been overcome by the development of special processing skills on the part of PVC fabricators, combined with the use of Paraloid acrylic processing aids developed by Rohm and Haas.

PARALOID PROCESSING AIDS improve the processing of PVC through:

- Faster, more controlled fusion
- Greater homogenization of the melt
- Elimination of plateout
- Increased melt strength and elongation
- Elimination of surface defects

These features result in numerous benefits to the PVC processor, both during manufacturing operations:

- Less sensitivity to fluctuations in processing conditions
- Broader processing temperature range
- More consistent melt characteristics
- Less downtime for cleaning
- Higher yields
- Easier use of lower cost powder blends

. . . . and in the finished product

- Improved surface quality
- Improved sparkle and clarity
- Improved physical properties

Depending on the actual process being used, these benefits are manifested in many different ways:

CALENDERING

- Higher and more uniform output from compounding extruders or intensive mixers
- Lower scrap rates due to less critical processing conditions, and quicker establishment of saleable quality production
- Film and sheet products with improved end-use fabrication properties, including improved thermoformability and higher consumer appeal

EXTRUSION

- Improved control of dimensional tolerances due to more consistent melt flow
- Lower scrap rates due to more rapid establishment of equilibrium conditions and elimination of melt fracture
- Increased output rates, resulting in lower unit costs

BLOW MOULDING

- Stronger, more consistent components through improved parison strength and better welds
- More uniform weight control
- More sparkling appearance and greater consumer appeal

INJECTION MOULDING

- Lower reject rates due to reduction of splash
- Higher quality parts, due to less obvious, stronger weld lines, and improved appearance
- Improved performance of parts due to lower moulding stress

Processing Aids

PARALOID K 120 N

Versatile, all purpose processing aid, the original standard of the industry. Usually chosen for medium and high K-value PVC and for use in plasticised PVC formulations.

PARALOID K 125

All purpose, high efficiency processing aid with similar properties to Paraloid K 120 N.

LUBRICATING - PROCESSING AID

PARALOID K 175

This product offers strong external lubricating action supplemented with acrylic processing aid functionality. This unique polymeric lubricant greatly facilitates PVC processing but without the detrimental side effects of conventional lubricants. The outstanding features of Paraloid K 175 are:

- Reduced sticking even at high temperatures
- Improved dynamic heat stability
- No impairment of clarity
- All the benefits of an acrylic processing aid

The key lubricating feature of Paraloid K 175 gives production cost savings through longer production runs between clean-ups of calender rolls or extrusion dies, higher output at higher operating temperatures, and lower scrap rates. Formulating versatility is gained especially in clear applications, resulting in easy processing of calendered film and blow moulded bottles. Paraloid K 175 is also effective in less critical industrial applications: for example it considerably broadens the range of injection speeds and temperatures tolerable in injection moulding; it can improve the binding of high filler levels in low cost formulations for pipe and other products; it eliminates lubricant migration in film and sheet giving improved print and ink acceptance.

PARALOID MBS IMPACT MODIFIERS

Impact modifiers are included in rigid PVC formulations to achieve improved toughness levels required for many commercial applications, for example in

Bottles

applications, for food container requirements, including mineral water, vegetable oils and fruit squashes and for non-food applications such as, cosmetics, household products, detergents and motor oils, etc.

Film and Sheet

for blister packs furniture films and extruded sheet, and for

Industrial applications and Building Products

such as pipes, profiles, injection mouldings, etc.

For each application Rohm and Haas offers a complete range of MBS products, both KM and BTA series, with specific properties specially developed to meet exactly the specifiers requirements.

PARALOID KM 611

- High impact strength
- Outstanding resistance to stress crazing by vegetable oils and other aggressive liquids
- Good clarity
- Excellent processability

A most versatile general-purpose impact modifier widely used in all major processes and applications; specially suited for calendered films and oil bottles. Unique stress-crazing resistance broadens range of PVC for packaging applications. In extrusion provides an excellent balance of processability and impact strength in the finished products.

PARALOID BTA III F

- Extremely low contribution to taste and odour
- High resistance to breakage
- Excellent clarity and high surface gloss
- Absence of colour reversal and fuorescence
- Good colour and thermal stability in calcium-zinc and tin stabilised PVC compounds

An impact modifier offering an outstanding balance of properties for PVC bottles, particularly for the packaging of mineral water and other products sensitive to taste and odour contribution. Also suitable for calendered film intended for food packaging applications.

PARALOID BTA III FF1

- Improved transparency
- Higher impact strength efficiency
- Improved thermal stability and better early colour
- Free flowing powder characteristics
- Extremely low contribution to taste and odour
- High resistance to breakage
- Excellent clarity and high surface gloss
- Absence of colour reversal and fluorescence
- Good colour and thermal stability in calcium-zinc and tin stabilised PVC compounds

Developed especially for the production of PVC bottles for mineral water and the packaging of taste-sensitive products. Its inherent free-flowing powder characteristics make it more easily handled in automated processing equipment, or for applications where performance improvements above Paraloid BTA III F are required.

PARALOID BTA III OR

- Very high resistance to stress-cracking in presence of vegetable oils and many other aggressive materials
- Good clarity and surface gloss
- Good colour and thermal stability in calcium-zinc and tin stabilised PVC compounds
- Low contribution to taste and odour

Specially developed impact modifier to minimise stress-cracking in PVC bottles used for packaging of vegetable oils and several other aggressive products.

PARALOID MBS IMPACT MODIFIERS

PARALOID BTA III N2

- Excellent transparency
- High impact strength
- Low degree of odour and taste transfer
- Excellent rheological performance under critical calendering conditions
- Good thermal stability
- Low contribution to crease whitening

An impact modifier designed for use in calendering and extrusion of PVC film and sheet for all applications particularly food packaging, requiring a high level of transparency, and extruded PVC applications requiring high levels of impact strength efficiency. Also suitable for PVC bottles intended for food and non food packaging applications.

PARALOID BTA 702

- Non-crease whitening characteristics
- Excellent clarity, gloss and sparkle
- High impact strength
- Excellent thermoforming properties

Paraloid BTA 702 represents a new generation of Methyl Methacrylate-Butadiene-Styrene modifier resins which provide a major advance in the balance of high impact strength with extremely low crease whitening. Applications include folding cartons, adhesive tapes, blister packaging, vacuum formed containers, credit cards, overlays, laminates, cups and trays.

PARALOID BTA 753

- Outstanding impact strength efficiency
- Unsurpassed low temperature impact resistance
- Excellent processability
- Good dynamic thermal stability

Paraloid BTA 753 is a highly efficient MBS copolymer, developed primarily to provide higher impact strength resistance, especially at low temperatures, than conventional MBS, CPE, EVA or ABS impact modifiers. Applications include industrial calendered or extruded film and sheet and blow moulded PVC bottles requiring extra high toughness. It is also a valuable additive at low concentrations for conduit, pipes and pipe fittings and other industrial goods.

PARALOID ACRYLIC IMPACT MODIFIERS

Unfortunately, MBS modifiers cannot maintain their toughness and efficiency under prolonged outdoor exposure. To satisfy the demands of the rapidly growing building products market, PVC formulators require impact modifiers which can preserve appearance and toughness through long-term weathering and exposure.

Rohm and Haas, the world leaders in acrylic chemistry, has developed impact modifier technology based on durable acrylic elastomers to meet this need.

PARALOID KM 323B

- High impact strength
- Excellent weatherability
- Easy, non-critical processing
- Little effect on stiffness or other physical properties

Proven performance in outdoor PVC applications requiring long term retention of toughness and appearance such as window-profiles, door and door frames cladding, rainwater goods, etc. Allows complete freedom in the formulation of PVC compounds because it is effective with almost all types of PVC resin and any desired toughness level can easily be achieved. Easy processing and effective performance over a wide range of conditions. Strength/stiffness properties of PVC are maintained after modification, as needed for building products.

PARALOID KM 334

- Outstanding impact strength, especially at low temperatures
- Excellent weatherability and impact retention
- Wide processing range
- Little effect on stiffness or other physical properties

A higher efficiency all acrylic impact modifier combining the benefits of Paraloid KM 323B with improved low temperature toughness. Building product applications containing Paraloid KM 334 exhibit excellent retention of impact strength and appearance after prolonged outdoor exposure. Paraloid Acrylic Impact Modifiers have also been found to be effective in improving the properties of thermoplastics other than PVC.

PARALOID SPECIALITY MODIFIERS

PARALOID KM 318 F

PVC FOAM MODIFIER

- Combination of processability, lubrication and impact
- Unsurpassed foam stability, cell uniformnity
- Low foam density
- Excellent surface quality
- High impact strength and resistance to nail-splitting

Specially developed modifier for PVC foam compounds used in extruding profiles for construction, mouldings, siding, pipes, fittings and panels.

PARALOID KF 710

GLOSS CONTROL AND ANTI-BLOCKING AGENT

- Stable matt finish in PVC and ABS
- Excellent blocking resistance; good slip and resistance to soiling
- High contact clarity
- No effect on impact strength or other performance properties

A fine-grind powder for easy dispersibility in soft PVC compounds or with low shear processes.

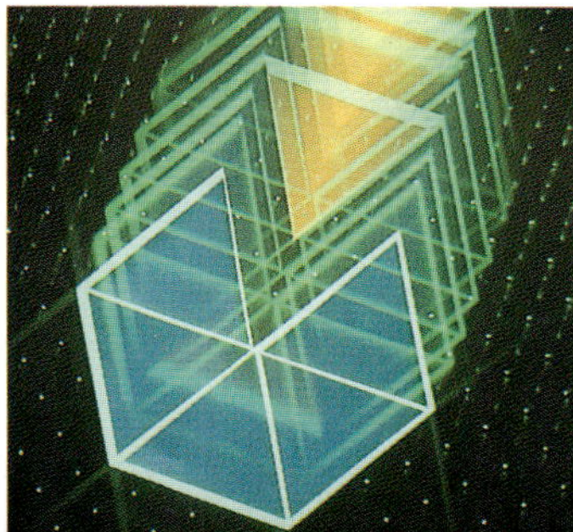

Henkel
Dehydag
New Lubricants for Calendering:
LOXIOL® G 70 S, LOXIOL® G 71 S
For the production of rigid-PVC-sheet of first class quality on modern calenders, the manufacturer needs highly effective lubricants that provide assured metal release with the highest possible compatibility at today's processing temperatures. For packaging wraps, the physiological safety of the lubricant is of the utmost importance.
With our new lubricants for calender formulations, the solid LOXIOL G 70 S and the liquid LOXIOL G 71 S, we invite sheet manufacturers to experience a further improvement in our products compared to our LOXIOL G 70 and LOXIOL HOB 7107 series. These products provide 30% better metal release at equal additive level.
A reduced additive level exhibits the same release. The advantages are particularly evident, when high transparency and compatibility with the polymer are necessary.
For food packaging, LOXIOL G 70 S is approved worldwide.
Our LOXIOL G 70 and HOB 7107 will continue to be available. Please contact:
Henkel KGaA
R-Organische Spezialchemikalien
Dehydag KLF-T
D-4000 Düsseldorf 1 · P.O. Box 1100
Telephone (211) 79 70
Telex 85 817 129 hd d
Telefax (211) 797 96 38
K 284 us - 2

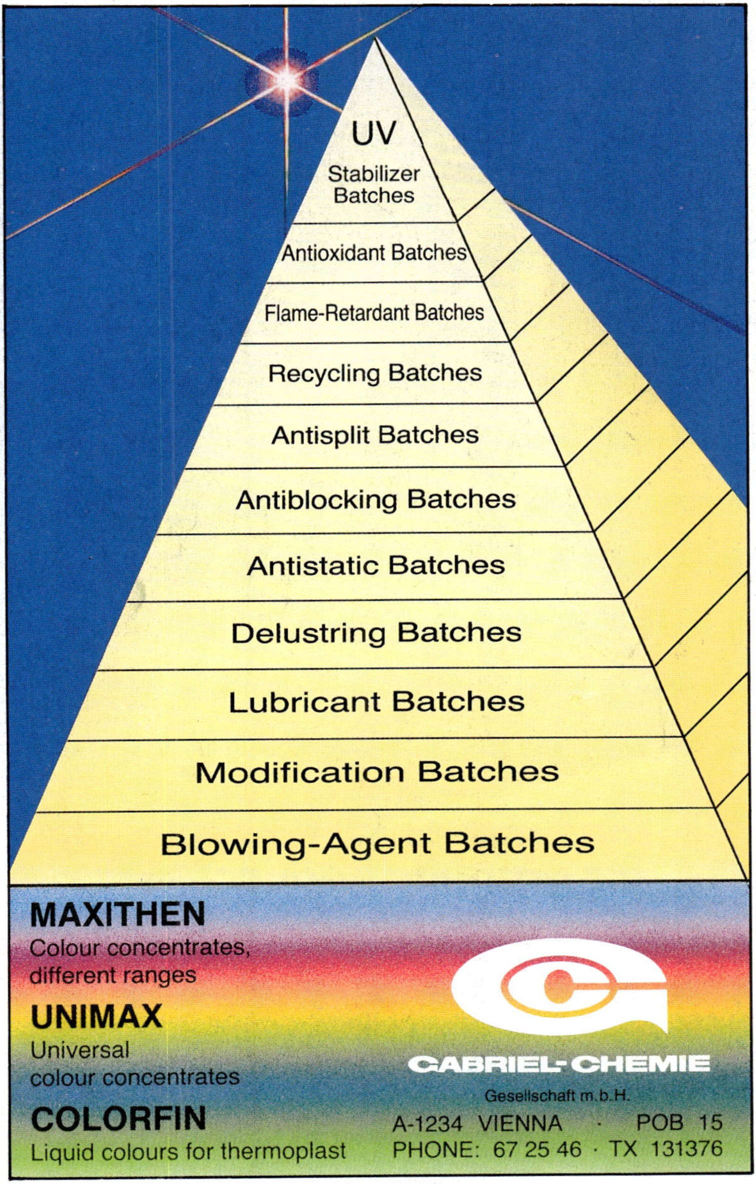

Index of Advertisers